# METHODS IN MOLECULAR BIOLOGY



*Series Editor*
**John M. Walker**
**School of Life and Medical Sciences**
**University of Hertfordshire**
**Hatfield, Hertfordshire, UK**

For further volumes:
http://www.springer.com/series/7651

For over 35 years, biological scientists have come to rely on the research protocols and methodologies in the critically acclaimed *Methods in Molecular Biology* series. The series was the first to introduce the step-by-step protocols approach that has become the standard in all biomedical protocol publishing. Each protocol is provided in readily-reproducible step-by-step fashion, opening with an introductory overview, a list of the materials and reagents needed to complete the experiment, and followed by a detailed procedure that is supported with a helpful notes section offering tips and tricks of the trade as well as troubleshooting advice. These hallmark features were introduced by series editor Dr. John Walker and constitute the key ingredient in each and every volume of the *Methods in Molecular Biology* series. Tested and trusted, comprehensive and reliable, all protocols from the series are indexed in PubMed.

# Legume Genomics

## Methods and Protocols

Edited by

**Mukesh Jain**

*School of Computational and Integrative Sciences, Jawaharlal Nehru University, New Delhi, India*

**Rohini Garg**

*Department of Life Sciences, School of Natural Sciences, Shiv Nadar University, Gautam Buddha Nagar, Uttar Pradesh, India*

Humana Press

*Editors*
Mukesh Jain
School of Computational and Integrative Sciences
Jawaharlal Nehru University
New Delhi, India

Rohini Garg
Department of Life Sciences
School of Natural Sciences
Shiv Nadar University
Gautam Buddha Nagar, Uttar Pradesh, India

ISSN 1064-3745 ISSN 1940-6029 (electronic)
Methods in Molecular Biology
ISBN 978-1-0716-0237-9 ISBN 978-1-0716-0235-5 (eBook)
https://doi.org/10.1007/978-1-0716-0235-5



This Humana imprint is published by the registered company Springer Science+Business Media, LLC, part of Springer Nature.
The registered company address is: 233 Spring Street, New York, NY 10013, U.S.A.

# Preface

Legumes contribute a major portion of dietary proteins for humans and animals. Unfortunately though, the full yield potential has not been achieved for legume crops due to several environmental constraints. Thus, there is a great need for improving legume crops to realize their full yield potential and meet the demands of the ever-increasing world population. In recent years, advancements and availability of various genomic methods and tools have provided impetus to legume research. These methods and tools can be harnessed to develop better legume crops via molecular breeding and/or genome engineering approaches, which can overcome various constraints. This book provides a compilation of a range of methods and protocols that allow researchers to better understand the biology of various cellular processes and agronomic traits and help devise better strategies to improve legume crops.

This book provides important reference material for genomicists, molecular breeders, plant molecular biologists, biotechnologists, computational biologists, and developmental biologists. The chapters are suitable for senior and young researchers/faculty, teachers, and graduate students involved in various molecular aspects of legume biology or in general plants.

The chapters are authored by reputed researchers and academicians working in the related area. The first two chapters provide updates on legume genomic resources followed by chapters describing methods of legume pangenome and organelle genome construction. Then, chapters on methods for transcriptome analysis, single-base resolution DNA methylation analysis, and genome-wide association studies using genotyping-by-sequencing and double-digest restriction site-associated DNA sequencing follow. In addition, chapters describing genomic selection in breeding, target enrichment sequencing via probe capture, and development of transposable element marker system for molecular breeding have been included. Two chapters provide methods for computational prediction of long non-coding RNAs and G-quadruplex sequences in legumes. One chapter describes a method for screening legume crops for abiotic stress tolerance and another illustrates an efficient method for mitochondrial DNA isolation. Three chapters deal with different plant transformation methods in legumes, including embryonic explant and plumular meristem transformation, hairy root transformation, and CRISPR/Cas9-based gene editing. Methods for genome-wide mining of disease resistance gene analogs and determination of bioactive compounds using LC-MS technique have also been included.

*We express sincere thanks and gratefulness to all of the esteemed authors. Without their strenuous efforts, this book would not have been possible.*

***New Delhi, India*** ***Mukesh Jain***
***Gautam Buddha Nagar, India*** ***Rohini Garg***

## Contents

*Preface* . . . . . . . . . . . . . . . . . . . . . . . . . . . . . . . . . . . . . . . . . . . . . . . *v*
*Contributors* . . . . . . . . . . . . . . . . . . . . . . . . . . . . . . . . . . . . . . . . . . . . . *ix*

1 Updates on Legume Genome Sequencing . . . . . . . . . . . . . . . . . . . . . 1
***Jungmin Ha and Suk-Ha Lee***

2 Updates on Genomic Resources in Chickpea for Crop Improvement . . . . . . . 19
***Rajesh Ghangal, Vikash K. Singh, Niraj K. Khemka, Mohan Singh Rajkumar, Rohini Garg, and Mukesh Jain***

3 Legume Pangenome Construction Using an Iterative Mapping and Assembly Approach . . . . . . . . . . . . . . . . . . . . . 35
***Haifei Hu, Yuxuan Yuan, Philipp E. Bayer, Cassandria T. Fernandez, Armin Scheben, Agnieszka A. Golicz, and David Edwards***

4 Methods and Tools for Plant Organelle Genome Sequencing, Assembly, and Downstream Analysis . . . . . . . . . . . . . . . . . . . . . 49
***Sanand Sandhya, Harsha Srivastava, Tanvi Kaila, Anshika Tyagi, and Kishor Gaikwad***

5 Methods of Gene Expression Profiling to Understand Abiotic Stress Perception and Response in Legume Crops . . . . . . . . . . . . . 99
***Meenu Bala, Ragini Sinha, M. A. Mallick, Tilak Raj Sharma, and Anil Kumar Singh***

6 Method for Bisulfite Sequencing Data Analysis for Whole-Genome Level DNA Methylation Detection in Legumes . . . . . . . . . . . . . 127
***Khushboo Gupta and Rohini Garg***

7 Method for Genome-Wide Association Study: A Soybean Example . . . . . . 147
***Robyn Anderson, Cassandria Tay Fernandez, Yuxuan Yuan, Agnieszka A. Golicz, David Edwards, and Philipp E. Bayer***

8 Genotyping for Species Identification and Diversity Assessment Using Double-Digest Restriction Site-Associated DNA Sequencing (ddRAD-Seq) . . . . . . . . . . . . . . . . . . . . . 159
***Anita A. Severn-Ellis, Armin Scheben, Ting Xiang Neik, Nur Shuhadah Mohd Saad, Aneeta Pradhan, and Jacqueline Batley***

9 Functional Genomic Selection in Crop Breeding . . . . . . . . . . . . . . . . 189
***Yongle Li***

10 Methods for Target Enrichment Sequencing via Probe Capture in Legumes . . . . . . . . . . . . . . . . . . . . . 199
***Ze Peng, Dev Paudel, Liping Wang, Ziliang Luo, Qian You, and Jianping Wang***

11 Developing Transposable Element Marker System for Molecular Breeding . . . . . . . . . . . . . . . . . . . . . 233
***R. S. Bhat, K. Shirasawa, Y. Monden, H. Yamashita, and M. Tahara***

12 Machine Learning-Based Annotation of Long Noncoding RNAs Using PLncPRO . . . . . 253
***Niraj K. Khemka, Urminder Singh, Anuj K. Dwivedi, and Mukesh Jain***

13 Methods for Identification and Validation of G-Quadruplex Sequences in Legumes . . . . . 261
***Surabhi Singh, Tanya Mathur, Khushboo Gupta, and Rohini Garg***

14 Method for Characterization of Root Traits in Chickpea Germplasm for Legume Genomics and Breeding . . . . . 269
***Yinglong Chen, Tao Zhou, and Kadambot H. M. Siddique***

15 Methods for Screening Legume Crops for Abiotic Stress Tolerance through Physiological and Biochemical Approaches . . . . . 277
***Ragini Sinha, Meenu Bala, Madan Kumar, Tilak Raj Sharma, and Anil Kumar Singh***

16 An Efficient Method of Mitochondrial DNA Isolation from *Vigna radiata* for Genomic Studies . . . . . 305
***Pooja Singh, Ranjan Kumar Sahoo, Mallesham Bulle, and Kapuganti Jagadis Gupta***

17 Embryonic Explant and Plumular Meristem Transformation Methods for Development of Transgenic Pigeon Pea . . . . . 317
***Shreeparna Ganguly, Arnab Purohit, Rituparna Kundu Chaudhuri, Sampa Das, and Dipankar Chakraborti***

18 Optimization of Hairy Root Transformation for the Functional Genomics in Chickpea: A Platform for Nodule Developmental Studies . . . . . 335
***Drishti Mandal, Deevita Srivastava, and Senjuti Sinharoy***

19 CRISPR/Cas9-Based Gene Editing in Soybean . . . . . 349
***Aili Bao, Lam-Son Phan Tran, and Dong Cao***

20 Genome-Wide Mining of Disease Resistance Gene Analogs Using Conserved Domains . . . . . 365
***Soodeh Tirnaz, Yueqi Zhang, and Jacqueline Batley***

21 Determination of Bioactive Compounds of Fenugreek (*Trigonella foenum-graecum*) Seeds Using LC-MS Techniques . . . . . 377
***Pratibha Singh, Vikas Bajpai, Vikaskumar Gond, Ashok Kumar, Narender Tadigoppula, and Brijesh Kumar***

22 Quantitative Phosphoproteomic Analysis of Legume Using $TiO_2$-Based Enrichment Coupled with Isobaric Labeling . . . . . 395
***Pragya Barua, Nilesh Vikram Lande, Sunil Kumar, Subhra Chakraborty, and Niranjan Chakraborty***

*Index* . . . . . *407*

## Contributors

ROBYN ANDERSON • *School of Biological Sciences and Institute of Agriculture, University of Western Australia, Perth, WA, Australia*
VIKAS BAJPAI • *Sophisticated Analytical Instrument Facility, CSIR-Central Drug Research Institute, Lucknow, Uttar Pradesh, India*
MEENU BALA • *ICAR-Indian Institute of Agricultural Biotechnology, Ranchi, India; Vinoba Bhave University, Hazaribagh, India*
AILI BAO • *Key Laboratory of Biology and Genetic Improvement of Oil Crops, Ministry of Agriculture and Rural Affairs, Oil Crops Research Institute, Chinese Academy of Agricultural Sciences, Wuhan, China*
PRAGYA BARUA • *National Institute of Plant Genome Research, Jawaharlal Nehru University Campus, New Delhi, India*
JACQUELINE BATLEY • *School of Biological Sciences, University of Western Australia, Crawley, WA, Australia*
PHILIPP E. BAYER • *School of Biological Sciences and Institute of Agriculture, University of Western Australia, Perth, WA, Australia*
R. S. BHAT • *Department of Biotechnology, University of Agricultural Sciences, Dharwad, Karnataka, India*
MALLESHAM BULLE • *National Institute of Plant Genome Research, New Delhi, India*
DONG CAO • *Key Laboratory of Biology and Genetic Improvement of Oil Crops, Ministry of Agriculture and Rural Affairs, Oil Crops Research Institute, Chinese Academy of Agricultural Sciences, Wuhan, China*
DIPANKAR CHAKRABORTI • *Department of Biotechnology, St. Xavier's College (Autonomous), Kolkata, West Bengal, India; Department of Genetics, University of Calcutta, Kolkata, West Bengal, India*
NIRANJAN CHAKRABORTY • *National Institute of Plant Genome Research, Jawaharlal Nehru University Campus, New Delhi, India*
SUBHRA CHAKRABORTY • *National Institute of Plant Genome Research, Jawaharlal Nehru University Campus, New Delhi, India*
RITUPARNA KUNDU CHAUDHURI • *Department of Botany, Krishnagar Govt. College, Krishnagar, West Bengal, India*
YINGLONG CHEN • *The UWA Institute of Agriculture, and School of Agriculture and Environment, The University of Western Australia, Perth, WA, Australia; Institute of Soil and Water Conservation, Northwest A&F University, and Chinese Academy of Sciences, Yangling, Shaanxi, China*
SAMPA DAS • *Division of Plant Biology, Bose Institute, Kolkata, West Bengal, India*
ANUJ K. DWIVEDI • *School of Computational and Integrative Sciences, Jawaharlal Nehru University, New Delhi, India*
DAVID EDWARDS • *School of Biological Sciences and Institute of Agriculture, University of Western Australia, Perth, WA, Australia*
CASSANDRIA T. FERNANDEZ • *School of Biological Sciences and Institute of Agriculture, University of Western Australia, Perth, WA, Australia*
KISHOR GAIKWAD • *ICAR-National Institute for Plant Biotechnology, New Delhi, India*

SHREEPARNA GANGULY • *Department of Biotechnology, St. Xavier's College (Autonomous), Kolkata, West Bengal, India*
ROHINI GARG • *Department of Life Sciences, School of Natural Sciences, Shiv Nadar University, Gautam Buddha Nagar, Uttar Pradesh, India*
RAJESH GHANGAL • *School of Computational & Integrative Sciences, Jawaharlal Nehru University, New Delhi, India*
AGNIESZKA A. GOLICZ • *Plant Molecular Biology and Biotechnology Laboratory, Faculty of Veterinary and Agricultural Sciences, University of Melbourne, Melbourne, VIC, Australia*
VIKASKUMAR GOND • *Sophisticated Analytical Instrument Facility, CSIR-Central Drug Research Institute, Lucknow, Uttar Pradesh, India*
KAPUGANTI JAGADIS GUPTA • *National Institute of Plant Genome Research, New Delhi, India*
KHUSHBOO GUPTA • *Department of Life Sciences, School of Natural Sciences, Shiv Nadar University, Gautam Buddha Nagar, Uttar Pradesh, India*
JUNGMIN HA • *Department of Plant Science and Research Institute of Agriculture and Life Sciences, Seoul National University, Seoul, Republic of Korea; Plant Genomics and Breeding Institute, Seoul National University, Seoul, Republic of Korea*
HAIFEI HU • *School of Biological Sciences and Institute of Agriculture, University of Western Australia, Perth, WA, Australia*
MUKESH JAIN • *School of Computational & Integrative Sciences, Jawaharlal Nehru University, New Delhi, India*
TANVI KAILA • *ICAR-National Institute for Plant Biotechnology, New Delhi, India*
NIRAJ K. KHEMKA • *School of Computational & Integrative Sciences, Jawaharlal Nehru University, New Delhi, India*
ASHOK KUMAR • *Academy of Scientific and Innovative Research (AcSIR), New Delhi, India; Medicinal and Process Chemistry, CSIR-Central Drug Research Institute, Lucknow, Uttar Pradesh, India*
BRIJESH KUMAR • *Sophisticated Analytical Instrument Facility, CSIR-Central Drug Research Institute, Lucknow, Uttar Pradesh, India; Academy of Scientific and Innovative Research (AcSIR), New Delhi, India*
MADAN KUMAR • *ICAR-Indian Institute of Agricultural Biotechnology, Ranchi, India*
SUNIL KUMAR • *National Institute of Plant Genome Research, Jawaharlal Nehru University Campus, New Delhi, India*
NILESH VIKRAM LANDE • *National Institute of Plant Genome Research, Jawaharlal Nehru University Campus, New Delhi, India*
SUK-HA LEE • *Department of Plant Science and Research Institute of Agriculture and Life Sciences, Seoul National University, Seoul, Republic of Korea; Plant Genomics and Breeding Institute, Seoul National University, Seoul, Republic of Korea*
YONGLE LI • *School of Agriculture, Food and Wine, The University of Adelaide, Adelaide, SA, Australia*
ZILIANG LUO • *Agronomy Department, University of Florida, Gainesville, FL, USA*
M. A. MALLICK • *Vinoba Bhave University, Hazaribagh, India*
DRISHTI MANDAL • *National Institute of Plant Genome Research, New Delhi, India*
TANYA MATHUR • *Department of Life Sciences, School of Natural Sciences, Shiv Nadar University, Gautam Buddha Nagar, Uttar Pradesh, India*
Y. MONDEN • *Graduate School of Environmental and Life Science, Okayama University, Okayama, Japan*

TING XIANG NEIK • *School of Biological Sciences, University of Western Australia, Crawley, WA, Australia*
DEV PAUDEL • *Agronomy Department, University of Florida, Gainesville, FL, USA*
ZE PENG • *Agronomy Department, University of Florida, Gainesville, FL, USA*
ANEETA PRADHAN • *School of Biological Sciences, University of Western Australia, Crawley, WA, Australia*
ARNAB PUROHIT • *Department of Biotechnology, St. Xavier's College (Autonomous), Kolkata, West Bengal, India*
MOHAN SINGH RAJKUMAR • *School of Computational & Integrative Sciences, Jawaharlal Nehru University, New Delhi, India*
NUR SHUHADAH MOHD SAAD • *School of Biological Sciences, University of Western Australia, Crawley, WA, Australia*
RANJAN KUMAR SAHOO • *Department of Biotechnology, Centurion University of Technology and Management, Bhubaneswar, Odisha, India*
SANAND SANDHYA • *ICAR-National Institute for Plant Biotechnology, New Delhi, India*
ARMIN SCHEBEN • *School of Biological Sciences and Institute of Agriculture, University of Western Australia, Perth, WA, Australia*
ANITA A. SEVERN-ELLIS • *School of Biological Sciences, University of Western Australia, Crawley, WA, Australia*
TILAK RAJ SHARMA • *ICAR-Indian Institute of Agricultural Biotechnology, Ranchi, India*
K. SHIRASAWA • *Department of Frontier Research and Development, Kazusa DNA Research Institute, Chiba, Japan*
KADAMBOT H. M. SIDDIQUE • *The UWA Institute of Agriculture, and School of Agriculture and Environment, The University of Western Australia, Perth, WA, Australia; Institute of Soil and Water Conservation, Northwest A&F University, and Chinese Academy of Sciences, Yangling, Shaanxi, China*
ANIL KUMAR SINGH • *ICAR-Indian Institute of Agricultural Biotechnology, Ranchi, India*
POOJA SINGH • *National Institute of Plant Genome Research, New Delhi, India*
PRATIBHA SINGH • *Sophisticated Analytical Instrument Facility, CSIR-Central Drug Research Institute, Lucknow, Uttar Pradesh, India; Academy of Scientific and Innovative Research (AcSIR), New Delhi, India*
SURABHI SINGH • *Department of Life Sciences, School of Natural Sciences, Shiv Nadar University, Gautam Buddha Nagar, Uttar Pradesh, India*
URMINDER SINGH • *School of Computational and Integrative Sciences, Jawaharlal Nehru University, New Delhi, India*
VIKASH K. SINGH • *School of Computational & Integrative Sciences, Jawaharlal Nehru University, New Delhi, India*
RAGINI SINHA • *ICAR-Indian Institute of Agricultural Biotechnology, Ranchi, India*
SENJUTI SINHAROY • *National Institute of Plant Genome Research, New Delhi, India*
DEEVITA SRIVASTAVA • *National Institute of Plant Genome Research, New Delhi, India*
HARSHA SRIVASTAVA • *ICAR-National Institute for Plant Biotechnology, New Delhi, India*
NARENDER TADIGOPPULA • *Academy of Scientific and Innovative Research (AcSIR), New Delhi, India; Medicinal and Process Chemistry, CSIR-Central Drug Research Institute, Lucknow, Uttar Pradesh, India*
M. TAHARA • *Graduate School of Environmental and Life Science, Okayama University, Okayama, Japan*
SOODEH TIRNAZ • *School of Biological Sciences, University of Western Australia, Crawley, WA, Australia*

LAM-SON PHAN TRAN • *Institute of Research and Development, Duy Tan University, Da Nang, Vietnam; Stress Adaptation Research Unit, RIKEN Center for Sustainable Resource Science, Yokohama, Japan*

ANSHIKA TYAGI • *ICAR-National Institute for Plant Biotechnology, New Delhi, India*

JIANPING WANG • *Agronomy Department, University of Florida, Gainesville, FL, USA; Plant Molecular and Cellular Biology Program, Genetics Institute, University of Florida, Gainesville, FL, USA*

LIPING WANG • *Agronomy Department, University of Florida, Gainesville, FL, USA*

H. YAMASHITA • *Graduate School of Environmental and Life Science, Okayama University, Okayama, Japan*

QIAN YOU • *Agronomy Department, University of Florida, Gainesville, FL, USA*

YUXUAN YUAN • *School of Biological Sciences and Institute of Agriculture, University of Western Australia, Perth, WA, Australia*

YUEQI ZHANG • *School of Biological Sciences, University of Western Australia, Crawley, WA, Australia*

TAO ZHOU • *College of Agronomy, Sichuan Agricultural University, Chengdu, China*

# Chapter 1

# Updates on Legume Genome Sequencing

**Jungmin Ha and Suk-Ha Lee**


## Abstract

Legumes (Fabaceae) are agronomically and economically one of the most important crops. Because legumes serve as a source of food, feed, and industrial materials, many studies in the field of legume genomics, including genome sequencing, have been conducted over the last decade. Here, we update the progress in genome sequencing of legume crops, including soybean (*Glycine max* [L.] Merr.), mung bean (*V. radiata* var. *radiata*), adzuki bean (*V. angularis* var. *angularis*), common bean (*Phaseolus vulgaris* L.), pigeon pea (*Cajanus cajan*), chickpea (*Cicer arietinum*), and peanut (*Arachis hypogaea*). Since the publication of the first reference genome sequence of each species, many accessions have been resequenced to study genetic diversity, speciation, and polyploidization in the legume lineage.


**Key words** Legume, Genome, Sequencing, Linkage map, Pseudomolecule, Resequencing, Genetic diversity

## 1 Introduction

Legumes are one of the most important crops worldwide, as they are a primary source of plant protein and oil in human diet and animal feed. In addition to protein and oil, legumes are rich in carbohydrates, vitamins, and essential nutrients, and depending on the species, legumes are used not only as food and feed but also as industrial materials [1–6]. Additionally, legume species improve soil quality by nitrogen fixation, and some legumes easily fit into crop rotation cycles with cereal crops [7–10]. Despite their agronomical and economical benefits, most legume species have not been studied intensively in genome sequencing and breeding programs.

The first legume species to be sequenced was soybean (*Glycine max* [L.] Merr.), a model legume crop. Since the publication of the reference genome sequence of soybean in 2010 [9], genomes of various other legume species have been published, including pigeon pea [11], chickpea [12], common bean [13], mung bean [14], adzuki bean [15], and peanut [16]. Moreover, diverse germplasm including wild relatives, ancestral species, and elite cultivars

Mukesh Jain and Rohini Garg (eds.), *Legume Genomics: Methods and Protocols*, Methods in Molecular Biology, vol. 2107,
https://doi.org/10.1007/978-1-0716-0235-5_1, 

of each of the abovementioned legume species have been sequenced using resequencing strategies [17–20].

With the increased economic significance of legume crops and demand for improved varieties, research in the area of legume genomics has increased. The accumulation of genomic information on legume crops will serve as a valuable resource in the near future for the development of improved cultivars capable of withstanding environmental challenges.

## 2 Materials and Methods

### 2.1 Soybean

Soybean is one of the most important crops worldwide. World production and growth area reached 353 tons and 124 million ha in 2017 (FAOstat, http://www.fao.org/faostat/). Generally, soybean seeds consist of 18% oil and 38% protein, and provide 30% and 69% of the oil and protein requirement in the human diet, respectively [2, 17]. Soybean meal is used as animal feed because of its high protein content [5, 9], and soybean oil is used not only as a food product but also as an industrial product in cosmetics, paint removers, and plastics [2]. Soybean is a palaeopolyploid that has undergone two cycles of whole genome duplication [9]. Analysis of the synonymous distance between paralogs in the soybean genome revealed that whole genome duplication of soybean occurred approximately 59 and 13 million years ago. Because of two whole genome duplications, 75% of the genes in the soybean genome exist as multiple copies [9].

The soybean genome was sequenced using the whole genome shotgun (WGS) sequencing approach. A total of 15,332,163 reads were generated by Sanger sequencing, and 3363 scaffolds covering 959.5 Mb of the soybean genome were generated after assembly using Arachne [21]. Anchoring the sequence reads to the genetic map generated 20 chromosome-scale pseudomolecules covering 950 Mb, which represented 85% of the soybean genome [9]. The authors identified 28 nodule synthesis genes and 24 nodule regulatory genes based on the comparison between soybean gene model and nodulation genes identified previously. In addition, 1127 genes with putative roles in lipid and oil biosynthesis were identified [9]. The number of genes involved in lipid and oil biosynthesis in soybean is twofold greater than that in *Arabidopsis thaliana*. Additionally, genes involved in lipid signaling, storage lipid degradation, and membrane lipid biosynthesis is two- to threefold higher in soybean than in Arabidopsis [9].

After the publication of the soybean reference genome, several soybean accessions have been resequenced. Lam et al. [17] showed genome-wide genetic variation between 17 wild and 14 cultivated soybean accessions that were resequenced using Illumina Genome Analyzer (Illumina GA) II platform. A total of 901.75 million 45 or

76 bp paired-end reads spanning 180 Gb were generated. One of the wild soybean accessions was sequenced at 80× genome coverage. The sequenced reads were aligned against the soybean reference genome, and 63,018,109 single nucleotide polymorphisms (SNPs) and 186,177 presence–absence variations (PAVs) were identified. Comparison between cultivated and wild soybean accessions showed that cultivated soybean accessions have lower genetic diversity than wild soybean genotypes, probably because of domestication bottleneck. Linkage disequilibrium (LD) analysis showed that cultivated soybeans had longer LD decay than that of wild soybeans (cultivated soybean: ~150 kb; wild soybean: ~75 kb). De novo assembly of the wild soybean genome sequence revealed 186,177 indels (<5 bp) in wild soybean, and 4444 and 1148 large PAVs (>500 bp) that were absent in the soybean reference genome and de novo assembly of wild soybean, respectively [17].

Additionally, a large number of plant species have been recently sequenced because of the rapid decline in sequencing cost. Resequencing of 62 wild soybeans, 130 landraces, and 110 improved cultivars using Illumina HiSeq 2000 identified genetic factors affected by domestication and breeding [19]. Each sample was sequenced at an average depth of 11×, generating a total of 33 billion 100 bp paired-end reads. Sequence analysis led to the identification of 9,790,744 SNPs, 876,799 indels (<6 bp), 1614 copy number variations (CNVs), and 6388 segmental deletions. A total of 230 selective sweeps were detected, of which 121 occurred during domestication and 109 during crop improvement. Among these, 53 domestication- and 43 improvement-selective sweeps contain quantitative trait loci (QTLs) for oil and 21 fatty acid biosynthesis genes. Morphological features such as seed size and seed coat color, and shine have been selected during domestication and improvement. In addition, some genetic factors, such as the *E1* gene, responsible for flowering time, and *T* locus, which regulates pubescence color, were selected because of geographic differentiation or local breeding. GWAS using sequencing data revealed that SNPs and CNVs, which were associated with traits such as flower color, seed coat color, pubescence form, cyst nematode resistance, hilum color, and plant height, were located within selective sweeps caused by domestication and crop improvement [19].

### 2.2 Wild Soybean

Wild soybean (*Glycine soja* Sieb. & Zucc.) is a close relative and an undomesticated ancestor of cultivated soybean [22]. Like cultivated soybean, wild soybean is also a primary source of plant protein and oil in the human diet and in animal feed. Compared with cultivated soybean, which has lost a substantial amount of genetic diversity because of breeding, wild soybean is genetically highly diverse and is adapted to various environmental conditions in a wide range of latitudes [18, 23, 24]. Therefore, genes or genetic variants identified in the wild soybean genome could be used to improve the environmental adaptation of cultivated soybean [18].

With the availability of the genome sequence of cultivated soybean [9], a number of studies have been conducted for the identification of novel sources of genetic variation for introgression into cultivated soybean [18, 25]. In 2010, Kim et al. generated 48.8 and 3.36 Gb of wild soybean genome sequence (var. IT182932) using Illumina GA and Roche Genome Sequencer FLX (GS FLX), respectively [25]. Mapping sequence reads to the reference genome sequence of cultivated soybean (var. Williams 82) revealed 2,167,728 and 337,257 SNPs in nongenic and genic regions, respectively. In addition, 32.4 Mb of cultivated soybean sequences were absent in the genome of wild soybean because of large deletions, and 32,262 contigs spanning 8.2 Mb, which were supposed to be wild soybean-specific large-scale insertions or unsequenced gaps in cultivated soybean, were identified.

However, because analysis of resequenced data is dependent on sequence similarity to the reference genome, resequencing is limited to the identification of large-scale structural variations including PAVs and CNVs. Moreover, a single genome is insufficient to represent the genomic content of wild soybean, as individual accessions of wild soybean are highly distinct because of the high level of genetic diversity and low level of genetic recombination. This suggested the need for a pangenome sequence comprising the majority of the genetic diversity in soybean. Li et al. [9] conducted de novo genome assembly of seven wild soybean accessions originating in the north, Huanghuai, and south of China, Japan, Korea, and Russia using sequence reads generated from Illumina HiSeq 2000 for three libraries with insert sizes of 180 bp, 500 bp, and 2 kb [18]. This generated contigs and scaffold N50 values ranging from 8 to 27 kb and 17–65.1 kb, respectively, with 94% coverage of the 54,175 reference genes in soybean. Comparison of pangenome sequences with the reference genome sequence resulted in 3.63–4.72 million SNPs, 0.50–0.77 million indels, 2.3–3.9 Mb wild soybean-specific PAVs, and 1978 CNV.

### 2.3 Mung Bean

Mung bean is a fast growing warm season legume species, mostly cultivated in South, East, and Southeast Asia for its edible seeds and sprouts [14]. It is a good source of dietary protein and contains higher levels of folate and iron than most other legumes [4, 14]. Additionally, mung bean fixes atmospheric nitrogen and fits easily into crop rotations with cereals [10]. Moreover, it has a short growth cycle (75–90 days), requires little water during cultivation, and grows well under arid and semiarid conditions [10].

The mung bean genome was sequenced using five libraries of 180 bp fragment, 5, 10, 40 kb mate-pair libraries, and a single linear library on two NGS platforms, Illumina HiSeq 2000 and GS FLX [14]. A total of 2748 scaffolds with an N50 length of 1.52 Mb were produced, covering 431 Mb and corresponding to 80% of the estimated genome size of 543 Mb obtained from 25 base kmer frequency distribution. To gain a better understanding of the domestication of mung bean, genome duplication and speciation were investigated by sequencing a wild relative of mung bean (*V. radiata* var. *sublobata*) and a tetraploid relative of mung bean (*V. reflexo-pilosa* var. *glabra*). *V. radiata* var. *sublobata* was sequenced using two types of libraries, a 180 bp paired-end library and 5 kb mate-pair library, and 8161 scaffolds with an N50 length of 214 kb were produced, covering 423 Mb and representing 84% of the estimated genome size of 501 Mb. *V. reflex-pilosa* var. *glabra* was sequenced using 180 bp paired-end and 5 kb mate-pair libraries, and the resulting assembly consisted of 29,166 scaffolds with an N50 length of 63 kb, covering 792 Mb and representing 82% of the estimated genome size of 968 Mb, which was almost twice the size of the mung bean genome [14]. Additionally, mung bean genetic map was constructed from an F6 population of 190 recombinant inbred lines (RILs) generated by single seed descent from a cross between *V. radiata* var. *radiata* VC1973A and the Korean landrace *V. radiata* var. *radiata* V2984 via genotyping-by-sequencing (GBS). Among 1993 SNPs identified from GBS, 1321 SNPs (covering 11 linkage groups [LGs]) were used to construct a genetic map. A total of 239 scaffolds were anchored to this genetic map, while 86 scaffolds remained unoriented. The resulting pseudochromosomes represented 11 LGs, with an N50 length of 35.4 Mb, and covered 314 Mb, corresponding to 73% of the total assembled genome sequence [14].

Information gained from the whole genome sequence of mung bean will facilitate further genomics-based research in *Vigna* species and provide a potential framework for future resequencing of other species in this genus. The whole genome sequence and high-density genetic map of mung bean will assist in efficient SNP discovery, thus improving genomics-assisted selection in mung bean breeding.

### 2.4 Adzuki Bean

Adzuki bean (diploid; $2n = 2x = 22$) is the second most important legume crop in East Asia [26]. It is cultivated in over 30 countries [27, 28] and consumed by more than a billion people [29]. Adzuki bean is known for its sweetness and is typically used as an ingredient in desserts. It is also known for its high nutritional value, as it is rich in protein, starch, minerals, and vitamins [3, 30]. Because of its high starch and low fat content, adzuki bean is categorized as a typical non-oil type legume crop [31]. Adzuki bean adapts well to soil with poor fertility and improves soil quality by nitrogen fixation [7, 8]. Because of the increased economic significance and demand for improved varieties of adzuki bean, the genome of adzuki bean has been sequenced [15, 31, 32]. Construction of a reference genome sequence of adzuki bean will not only be helpful for devising efficient breeding methods but also for comparative genomics of adzuki bean and other important legume crops.

The first genome sequence draft of adzuki bean was generated by sequencing a widely adopted Korean cultivar, Gyeongwon (IT213134) [15]. For the de novo assembly of adzuki bean genome sequence, Kang et al. [14] sequenced two 180 bp paired-end libraries and two 5 kb mate pair, and one 10 kb mate pair libraries on Illumina HiSeq 2000 and a linear library on Roche GS FLX+. Short and long reads produced by HiSeq 2000 and Roche GS FLX+, respectively, were assembled using the ALLPATHS-LG assembler and Newbler software, respectively [33]. Using the Jellyfish software [34] at 22 k-mer frequency, the genome size of adzuki bean was estimated as 591 Mb. A total of 3883 scaffolds with an N50 length of 703 kb were generated from the assembly, and the total length of all scaffolds (443 Mb) covered 75% of the predicted genome size.

The super-scaffolding strategy was used to understand the syntenic relationship between the genomes of adzuki bean and two other warm season legume crops, common bean and mung bean [15]. Conserved collinear genomic blocks produced by the McScanX software [35] were extracted and used for linking scaffolds into super-scaffolds. Consequently, this strategy improved the N50 length to 1.5 Mb and maximum scaffold length to 11.1 Mb. To assemble pseudomolecules, Kang et al. adopted the GBS method [36] and prepared a high-quality genetic map using 133 F4 lines derived from a cross between the adzuki bean accession Gyeongwon and wild species *V. nakashimae* (IT178530) [15]. After discarding cosegregating SNPs and SNPs with segregation distortions, 11 LGs with 814 SNPs were generated. A total of 158 scaffolds were anchored to the genetic map, yielding 11 pseudochromosomes. The total scaffold length and N50 length were 210 and 25 Mb, respectively.

In an independent study, Yang et al. [31] sequenced the genome of a Chinese adzuki bean cultivar, Jingnong 6, using the WGS sequencing method on Illumina HiSeq 2000 platform. High-quality reads of 90.88 Gb were produced and assembled using the SOAPdenovo software [37]. The genome size of adzuki bean was predicted as 542 Mb, according to k-mer analysis, and the assembled scaffolds of 466.7 Mb represented 86.11% of the genome. Yang et al. [31] subsequently constructed a high-density SNP genetic map with restriction-site associated DNA-sequencing (RAD-Seq) using 150 F2 individuals derived from a cross between the adzuki bean cultivar Ass001 and wild adzuki bean accession CWA108. Using mapped SNPs, a total of 372.9 Mb of scaffolds were assigned to 11 pseudochromosomes, covering 79.9% of the genome.

Recently, to enhance the contiguity and coverage of adzuki bean genome compared with the currently assembled legume genomes, single molecule real-time (SMRT) sequencing technology [38] was implemented by Sakai et al. [32]. For de novo assembly, Sakai et al. used PacBio RS II platform to produce 27.6 Gb of reads and Celera Assembler ver. 8.2beta [39] to assemble error-corrected reads. Compared with the assembly using second generation sequencers, SMRT-based sequencing resulted in 100-fold longer contigs with 100-fold smaller gaps, producing 279 anchored scaffolds that covered 85.6% of the genome [32].

### 2.5 Common Bean

Common bean, one of the most widely cultivated legumes in the world, is rich in protein, carbohydrates, some vitamins, and essential minerals including iron and zinc [6, 40, 41]. Because of the ease of cultivation, common bean is a staple food crop in developing and developed countries. Common bean provides 15% and 36% of the total daily calorie and protein requirement, respectively, in the human diet in parts of Africa and America. In sub-Saharan Africa, common bean is responsible for the daily nutrition of 200 million people.

Wild common bean consists of two distinct gene pools, Mesoamerican and Andean, which are thought to have diverged from a common ancestor more than 100,000 years ago [42]. These gene pools underwent domestication independently approximately 8000 years ago, resulting in landraces with different characteristics [43, 44].

Given that common bean is a staple food crop worldwide, a reference genome of common bean is required for the identification of genes responsible for increasing nutrient concentrations in common bean. Moreover, because of the presence of two distinct gene pools within a single species, common bean serves as an excellent material for studying parallel evolution and introgression.

An inbred landrace of common bean (G19833) derived from the Andean gene pool (Race Peru) was sequenced for constructing the reference genome of common bean [13]. Two different types of WGS sequencing platforms were used: Roche 454 and Illumina HiSeq 2000. Two types of linear 454 data (10.7 Gb of XLR data and 5.6 Gb of FLX+ data) were collected. Six 454 linear paired-end libraries with different insert sizes (2.8–4.8, 8.0, 9.2, 11.9, and 12.2 kb) were sequenced using standard XLR. Additionally, 24.1 Gb reads were produced from 400 bp insert libraries. For longer-range linkage, two fosmid libraries (average insert sizes: 35.0 and 36.0 kb) and three bacterial artificial chromosome (BAC) libraries (average insert sizes: 127.0, 135.3, and 122.0 kb) were sequenced on the Sanger platform, resulting in 591,289 reads and 402.5 Mb of high-quality sequence.

Prior to assembly, reads that met any of the following conditions were eliminated: (1) organelle DNA (mitochondria, chloroplast, and ribosomal DNA [rDNA]), (2) read length <200 bp among 454 linear reads, and (3) read length <50 bp of at least one 454 paired-end read. Additionally, to avoid inadequate assembly of transposon sequences, >80% 24-mers, which occurred at least 400 times in the data set, were also removed. Arachne v.20071016 [45] was used for genome assembly, and contigs <300 bp in length or supporting depth of 4 or less were removed as additional filtering. Consequently, 1627 scaffolds, with a scaffold L50 value of 6.0 Mb, were obtained. Subsequently, scaffolds consisting of >95% 24-mers that (1) occurred four other times in scaffolds greater than 50 kb in length, (2) contained only unanchored RNA sequences, or (3) were < 1 kb in length were filtered again. To obtain the common bean reference genome sequence, 19,619 Mb of 121 bp paired-end reads collected from a diverse set of genotypes were aligned with 14× coverage, and SNPs were called using the CASAVA1.7 software (Illumina, 2010). Illumina Infinium BeadChip (BARCBEAN6K_1 with 5232 SNPs) was designed using the obtained SNPs, and another Illumina BeadChip (BARCBEAN6K_2 with 5514 SNPs) was also designed using the same steps as those used for the *P. vulgaris* v0.9 assembly. Using Illumina BeadChip and 261 simple sequence repeat (SSR) markers, 267 F2 progeny and 88 F5 RILs derived from a cross between the common bean cultivars Stampede and Red Hawk were genotyped to construct a linkage map [46]. The final linkage map includes 7276 SSR and SNP markers in 11 LGs. Of these, 7015 markers from the genetic map were aligned to the reference genome assembly using BLAT [47], generating 11 pseudomolecules.

The final assembly of common bean reference genome contained 708 scaffolds with 41,391 contigs, which covered 472.5 Mb of the 587 Mb genome, with a contig N50 value of 39.5 Mb and a scaffold N50 value of 50.4 Mb. Pseudomolecules were refined once again on the basis of synteny with soybean. For annotation, Schmutz et al. [13] built seven transcript assemblies from approximately 727 million reads of paired-end Illumina RNA-Seq data using PERTRAN. Schmutz et al. also constructed 47,464 transcript assemblies using PASA [48] from 79,630 common bean Sanger expressed sequence tags (ESTs) and RNA-Seq transcript assemblies. Loci were identified using transcript assembly alignments and/or EXONERATE alignment. Gene models were predicted using homology-based predictors FGENESH+ [49], FGENESH_EST, and GenomeScan [50]. Selected gene predictions were improved by PASA, and corresponding peptides were subjected to peptide homology analysis. Subsequently, Pfam analysis was conducted, resulting in 27,197 and 31,638 protein-coding genes and transcripts, respectively. Repeat elements were also studied using the common bean genome sequence and 15 BAC clone sequences of common bean publicly available in GenBank. Schmutz et al. [13] identified and classified 2668 complete long terminal repeat (LTR) retrotransposons in 165 families.

### 2.6 Pigeon Pea

Pigeon pea is a diploid legume species ($2n = 2x = 22$) with an estimated genome size of 853 Mb [51]. It is an orphan crop and has not been thoroughly investigated, despite its importance for food security in poor countries. Since the domestication of pigeon pea approximately 3500 years ago [52], the tropical legume crop with low input requirements has been grown in many semiarid tropical regions as a main source of protein. However, there is a large gap between potential and observed yield of pigeon pea (866.2 kg/ha in Asia and 736.2 kg/ha in Africa) [53]. In the last few decades, low genetic diversity in the primary gene pool has been a bottleneck for crop improvement in breeding programs. Since the publication of pigeon pea reference genome sequence in 2012, breeding lines, landraces, and wild species of the genus *Cajanus* have been used for genome-wide association study (GWAS) to discover genomic variants associated with important agronomic traits [20].

Pigeon pea variety 'Asha' (ICPL87119) was selected for whole genome sequencing [11, 54]. Varshney et al. constructed 11 small-insert (180–800 bp) and 11 large-insert (2–20 kb) libraries, and a total of 237.2 Gb paired-end reads, ranging from 50 to 100 bp in size, were generated on Illumina GA and HiSeq 2000 sequencing platforms [11]. After quality control (QC) and correction, 130.7 Gb of high-quality sequences covering more than 163.4× of pigeon pea genome remained. An additional set of 88,860 BAC-end sequences (BESs) was produced from Sanger sequencing of two BAC libraries (69,120 clones) constructed using *Hin*dIII (34,560 clones) and *Bam*HI (34,560 clones) restriction enzymes. De novo assembly was performed using SOAPdenovo, generating 605.78 Mb of the pigeon pea genome with a contig N50 value of 21.95 kb and the longest contig of 185.39 kb. The paired-BES reads and a genetic map improved the assembly results with a scaffold N50 value of 516.06 kb and the longest scaffold length of 48.97 Mb. The pigeon pea genome size, using k-mer statistics, was estimated as 833.07 Mb. Thus, 72.7% of the total pigeon pea genome (605.78 Mb) was assembled, and 48,680 genes were predicted using the GLEAN algorithm [55].

Singh et al. [54] sequenced 19 plates of short DNA fragments from WGS libraries using GS FLX Phase D chemistry and three plates of paired-end sequences from 20 kb long fragment library using GS FLX Titanium chemistry [54, 56]. A total of 25,489,474 reads, representing 10.1 Gb sequence, were generated. To assemble the pigeon pea reference genome with filtered high-quality sequences, Newbler GS De Novo assembler version 2.5.3 (Roche Inc., Germany) was used with the following options: overlap minimum match length = 25 bp, large genome = true, number of CPU used = 0 (all), exclude contigs <500 bp. The size of the final assembled genome was 510,809,477 bp, with 191,705 QC-passed large contigs arranged in 59,681 scaffolds (average scaffold size = 7678 bp, N50 length = 13,989 bp, longest scaffold length = 177,971 bp); this is a first draft of pigeon pea genome sequence generated using only 454 GS FLX sequences.

Based on the reference genome of pigeon pea, whole genome resequencing of 292 *Cajanus* accessions, including breeding lines, landraces and wild species, was performed for the identification of genomic variants associated with domestication and agronomic traits [20]. A total of 21.7 billion 101 bp paired-end reads were generated and mapped to the pigeon pea reference genome [11] using BWA [57]. A total of 17.2 million variants including 15.1 million SNPs, 0.9 million small insertions and 1.2 million small deletions were identified. Furthermore, 19 genomic regions, comprising a selective sweep, were identified based on nonsynonymous to synonymous substitution ratio (Ka/Ks) greater than 2.5 by scanning the genome in nonoverlapping windows of 10 kb and 1 Mb in length. These regions, comprising 1749 genes, were considered the result of domestication and artificial selection in modern breeding programs.

## 2.7 Chickpea

Chickpea ($2n = 2x = 16$) is one of the most important staple food crops and the third most highly grown legume crop worldwide [12]. In 2017, the global production area of chickpea was approximately 260,000 ha, and the average yield was approximately 1.5 tons/ha, with a global production of 12 million tons. The yield and area of cultivation of chickpea were the highest in India at 9.1 million tons and 10 million ha, respectively (FAOSTAT). The gap between average and observed maximum yield in the same year suggests that there is extensive scope for crop improvement. Chickpea is rich in essential nutrients and a member of "founder crop package"; however, unlike other crops in this group including wheat and barley, chickpea has received far less attention [58]. The wild progenitors of chickpea are believed to be *C. reticulatum* (primary gene pool) and *C. echinospermum* (secondary gene pool). These wild progenitors are crossable with the cultivated chickpea but the hybrids tend to be sterile. Throughout the 7500 years of cultivation, chickpea has suffered severe evolutionary bottlenecks, causing a narrow genetic base and lowering biotic stress resistance [59, 60].

The WGS sequence of CDC Frontier, a Canadian *kabuli* chickpea variety, has been produced [12]. This cultivar exhibits several important disease and insect resistance traits. In addition, 29 other elite cultivars have been sequenced [12]. A total of 153 Gb paired-end raw sequence reads (207.32× genome) were generated using Illumina HiSeq 2000, and 88 Gb sequence was retained after quality filtering. The high-quality raw reads were assembled into scaffolds of at least 1 kb with 544 Mb of genomic sequence. The N50 length of scaffolds was 645.3 kb, and the longest scaffold was 6.17 Mb. Using k-mer statistics, the chickpea genome was estimated as 738.09 Mb, indicating that 73.8% of the genome was captured. The unassembled region was enriched for repetitive sequences. With the aid of 46,270 repeat masked BESs, 532.29 Mb sequence containing 7163 scaffolds was produced. Using ~1300 genetic markers, 65% of the assembly was anchored to eight genetic LGs. This allowed the identification of eight pseudomolecules. Additionally, resequencing of 90 cultivated and wild chickpea accessions allowed the detection of a selective sweep related to breeding and balancing selection [12].

Another study produced 13 Gb of high-quality sequence data of chickpea ICC4958 by sequencing WGS libraries and mate-pair libraries on the 454/Roche GS FLX Titanium platform. The Illumina GA IIx was used to produce 43.7 Gb of additional paired-end data [60]. Initial contig assembly of 304.9 Mbp generated using ABYSS was used in tandem with quality filtered GS FLX data to generate the primary assembly, which was further assembled using BESs. The scaffolds were anchored to the linkage group based on a previously constructed genetic map from the mapping population of *C. arietinum* ICC4958 × *C. reticulatum* PI489777 [61]. Overall, a 519,846,222 bp sequence was assembled into scaffolds with an N50 length of 77 kb. A total of 532 scaffolds of 124.4 Mb were allocated to eight LGs with the largest pseudomolecule, chromosome 3, of 23.4 Mb. The draft assembly covered 70% of the predicted chickpea genome size of 740 Mb at an average of 15× GS FLX read coverage. The assembled bases had a quality of at least 40. Mapping high-quality Illumina reads to the assembly showed a mismatch error frequency of 1.2 bp/10 kb and an indel frequency of 1.6/10 kb, which could be attributed to heterozygosity. The chickpea genome showed signs of segmental duplication but no recent whole genome duplication [60].

Based on the genome assembly by Jain et al. [60], a more advanced draft genome has been published [62]. The authors generated two additional raw data sets, including 1.01 Gb from mate-pair libraries and 23.93 Gb from paired-end libraries, using 454/Roche and Illumina Hiseq1000 platforms technology, respectively [62]. NEWBLER v2.6 tool allowed the researchers to improve the scaffold size, and any scaffold size ≤1 kb was not used for genome assembly. Overall, the number of scaffolds was reduced from 182,734 to 42,978, and the gaps between scaffolds were filled with paired-end reads using SOAPdenovo gap closure tool version 1.1. The gap filling process lowered the ambiguous nucleotides (Ns) from 13.8% to 8.68%, which is much lower than the genome assembly of CDC Frontier [12, 60, 62]. The biparental mapping population of *C. arietinum* cv. ICC 4958 × *C. reticulatum* cv. PI 489777 and its genetic linkage map allowed linking 511 Mbp to the corresponding linkage group. Overall, 334 Mbp sequence (66%) was assigned to the eight LGs; the largest linkage group spanned 54.99 Mb. Additionally, more than 4000 intraspecific SNP markers between *desi* and *kabuli* types were discovered [60].

*Cicer reticulatum* L. is the wild progenitor of chickpea. Gupta et al. [59] assembled the genome of the chickpea progenitor using three paired-end and two mate-pair libraries. Using ALLPATHS-LG, 1,198,234,588 filtered reads of *C. reticulatum* were de novo assembled, generating a total scaffold length of 16,562,993 bp with 49× coverage over the predicted genome size of 817.46 Mb [59]. A total of 7383 scaffolds, with an N50 value of 216.8 kb and average size of 56 kb, were created. Using genetic markers identified from a linkage map constructed using a RIL mapping population of *C. arietinum* cv. ICC4958 × *C. reticulatum* cv. PI489777, 1651 scaffolds were anchored to eight LGs, resulting in eight pseudomolecules of 327.07 Mb, with the largest linkage group (LG6) being 58.23 Mb. Out of the anchored sequences, around 65% were oriented successfully. The genome assembly of the wild progenitor showed high synteny to the model plant *Medicago truncatula*. Interestingly, analysis of gene sequence and nucleotide diversity suggested that *desi* type chickpea is genetically more similar to the wild species than the *kabuli* type [59].

### 2.8 Peanut

Peanut (groundnut) is a grain legume and a staple food crop in tropical and subtropical regions. Peanut seeds comprise 45–50% edible oil and play a key role in human nutrition [1]. The genus *Arachis* originated in South America and comprises ~80 species, most of which are diploid ($2n = 2x = 20$) [63]. Because cultivated peanut is an allopolyploid (AABB; $2n = 4x = 40$) with a large proportion of repetitive sequences (64%) in a relatively large genome (~2.7 Gb), the assembly of chromosomal pseudomolecules has been challenging [64]. To assemble the chromosomal pseudomolecules of peanut, two diploid ancestors, with genomes similar to the A and B subgenomes of cultivated peanut, were sequenced. *A. duranensis* Krapov. & W.C. Greg. and *A. ipaensis* Krapov. & W.C. Greg. are considered as the donors of the A and B subgenomes of cultivated peanut, respectively, based on cytogenetic, phylogeographic, and molecular data [65, 66]. In situ hybridization using genomic DNA of the two diploid species as probes showed that A and B subgenomes are clearly distinguishable from each other without large A–B mosaics [65, 67]. Additionally, the sum of genome sizes of *A. duranesis* (1.25 Gb) and *A. ipaensis* (1.56 Gb) is approximately equal to the genome size of peanut (~2.7 Gb), suggesting that no substantial changes have occurred in the peanut genome since polyploidization [68].

Considering that *A. duranensis* and *A. ipaensis* are good representatives of the ancestral diploid species of the cultivated tetraploid peanut, genomes of these two diploid ancestors were sequenced as a first step toward characterizing the genome of cultivated peanut [16, 69, 70].

To sequence the ancestral diploid species of peanut, *Arachis duranensis* and *A. ipaensis*, Bertioli et al. [16] constructed and sequenced 14 and 19 libraries of different insert sizes (250 bp, 500 bp, 2 kb, 5 kb, 10 kb, 20 kb, and 40 kb), generating 325.73 and 416.59 Gb raw data, respectively. During QC, the following reads were filtered out: reads with more than 5% Ns or polyadenylated termini; reads from short-insert libraries (170–800 bp) with 20 or more bases having quality scores $\leq$7; reads from large-insert libraries (2–40 kb) with 40 or more bases having quality score $\leq$ 7; reads with adaptor contamination; reads with $\geq$10 bp overlap between read 1 and read 2; reads identical to each other at both ends; and reads where the quality of the bases at the head or tail was $\leq$7.

Genome sizes were estimated by dividing the total numbers of k-mers by the depths of the major peaks. Sequencing errors in the 17-mers, with frequencies lower than 3, were corrected for 250 and 500 bp insert libraries. Genome assembly was conducted using SOAPdenovo v. 2.05 with parameters --K 81 --R [37]. Gaps were closed using KGF and GAPcloser [71]. Contigs were scaffolded using SSPACE, where at least five supporting reads exist [72]. Paired-end reads from the 250 bp insert library were joined into single longer reads using COPE [73].

Genetic maps were generated through GBS of two diploid RIL populations. Overall, 1692 and 459 scaffolds, representing 96.0% and 99.2% of the sequences in contigs >10 kb in length, were ordered into ten chromosomal pseudomolecules each for *A. duranensis* and *A. ipaensis*, respectively.

To perform genome-wide comparison and assembly validation, 5.74 Gb of long-sequence Moleculo reads were generated from TruSeq synthetic long-read sequencing libraries in *A. hypogaea* cv. Tifrunner [74], and 100 bp paired-end reads generated on HiSeq2500 were assembled into TruSeq (Moleculo) synthetic long reads. In total, 1,263,111 Moleculo long reads with N50 value of 6137 bp, providing 2.3× genome coverage, were mapped to the diploid pseudomolecules. Moleculo reads of peanut and pseudomolecules of *A. duranensis* and *A. ipaensis* were 98.36% and 99.96%, respectively.

Chen et al. [69] sequenced *A. duranensis* using the WGS sequencing strategy. Paired-end sequencing was performed using Illumina HiSeq 2500 producing 229.95 Gb reads from short-insert libraries (250, 500, and 800 bp) and long-insert libraries (2, 5, 10, and 20 kb). Low-quality reads were filtered out as follows: reads with four low-quality bases at both ends in short-insert libraries; reads with three low-quality bases in long-insert libraries; duplicated reads from long-insert libraries; three reads with at least 10 Ns and low quality bases.

Initial assembly of *A. duranensis* was performed using SOAPdenovo2 v. 2.04.4 (pregraph -K 79 -p 16 -d 60 5; scaff -F -b 1.5) using 159.07 Gb filtered reads [71]. SSPACE was used for scaffolding [72]. A draft genome of *A. duranensis* consisting of 8173 scaffolds and covering 1,051,523,805 bp with an N50 value of 649,840 was constructed. The quality of assembly was validated by PCR amplification of randomly selected regions, sequence-depth distribution, and expressed sequence tag [69].

*A. ipaensis* was sequenced using a shotgun approach on the Illumina HiSeq2500 by Lu et al. [70]. In total, 11 paired-end sequencing libraries consisting of 250, 500, 800 bp, 2, 5, 10 and 20 Kb insert sizes. SOAPdenovo2 and SSPACE were used to construct contigs and to link the contigs into scaffolds [71, 72]. To calculate sequencing depth, the frequency of each 17-mer were evaluated from the raw reads and the genome size was estimated using the total read length divided by sequencing depth [70]. High-quality reads of 250.4 Gb were generated representing 149.5× genome coverage. The *A. ipaensis* genome sequence of 1391.7 Mb was assembled with a contig N50 of 8067 bp and a scaffold N50 of 170,050 bp.

## 3 Notes

Because of their nutritional and agronomical importance, reference genome sequences of legume crops have been published. With the increased demand for improved varieties, resequencing of wild relatives, ancestral species, and elite cultivars have been conducted in each legume species. In this chapter, we updated genomic information on legume crops to be served as a valuable resource to develop improved cultivars against environmental challenges.

## Appendix

| Species | Sequencing platform | Strategy | Reference |
|---|---|---|---|
| Soybean | Sanger<br>Illumina GA II<br>Illumina Hiseq2000 | De novo<br>Resequencing<br>Resequencing | Schmutz et al. [13]<br>Lam et al. [17]<br>Zhou et al. [19] |
| Wild soybean | Illumina GA<br>GS FLX<br>Illumina Hiseq2000 | Resequencing<br><br>De novo | Kim et al. [25]<br><br>Li et al. [18] |
| Mung bean | Illumina Hiseq2000<br>GS FLX | De novo | Kang et al. [14] |
| Adzuki bean | Illumina Hiseq2000<br>GS FLX+<br>Illumina Hiseq2000<br>Pacbio RS II | De novo<br><br>De novo<br>De novo | Kang et al. [15]<br><br>Yang et al. [15]<br>Sakai et al. [32] |

(continued)

| Species | Sequencing platform | Strategy | Reference |
|---|---|---|---|
| Common bean | GS FLX+<br>GS XLR<br>Illumina Hiseq2000<br>Sanger | De novo | Schmutz et al. [13] |
| Pigeon pea | Illumina GA<br>Illumina Hiseq2000 | De novo | Varshney et al. [11] |
| | GS FLX | De novo | Singh et al. [54] |
| | Illumina Hiseq2000 | Resequencing | Varshney et al. [20] |
| Chickpea | Illumina Hiseq2000 | De novo resequencing | Varshney et al. [12] |
| | GS FLX<br>Illumina GA IIx | De novo | Jain et al. [60] |
| | GS FLX+<br>Illumina HIseq1000 | De novo | Parween et al. [62] |
| | Illumina HIseq1000 | De novo | Gupta et al. [59] |
| Peanut | Illumina Hiseq2000<br>Illumina Moleculo | De novo | Bertioli et al. [16] |
| | Illumina Hiseq2500 | De novo | Chen et al. [69] |
| | Illumina Hiseq2500 | De novo | Lu et al. [70] |

## References


1. Parthasarathy S, Khoo JC, Miller E et al (1990) Low density lipoprotein rich in oleic acid is protected against oxidative modification: implications for dietary prevention of atherosclerosis. Proc Natl Acad Sci 87:3894–3898
2. Liu K (2008) Food use of whole soybeans. In: Johnson L, White P, Galloway R (eds) Soybeans. Elsevier, Amsterdam, pp 441–481
3. Xu N, Cheng X-Z, Wang S-H et al (2008) Establishment of an adzuki bean (*Vigna angularis*) core collection based on geographical distribution and phenotypic data in China. Acta Agron Sin 34:1366–1373
4. Keatinge JDHD, Easdown WJ, Yang RY et al (2011) Overcoming chronic malnutrition in a future warming world: the key importance of mungbean and vegetable soybean. Euphytica 180:129–141. https://doi.org/10.1007/s10681-011-0401-6
5. Hartman GL, West ED, Herman TK (2011) Crops that feed the world 2. Soybean—worldwide production, use, and constraints caused by pathogens and pests. Food Secur 3:5–17. https://doi.org/10.1007/s12571-010-0108-x
6. Celmeli T, Sari H, Canci H et al (2018) The nutritional content of common bean (*Phaseolus vulgaris* L.) landraces in comparison to modern varieties. Agronomy 8:166. https://doi.org/10.3390/agronomy8090166
7. Duan H (1989) Small bean edible bean crops. Science 1989:160–171
8. Sikkema P, Soltani N, Shropshire C, Robinson D (2006) Response of adzuki bean to pre-emergence herbicides. Can J Plant Sci 86:601–604
9. Schmutz J, Cannon SB, Schlueter J et al (2010) Genome sequence of the palaeopolyploid soybean. Nature 463:178–183. https://doi.org/10.1038/nature08670
10. Dahiya PK, Linnemann AR, Boekel MAJSV et al (2015) Mung bean: technological and nutritional potential. Crit Rev Food Sci Nutr 55:670–688. https://doi.org/10.1080/10408398.2012.671202
11. Varshney RK, Chen W, Li Y et al (2012) Draft genome sequence of pigeonpea (*Cajanus cajan*), an orphan legume crop of resource-poor farmers. Nat Biotechnol 30:83–89. https://doi.org/10.1038/nbt.2022
12. Varshney RK, Song C, Saxena RK et al (2013) Draft genome sequence of chickpea (*Cicer arietinum*) provides a resource for trait improvement. Nat Biotechnol 31:240–246. https://doi.org/10.1038/nbt.2491

13. Schmutz J, McClean PE, Mamidi S et al (2014) A reference genome for common bean and genome-wide analysis of dual domestications. Nat Genet 46:707–713. https://doi.org/10.1038/ng.3008
14. Kang YJ, Kim SK, Kim MY et al (2014) Genome sequence of mungbean and insights into evolution within *Vigna* species. Nat Commun 5:5443. https://doi.org/10.1038/ncomms6443
15. Kang YJ, Satyawan D, Shim S et al (2015) Draft genome sequence of adzuki bean, *Vigna angularis*. Sci Rep 5:8069. https://doi.org/10.1038/srep08069
16. Bertioli DJ, Cannon SB, Froenicke L et al (2015) The genome sequences of Arachis duranensis and Arachis ipaensis, the diploid ancestors of cultivated peanut. Nat Genet 47:438
17. Lam H-M, Xu X, Liu X et al (2010) Resequencing of 31 wild and cultivated soybean genomes identifies patterns of genetic diversity and selection. Nat Genet 42:1053–1059. https://doi.org/10.1038/ng.715
18. Li Y, Zhou G, Ma J et al (2014) *De novo* assembly of soybean wild relatives for pan-genome analysis of diversity and agronomic traits. Nat Biotechnol 32:1045–1052. https://doi.org/10.1038/nbt.2979
19. Zhou Z, Jiang Y, Wang Z et al (2015) Resequencing 302 wild and cultivated accessions identifies genes related to domestication and improvement in soybean. Nat Biotechnol 33:408–414. https://doi.org/10.1038/nbt.3096
20. Varshney RK, Saxena RK, Upadhyaya HD et al (2017) Whole-genome resequencing of 292 pigeonpea accessions identifies genomic regions associated with domestication and agronomic traits. Nat Genet 49:1082–1088. https://doi.org/10.1038/ng.3872
21. Jaffe DB (2003) Whole-genome sequence assembly for mammalian genomes: arachne 2. Genome Res 13:91–96. https://doi.org/10.1101/gr.828403
22. Nawaz MA, Yang SH, Rehman HM et al (2017) Genetic diversity and population structure of Korean wild soybean (*Glycine soja* Sieb. and Zucc.) inferred from microsatellite markers. Biochem Syst Ecol 71:87–96. https://doi.org/10.1016/j.bse.2017.02.002
23. Qiu L-J, Xing L-L, Guo Y et al (2013) A platform for soybean molecular breeding: the utilization of core collections for food security. Plant Mol Biol 83:41–50. https://doi.org/10.1007/s11103-013-0076-6
24. He S-L, Wang Y-S, Li D-Z, Yi T-S (2016) Environmental and historical determinants of patterns of genetic differentiation in wild soybean (*Glycine soja* Sieb. et Zucc). Sci Rep 6:22795. https://doi.org/10.1038/srep22795
25. Kim MY, Lee S, Van K et al (2010) Whole-genome sequencing and intensive analysis of the undomesticated soybean (*Glycine soja* Sieb. and Zucc.) genome. Proc Natl Acad Sci 107:22032–22037. https://doi.org/10.1073/pnas.1009526107
26. Vaughan D, Tomooka N, Kaga A (2005) Azuki bean [*Vigna angularis* (Wild.) Ohwi & Ohashi]. In: Singh RJ, Jauhar PP (eds) Genetic resources, chromosome engineering, and crop improvement: grain legumes. Taylor Francis, Boca Raton, FL, pp 347–359
27. Tomooka N, Vaughan D, Moss H, Maxted N (2002) The Asian Vigna: the genus Vigna subgenus Ceratotropis genetic resouces. Springer, New York, NY, p 270
28. Kramer C, Soltani N, Robinson DE et al (2012) Control of volunteer adzuki bean in soybean. Agric Sci 03:501–509. https://doi.org/10.4236/as.2012.34059
29. Lestari P, Kang YJ, Han K-S et al (2014) Genome-wide single nucleotide polymorphism discovery and validation in adzuki bean. Mol Breed 33:497–501. https://doi.org/10.1007/s11032-013-9962-5
30. Lin R, Chai Y, Liao Q et al (2002) Minor grain crops in china. China Agricultural Science and Technology Press, Beijing, pp 210–223
31. Yang K, Tian Z, Chen C et al (2015) Genome sequencing of adzuki bean (*Vigna angularis*) provides insight into high starch and low fat accumulation and domestication. Proc Natl Acad Sci 112:13213–13218. https://doi.org/10.1073/pnas.1420949112
32. Sakai H, Naito K, Ogiso-Tanaka E et al (2015) The power of single molecule real-time sequencing technology in the *de novo* assembly of a eukaryotic genome. Sci Rep 5:16780. https://doi.org/10.1038/srep16780
33. Gnerre S, MacCallum I, Przybylski D et al (2011) High-quality draft assemblies of mammalian genomes from massively parallel sequence data. Proc Natl Acad Sci 108:1513–1518. https://doi.org/10.1073/pnas.1017351108
34. Marçais G, Kingsford C (2011) A fast, lock-free approach for efficient parallel counting of occurrences of k-mers. Bioinformatics 27:764–770. https://doi.org/10.1093/bioinformatics/btr011
35. Wang Y, Tang H, DeBarry JD et al (2012) MCScanX: a toolkit for detection and

evolutionary analysis of gene synteny and collinearity. Nucleic Acids Res 40:e49–e49
36. Elshire RJ, Glaubitz JC, Sun Q et al (2011) A robust, simple genotyping-by-sequencing (GBS) approach for high diversity species. PLoS One 6:e19379. https://doi.org/10.1371/journal.pone.0019379
37. Li R, Zhu H, Ruan J et al (2010) *De novo* assembly of human genomes with massively parallel short read sequencing. Genome Res 20:265–272. https://doi.org/10.1101/gr.097261.109
38. Eid J, Fehr A, Gray J et al (2009) Real-time DNA sequencing from single polymerase molecules. Science 323:133–138. https://doi.org/10.1126/science.1162986
39. Miller JR, Delcher AL, Koren S et al (2008) Aggressive assembly of pyrosequencing reads with mates. Bioinformatics 24:2818–2824. https://doi.org/10.1093/bioinformatics/btn548
40. Singh J et al Cotyledon cell structure and in vitro starch digestion in navy beans. ScienceDirect. https://www.sciencedirect.com/science/article/pii/B9780124046108000086. Accessed 7 Jan 2019
41. Geil PB, Anderson JW (1994) Nutrition and health implications of dry beans: a review. J Am Coll Nutr 13:549–558. https://doi.org/10.1080/07315724.1994.10718446
42. Mamidi S, Rossi M, Moghaddam S et al (2013) Demographic factors shaped diversity in the two gene pools of wild common bean *Phaseolus vulgaris* L. Heredity 110:267–276. https://doi.org/10.1038/hdy.2012.82
43. Bitocchi E, Nanni L, Bellucci E et al (2012) Mesoamerican origin of the common bean (*Phaseolus vulgaris* L.) is revealed by sequence data. Proc Natl Acad Sci 109:E788–E796. https://doi.org/10.1073/pnas.1108973109
44. Bitocchi E, Bellucci E, Giardini A et al (2013) Molecular analysis of the parallel domestication of the common bean (*Phaseolus vulgaris*) in Mesoamerica and the Andes. New Phytol 197:300–313. https://doi.org/10.1111/j.1469-8137.2012.04377.x
45. Jaffe DB, Butler JL, Gnerre S et al (2003) Whole-genome sequence assembly for mammalian genomes: Arachne 2. Genome Res 13:91–96
46. Van Ooijen JW (2006) JoinMap® 4.0: software for the calculation of genetic linkage maps in experimental population. Kyazma BV, Wageningen
47. James Kent W (2002) BLAT – the BLAST-like alignment tool. Genome Res 12:656–664. https://doi.org/10.1101/gr.229202
48. Haas BJ, Delcher AL, SMM Sm et al (2003) Improving the Arabidopsis genome annotation using maximal transcript alignment assemblies. Nucleic Acids Res 31:5654–5666. https://doi.org/10.1093/nar/gkg770
49. Salamov A, Solovyev VV (2000) Ab initio gene finding in Drosophila genomic DNA. Genome Res 10:516–522
50. Yeh RF, Lim LP, Burge CB (2001) Computational inference of homologous gene structures in the human genome. Genome Res 11:803–816. https://doi.org/10.1101/gr.175701
51. Greilhuber J, Obermayer R (1998) Genome size variation in *Cajanus cajan* (Fabaceae): a reconsideration. Plant Syst Evol 212:135–141
52. Royes WV (1976) Pigeonpea. Evol Crop Plan Theory 1976:154–156
53. Mula M, Saxena K (2010) Lifting the level of awareness on pigeonpea-a global perspective. International Crops Research Institute for the Semi-Arid Tropics, Patancheru
54. Singh NK, Gupta DK, Jayaswal PK et al (2012) The first draft of the pigeonpea genome sequence. J Plant Biochem Biotechnol 21:98–112. https://doi.org/10.1007/s13562-011-0088-8
55. Elsik CG, Mackey AJ, Reese JT et al (2007) Creating a honey bee consensus gene set. Genome Biol 8:R13. https://doi.org/10.1186/gb-2007-8-1-r13
56. Margulies M, Egholm M, Altman WE et al (2005) Genome sequencing in microfabricated high-density picolitre reactors. Nature 437:376
57. Li H, Durbin R (2009) Fast and accurate short read alignment with Burrows–Wheeler transform. Bioinformatics 25:1754–1760
58. Jukanti AK, Gaur PM, Gowda CLL, Chibbar RN (2012) Nutritional quality and health benefits of chickpea (*Cicer arietinum* L.): a review. Br J Nutr 108(Suppl 1):S11–S26. https://doi.org/10.1017/S0007114512000797
59. Gupta S, Nawaz K, Parween S et al (2017) Draft genome sequence of Cicer reticulatum L., the wild progenitor of chickpea provides a resource for agronomic trait improvement. DNA Res Int J Rapid Publ Rep Genes Genomes 24:1–10. https://doi.org/10.1093/dnares/dsw042
60. Jain M, Misra G, Patel RK et al (2013) A draft genome sequence of the pulse crop chickpea (Cicer arietinum L.). Plant J 74:715–729. https://doi.org/10.1111/tpj.12173
61. Gaur R, Azam S, Jeena G et al (2012) High-throughput SNP discovery and genotyping for constructing a saturated linkage map of

chickpea (*Cicer arietinum* L.). DNA Res Int J Rapid Publ Rep Genes Genomes 19:357–373. https://doi.org/10.1093/dnares/dss018
62. Parween S, Nawaz K, Roy R et al (2015) An advanced draft genome assembly of a *desi* type chickpea (*Cicer arietinum* L.). Sci Rep 5:12806. https://doi.org/10.1038/srep12806
63. Kochert G, Stalker HT, Gimenes M et al (1996) RFLP and cytogenetic evidence on the origin and evolution of allotetraploid domesticated peanut, *Arachis hypogaea* (Leguminosae). Am J Bot 83:1282–1291
64. Dhillon SS, Rake AV, Miksche JP (1980) Reassociation kinetics and cytophotometric characterization of peanut (*Arachis hypogaea* L.) DNA. Plant Physiol 65:1121–1127
65. Seijo G, Lavia GI, Fernández A et al (2007) Genomic relationships between the cultivated peanut (*Arachis hypogaea*, Leguminosae) and its close relatives revealed by double GISH. Am J Bot 94:1963–1971
66. Moretzsohn MC, Gouvea EG, Inglis PW et al (2012) A study of the relationships of cultivated peanut (*Arachis hypogaea*) and its most closely related wild species using intron sequences and microsatellite markers. Ann Bot 111:113–126
67. Ramos ML, Fleming G, Chu Y et al (2006) Chromosomal and phylogenetic context for conglutin genes in Arachis based on genomic sequence. Mol Gen Genomics 275:578–592
68. Samoluk SS, Robledo G, Podio M et al (2015) First insight into divergence, representation and chromosome distribution of reverse transcriptase fragments from L1 retrotransposons in peanut and wild relative species. Genetica 143:113–125
69. Chen X, Li H, Pandey MK et al (2016) Draft genome of the peanut A-genome progenitor (Arachis duranensis) provides insights into geocarpy, oil biosynthesis, and allergens. Proc Natl Acad Sci 113:6785–6790
70. Lu Q, Li H, Hong Y et al (2018) Genome sequencing and analysis of the peanut B-genome progenitor (Arachis ipaensis). Front Plant Sci 9:604
71. Luo R, Liu B, Xie Y et al (2012) SOAPdenovo2: an empirically improved memory-efficient short-read de novo assembler. GigaScience 1:18
72. Boetzer M, Henkel CV, Jansen HJ et al (2010) Scaffolding pre-assembled contigs using SSPACE. Bioinformatics 27:578–579
73. Liu B, Yuan J, Yiu S-M et al (2012) COPE: an accurate k-mer-based pair-end reads connection tool to facilitate genome assembly. Bioinformatics 28:2870–2874
74. McCoy RC, Taylor RW, Blauwkamp TA et al (2014) Illumina TruSeq synthetic long-reads empower de novo assembly and resolve complex, highly-repetitive transposable elements. PLoS One 9:e106689

# Chapter 2

# Updates on Genomic Resources in Chickpea for Crop Improvement

**Rajesh Ghangal, Vikash K. Singh, Niraj K. Khemka, Mohan Singh Rajkumar, Rohini Garg, and Mukesh Jain**

## Abstract

In recent years, rapid advancement has been done in generation of genomic resources for the important legume crop chickpea. Here, we provide an update on important advancements made on availability of genomic resources for this crop. The availability of reference genome and transcriptome sequences, and resequencing of several accessions have enabled the discovery of gene space and molecular markers in chickpea. These resources have helped in elucidating evolutionary relationship and identification of quantitative trait loci for important agronomic traits. Gene expression in different tissues/organs during development and under abiotic/biotic stresses has been interrogated. In addition, single-base resolution DNA methylation patterns in different organs have been analyzed to understand gene regulation. Overall, we provide a consolidated overview of available genomic resources of chickpea that may help in fulfilling the promises for improvement of this important crop.

**Key words** Chickpea, Genome, Transcriptome, Web resource, Molecular markers, Linkage maps, QTLs

## 1 Introduction

Chickpea (*Cicer arietinum*) is self-pollinating diploid ($2n = 2x = 16$) plant with genome size of 740 Mb. It is world's second most widely grown legume crop and excellent source of proteins, carbohydrates, fiber and minerals. It is currently cultivated in more than 50 countries, covering approximately 14.8 million ha area [1]. Across the world, Southern and South-Eastern Asia account for 80% of the total chickpea production, whereby India leads the list accounting for approximately 70% of total world chickpea production.

Chickpea contributes to sustainable agriculture by virtue of symbiotic nitrogen fixation. Its productivity is 0.962 t/ha globally

---

Rajesh Ghangal and Vikash K. Singh contributed equally to this work.

Mukesh Jain and Rohini Garg (eds.), *Legume Genomics: Methods and Protocols*, Methods in Molecular Biology, vol. 2107, https://doi.org/10.1007/978-1-0716-0235-5_2, © Springer Science+Business Media, LLC, part of Springer Nature 2020

[1], which is very less than its potential (6 t/ha) under optimum growing conditions [2]. Chickpea production is affected by several abiotic and biotic stress conditions. Among abiotic stress factors, drought stress alone results in about 50% global reduction in yield, as it mainly occurs at the terminal stage of seed development in chickpea [3]. Besides this, salinity stress has also emerged as another abiotic stress factor affecting chickpea production globally. In addition, low temperature hampers seed development and reduces crop productivity remarkably [2]. Prominently, heat stress and terminal drought present major constraints for its productivity [4]. Several biotic stress factors also negatively affect growth and yield. Fungal attack has been reported to be one of the most severe causes of crop loss. Fungi that severely infect chickpea are *Fusarium oxysporum* f. sp. *ciceri* (causes plant to wilt), *Rhizoctonia bataticola* (results in dry root rot), and *Sclerotium rolfsii* (causing collar rot). Ascochyta blight, caused by *Ascochyta rabiei*, is one of the most destructive diseases in chickpea. Another disease, Botrytis grey mold (BGM) caused by *Botrytis cineria* Pres., also results in greater yield loss [5, 6].

Chickpea growth and development is affected greatly due to various abiotic and biotic factors. Hence, there is a need to develop improved chickpea varieties or adapt alternate strategies to enhance crop productivity under challenging environmental conditions. To develop stress-tolerant chickpea varieties, better understanding of crop responses to such stresses is required. Under stress conditions, networks of several genes are regulated by specific transcription factors to control stress responses in plants. Therefore, combining genomics and molecular breeding approaches together is expected to provide more efficient methods to develop stress-tolerant cultivars.

Resources for genomic research have been increasing rapidly for chickpea in recent years, including assembly of genome and transcriptome sequences, annotation of genes encoding functional proteins, construction of genetic linkage maps and gene expression profiling in different biological contexts [7–12]. Several efforts have been made to identify molecular markers using high-throughput sequencing technologies and many of them have been associated with agronomically important traits [10, 13–15]. This chapter describes current status of chickpea genomic resources and their potential applications toward development of improved cultivars (Table 1).

## 2 Genomic Resources for Chickpea

### 2.1 Genome Sequence

Genome sequences of two cultivated chickpea types, desi [7] and kabuli [8], are now available. Jain et al. [7] assembled 520 Mb of desi type chickpea (ICC 4958) genome representing 70% of the total genome (~740 Mb) and more than 80% of the genes encoding

**Table 1**
**Web resources available for chickpea**

| Resource | Remarks | Reference |
|---|---|---|
| Desi chickpea genome sequence | *Cicer arietinum* ICC 4958 (desi)<br>Sequenced genome size (520 Mb)<br>Weblink: http://nipgr.res.in/CGAP/home.php | [7] |
| Kabuli chickpea genome sequence | *Cicer arietinum* CDC Frontier (kabuli)<br>Sequenced genome size (532 Mb)<br>The web page has a wide array of features like genetic and genomic resources, trait mapping, functional analysis, and transcriptomics which can be helpful for researchers<br>Weblink: http://cegsb.icrisat.org/gt-bt/ICGGC/ICGSC.htm | [8] |
| Desi chickpea genome sequence (v2) | *Cicer arietinum* ICC 4958 (desi)<br>Sequenced genome size (511 Mb)<br>Weblink: http://nipgr.res.in/CGAP2/home.php | [16] |
| Wild chickpea genome sequence | *Cicer reticulatum* PI 489777 (wild)<br>Sequenced genome size (327 Mb)<br>Wild chickpea genome analysis project (WCGAP)<br>Genome sequence and its annotation can be downloaded<br>The webpage also have linkage maps, marker sequences, SSRs, SNPs as genomic resource<br>Weblink: http://nipgr.res.in/WCGAP/ | [17] |
| Chickpea Transcriptome Database (CTDB) | An integrated chickpea transcriptome database for functional and applied genomics transcriptome sequence from cultivated desi (ICC 4958) and kabuli (ICC V2)) and wild (PI 489777) chickpea genotypes<br>The tools available include basic search function, gene expression profiling in flower development, molecular markers like microsatellites and SNPs, and comparative genomics<br>Weblink: http://nipgr.res.in/ctdb.html | [22, 82] |
| Chickpea Genomic Web Resource (CGWR) | Visualization of desi (ICC 4958) chickpea genome and comparative analysis with wild and cultivated chickpea genotypes, and various other leguminous and nonleguminous plants<br>CGWR offers functional genomic resource along with customized maps and gene-clustering options<br>Weblink: http://www.nipgr.res.in/CGWR/home.php | [83] |
| Chickpea microsatellite database (CicArMiSatDB) | Provides detailed marker information on chickpea SSRs and their genomic features<br>The database archives 48,298 SSRs with integrated BLAST search<br>Weblink: https://cegresources.icrisat.org/CicArMiSatDB/ | [84] |

(continued)

**Table 1**
**(continued)**

| Resource | Remarks | Reference |
|---|---|---|
| CicerTransDB (Cicer Transcription Factor Database) | Classification of chickpea transcription factors in families<br>Provides information of the gene, gene ontology, and domain and motif architecture<br>Weblink: http://www.cicertransdb.esy.es/ | [88] |
| Chickpea Microsatellite Database (CMsDB) | An informative polymorphic SSR marker database generated using desi and kabuli chickpea<br>Provides microsatellites annotation with size, and genomic location (genic or intergenic) including their physical position on chickpea chromosome, pseudomolecules/ linkage groups<br>Weblink: http://www.nipgr.res.in/CMsDB.html | [85] |
| CicArVarDB | SNP and InDel database for advancing genetics research, molecular genetics, gene mapping, and breeding applications in chickpea<br>The database contains more than 1.9 million variations (substitutions, small length insertions, and deletions) spanning across (upstream and downstream regions of genes, intronic, exonic, and intergenic regions) the chickpea genome<br>Weblink: https://cegresources.icrisat.org/cicarvardb/ | [86] |
| Chickpea ISM-ILP Marker Database | This marker database contains genome-wide 119,169 and 110,491 ISMs from 23,129 desi and 20,386 kabuli protein-coding genes<br>It also catalogues 7454 in silico InDel, (1–45-bp)-based ILP markers from 3283 genes<br>Its use will accelerate the identification of molecular tags associated with important agronomic traits for genomics-assisted chickpea breeding<br>Weblink:<br>http://webapp.cabgrid.res.in/chickpea/index.html | [87] |

functional proteins. Parween et al. [16] reported an improved genome assembly (version 2.0) of ICC 4958 representing more than 94% of the annotated genes and substantial reduction in sequence gaps with 2.7-fold increase in length of pseudomolecules. Varshney et al. [8] reported 544.73 Mb assembled genome of kabuli type chickpea that covers 74% of chickpea genome. Approximately half (49.41%) of the kabuli chickpea genome is represented by transposable elements and unclassified repeats. Segmental duplication analysis of genome sequences revealed presence of 110 syntenic blocks that include 5–62 gene pairs [8]. Synteny analysis carried out till date have illustrated extensive conservation between chickpea and other legumes [7, 8]. Recently, a draft genome assembly (416 Mb) of a wild chickpea [*Cicer reticulatum* L. (PI 489777)] covering about 90% of the annotated genes has been reported (*see* **Note 1**) [17].

**2.2 Transcriptome Sequence**

Earlier, expressed sequence tag (EST) based transcriptome sequencing generated thousands of single-pass sequences for chickpea [13, 18–21], mostly from tissues under various abiotic stress conditions. Later, the transcriptome sequences of cultivated and wild chickpea types were sequenced via high-throughput deep sequencing technologies (*see* **Note 1**) [15, 22–25]. Garg and colleagues [23] presented high-quality transcriptome sequence of desi type chickpea (genotype ICC 4958) using Roche 454 and Illumina next generation sequencing (NGS) platforms, and highlighted several conserved biological processes and pathways. In addition, they identified lineage- and species-specific genes in chickpea that can facilitate understanding of evolutionary relationships among legumes. Jhanwar et al. [15] reported 37, 265 transcripts in wild chickpea, *C. reticulatum* (genotype PI 489777), representing a total transcriptome size of 35.2 Mb. Following the availability of desi and wild chickpea transcriptome sequences, Agarwal et al. [25] reported the transcriptome sequence (43,389 transcripts) of a kabuli type chickpea (genotype ICC V2), which has high commercial value and protein content as compared to the desi type. Taking 22 different tissue samples of chickpea (genotype ICC 4958), Hiremath et al. [24] generated a total of 103,215 unique transcripts. *Cicer arietinum* Transcriptome Assembly version 2 (CaTA v2), a hybrid transcriptome assembly generated from sequences obtained using NGS platforms (Illumina and FLX/454) and Sanger sequencing method, comprised 46,369 transcript contigs [26].

**2.3 Molecular Markers**

Discovery of large number of different molecular markers help plant breeders to overcome several problems of conventional breeding and assist in identification of candidate genes that are linked with agronomic traits (*see* **Note 2**). Using molecular markers, several genetic and physical maps have been constructed in chickpea [10, 27]. Isozyme markers were the first to be employed for construction of map in chickpea [28]. Subsequently, restriction fragment length polymorphism (RFLP) [29], random amplification of polymorphic DNA (RAPD) [30], amplified fragment polymorphism (AFLP) [31], and simple sequence repeats (SSRs) [32] were exploited for genetic mapping and diversity studies. However, only the SSR markers have been used extensively in chickpea genetic and breeding programs. Numerous SSR markers were generated using genomic DNA libraries [33–35]. Buhariwalla et al. [18] developed functional markers using ESTs. Kudapa et al. [26] developed intron spanning regions (ISRs) molecular markers in chickpea using hybrid transcriptome assembly. Gene-based molecular markers, including 2088 intron-spanning region (ISR), 728 SSRs, 495 SNPs, and 387 conserved orthologous sequence (COS) markers were identified by Hiremath et al. [24] in their transcriptome study. Varshney et al. [13] developed 177 drought and salinity responsive EST-SSR markers. In addition, transcriptome sequences have been utilized to identify markers that might have association with abiotic stress responses [22, 24].

Single nucleotide polymorphisms (SNPs) have been revealed from ESTs, transcriptome and genomic sequences. Jhanwar et al. [15] revealed 37,954 SNPs between desi-type/kabuli-type using transcriptome sequences. Agarwal et al. [25] detected 1986 SNPs between kabuli/desi genotypes and 37,954 SNPs between kabuli/wild genotypes. Genetic variations were found to be lesser between the cultivated (kabuli/desi) as compared to those between cultivated and wild type [15, 25]. The genomes of different chickpea species/accessions have been re/sequenced to identify SNPs from coding and noncoding regions [7, 8, 27, 36]. A total of 44,844 high-quality SNPs were discovered from 93 diverse wild and cultivated accessions [10]. Gaur et al. [37] integrated 6698 SNPs on eight linkage groups of interspecific recombinant inbred line (RIL) population. To identify SNPs responsible for contrasting seed size in chickpea, Garg et al. [11] screened the promoter region of the genes controlling seed weight/size in ten cultivars/genotypes comprising four small- and six large-seeded chickpea. Overall, SNPs have garnered considerable attention in past years due to their higher abundance and ease of identification via high-throughput approaches.

### 2.4 Linkage Maps

Interspecific crosses of *C. arietinum* with *C. reticulatum* and *Cicer echinospermum* and isozyme markers were used to create the first linkage map of chickpea [28]. Subsequently, an integrated genetic linkage map of chickpea was developed consisting of nine morphological, 27 isozymes, and 10 RFLP and 45 RAPD markers [38]. RIL mapping population of *C. arietinum* (ICC 4958) × *C. reticulatum* (PI 489777) was taken as reference to construct high density genetic map of other chickpea cultivars [33, 35, 39, 40]. This reference mapping population was used to develop a comprehensive map with 521 loci comprising mainly SSR markers [33]. This reference genetic map was further enriched by integration of BES-SSRs, DArT, and gene-based markers having 1291 loci [35]. Hiremath et al. [41] constructed genetic map for this population using KASPar assays for 1328 loci. Among them, 625 loci were novel chickpea KAspar assay markers (CKAMs), 314 tentative orthologous gene-SNPs (TOG-SNPs), and 389 published loci.

Screening of molecular markers from transcriptome sequences have been used to map loci in the same parental lines ICC 4958 and PI 489777 [39, 41, 42]. The first intraspecific map for cultivated chickpea from ICCV2 × JG62 RILs was developed by Cho et al. [43]. This map was utilized for recognizing genes responsible for vital morphological traits together with double-podded character. STMS markers were also mapped on the linkage map of this population by Gaur et al. [34]. An intraspecific genetic map of chickpea comprising STMS and ISSRs along with 12 resistance gene analogs was generated using F2 population of chickpea [44], whereas another map was generated through a Kabuli × Desi cross [45]. Linkage maps have also been generated via specific mapping populations possessing diverse agronomical traits and molecular markers [10, 14, 46, 47].

A comprehensive genetic map through both inter- and intra-specific populations, including five interspecific (*C. arietinum* × *C. reticulatum*) and five intraspecific (desi × kabuli genotypes) populations has also been generated [48]. Upadhyaya et al. [49] studied the inheritance of seed size in chickpea via crossing a normal seed size parent (ICC 11255) and a small seed size parent (ICC 5002).

### 2.5 QTLs for Agronomically Important Traits

Several studies have been performed to associate genomic regions with agronomically important traits, like resistance to Fusarium wilt [50], Ascochyta blight [51], rust [52], botrytis grey mold [53], drought tolerance [54, 55], salinity stresses [56], and seed traits and yield (*see* **Note 3**) [9]. Paul et al. [57] identified QTLs associated with heat tolerance in chickpea using RILs developed by crossing ICC 4567 (heat sensitive) and ICC 15614 (heat tolerant) genotypes. A genetic map consisting of 271 SNP markers was constructed using genotyping by sequencing (GBS) and 25 putative heat-stress candidate genes were identified in the two major genomic regions (CaLG05 and CaLG06) [57].

Using genome-wide association studies (GWAS), Li et al. [58] fine mapped QTLs related to ascochyta blight resistance in chickpea. Efforts toward resequencing the whole genome revealed genomic regions that were under positive selection for yield, nitrogen fixation, and abiotic stress traits in chickpea [59]. Using the pool-based DNA analysis, Kujur et al. [60] identified seven significant QTLs related to seed weight, five of which were present on the linkage group (LG) 1 and two were present on LG 2. Bajaj et al. [61] identified two QTLs/eQTLs on LG 2 and LG 7 for seed weight, whereas another study identified five SNPs on LG 1, 2, and 4, which were significantly and consistently associated with seed weight QTLs [62]. In a separate study, Hossain et al. [63] identified two major QTLs (on LG 4 and LG 1) that together accounted for 20% of the seed size variation. Both loci accounted for 37% of the phenotypic variance indicating close genetic relationship between seed size and weight. They also assessed the role of environment in seed size by setting up the experiment at two locations with contrasting rainfall, soil type, and temperature and found limited environmental influence on this trait. Using SSR and GBS markers, Varshney et al. [54] delineated a region on chickpea chromosome 4 as "QTL-hotspot." Kale et al. [55] narrowed down this region to a ~300 kb region harboring 26 genes. Taking 189 RILs, derived from a cross between ICC 995 (medium-protein content) and ICC 5912 (high-protein content) genotypes of the desi chickpea, QTL at LG 1 and 3 were found to be associated with seed filling and seed coat development [64].

### 2.6 Gene Expression Datasets

Genes that are differentially expressed under biotic and abiotic stress conditions, and/or across various tissues/stages of development are the candidates that have potential role in adaptation of plants to stress conditions, and development of organs/tissues, respectively (*see* **Note 4**). Many genes/ESTs engaged in regulation of diverse stress responses have been detected through transcriptome studies in chickpea [13, 65–67]. Various methods, such as suppression subtractive hybridization, EST sequencing, microarray, super serial analysis of gene expression (SuperSAGE), and RNA-seq, have been employed to detect abiotic stress-responsive genes in chickpea [13, 18, 67]. The transcriptome analysis of roots and nodules was performed under salt stress by Molina et al. [65] using deep SuperSAGE. Jain and Chattopadhyay [20] analyzed two chickpea genotypes (BGD 72 and ICC V2) for differential regulation of genes under drought stress. Further, Deokar et al. [21] via a comparative study of drought-tolerant (ICC 4958) and susceptible (ICC 1882) genotypes determined genes that are significantly up- or downregulated under drought condition. Garg et al. [66] reported 3536 differentially expressed genes under different abiotic stress conditions (desiccation, salinity, and cold) in chickpea (ICC 4958), and reported involvement of these genes in transcription regulation, photosynthesis, energy metabolism, secondary metabolite biosynthesis, hormonal responses and osmoprotectant metabolism under stress conditions. In an another study, Garg et al. [67] reported 4954 genes that were specifically upregulated in the drought-tolerant genotype (ICC 4958) and 5545 genes that were specifically upregulated in the salinity-tolerant genotype (JG 62) under drought and salinity stress, respectively. Two contrasting chickpea genotypes, wilt-susceptible (JG 62) and wilt-resistant (WR 315), in disease and immune state were profiled using microarray analysis to understand gene expression switches in host specific resistance [68]. Via global network and subcellular analysis, 389 coregulated genes and 76 core genes related to disease/immunity, respectively, were identified.

Moreover, few studies have been performed to profile gene expression across various tissue types or stages of development in chickpea. Singh et al. [69] exploited high-throughput sequencing to measure the transcript abundance during chickpea flower development. Among transcription factors, several members of MADS-box gene family were found upregulated at different stages of flower development [69]. A gene expression resource, *Cicer arietinum* Gene Expression Atlas (CaGEA), consisting of 27 tissue samples from different organs and five developmental stages (germination and seedling stages, vegetative, reproductive, and senescence) resulted in identification of differentially expressed genes involved in flowering, nodulation, and seed and root development [12]. They also identified drought-responsive genes in chickpea and validated nine candidate genes present in "QTL-hotspot" region that were associated with drought stress response. Garg et al. [11] analyzed gene expression profiles across different stages of seed development in small-seeded (Himchana 1) and large-seeded (JGK 3) chickpea cultivars. The transcriptome study revealed differential gene expression programs between these two cultivars during seed development. Gene expression analysis revealed extended period of cell division during embryogenesis along with higher level of endoreduplication during the later stages of seed maturation determine large seed size/weight in JGK 3 cultivar [11].

### 2.7 Noncoding RNAs

Noncoding RNAs do not code for functional proteins, but regulate the expression of protein coding genes. In chickpea, several microRNAs (miRNAs) that target mRNAs involved in various processes of plant development, biotic and abiotic stress responses, have been identified (*see* **Note 5**) [70–75]. Employing NGS platform for genome-wide discovery of miRNAs, Jain et al. [72] identified 440 conserved and 178 novel miRNAs from seven different tissues/organs in chickpea. The roles of miRNAs have been analyzed in various tissues/organs in chickpea [72, 74]. These studies suggest the involvement of miRNAs through regulation of transcription factor encoding genes, such as AP2, MYB, SBP, and GRF families [72, 74]. In addition, tissue/organ/developmental stage specific expression of members of few miRNA families have also been reported [72]. Another study revealed the importance of miRNAs in root development during water deficiency and salinity stress [73]. By sequencing three chickpea libraries, fungal-infected (fusarium wilt), salt-treated, and untreated seedlings, Kohi et al. [75] identified 59 novel and 122 conserved miRNAs of 25 different families. MicroRNA, miR530, was found to be upregulated in response to fungal infection causing wilt disease in chickpea [75]. In a recent study, of the 651 miRNAs (478 known and 173 novel), differential expression of 297 miRNAs was observed between different genotypes, conditions and time points in response to ascochyta blight infection [76].

Long noncoding RNAs (lncRNAs) are represented by >200 bp long transcripts, but lack ability to encode functional proteins. From RNA-Seq data of 11 tissues, Khemka et al. [77] identified the first set of 2248 lincRNAs in chickpea. Network propagation algorithm was employed to functionally annotate these lincRNAs. A comprehensive expression profiling revealed preferential expression of several lincRNAs in flower developmental stages and tissues representing actively dividing cells. Out of 2248 total lincRNAs identified, 79% and 13% were assigned with a putative function and tissue-specific expression during various stages of flower development, respectively [77]. LncRNAs have also been detected in different cultivars of chickpea under drought and salinity stress responses using a novel tool PLncPRO [78]. These studies suggest tissue-specific functions of lncRNAs in chickpea.

### 2.8 DNA Methylation

There is a rapid surge in the studies involving epigenetic regulation of gene expression. DNA methylation and histone modifications are the important epigenetic factors that are widely known to be involved in regulation of eukaryotic gene expression (*see* **Note 5**). Few studies have been performed via methylation-sensitive amplified polymorphism (MSAP) method to profile DNA methylation changes in chickpea genotypes subjected to biotic [79] and abiotic stresses [80]. Recently, Bhatia et al. [81] reported a single base resolution methylome analysis in different tissues/organs of chickpea. A larger number of CG context differentially methylated regions (DMRs) were identified between cultivated and wild chickpeas, whereas CHH context DMRs were more abundant among the tissues/organs of cultivated chickpea [81]. Higher CG and CHH methylation was observed within gene-body and intergenic (promoter/downstream) regions, respectively. Many more such studies are expected in near future to dissect the epigenetic regulation in different biological contexts in chickpea.

### 2.9 Web Resources

Garg et al. [22] constructed Chickpea Transcriptome Database (CTDB), as a searchable interface for transcriptome sequence of chickpea (genotype ICC 4958). An updated release of the database (CTDB v2.0) hosts transcriptome sequence from desi (ICC 4958), kabuli (ICC V2), and wild (PI 489777) chickpea genotypes [82]. This database can be used to perform similarity and functional annotation search along with expression profiles of chickpea transcripts in major tissues/organs and various flower development stages. Chickpea Genomic Web Resource (CGWR) offers visual and functional genomic analysis of the desi chickpea genome [83]. Doddamani et al. [84] constructed chickpea microsatellite database (CicArMiSatDB) having detailed information on chickpea SSRs and their genomic features. The polymorphic SSR markers between desi and kabuli chickpea types have been cataloged in the Chickpea Microsatellite Database (CMsDB), which can be used for efficient large-scale genotyping applications [85]. CicArVarDB has an entry of 1.9 million SNPs and InDels. The repository has features like webBLAST search and JBrowse visualization which can be used to select variations around specific regions associated with QTLs [86]. Srivastava et al. [87] used intron-spanning markers (ISM), insertion-deletion (InDel), and intron-length polymorphism (ILP) markers from protein-coding genes of desi and kabuli for the construction of "Chickpea ISM-ILP Marker Database." Gayali et al. [88] developed Cicer Transcription Factor Database (CicerTransDB) for systematic studies like gene information, gene ontology, and domain and motif architecture of transcription factors in chickpea.

## 3 Conclusion

Chickpea is one of the most important legume crops for global food and nutritional security. In the past decade, there has been a huge paradigm shift that once considered to be an orphan legume, chickpea is now a genomic resource-rich legume. Here, we summarized the available genomic resources to assist in designing robust strategies to improve chickpea productivity. Although research on chickpea was initiated at a later time as compared with other model crops/plants, most of the necessary genomic resources are now available. Browsers and web resources dedicated to chickpea are also available publicly. These resources can provide ingredients to identify candidate genes for the crop improvement.

## 4 Notes

1. Genome and transcriptome sequences of various chickpea types/genotypes are important resources available which can be used for various structural and functional analyses.
2. Large number of molecular markers are available for chickpea which have been used for creating linkage maps, studying genetic diversity, and genotyping applications.
3. Various mapping population studies have resulted in marking of QTLs associated with important agronomical traits.
4. Differential gene expression datasets comprised of different tissues and various development stages have shown tissue- and stage-specific expression of few of the genes that can help in prioritizing candidate genes for functional studies.
5. Studies pertaining to regulation of gene expression via ncRNAs (miRNAs and lncRNAs) and DNA methylation have revealed important insights into gene regulation in chickpea.

## Acknowledgments

Funding from the Department of Biotechnology, Government of India, New Delhi under the Challenge Programme on Chickpea Functional Genomics (BT/AGR/CG-PhaseII/01/2014) and Tata Innovation Fellowship Scheme (BT/HRD/35/01/04/2019) is gratefully acknowledged.

## References


1. FAOSTAT (2014). http://faostat.fao.org/beta/en/#data/QC
2. Singh KB (1987) Chickpea breeding. In: Saxena MC, Singh KB (eds) The chickpea. C.A.B. International, Oxon, pp 127–162
3. Johansen C, Krishnamurthy L, Saxena NP, Sethi SC (1994) Genotypic variation in moisture response of chickpea grown under line-source sprinklers in a semi-arid tropical environment. Field Crops Res 37:103–112
4. Singh KB, Malhotra RS, Halila MH, Knights EJ, Verma MM (1994) Current status and future strategy in breeding chickpea for resistance to biotic and abiotic stresses. Euphytica 73:137–149
5. Smithson JB, Thompson JA, Summerfield RJ (1985) Chickpea (*Cicer arietinum* L.). In: Summerfield RJ, Roberts EH (eds) Grain legume crops. Collins Publications, London, pp 312–390
6. Kaiser WJ (1992) Epidemiology of *Ascochyta rabiei*. In: Singh KB, Saxena MC (eds) Disease resistance breeding in kabuli chickpeas. ICARDA, Aleppo, pp 117–143
7. Jain M, Misra G, Patel RK, Priya P, Jhanwar S et al (2013) A draft genome sequence of the pulse crop chickpea (*Cicer arietinum* L.). Plant J 74:715–729
8. Varshney RK, Song C, Saxena RK, Azam S, Yu S et al (2013) Draft genome sequence of chickpea (*Cicer arietinum*) provides a resource for trait improvement. Nat Biotechnol 31:240–246
9. Karami E, Talebi R, Kharkesh M, Saidi A (2015) Linkage map of chickpea (*Cicer arietinum* L.) based on population from ILC3279×ILC588 crosses: location of genes for time to flowering, seed size and plant height. Genetika 47:253–263
10. Kujur A, Bajaj D, Upadhyaya HD, Das S, Ranjan R, Shree T et al (2015) A genome-wide SNP scan accelerates trait-regulatory genomic loci identification in chickpea. Sci Rep 5:11166
11. Garg R, Singh VK, Rajkumar MS, Kumar V, Jain M (2017) Global transcriptome and coexpression network analyses reveal cultivar-specific molecular signatures associated with seed development and seed size/weight determination in chickpea. Plant J 91(6):1088–1107
12. Kudapa H, Garg V, Chitikineni A, Varshney RK (2018) The RNA-Seq-based high resolution gene expression atlas of chickpea (*Cicer arietinum* L.) reveals dynamic spatio-temporal changes associated with growth and development. Plant Cell Environ 41(9):2209–2225
13. Varshney RK, Hiremath PJ, Lekha P, Kashiwagi J, Balaji J et al (2009) A comprehensive resource of drought- and salinity-responsive ESTs for gene discovery and marker development in chickpea (*Cicer arietinum* L.). BMC Genomics 10:523
14. Verma S, Gupta S, Bandhiwal N, Kumar T, Bharadwaj C et al (2015) High-density linkage map construction and mapping of seed trait QTLs in chickpea (*Cicer arietinum* L.) using genotyping-by-sequencing (GBS). Sci Rep 5:17512
15. Jhanwar S, Priya P, Garg R, Parida SK, Tyagi AK et al (2012) Transcriptome sequencing of wild chickpea as a rich resource for marker development. Plant Biotechnol J 10:690–702
16. Parween S, Nawaz K, Roy R, Pole AK, Venkata Suresh B et al (2015) An advanced draft genome assembly of a desi type chickpea (*Cicer arietinum* L.). Sci Rep 5:1–14
17. Gupta S, Nawaz K, Parween S, Roy R, Sahu K et al (2017) Draft genome sequence of *Cicer reticulatum* L., the wild progenitor of chickpea provides a resource for agronomic trait improvement. DNA Res 24:1–10
18. Buhariwalla HK, Jayashree B, Eshwar K, Crouch JH (2005) Development of ESTs from chickpea roots and their use in diversity analysis of the *Cicer* genus. BMC Plant Biol 5:16
19. Ashraf N, Ghai D, Barman P, Basu S, Gangisetty N et al (2009) Comparative analyses of genotype dependent expressed sequence tags and stress-responsive transcriptome of chickpea wilt illustrate predicted and unexpected genes and novel regulators of plant immunity. BMC Genomics 10:415
20. Jain D, Chattopadhyay D (2010) Analysis of gene expression in response to water deficit of chickpea (*Cicer arietinum* L.) varieties differing in drought tolerance. BMC Plant Biol 10:24
21. Deokar AA, Kondawar V, Jain PK, Karuppayil SM, Raju NL et al (2011) Comparative analysis of expressed sequence tags (ESTs) between drought-tolerant and -susceptible genotypes of chickpea under terminal drought stress. BMC Plant Biol 11:70
22. Garg R, Patel RK, Jhanwar S, Priya P, Bhattacharjee A et al (2011) Gene discovery and tissue specific transcriptome analysis in chickpea with massively parallel pyrosequencing and web

resource development. Plant Physiol 156:1661–1678
23. Garg R, Patel RK, Tyagi AK, Jain M (2011) De novo assembly of chickpea transcriptome using short reads for gene discovery and marker identification. DNA Res 18:53–63
24. Hiremath PJ, Farmer A, Cannon SB, Woodward J, Kudapa H et al (2011) Large-scale transcriptome analysis in chickpea (*Cicer arietinum* L.), an orphan legume crop of the semi-arid tropics of Asia and Africa. Plant Biotechnol J 9:922–931
25. Agarwal G, Jhanwar S, Priya P, Singh VK, Saxena MS et al (2012) Comparative analysis of kabuli chickpea transcriptome with desi and wild chickpea provides a rich resource for development of functional markers. PLoS One 7:e52443
26. Kudapa H, Azam S, Sharpe AG, Taran B, Li R et al (2014) Comprehensive transcriptome assembly of chickpea (*Cicer arietinum* L.) using sanger and next generation sequencing platforms: development and applications. PLoS One 9(1):e86039
27. Deokar AA, Ramsay L, Sharpe AG, Diapari M, Sindhu A et al (2014) Genome wide SNP identification in chickpea for use in development of a high density genetic map and improvement of chickpea reference genome assembly. BMC Genomics 15:708
28. Gaur PM, Slinkard AE (1990) Genetic control and linkage relations of additional isozymes markers in chickpea. Theor Appl Genet 80:648–653
29. Udupa SM, Sharma A, Sharma RP, Pai RA (1993) Narrow genetic variability in *Cicer arietinum* L. as revealed by RFLP analysis. J Plant Biochem Biotechnol 2:83
30. Iruela M, Rubio J, Cubero JI, Gil J, Millan T (2002) Phylogenetic analysis in the genus *Cicer* and cultivated chickpea using RAPD and ISSR markers. Theor Appl Genet 104:643–651
31. Nguyen TT, Taylor PWJ, Redden RJ, Ford R (2004) Genetic diversity estimates in *Cicer* using AFLP analysis. Plant Breed 123:173–179
32. Huttel B, Winter P, Weising K, Choumane W, Weigand F et al (1999) Sequence tagged microsatellite markers for chickpea (*Cicer arietinum* L.). Genome 42:210–217
33. Nayak SN, Zhu H, Varghese N, Datta S, Choi HK et al (2010) Integration of novel SSR and genebased SNP marker loci in the chickpea genetic map and establishment of new anchor points with *Medicago truncatula* genome. Theor Appl Genet 120:1415–1441
34. Gaur R, Sethy NK, Choudhary S, Shokeen B, Gupta V et al (2011) Advancing the STMS genomic resources for defining new locations on the intraspecific genetic linkage map of chickpea (*Cicer arietinum* L.). BMC Genomics 12:117
35. Thudi M, Bohra A, Nayak SN, Varghese N, Shah TM et al (2011) Novel SSR markers from BAC end sequences, DArT arrays and a comprehensive genetic map with 1,291 marker loci for chickpea (*Cicer arietinum* L.). PLoS One 6:e27275
36. Rajkumar MS, Garg R, Jain M (2018) Genome-wide discovery of DNA polymorphisms among chickpea cultivars with contrasting seed size/weight and their functional relevance. Sci Rep 8(1):16795
37. Gaur R, Jeena G, Shah N, Gupta S, Pradhan S et al (2015) High density linkage mapping of genomic and transcriptomic SNPs for synteny analysis and anchoring the genome sequence of chickpea. Sci Rep 5:13387
38. Simon CJ, Muehlbauer FJ (1997) Construction of a chickpea linkage map and its comparison with maps of pea and lentil. J Hered 88:115–119
39. Gujaria N, Kumar A, Dauthal P, Dubey A, Hiremath P et al (2011) Development and use of genic molecular markers (GMMs) for construction of a transcript map of chickpea (*Cicer arietinum* L.). Theor Appl Genet 122:1577–1589
40. Gaur R, Azam S, Jeena G, Khan AW, Choudhary S et al (2012) High-throughput SNP discovery and genotyping for constructing a saturated linkage map of chickpea (*Cicer arietinum* L.). DNA Res 19(5):357–373
41. Hiremath PJ, Kumar A, Penmetsa RV, Farmer A, Schlueter JA et al (2012) Large-scale development of cost-effective SNP marker assays for diversity assessment and genetic mapping in chickpea and comparative mapping in legumes. Plant Biotechnol J 10:716–732
42. Choudhary S, Gaur R, Gupta S, Bhatia S (2012) EST-derived genic molecular markers: development and utilization for generating an advanced transcript map of chickpea. Theor Appl Genet 124:1449–1462
43. Cho S, Kumar J, Shultz JL, Anupama K, Tefera F et al (2002) Mapping genes for double podding and other morphological traits in chickpea. Euphytica 128:285–292
44. Flandez-Galvez H, Ades PK, Ford R, Pang ECK, Taylor PWJ (2003) QTL analysis for ascochyta blight resistance in an intraspecific population of chickpea (*Cicer arietinum* L.). Theor Appl Genet 107:1257–1265

45. Cobos MJ, Fernández MJ, Rubio J, Kharrat M, Moreno MT et al (2005) A linkage map of chickpea (*Cicer arietinum* L.) based on populations from Kabuli x Desi crosses: location of genes for resistance to fusarium wilt race o. Theor Appl Genet 110:1347–1353
46. Kujur A, Bajaj D, Saxena MS, Tripathi S, Upadhyaya HD et al (2013) Functionally relevant microsatellite markers from chickpea transcription factor genes for efficient genotyping applications and trait association mapping. DNA Res 20:355–374
47. Kujur A, Upadhyaya HD, Bajaj D, Gowda CL, Sharma S et al (2016) Identification of candidate genes and natural allelic variants for QTLs governing plant height in chickpea. Sci Rep 6:27968
48. Millan T, Winter P, Jüngling R, Gil J, Rubio J et al (2010) A consensus genetic map of chickpea (*Cicer arietinum* L.) based on 10 mapping populations. Euphytica 175:175–189
49. Upadhyaya HD, Kumar S, Gowda CLL, Singh S (2006) Two major genes for seed size in chickpea (*Cicer arietinum* L.). Euphytica 147:311
50. Jain M, Pole AK, Singh VK, Ravikumar RL, Garg R (2015) Discovery of molecular markers for Fusarium wilt via transcriptome sequencing of chickpea cultivars. Mol Breed 35:198
51. Anbessa Y, Taran B, Warkentin TD, Tullu A, Vandenberg A (2009) Genetic analyses and conservation of QTL for ascochyta blight resistance in chickpea (*Cicer arietinum* L.). Theor Appl Genet 119:757–765
52. Madrid E, Rubiales D, Moral A, Moreno MT, Millán T et al (2008) Mechanism and molecular markers associated with rust resistance in a chickpea interspecific cross (*Cicer arietinum* x *Cicer reticulatum*). Eur J Plant Pathol 121:43–53
53. Anuradha C, Gaur PM, Pande S, Gali KK, Ganesh M et al (2011) Mapping QTL for resistance to botrytis grey mould in chickpea. Euphytica 182:1–9
54. Varshney RK, Thudi M, Nayak SN, Gaur PM, Kashiwagi J et al (2014) Genetic dissection of drought tolerance in chickpea (*Cicer arietinum* L.). Theor Appl Genet 127:445–462
55. Kale SM, Jaganathan D, Ruperao P, Chen C, Punna R et al (2015) Prioritization of candidate genes in "QTL-hotspot" region for drought tolerance in chickpea (*Cicer arietinum* L.). Sci Rep 5:15296
56. Vadez V, Krishnamurthy L, Thudi M, Anuradha C, Colmer TD et al (2012) Assessment of ICCV 2 × JG 62 chickpea progenies shows sensitivity of reproduction to salt stress and reveals QTL for seed yield and yield components. Mol Breed 30:9–21
57. Paul PJ, Samineni S, Thudi M, Sajja SB, Rathore A et al (2018) Molecular mapping of QTLs for heat tolerance in chickpea. Int J Mol Sci 19(8):2166
58. Li Y, Ruperao P, Batley J, Edwards D, Davidson J et al (2017) Genome analysis identified novel candidate genes for ascochyta blight resistance in chickpea using whole genome re-sequencing data. Front Plant Sci 8:1–13
59. Thudi M, Khan AW, Kumar V, Gaur PM, Katta K et al (2016) Whole genome re-sequencing reveals genome-wide variations among parental lines of 16 mapping populations in chickpea (*Cicer arietinum* L.). BMC Plant Biol 16:10
60. Kujur A, Bajaj D, Saxena MS, Tripathi S, Upadhyaya HD et al (2014) An efficient and cost-effective approach for genic microsatellite marker-based large-scale trait association mapping: identification of candidate genes for seed weight in chickpea. Mol Breed 34:241–265
61. Bajaj D, Upadhyaya HD, Khan Y, Das S, Badoni S et al (2015) A combinatorial approach of comprehensive QTL-based comparative genome mapping and transcript profiling identified a seed weight-regulating candidate gene in chickpea. Sci Rep 5:1–14
62. Bajaj D, Das S, Badoni S, Kumar V, Singh M et al (2015) Genome-wide high-throughput SNP discovery and genotyping for understanding natural (functional) allelic diversity and domestication patterns in wild chickpea. Sci Rep 5:1–17
63. Hossain S, Ford R, McNeil D, Pittock C, Panozzo JF (2010) Inheritance of seed size in chickpea (*Cicer arietinum* L.) and identification of QTL based on 100-seed weight and seed size index. Aust J Crop Sci 4(2):126
64. Wang R, Gangola MP, Irvine C, Gaur PM, Båga M et al (2019) Co-localization of genomic regions associated with seed morphology and composition in a desi chickpea (*Cicer arietinum* L.) population varying in seed protein concentration. Theor Appl Genet 132 (4):1263–1281
65. Molina C, Zaman-Allah M, Khan F, Fatnassi N, Horres R et al (2011) The salt responsive transcriptome of chickpea roots and nodules via deepSuperSAGE. BMC Plant Biol 11:31
66. Garg R, Bhattacharjee A, Jain M (2015) Genome-scale transcriptomic insights into molecular aspects of abiotic stress responses in chickpea. Plant Mol Biol Rep 33:388–400
67. Garg R, Shankar R, Thakkar B, Kudapa H, Krishnamurthy L et al (2016) Transcriptome

analyses reveal genotype- and developmental stage-specific molecular responses to drought and salinity stresses in chickpea. Sci Rep 6:19228

68. Ashraf N, Basu S, Narula K, Ghosh S, Tayal R et al (2018) Integrative network analyses of wilt transcriptome in chickpea reveal genotype dependent regulatory hubs in immunity and susceptibility. Sci Rep 8(1):6528
69. Singh VK, Garg R, Jain M (2013) A global view of transcriptome dynamics during flower development in chickpea by deep sequencing. Plant Biotechnol J 11:691–701
70. Hajyzadeh M, Turktas M, Khawar KM, Unver T (2015) miR408 overexpression causes increased drought tolerance in chickpea. Gene 555:186–193
71. Hu J, Sun L, Ding Y (2013) Identification of conserved microRNAs and their targets in chickpea (*Cicer arietinum* L.). Plant Signal Behav 8(4):e23604
72. Jain M, Chevala VVSN, Garg R (2014) Genome-wide discovery and differential regulation of conserved and novel microRNAs in chickpea via deep sequencing. J Exp Bot 65:5945–5958
73. Khandal H, Parween S, Roy R, Meena MK, Chattopadhyay D (2017) MicroRNA profiling provides insights into post-transcriptional regulation of gene expression in chickpea root apex under salinity and water deficiency. Sci Rep 7:4632
74. Srivastava S, Zheng Y, Kudapa H, Jagadeeswaran G, Hivrale V et al (2015) High throughput sequencing of small RNA component of leaves and inflorescence revealed conserved and novel miRNAs as well as phasiRNA loci in chickpea. Plant Sci 235:46–57
75. Kohli D, Joshi G, Deokar AA, Bhardwaj AR, Agarwal M et al (2014) Identification and characterization of wilt and salt stress-responsive microRNAs in chickpea through high-throughput sequencing. PLoS One 9:e108851
76. Garg V, Khan AW, Kudapa H, Kale SM, Chitikineni A et al (2019) Integrated transcriptome, small RNA and degradome sequencing approaches provide insights into Ascochyta blight resistance in chickpea. Plant Biotechnol J 17:914–931
77. Khemka N, Singh VK, Garg R, Jain M (2016) Genome-wide analysis of long intergenic non-coding RNAs in chickpea and their potential role in flower development. Sci Rep 6:33297
78. Singh U, Khemka N, Rajkumar MS, Garg R, Jain M (2017) PLncPRO for prediction of long non-coding RNAs (lncRNAs) in plants and its application for discovery of abiotic stress-responsive lncRNAs in rice and chickpea. Nucleic Acids Res 45:e183
79. Mohammadi P, Bahramnejad B, Badakhshan H, Kanouni H (2015) DNA methylation changes in fusarium wilt resistant and sensitive chickpea genotypes (*Cicer arietinum* L.). Physiol Mol Plant Pathol 91:72–80
80. Rakei A, Maali-Amiri R, Zeinali H, Ranjbar M (2016) DNA methylation and physio-biochemical analysis of chickpea in response to cold stress. Protoplasma 253(1):61–76
81. Bhatia H, Khemka N, Jain M, Garg R (2018) Genome-wide bisulphite-sequencing reveals organ-specific methylation patterns in chickpea. Sci Rep 8(1):9704
82. Verma M, Kumar V, Patel RK, Garg R, Jain M (2015) CTDB: an integrated chickpea transcriptome database for functional and applied genomics. PLoS One 10:e0136880
83. Misra G, Priya P, Bandhiwal N, Bareja N, Jain M et al (2014) The chickpea genomic web resource: visualization and analysis of the desi-type *Cicer arietinum* nuclear genome for comparative exploration of legumes. BMC Plant Biol 14(1):315
84. Doddamani D, Katta MA, Khan AW, Agarwal G, Shah TM et al (2014) CicArMiSatDB: the chickpea microsatellite database. BMC Bioinformatics 15(1):212
85. Parida SK, Verma M, Yadav SK, Ambawat S, Das S et al (2015) Development of genome-wide informative simple sequence repeat markers for large-scale genotyping applications in chickpea and development of web resource. Front Plant Sci 6:645
86. Doddamani D, Khan AW, Katta MA, Agarwal G, Thudi M et al (2015) CicArVarDB: SNP and InDel database for advancing genetics research and breeding applications in chickpea. Database 2015:pii: bav078
87. Srivastava R, Bajaj D, Sayal YK, Meher PK, Upadhyaya HD et al (2016) Genome-wide development and deployment of informative intron-spanning and intron-length polymorphism markers for genomics-assisted breeding applications in chickpea. Plant Sci 252:374–387
88. Gayali S, Acharya S, Lande NV, Pandey A, Chakraborty S et al (2016) CicerTransDB 1.0: a resource for expression and functional study of chickpea transcription factors. BMC Plant Biol 16(1):169

Chapter 3

# Legume Pangenome Construction Using an Iterative Mapping and Assembly Approach

Haifei Hu, Yuxuan Yuan, Philipp E. Bayer, Cassandria T. Fernandez, Armin Scheben, Agnieszka A. Golicz, and David Edwards

## Abstract

A pangenome is a collection of genomic sequences found in the entire species rather than a single individual. It allows for comprehensive, species-wide characterization of genetic variations and mining of variable genes which may play important roles in phenotypes of interest. Recent advances in sequencing technologies have facilitated draft genome sequence construction and have made pangenome constructions feasible. Here, we present a reference genome-based iterative mapping and assembly method to construct a pangenome for a legume species.

**Key words** Pangenome, Legume, De novo, Mapping, Assembly

## 1 Introduction

Advances in DNA sequencing and algorithms, as well as the decreasing DNA sequencing cost, have facilitated the construction of genome assemblies for different species. However, currently, most genomic studies conducted are based on a comparison with a single reference genome. A reference genome assembled from one individual does not capture the whole genomic information in a species owing to genetic variations between individuals [1]. Therefore, a more comprehensive reference genome containing all genomic information in a species is necessary.

The term "pangenome" was coined to describe the complete gene set of a species [2]. A pangenome comprises core genes shared by all individuals of a species and dispensable or variable genes which are absent in one or more individuals [3]. Pangenomes were initially constructed and annotated in bacteria, benefiting from their relatively simple and small genomes [2, 4]. The assembly and annotation of eukaryotic pangenomes, however, is more

Mukesh Jain and Rohini Garg (eds.), *Legume Genomics: Methods and Protocols*, Methods in Molecular Biology, vol. 2107, https://doi.org/10.1007/978-1-0716-0235-5_3, © Springer Science+Business Media, LLC, part of Springer Nature 2020

challenging, especially for large, complex, and repetitive genomes, which are common in plants.

Two main methods have been applied for assembling plant pangenomes: the de novo assembly and comparison approach [5–8] and the iterative mapping and assembly approach [9–11]. The de novo assembly approach builds a pangenome by de novo assembling individual genomes followed by whole genome sequence comparisons to identify core and variable genes [2]. This approach can provide positional information for variable genes as well as copy number variation information. The iterative mapping and assembly approach works by aligning reads to a reference genome and assembling reads that do not align, adding the annotated new contigs to the pangenome reference. Remapping all the data back to the pangenome allows for the calling of gene presence/absence for all individuals. This method requires lower sequencing coverage than required for de novo assembly and thus is cost-effective for a greater number of individuals.

Although iterative and de novo assembly approaches both offer advantages for pangenome construction, both have limitations. For instance, in the comparative de novo approach, the read coverage requirement is relatively high, which increases the sequencing cost [12]. Errors in genome assembly and annotation are common and these also lead to gene presence/absence calling errors [13]. In a recent iterative pangenome assembly for wheat, sequencing coverages of 8.4–20.0× were used [11]. In contrast, de novo assembly of accessions for a rice pangenome used an average sequencing coverage of 115× [6]. Additionally, while iterative mapping has a moderate to low requirement for computational resources, de novo assembly is highly computationally intensive and can require extensive high-performance computing resources when many individuals are assembled.

Currently, we believe that both pangenome construction methods are complementary. Whole genome assembly and comparison permits the placement of the variable genes, while the iterative mapping and assembly approach permits the calling of presence/absence variation (PAV) for much larger numbers of individuals supporting population scale analysis.

Legumes are important agronomic crops, which can provide food for both animals and humans. In this chapter, we describe how to construct a legume pangenome using the iterative mapping and assembly approach [14]. An overview of the workflow is given in Fig. 1.

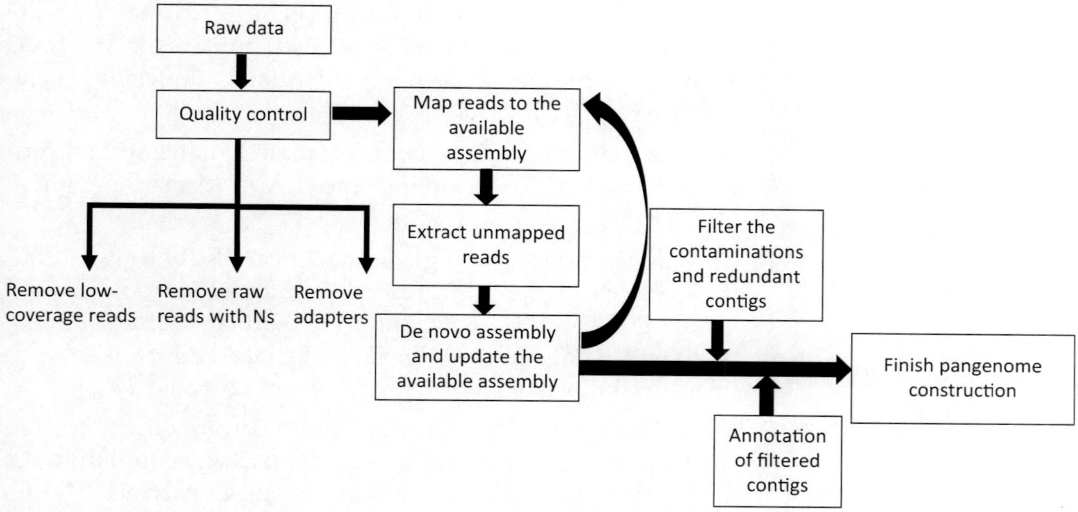


**Fig. 1** Overall workflow of the reference genome-based iterative mapping and assembly approach. In this approach, the data are firstly quality controlled by removing adapters in the reads and filtering reads with unknown sequences and samples with lower *coverages*. After that, mapping cleaned sequencing reads of the first dataset to the targeted reference genome followed by extracting both paired reads and singletons that do not align to the reference genome. And then de novo assembling the unmapped reads and updating the available assembly by integration with the newly assembled contigs. Repeating the mapping, extraction and assembly steps using the updated references for the remaining datasets until a primary pangenome is obtained for the whole group of individuals. A final pangenome is obtained after removing contigs from contaminations and contigs that are redundant. Pangenome annotation is also performed to check how many novel genes captured

## 2 Materials

### 2.1 Sequencing Data

Illumina short sequencing reads with a minimum coverage depth of 10× are required in the reference genome-based iterative assembly approach (*see* **Note 1**).

### 2.2 Bioinformatics Tools and Scripts

This pangenome pipeline is recommended to run on a server with 24 CPU cores and at least 128 GB RAM. A set of publicly available bioinformatics tools and custom Python scripts are used for the analysis.

1. Trimmomatic (http://www.usadellab.org/cms/?page=trimmomatic) is a flexible and multithreaded trimming tool for removing adapter sequences and low-quality sequencing reads for Illumina NGS data [15].
2. FastQC (https://www.bioinformatics.babraham.ac.uk/projects/fastqc/) is a quality control tool for Illumina NGS data [16].
3. Bowtie2 (http://bowtie-bio.sourceforge.net/bowtie2/index.shtml) is an ultrafast short sequencing read aligning tool with low memory consumption [17] (*see* **Note 2**).

4. Samtools (http://samtools.sourceforge.net/) contains a suite of utilities designed for manipulating alignments in the SAM format, including sorting, merging, indexing, and compressing [18].
5. Bamtools (https://github.com/pezmaster31/bamtools) provides a toolkit for handling the BAM format alignment files [19].
6. BBtools (https://jgi.doe.gov/data-and-tools/bbtools/) contains a suite of fast, multithreaded bioinformatics tools that can handle DNA and RNA data in various formats including SAM, fastq, and BAM. In this pipeline, BBtools' BBmap is used to calculate the average insert size of mapping reads [20].
7. MaSuRCA (http://www.genome.umd.edu/masurca.html) is an efficient de novo assembly software package combining the de Bruijn graph and Overlap-Layout-Consensus (OLC) approaches [21] (*see* **Note 3**).
8. NCBI-BLAST+ (https://www.ncbi.nlm.nih.gov/books/NBK279671/) consists of a number of utilities that can be used for searching sequence databases, comparing the similarity of two sequences, and identifying homologous genome sequence candidates [22].
9. In-house Python scripts (https://github.com/appliedbioinformatics/pangenome) were written to perform various tasks associated with data processing, remove reads with Ns and to extract the unmapped reads based on alignment in BAM files (1). These scripts are open source and freely available.

## 3 Methods

The workflow for building a soybean pangenome using the iterative mapping and assembly method is shown in Fig. 1. The workflow can be divided into five main steps: data quality control, mapping cleaned reads to the available assembly and extracting the unmapped reads, de novo assembling the unmapped reads, updating the starting reference, repeating the steps, and filtering the contamination and redundant contigs until obtaining a primary pangenome.

### 3.1 Data Quality Control

The example published soybean dataset can be downloaded from NCBI Sequence Read Archive (SRA) (https://www.ncbi.nlm.nih.gov/sra); the accession is SRP045129. Data quality control is an essential step for performing the iterative mapping and assembly pangenome construct pipeline and failure in data quality control can produce an incorrect pangenome assembly and mislead gene PAV detection, pangenome-wide SNP calling and other

**Table 1**
**List of the utility of Python script designed to perform tasks associated with the pangenome construction method of iterative mapping and assembly**

| Python scripts | Description |
|---|---|
| Remove_reads_with_Ns.py | This script is used to remove paired raw reads with unknown bases (Ns) as these reads will affect the performance of mapping and end up in the pool of unmapped reads—the assumption being that the sequencing run's quality was good enough to have only few reads with Ns |
| splitUp.py | This script is used to split the unmapped fastq files into R1, R2, and singlets for assembly by MaSuRCA |

pangenome-based downstream analyses. To solve this potential problem, data quality control is needed.

Data quality control is performed to remove reads with low coverage, adapters, and multiple N bases. All steps presented here are run on the command line.

1. `fastqc *.fastq.gz`

   FastQC [16] is employed to check the read length, completeness, and consistency of the Illumina paired-end reads. Problematic reads in the samples can also be reported by FastQC (*see* **Note 4**).

2. `multiqc /path/to/fastqc/output && multiqc output`

   After FastQC, MultiQC [23] is used to summarize multiple FastQC runs and to generate a visualized report of FastQC results.

3. `java -jar trimmomatic-0.33.jar PE R_1.fastq R_2.fastq R_1_Paired R_1_Unpaired R_2_Paired R_2_Unpaired ILLUMINACLIP:All.fa:2:30:10 MINLEN:99 TOPHRED33`

   Trimmomatic v0.33 [15] is used to remove those sequencing reads containing adapters and low-quality sequencing reads. To keep consistency, on the command line, TOPHRED33 [15] is applied to convert raw data to "Phred +33" encoding format, which can be directly interpreted by Bowtie2 [17] for read mapping, as by default bowtie2 assumes all input data are "Phred+33" encoded. ILLUMINACLIP requires a fasta-formatted file listing potential adapter sequence. Here "All.fa" generated from all adapter sequences provided in the Trimmomatic package is used. MINLEN can help to remove reads which are lower than a certain length. Here we limit the minimum read length to 99 bp (*see* **Note 5**).

4. `python Remove_reads_with_Ns.py R1.fq R2.fq`

   A custom Python script (Remove_reads_with_Ns.py) is used to remove the sequence reads containing multiple N bases (reads with "Ns" in the sequencing data) (Table 1, *see* **Note 6**). Eventually, the adapter-free and "N"-free clean paired-end resequencing data are ready for the next stage of the analysis: iterative mapping and assembly.

### 3.2 Mapping and Extraction of the Unmapped Reads

The key of the iterative mapping and assembly pangenome construction pipeline is to extract and assemble the unmapped reads. In this step, clean paired-end reads generated from the steps above are mapped to the reference genome to identify the unmapped reads.

The example published soybean genome is downloaded from Phytozome [24, 25] and used as the starting reference for read mapping by Bowtie2 [17] (*see* **Note 7**).

1. `bowtie2-build Glycine_max_Wm82.a2.v1.fa Glycine_max_Wm82.a2.v1`

   Bowtie2-build [17] is used to index the reference. On this command line, a FASTA format reference genome "Glycine_max_v2.1.fa" is used as an example. The prefix "Glycine_max_v2.1" is given to name all index files generated.

2. `bowtie2 -I 0 -X 1000 -x Glycine_max_Wm82.a2.v1 -1 R1.fastq -2 R2.fastq --end-to-end --sensitive -S mapped.sam`

   Bowtie 2 is applied to aligned paired reads ("-1 and -2") to the reference. Bowtie2 only accepts the indexed reference ("-x") from bowtie2-build. To ensure all the reads can be aligned confidently to the reference without soft-clipping, the "end-to-end" mode and "sensitive" option are used. As the mapping accuracy of mate pair reads is relatively low [26], we limit the insert size from 0 to 1000 bp ("-I and -X") to reduce the potential of getting false positive mappings. "-S" is used to ensure that the mapping information is encoded in a standardized SAM file.

3. `reformat.sh in=mapped.sam ihist=ihist.txt`

   BBmap reformat.sh [20] is applied to calculate the average insert size of the mapped paired-end reads for the MaSuRCA assembly. reformat.sh requires the read mapping information ("in") to calculate the insert size ("ihist").

4. `samtools view -bS -f 68 mapped.sam>out_pfx.forward.bam # forward reads`

5. `samtools view -bS -f 132 mapped.sam>out_pfx.reverse.bam # reverse reads`

   The unmapped reads are identified from the SAM file generated from bowtie2 using SAMTtools. Flag ("-f") 68 is used to extract forward unmapped reads and flag 132 is used to extract reverse unmapped reads. "-b" is used to converted the SAM output into BAM format [18].

6. `samtools merge out_pfx.merged.bam out_pfx.forward.bam out_pfx.reverse.bam`
7. `samtools sort -n -o out_pfx.sorted.bam out_pfx.merged.bam`

   The extracted forward and reverse unmapped reads are merged and stored in to a BAM file using "samtools merge" followed by reads sorting by sequence IDs ("samtools sort -n"). The purpose to sort reads by ID is that forward reads and reverse reads can be sorted next to each other.
8. `bamtools convert -in out_pfx.sorted.bam --out out_pfx.merge.fastq -format fastq`

   Using bamtools [19], we then convert the sorted unmapped reads in BAM format ("-in") into an interleaved FASTQ format ("-format fastq -out out_pfx.merge.fastq").
9. `python splitUP.py out_pfx.merge.fastq`

   The paired and solitary unmapped reads are further extracted from the interleaved FASTQ file by using a custom Python script ("splitUp.py") (Table 1). The extraction is based on the matching of reads ID and singletons in the interleaved FASTQ file.

### 3.3 De Novo Assembling Unmapped Reads

De novo assembly of unmapped reads is performed by MaSuRCA [21] using default parameters. MaSuRCA [21] requires a configuration file to set up the parameter and input files. A sample configuration file based on the v3.3.0 standard is presented as below:

```
1. # example configuration file
2. DATA
3. PE= aa 456 68 R1.PE.fastq R2.PE.fastq  ##Here, 456 is the average insert size of
   mapping reads, and 68 is the standard deviation of the average insert size
4. PE= ab 456 68 SE.fastq
5. END
6. PARAMETERS ## Users can modify the parameters based on their datasets and
   computing resource
7. EXTEND_JUMP_READS=0
8. GRAPH_KMER_SIZE = auto
```

```
9.  USE_LINKING_MATES = 0
10. LHE_COVERAGE=25
11. MEGA_READS_ONE_PASS=0
12. LIMIT_JUMP_COVERAGE=300
13. CLOSE_GAPS=0
14. CA_PARAMETERS = cgwErrorRate=0.15 ovlMemory=8GB
15. KMER_COUNT_THRESHOLD = 1
16. NUM_THREADS = 16
17. JF_SIZE = 2300000000
18. DO_HOMOPOLYMER_TRIM = 0
19. SOAP_ASSEMBLY=0
20. END
```

A pangenome study on *Brassica oleracea* [9] suggested that the order of individuals does not greatly influence the final assembly size and continuity. Therefore, the overall iterative processing order is based on the numerical order of the data sample name under the same ecotype (*see* **Note 8**). To eliminate misassembled contigs, those newly assembled contigs with a length smaller than 200 bp are discarded (*see* **Note 9**). The current reference assembly is updated by concatenating novel contigs to the reference sequence. Afterward, the steps mentioned in Subheadings 3.2 and 3.3 are reiterated for all remaining data until a preliminary pangenome is constructed.

### 3.4 Contamination Removal

NCBI-BLAST+ [22] is used to search for potential contamination by aligning the primary pangenome with the NCBI nucleotide database [27] to identify and remove contigs with non-plant species hit (e.g., hits with bacterial, fungal, and human sequences). Additionally, NCBI-BLAST+ [22] is used to search for redundant contigs by aligning the newly assembled contigs with the starting reference genome. Those redundancies are likely caused by incorrect assembly. During the processing, a 95% similarity and 90% sequence coverage are set to identify redundant contigs. After filtering contaminants and redundant contigs, the remained newly assembled contigs are used for gene prediction and functional annotation (*see* **Note 10**).

### *3.5 Gene Prediction and Functional Annotation*

The MAKER [28] pipeline uses gene prediction evidence, transcript evidence, and protein evidence for gene prediction and functional annotation (*see* **Note 11**). The detailed procedure can be found in the wiki tutorial page of MAKER (http://weatherby.genetics.utah.edu/MAKER/wiki/index.php/MAKER_Tutorial_for_WGS_Assembly_and_Annotation_Winter_School_2018). Finally, the filtered and removed redundant newly assembled contigs and their annotations are added to the reference genome and annotation to construct the species' entire pangenome.

### *3.6 Gene Presence and Absence Calling*

1. `bowtie2-build Glycine_max_pangenome.fa Glycine_max _pangenome`

   Bowtie2-build [17] is used to index the constructed pangenome as mentioned in Subheading 3.2.

2. `bowtie2 -I 0 -X 1000 -x Glycine_max _pangenome -1 R1.fastq -2 R2.fastq --end-to-end --sensitive -S line1_mapped.sam`

   The cleaned sequencing reads of each line generated from Subheading 3.1 are mapped to the constructed pangenome assembly using Bowtie2 [17] with the same settings mentioned in Subheading 3.2.

3. `samtools view -bS line1_mapped.sam > line1_mapped.bam`

4. `samtools sort -n line1_mapped.bam > line1_mapped.sorted.bam`

   The generated SAM files are converted to sorted BAM files using samtools [18].

5. `samtools view -f 2 line1_mapped.bam >line1_paired_mapped.bam`

6. `samtools index line1_paired_mapped.bam`

   Reads mapped in proper pairs are extracted using samtools with flag ("-f") 2 followed by indexing using "samtools index" [18]. The aim is to ensure that only correctly mapped reads are used in gene PAV detection.

7. `java -Xmx16g -jar SGSGeneLoss.v0.1.jar minCov=2 chromosomeList=all lostCutoff=0.05 bamPath=/path/to/bam bamFileList= line1_paired_mapped.bam gffFile=/path/to/gff_file chromosomeList=all`

Gene PAV detection is performed by SGSGeneLoss v0.1 [29] for each accession. On the command line, "bamFileList" is used to specify which bam file should be used to detect PAVs. A gff file is required ("gffFile") by SGSGeneLoss to identify gene positions on the chromosomes. In the example, genes with PAV are identified on all chromosomes and novel contigs in the pangenome ("chromosomeList=all"). To reduce the biases caused by cross-mapping, 2 is set as the minimum coverage ("minCov") to consider an exon as covered by reads. "lostCutoff" is used to assess whether a gene is lost or not. Here we set it to 0.05, which means if at least 5% of a gene is not covered by reads, we consider this gene is lost in the corresponding individual [9].

To further understand the biological function of gene PAV, gene ontology enrichment analysis can be performed by BLAST2GO [30] and TopGO package [31] to predict the functions they may play.

## 4 Notes

1. A minimum coverage of 10× has been calculated to accurately call PAV of genes for wheat, soybean, and *Brassica* species. For other species, this minimum coverage may be determined empirically by mapping reference data to the reference genome at different coverage.
2. BWA can be used as an alternative alignment tool for this application [32], but by default BWA soft clips reads, which can lead to false-positive alignments when, for example, reads from a slightly different paralogous gene copy not found in the reference assembly align with the copy in the reference after soft clipping.
3. Other software such as Velvet [33], SOAPdenovo2 [34], and ABySS [35] are also available for de novo assembly. However, in our experience, MaSuRCA has generated longer and more complete genome sequences than older assemblers such as Velvet or SOAPdenovo2.
4. FastQC is designed to analyze only one read library, which can become cumbersome when working with many sequencing runs for many individuals.
5. The order of arguments in Trimmomatic is important—if MINLEN is used before the ILLUMINACLIP argument, reads with adapters that are shorter than the given minimum length will be retained.
6. The "Ns" in sequencing reads will result in the Bowtie2 aligner producing false-positive unmapped reads and novel contigs highly similar to the reference genome.
7. A high-quality assembly and annotation are preferred to be the starting reference in pangenome construction, as it could reduce the biases of incorrect and inaccurate mapping.

8. To reduce the requirement of computational power and increase the pipeline efficiency, reads with the same ecotype are better pooled as a hybrid dataset and mapped to the same reference genome.
9. 200 bp is used as a criterion for filtering contigs after assembly for the majority of genome assembly studies.
10. Contigs with sequence length equal to or larger than 1000 bp are retained for gene prediction, as it is rare to predict genes smaller than 1000 bp. Those shorter contigs are highly likely repeats.
11. Other annotation tools such as EVidenceModeler (EVM) [36] and BRAKER [37] can also be used for gene annotation.

## References


1. Hurgobin B, Edwards D (2017) SNP discovery using a pangenome: has the single reference approach become obsolete? Biology (Basel) 6 (1):pii: E21
2. Tettelin H, Riley D, Cattuto C, Medini D (2008) Comparative genomics: the bacterial pan-genome. Curr Opin Microbiol 11 (5):472–477
3. Tettelin H, Masignani V, Cieslewicz MJ, Donati C, Medini D, Ward NL, Angiuoli SV, Crabtree J, Jones AL, Durkin AS, Deboy RT, Davidsen TM, Mora M, Scarselli M, Margarit-Ros I, Peterson JD, Hauser CR, Sundaram JP, Nelson WC, Madupu R, Brinkac LM, Dodson RJ, Rosovitz MJ, Sullivan SA, Daugherty SC, Haft DH, Selengut J, Gwinn ML, Zhou L, Zafar N, Khouri H, Radune D, Dimitrov G, Watkins K, O'Connor KJ, Smith S, Utterback TR, White O, Rubens CE, Grandi G, Madoff LC, Kasper DL, Telford JL, Wessels MR, Rappuoli R, Fraser CM (2005) Genome analysis of multiple pathogenic isolates of Streptococcus agalactiae: implications for the microbial "pan-genome". Proc Natl Acad Sci U S A 102(39):13950–13955
4. Plissonneau C, Hartmann FE, Croll D (2018) Pangenome analyses of the wheat pathogen Zymoseptoria tritici reveal the structural basis of a highly plastic eukaryotic genome. BMC Biol 16(1):5
5. Li Y-h, Zhou G, Ma J, Jiang W, L-g J, Zhang Z, Guo Y, Zhang J, Sui Y, Zheng L (2014) De novo assembly of soybean wild relatives for pan-genome analysis of diversity and agronomic traits. Nat Biotechnol 32 (10):1045–1052
6. Zhao Q, Feng Q, Lu H, Li Y, Wang A, Tian Q, Zhan Q, Lu Y, Zhang L, Huang T, Wang Y, Fan D, Zhao Y, Wang Z, Zhou C, Chen J, Zhu C, Li W, Weng Q, Xu Q, Wang Z-X, Wei X, Han B, Huang X (2018) Pan-genome analysis highlights the extent of genomic variation in cultivated and wild rice. Nat Genet 50 (2):278–284
7. Gordon SP, Contreras-Moreira B, Woods DP, Des Marais DL, Burgess D, Shu S, Stritt C, Roulin AC, Schackwitz W, Tyler L, Martin J, Lipzen A, Dochy N, Phillips J, Barry K, Geuten K, Budak H, Juenger TE, Amasino R, Caicedo AL, Goodstein D, Davidson P, Mur LAJ, Figueroa M, Freeling M, Catalan P, Vogel JP (2017) Extensive gene content variation in the Brachypodium distachyon pan-genome correlates with population structure. Nat Commun 8(1):2184
8. Schatz MC, Maron LG, Stein JC, Hernandez Wences A, Gurtowski J, Biggers E, Lee H, Kramer M, Antoniou E, Ghiban E, Wright MH, Chia JM, Ware D, McCouch SR, McCombie WR (2014) Whole genome de novo assemblies of three divergent strains of rice, Oryza sativa, document novel gene space of aus and indica. Genome Biol 15(11):506
9. Golicz AA, Bayer PE, Barker GC, Edger PP, Kim H, Martinez PA, Chan CKK, Severn-Ellis-A, McCombie WR, Parkin IA (2016) The pangenome of an agronomically important crop plant Brassica oleracea. Nat Commun 7:13390
10. Hurgobin B, Golicz AA, Bayer PE, Chan CKK, Tirnaz S, Dolatabadian A, Schiessl SV, Samans B, Montenegro JD, Parkin IA (2017) Homoeologous exchange is a major cause of gene presence/absence variation in the amphidiploid Brassica napus. Plant Biotechnol J 16 (7):1265–1274

11. Montenegro JD, Golicz AA, Bayer PE, Hurgobin B, Lee H, Chan CKK, Visendi P, Lai K, Doležel J, Batley J (2017) The pangenome of hexaploid bread wheat. Plant J 90 (5):1007–1013
12. Tao Y, Zhao X, Mace E, Henry R, Jordan D (2018) Exploring and exploiting pan-genomics for crop improvement. Mol Plant 12 (2):156–169
13. Bayer PE, Hurgobin B, Golicz AA, Chan CKK, Yuan YX, Lee H, Renton M, Meng JL, Li RY, Long Y, Zou J, Bancroft I, Chalhoub B, King GJ, Batley J, Edwards D (2017) Assembly and comparison of two closely related Brassica napus genomes. Plant Biotechnol J 15 (12):1602–1610
14. Golicz AA, Batley J, Edwards D (2016) Towards plant pangenomics. Plant Biotechnol J 14(4):1099–1105
15. Bolger AM, Lohse M, Usadel B (2014) Trimmomatic: a flexible trimmer for Illumina sequence data. Bioinformatics 30 (15):2114–2120
16. Andrews S (2010) FastQC: a quality control tool for high throughput sequence data.
17. Langmead B, Salzberg SL (2012) Fast gapped-read alignment with Bowtie 2. Nat Methods 9 (4):357–359
18. Li H, Handsaker B, Wysoker A, Fennell T, Ruan J, Homer N, Marth G, Abecasis G, Durbin R (2009) The sequence alignment/map format and SAMtools. Bioinformatics 25 (16):2078–2079
19. Barnett DW, Garrison EK, Quinlan AR, Stromberg MP, Marth GT (2011) BamTools: a C++ API and toolkit for analyzing and managing BAM files. Bioinformatics 27 (12):1691–1692
20. Bushnell B (2014) BBMap: a fast, accurate, splice-aware aligner. Conference: 9th annual genomics of energy & environment meeting, Walnut Creek, CA, 17–20 Mar 2014
21. Zimin AV, Marçais G, Puiu D, Roberts M, Salzberg SL, Yorke JA (2013) The MaSuRCA genome assembler. Bioinformatics 29 (21):2669–2677
22. Camacho C, Coulouris G, Avagyan V, Ma N, Papadopoulos J, Bealer K, Madden TL (2009) BLAST+: architecture and applications. BMC Bioinformatics 10:421
23. Ewels P, Magnusson M, Lundin S, Käller M (2016) MultiQC: summarize analysis results for multiple tools and samples in a single report. Bioinformatics 32(19):3047–3048
24. Schmutz J, Cannon SB, Schlueter J, Ma J, Mitros T, Nelson W, Hyten DL, Song Q, Thelen JJ, Cheng J, Xu D, Hellsten U, May GD, Yu Y, Sakurai T, Umezawa T, Bhattacharyya MK, Sandhu D, Valliyodan B, Lindquist E, Peto M, Grant D, Shu S, Goodstein D, Barry K, Futrell-Griggs M, Abernathy B, Du J, Tian Z, Zhu L, Gill N, Joshi T, Libault M, Sethuraman A, Zhang XC, Shinozaki K, Nguyen HT, Wing RA, Cregan P, Specht J, Grimwood J, Rokhsar D, Stacey G, Shoemaker RC, Jackson SA (2010) Genome sequence of the palaeopolyploid soybean. Nature 463(7278):178–183
25. Goodstein DM, Shu S, Howson R, Neupane R, Hayes RD, Fazo J, Mitros T, Dirks W, Hellsten U, Putnam N, Rokhsar DS (2012) Phytozome: a comparative platform for green plant genomics. Nucleic Acids Res 40(Database issue):D1178–D1186
26. Thankaswamy-Kosalai S, Sen P, Nookaew I (2017) Evaluation and assessment of read-mapping by multiple next-generation sequencing aligners based on genome-wide characteristics. Genomics 109(3-4):186–191
27. Benson DA, Karsch-Mizrachi I, Lipman DJ, Ostell J, Wheeler DL (2008) GenBank. Nucleic Acids Res 36(Database issue):D25
28. Cantarel BL, Korf I, Robb SM, Parra G, Ross E, Moore B, Holt C, Sanchez Alvarado A, Yandell M (2008) MAKER: an easy-to-use annotation pipeline designed for emerging model organism genomes. Genome Res 18(1):188–196
29. Golicz AA, Martinez PA, Zander M, Patel DA, Van De Wouw AP, Visendi P, Fitzgerald TL, Edwards D, Batley J (2015) Gene loss in the fungal canola pathogen Leptosphaeria maculans. Funct Integr Genomics 15(2):189–196
30. Conesa A, Gotz S, Garcia-Gomez JM, Terol J, Talon M, Robles M (2005) Blast2GO: a universal tool for annotation, visualization and analysis in functional genomics research. Bioinformatics 21(18):3674–3676
31. Alexa A, Rahnenfuhrer J (2010) topGO: enrichment analysis for gene ontology. R package version 2(0):2010
32. Li H, Durbin R (2009) Fast and accurate short read alignment with Burrows-Wheeler transform. Bioinformatics 25(14):1754–1760. https://doi.org/10.1093/bioinformatics/btp324
33. Zerbino D, Birney E (2008) Velvet: algorithms for de novo short read assembly using de Bruijn graphs. Genome Res 2008:074492.074107
34. Luo R, Liu B, Xie Y, Li Z, Huang W, Yuan J, He G, Chen Y, Pan Q, Liu Y, Tang J, Wu G, Zhang H, Shi Y, Liu Y, Yu C, Wang B, Lu Y, Han C, Cheung DW, Yiu SM, Peng S, Xiaoqian Z, Liu G, Liao X, Li Y, Yang H,

Wang J, Lam TW, Wang J (2015) Erratum: SOAPdenovo2: an empirically improved memory-efficient short-read de novo assembler. Gigascience 4:30

35. Simpson JT, Wong K, Jackman SD, Schein JE, Jones SJ, Birol I (2009) ABySS: a parallel assembler for short read sequence data. Genome Res 2009:089532.089108

36. Haas BJ, Salzberg SL, Zhu W, Pertea M, Allen JE, Orvis J, White O, Buell CR, Wortman JR (2008) Automated eukaryotic gene structure annotation using EVidenceModeler and the Program to Assemble Spliced Alignments. Genome Biol 9(1):R7

37. Hoff KJ, Lange S, Lomsadze A, Borodovsky M, Stanke M (2016) BRAKER1: unsupervised RNA-seq-based genome annotation with GeneMark-ET and AUGUSTUS. Bioinformatics 32(5):767–769

# Chapter 4

# Methods and Tools for Plant Organelle Genome Sequencing, Assembly, and Downstream Analysis

**Sanand Sandhya, Harsha Srivastava, Tanvi Kaila, Anshika Tyagi, and Kishor Gaikwad**

## Abstract

Organelles play an important role in a eukaryotic cell. Among them, the two organelles, chloroplast and mitochondria, are responsible for the critical function of photosynthesis and aerobic respiration. Organellar genomes are also very important for plant systematic studies. Here we have described the methods for isolation of the mitochondrial and plastid DNA and its subsequent sequencing with the help of NGS technology. We have also discussed in detail the various tools available for assembly, annotation, and visualization of the organelle genome sequence.

**Key words** Chloroplast, Mitochondria, NGS, Assembly, Annotation, Visualization

## 1 Introduction

Among several organelles present in plant cells, plastid/chloroplast and mitochondria are unique and essential organelles that possess complex modes of gene expression as they maintain their own genomes. Both genomes are quite informative and dynamic in terms of plant evolution, origins and diversification of eukaryotic life. The plastid genome generally appear to be highly conserved with a circular configuration and similar sizes (100–220 kb) across genera, but the mitochondrial genomes are present in multiple configurations with large variation in genome sizes (100–10,000 kb).

In 2006, chloroplast genome was sequenced using NGS technology by Moore and colleagues. Among several platforms available, Illumina is currently the most favored NGS platform due to the fact that it allows the use of rolling circle amplification products. De novo assembly is attained using bioinformatics programs which do not require reference genomes. However, newer systems like PacBio and Oxford Nanopore are facilitating easy capture of these

Mukesh Jain and Rohini Garg (eds.), *Legume Genomics: Methods and Protocols*, Methods in Molecular Biology, vol. 2107,
https://doi.org/10.1007/978-1-0716-0235-5_4, 

genomes due to their ability to sequence large reads. Combining these together along with easy to use scripts, it has now become much easier to decode the organelle genomes of plants and identify their role in different aspects of plant growth and development. Small size and high copy numbers of these organelles per cell presents relative ease for NGS sequencing when compared to larger genomes. Organelle genome sequencing has seen rapid strides in the recent years with a total 177,000 mitochondrial sequences and 41,000 chloroplast new sequences now deposited in the GOMP (an organelle genome database) till date.

## 2 Organelle Genomics: Historical Overview

The history of organelle genomes traces back to a series of endosymbiotic events about 1–1.5 billion years ago, followed by several structural, genetic, and biochemical modifications, that eventually culminated into the present-day autonomous organelles. From evolutionary point of view, both mitochondria and chloroplasts have been thought of as primitive cells that were ingested by the nucleus containing cells with their size getting reduced significantly. Interestingly, due to the expansion of noncoding part, some of the organelle genome sizes are larger than the bacterial genomes. The prokaryotic signature is still present in these organelles as evident through their metabolic mechanisms, genetics, protein transport, and so on. During the course of evolution, there have been several genetic exchanges between mitochondrial and chloroplast genomes in plants which also led to considerable changes in the sizes of these organelles.

### 2.1 Why Organelle Genome Sequencing

Mitochondria and chloroplasts are the only organelles in the biological system that have their own distinct DNA. The DNA can be circular or linear and mode of replication is also different from the nuclear genome. Besides, the structural and genomic variations in these organelles have helped in propounding evolutionary theories.

Some of the reasons which make organelle sequencing interesting and important include the following:

- Prokaryotic inheritance of these organelles.
- Mitochondria acting as the power house of the cell, and plastids performing photosynthesis.
- Aiding in evolution of plant genomes, due to constant exchange of genetic material with the nucleus.
- Involvement in important agricultural traits such as cytoplasmic male sterility as a result of mutation in mitochondrial genome.

## 3 Unique Features of Mitochondrial and Chloroplast Genomes

As stated earlier, mitochondria and chloroplast act as power houses, generating ATP in respective cells. Both the organelles are surrounded by two membranes—the outer and the inner, the latter being known as thylakoids in chloroplasts and cristae in mitochondria. The following characteristics are unique to mitochondria and chloroplasts:

- Their genomes are prone to genetic recombination, posttranscriptional modifications as well as DNA exchange with nuclear genome.
- Huge variations in their genetic makeup and copy number exist in different cell types. Further, most of the genes code for either RNA or protein and proportion of regulatory DNA sequence is quite less.
- Codon usage also differs from nuclear genome. In mitochondria, only 22 tRNAs code for amino acids required for protein synthesis. Many tRNAs recognize all the four nucleotides which justifies the requirement of only 22 tRNAs as compared to more than 30 for cytosolic protein synthesis.
- Variations in genetic code. Some codons in mitochondrial genetic code have different meaning from those present in nuclear genome.
- Mitochondrial/plastid genomes are inherited in a non-Mendelian manner.

### 3.1 Mitochondrial Genomes Structure: Genome Organization and Gene Content

Mitochondria are one of the vital organelles present in a cell. Structurally, they are either filamentous or granular like, surrounded by two membranes and carry their own genome. Plants also possess mitochondrial genomes that are larger and complex (from around 100 kb to more than 10 Mb) as compared to humans and animals (~16.5-kb) [1–5], most of which consists of large introns, repeated region (between 50 and 500 bp) and AT-rich noncoding regions [6–8]. They also exist either as linear, circular, or both, and sometimes a cell can harbor different subgenomic circles of variable sizes indicating a frequent capacity for inter and intramolecular recombination. Besides, the plant mitochondrial genomes have also inherited significant traces of short nuclear and chloroplast genomic sequences that might have crept in during evolution [9–12]. Till date, several mitochondrial genomes belonging to all major clades of plants (rhodophytes, chlorophytes, charophytes, hornworts, liverworts, mosses, ferns, lycophytes, gymnosperms, angiosperms) have been sequenced showing a significant diversity among genome sizes (e.g., 13 Kb in chlorophytes and up to 11.3 Mb in angiosperms [1, 13]). The mitochondrial genome codes for five complexes that are part of oxidative phosphorylation viz. Complex I (NADH dehydrogenase), Complex II (succinate dehydrogenase), Complex III (ubiquinol cytochrome c reductase), Complex IV (cytochrome c oxidase), and Complex V (ATP synthase). RNA editing is also a unique feature of mtDNA-encoded genes where specific C's in the RNA transcripts are deaminated to U's.

### 3.2 Chloroplast Genome Structure: Genome Organization and Gene Content

Chloroplast is a structure unique to kingdom Plantae. Their genome organization is highly conserved among different eukaryotic plants ranging from 107 kb (*Cathaya argyrophylla*) to 218 kb (*Pelargonium*). Some species (*Acetabularia*) contain up to 1.5 Mb [14]. Generally, the genome size ranges from 120 to 160 kb in length with GC content ranging from 30% to 40% in flowering plants. Chloroplast genome is organized as a large-single-copy region (LSC; 80–90 kb), a small-single-copy region (SSC; 16–27 kb), and two copies of inverted repeats (IRs) of ~20–28 kb in size. It contains around 120–130 genes participating in photosynthesis, transcription, and translation [15–17]. Like mitochondrial genomes, chloroplast genomes also display significant variations especially in terms of structural rearrangements (e.g., loss of IR, several inversions, and an expansion/contraction of the inverted repeat [18]) in related ancestries. Various fruits (total 24), vegetable (total 21), ornamental (total 38), medicinal (total 28), cereal (total 10), timber (total 24), fiber (total 3), and biofuel (total 7) plants have been sequenced and submitted in NCBI genome database for complete chloroplast genome analysis using NGS (https://www.ncbi.nlm.nih.gov/). The core genes present in chloroplast genome are essentially comprised of photosynthesis related genes with two subunits (PS I and PS II), the cytochrome b6f complex, ATP synthase gene with large subunit of ribulose bis-phosphate carboxylase (RUBISCO), component of both ribosomal proteins (LSU and SSU), NADH dehydrogenase gene, a bacterial-type RNA polymerase, introns, rRNAs, and tRNAs.

## 4 Progress in Organelle Genomics

Owing to the significant advancement of NGS technology, the sequencing of chloroplast and mitochondrial genomes has seen a significant rise over the past decade or so. Most of the genes present in these organelles are involved in the structure or assembly of multimeric complexes of the electron transport chain. Further, the sequencing of these organelles has provided insights into the unknown facets of organelle genome evolution. Recent data on chloroplast and mitochondrial genome assembly for the last 3 years has been compiled in Table 1.

**Table 1**
**List of some of the recently sequenced organelle genomes using different sequencing platforms**

| Crop | Organelle | Assembler | Annotation | Platform | Size (bp) | Reference |
|---|---|---|---|---|---|---|
| *Metasequoia glyptostroboides* | Chloroplast | Velvet Assembler version v1.2.07 | DOGMA | Illumina Miseq | 131,887 | [19] |
| Globe artichoke | Chloroplast | CLC Genomics Workbench v6.0.1 | DOGMA | Illumina GAIIx | 152,529 | [20] |
| *Vigna radiata* | Chloroplast | MIRA 4 and MITObim v1.7 | DOGMA | Illumina Hiseq 2000 | 151,283 | [21] |
| *Ananas comosus* | Chloroplast | MIRA v4.0.2 | DOGMA | Illumina and PacBio RSII | 159,636 | [22] |
| *Acacia ligulata* | Chloroplast | Velvet version 1.2.08 | DOGMA | Illumina Hiseq 2000 | 158,724 | [23] |
| *Ipomoea batata* | Chloroplast | Edena v2.1.1, SOAPdenovo2-r240 and Velvet v1.0.12 | DOGMA | Illumina Hiseq 2000 | 161,303 | [24] |
| *Actinidia chinensis* | Chloroplast | Velvet v1.2.07 | DOGMA | Illumina Hiseq 2000 | 156,346 | [25] |
| *Panax quinquefolius* | Chloroplast | CLC Genomic Workbench v3.6 | DOGMA | Roche 454 GS FLX and Illumina short-read | 156,359 | [26] |
| *Euonymus japonicus* | Chloroplast | Geneious v7.1.3 | DOGMA | Illumina Miseq | 157,637 | [27] |
| *Cajanus cajan* (*L.*) *Millspaugh* | Chloroplast | Newbler v2.6 and CLC Genomics Workbench v7.5.1 | DOGMA | Roche 454 GS FLX | 152,201 | [28] |
| *Cajanus scarabaeoides* (*L.*) *Thouars* | Chloroplast | Newbler v2.6 and CLC Genomics Workbench 7.5.1 | DOGMA | Roche 454 GS FLX | 152,242 | [28] |
| *Cyamopsis tetragonoloba* L. | Chloroplast | CLC genomics workbench v9.5.1 | DOGMA | Illumina Hiseq 1000 | 152,530 | [29] |
| *Lupinus luteus* | Chloroplast | The organelle assembler | DOGMA | Illumina HiSeq 2000 | 151,894 | [30] |
| *Lupinus albus* | Chloroplast | The organelle assembler | DOGMA | Illumina HiSeq 2000 | 151,921 | [30] |

(continued)

**Table 1**
**(continued)**

| Crop | Organelle | Assembler | Annotation | Platform | Size (bp) | Reference |
|---|---|---|---|---|---|---|
| *Lupinus atlanticus* | Chloroplast | The organelle assembler | DOGMA | Illumina HiSeq 2000 | 152,272 | [30] |
| *Lupinus princei* | Chloroplast | The organelle assembler | DOGMA | Illumina HiSeq 2000 | 152,243 | [30] |
| *Fragaria* × *ananassa* 'Benihoppe' | Chloroplast | MITObim v1.8 | DOGMA | Illumina HiSeq 2500 | 155,549 | [31] |
| *Ailanthus altissima* | Chloroplast | NOVOPlasty and GENEIOUS v8.1 | DOGMA | Illumina Hiseq 2500 | 160,815 | [32] |
| *Capsicum pubescens* | Chloroplast | Velvet v1.2.10 | DOGMA | Illumina Hiseq 2500 | 157,390 | [33] |
| *Panax stipuleanatus* | Chloroplast | SPAdes v3.10.1 | DOGMA | Illumina HiSeq 2500 | 156,069 | [34] |
| *Eucalyptus grandis* | Chloroplast | NOVOPlasty v1.1 | DOGMA, CpGAVAS, Geneious v10.0.5, and MFannot | Illumina Hiseq 2000 | 160,109 | [35] |
| *Prangos fedtschenkoi* | Chloroplast | Bowtie 2 v2.2.3 | DOGMA | Illumina Hiseq 2000 | 153,626 | [36] |
| *Prangos lipskyi* | Chloroplast | Bowtie 2 v2.2.3 | DOGMA | Illumina Hiseq 2000 | 154,143 | [36] |
| *Hordeum vulgare* | Mitochondria | Newbler v2.6 | BLASTn and Mitofy | Roche GS-FLX 454 and Illumina Miseq | 525,599 | [37] |
| *Cocos nucifera* | Mitochondria | Newbler v2.6 | ORF finder, Rfam and tRNAscan-SE | Roche/454 GS FLX and Illumina HiSeq 2000 | 678,653 | [38] |
| *Ophioglossum californicum* | Mitochondria | Velvet v1.210 and Newbler v2.3 | Bowtie v2 | Illumina HiSeq 2000 | 372,339 | [39] |
| *Gossypium raimondii* | Mitochondria | Newbler v3.0 | NCBI-BLASTn and tRNAscan-SE | 454 GS FLX and Illumina short-read | 676,078 | [40] |

| | | | | | | |
|---|---|---|---|---|---|---|
| *Psilotum nudum* | Mitochondria | Velvet v1.210 and Newbler v2.3 | Bowtie v2 | Illumina HiSeq 2000 | 628,553 | [39] |
| *Allium cepa L.* | Mitochondria | CLC genome assembler ver. v4.06 | Mitofy | Illumina NextSeq500 | 316,363 | [41] |
| *Utricularia reniformis* | Mitochondria | CLC Genomics Workbench v9 | Mitofy | Illumina MiSeq | 857,234 | [42] |
| *Nymphaea colorata* | Mitochondria | Canu | Geneious software v.10.0.2 | PacBio RSII | 617,195 | [43] |
| *Chrysanthemum boreale* | Mitochondria | Canu v1.4 | Mitofy | PacBio's SMRT | 211,002 | [44] |
| *Sophora japonica L.* | Mitochondria | Canu v1.7 | BLASTn and Geneious v10.0.2 | Illumina Hiseq 2000 and PacBio RSII | 484,916 | [45] |
| *Hibiscus cannabinus* | Mitochondria | Hierarchical Genome Assembly Process | BLASTn and Mitofy | PacBio RS II SMRT | 569,915 | [46] |
| *Caulerpa lentillifera* | Mitochondria | SOAP denovo2 | tRNAscan-SE 1.21 and tRNADB-CE | Illumina Hiseq 2500 | 209,034 | [47] |
| *Eucalyptus grandis* | Mitochondria | MITObim v1.6 | Mitofy, MFannot, and Geneious v10.0.5 | Illumina Hiseq 2000 | 478,813 | [35] |

## 5 Checklist: Important Points to Be Considered Before Undertaking Organelle Genome Sequencing

(a) *Choice of sample*: Tissues taken for DNA extraction should be fresh and rich in respective organelle (e.g., leaf tissue for chloroplast DNA and bud tissues for mitochondria) (*see* **Note 1**).

(b) *Choice of DNA extraction method*: Although the kits available for organelle DNA extraction are limited but, organelle specific DNA extraction methods are available for isolating abundant and good quality DNA. The available protocols can be modified according to one's need.

(c) *DNA quality requirements for de novo sequencing*: Biochemical purity and structural integrity of the DNA are the two major key parameters that are of highest concern. Long read sequencing technologies require best quality high molecular weight (HMW) DNA (*see* **Note 2**).

Second important parameter for accurate sequencing is structural integrity of DNA. Prolonged DNA storage in water and above −20 °C is not recommended, as it increases the DNA degradation risk due to hydrolysis (*see* **Note 3**).

Further, RNA contamination of DNA samples must be avoided. Majority of the NGS DNA library preps available can accurately utilize double-stranded DNA only (*see* **Note 4**).

(d) *Choice of a suitable sequencing platform*:

While selecting a sequencing platform, one must consider the objectives of the investigation, genome size, and its complexity, as well as the information available in the public domain regarding quality and type of data. Quality read, yield, read length, and cost heavily influence the selection of the sequencing method. Illumina's NGS sequencing platforms are quite popular and cost effective. It provides the greatest output at the lowest reagent cost and offers a lot of flexibility (e.g., different runs with varying read length, paired-end and single-end sequencing, and/ or multiplexing many samples). If longer read lengths are required, then one must consider Pacific Bioscience's RS II/ Sequel system or Oxford Nanopore MinION/GridION.

## 6 Isolation of Chloroplast/Mitochondrial DNA

Both kit based and biochemical methods are available for DNA isolation. Whole chloroplast genome amplification from total DNA using long polymerase chain reaction (PCR), and the direct isolation of cpDNAs using sucrose gradient centrifugation are the two most common methods for separation of nuclear DNAs from cpDNAs, the latter being the most frequently used. The cpDNA isolation comprises of three basic steps: Isolation of chloroplast, lysis, and purification of cpDNA. However, while using sucrose gradient centrifugation only, a small amount of chloroplast pellet is collected, thereby resulting in very less amount of cpDNA. But, since the library preparation requires a significant amount and quality of input DNA, repeated cpDNA isolation is required. Later two alternative methods, DNase I treatment and high salt method were introduced as a substitute for the sucrose gradient centrifugation method. The DNase I treatment method uses DNase I to digest nuclear DNA that adheres to the outer chloroplast membrane, but again this method is not very precise as DNase I digests not only the contaminating DNA but also cpDNA within chloroplasts [48].

Various methods for the isolation of functional mitochondria from different crop species and plant cell culture, have been reported in the past like *Arabidopsis* [49–52], pea [53], wheat [54], or plant cell culture [55]. Protocols for extracting mtDNAs and cpDNAs form different crop species also exist for nonfunctional analysis of mitochondria. Different methods for isolating chloroplast and mitochondrial DNA have been described in Tables 2 and 3.

## 7 Library Preparation

Efficient library preparation depends on quality and accurate quantification of input DNA. Fluorometric-based methods are available for quantification, such as Qubit or PicoGreen. DNA library is validated quantitatively and qualitatively prior to sequencing to ensure that sufficient amount of good quality DNA is present in the prepared library. The quality of the sample library is assessed with the help of Agilent Technologies 2100 Bioanalyzer.

Both quality and quantity play important roles in generating quality data (*see* **Note 5**). In terms of quality, a good quality library is the one that has a diverse set of DNA fragments with minimal duplicate fragments (*see* **Note 6**). Library quantification is performed using either qPCR or a fluorometric method like Qubit.

The Agilent Bioanalyzer checks for the size distribution of the library before the sequencing reaction. It verifies the presence of DNA fragments of desired sizes. Markers, ladders, and samples (either diluted or undiluted) are loaded onto the chip. The chip is then vortexed for 1 min at 650 × *g* before it is loaded onto the Bioanalyzer. The machine will monitor each well for the sample; this is visualized with peaks on a graph. The location of the peaks

**Table 2**
**Chloroplast isolation methods**

| | Liquid nitrogen sucrose density gradient centrifugation (LN) method | Modified high salt (HS) method | Percoll gradient (PG) centrifugation method | Sucrose gradient method |
|---|---|---|---|---|
| Reagent required | *Isolation buffer* (50 mM Tris–HCl pH 8.0, 0.35 M sucrose, 7 mM EDTA, 5 mM 2-mercaptoethanol, 0.1% BSA) | *Buffer A (PH 3.8)*<br>1.25 M NaCl, 0.25 M ascorbic acid, 10 mM sodium metabisulfite, 0.0125 M Borax, 50 mM Tris–HCl (pH 8.0), 7 mM EDTA, 1% PVP-40 (w/v), 0.1% BSA (w/v), 1 mM DTT<br>*Buffer B (PH 8.0)*<br>1.25 M NaCl, 0.0125 M borax, 1% PVP-40 (w/v), 50 mM, Tris–HCl (PH 8.0), 25 mM EDTA, 0.1% BSA (w/v), 1 mM DTT<br>*Buffer C*<br>100 mM NaCl, 100 mM Tris–HCl (PH 8.0), 50 mM EDTA, 1 mM DTT Cold isolation buffer: 1.25 M NaCl, 50 mM Tris–HCl (pH 8.0), 5 mM EDTA, 0.1% BSA (w/v), 0.1% β-mercaptoethanol (v/v) | *Isolation buffer*<br>50 mM HEPES–KOH pH 7.5, 0.33 M sorbitol, 5 mM $MgCl_2$, 5 mM $MnCl_2$, 5 mM EDTA, and 50 mM sodium ascorbate)<br>80% (v/v) Percoll (Sigma) | *Isolation buffer* 0.35 M sorbitol, 50 mM Tris–HCl (pH 8.0), 5 mM EDTA, 0.1% BSA, 0.1% β-mercaptoethanol (v/v); wash buffer: 0.35 M sorbitol, 50 mM Tris–HCl (pH 8.0), 25 mM EDTA |
| Tissue used | 50 g leaves | 20 g leaves | 20 g leaves | 20 g leaves |
| Homogenization | 1000 × *g* for 10 min | 3500 × *g* for 20 min | 2000 × *g* for 4 min | 1000 × *g* for 20 min |
| Resuspension | 3000 × *g* for 10 min | 10,000 × *g* centrifuge 20 min | 4000 × *g* for 10 min | 16,000 × *g* for 60 min |
| DNase I treatment | No | No | No | Yes |
| References | [56] | [48, 57] | [58] | [59] |

**Table 3**
**Mitochondria isolation methods**

| | Method 1 | Method 2 | Method 3 | Method 4 |
|---|---|---|---|---|
| Reagent required | *Isolation buffer (pH 7.5)* (0.5 M sucrose, 1 mM EDTA, 70 mM $KH_2PO_4$, 0.80% BSA, and 0.1% β-mercaptoethanol) | *Isolation buffer (pH7.5)* (0.5 M sucrose, 1 mM EDTA, 70 mM $KH_2PO_4$, 0.80% BSA, 0.1% β-mercaptoethanol, and 0.6% PVP<br>7% Percoll | *Isolation buffer (pH 7.5)* (0.5 M sucrose, 1 mM EDTA, 70 mM $KH_2PO_4$, 0.80% BSA, 0.1% β-mercaptoethanol, and 0.7 M sorbitol)<br>0.7% (v/v) Percoll (Sigma) | *Homogenization buffer* (0.4 M sucrose, 50 mM Tris–HCl pH 7.5, 50 mM EDTA, 1% bovine serum albumin, and 10 mM Β-mercaptoethanol)<br>*DNase I Buffer* (0.4 M sucrose, 10 mM Tris–HCl pH 7.5, and 50 mM $MgCl_2$). 2.5 U of DNase I)<br>*Lysis buffer* (100 mM Tris pH 8, 10 mM EDTA, 2% SDS, 15 μL of 20 mg/mL Proteinase K, 5 M NaCl, 2% CTAB) |
| Tissue used | 30 (no.) Dry starchy seeds | 2 g leaves | 50 mg oily seeds | 2–5 g leaves |
| Homogenization | 1800 × *g* for 7 min | 2000 × *g* for 5 min | 2000 × *g* for 4 min | 1000–3000 × *g* for 6 min |
| Resuspension | 20,000 × *g* for 17 min | 20,000 × *g* for 20 min | 20,000 × *g* for 20 min | 14,000 × *g* for 15 min |
| DNase I treatment | Yes | No | No | Yes |
| References | [60, 61] | [62] | [62] | [62] |

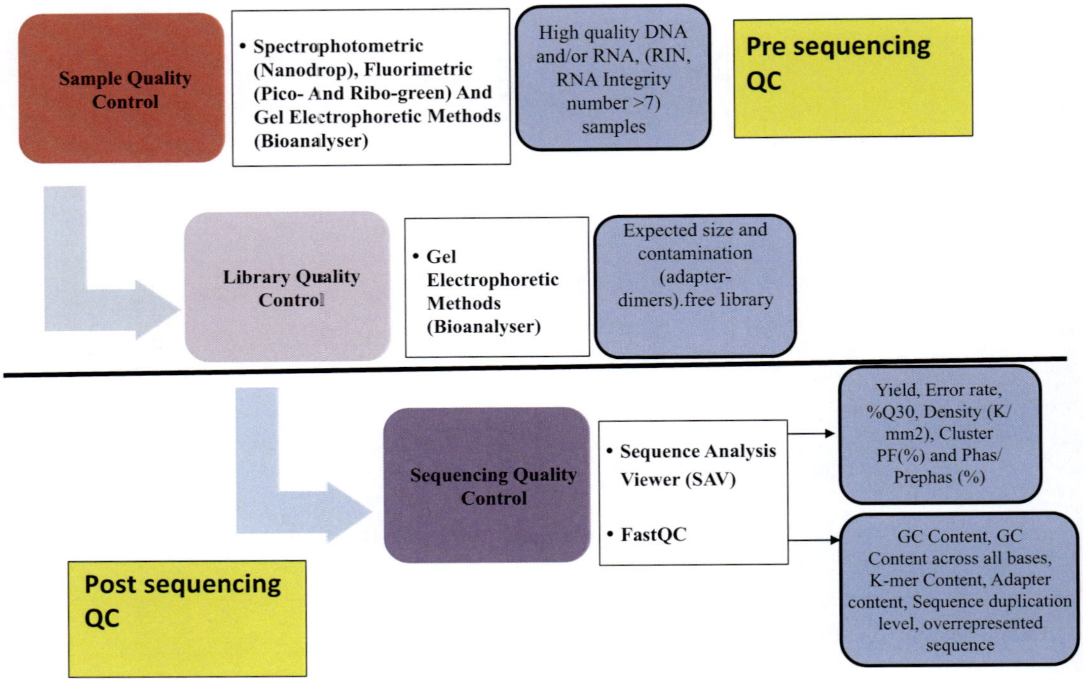


**Fig. 1** General flowchart for quality control process: Pre- and postsequencing processes necessary to ensure the result for quality control in each step

will indicate the markers and the sample size distribution of the library, while the peak height shows the number of fragments at a specific size. Figure 1 describes the quality control steps involved pre and post sequencing.

### 7.1 Quality Control

FastQC checks for high quality of the sequence data obtained (Fig. 2). First it checks the raw data and verifies the quality of sequencing data. Figure 3 shows the quality control report for different data. Analysis of GC content is done using Perl script. Detection of repeat sequences is another important quality control in characterizing mitochondrial genomes.

## 8 Organelle Genome Assembly

### 8.1 Reads Processing

NGS results are generated in the form of FASTQ files containing several small sequences known as reads and associated sequencing quality data. Figure 4, gives an overview of the organelle genome sequencing. Raw reads are first assembled into a contiguous sequence using difference bioinformatics programs. Two approaches are currently used to assemble reads. First is de novo assembly in which several short reads are arranged into long

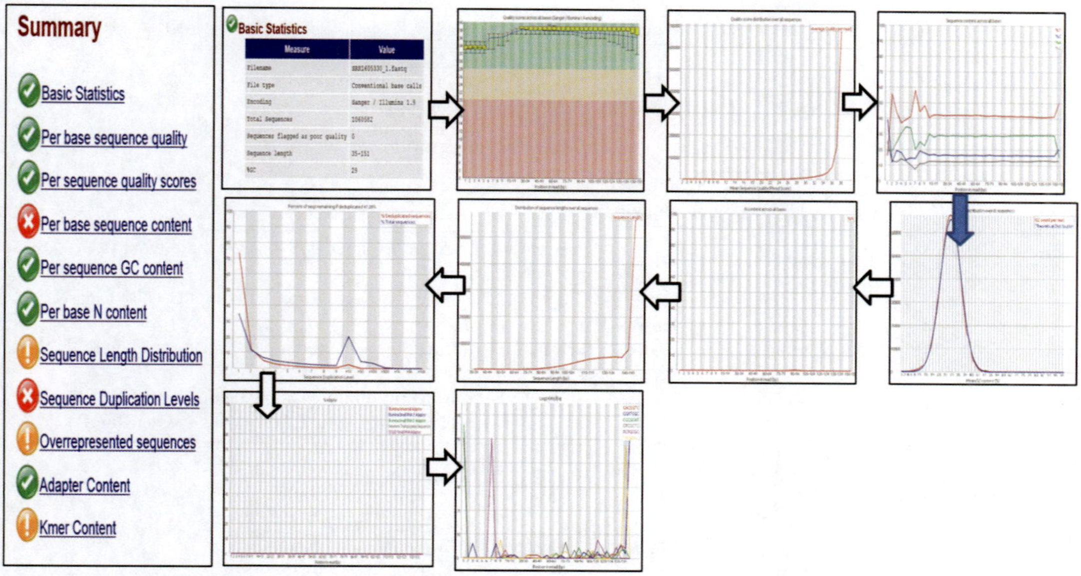


**Fig. 2** Graphical presentation of FastQC Report and the graph pattern for each parameter. Green tick (entirely normal) means QC is pass, while red cross (very unusual) means further processing required, and orange circle (slightly abnormal) means warning sign

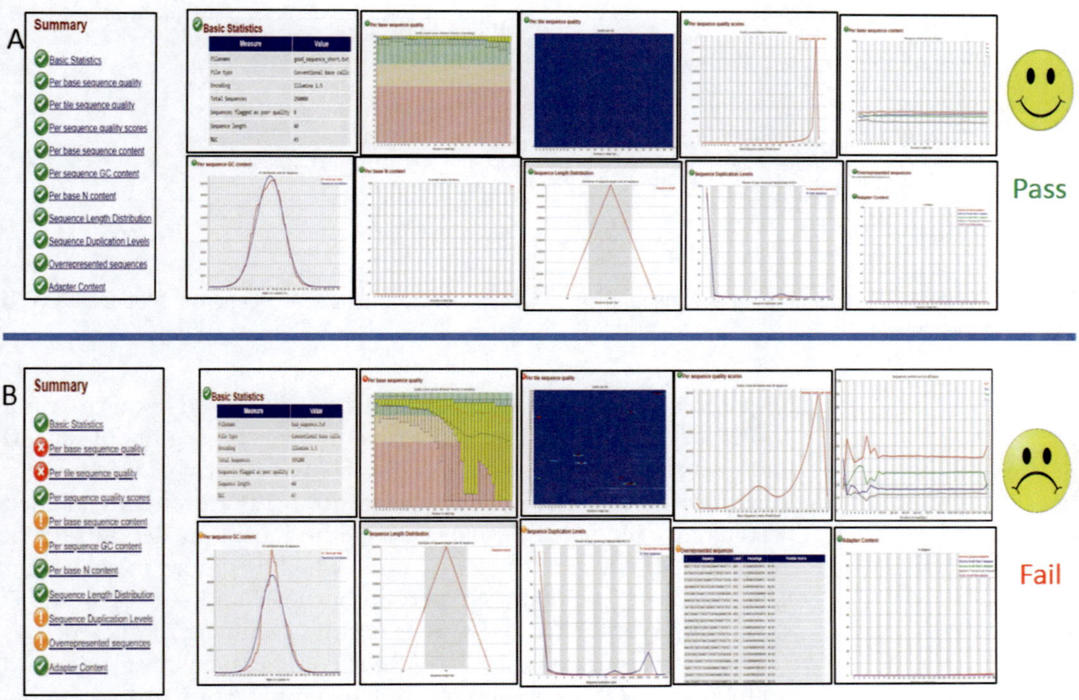


**Fig. 3** Quality control report generated by FastQC for understanding the quality of reads (**a**) Good data; (**b**) Bad data http:/www.bioinformatics.babraham.ac.uk/projects/fastqc/

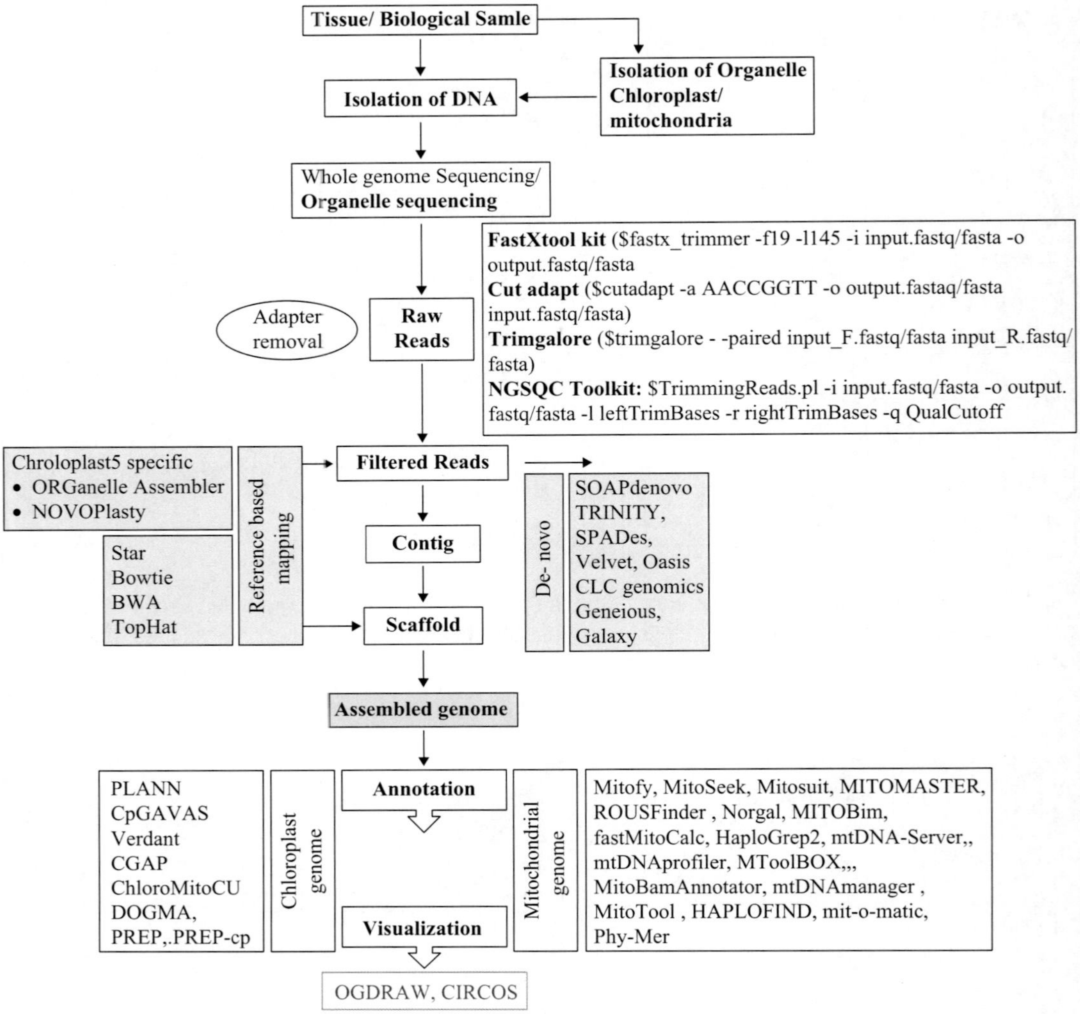


**Fig. 4** A flowchart for illustrating organelle DNA isolation, library preparation, genome sequencing, and bioinformatics pipeline

sequences by overlapping reads. The algorithm used in this method looks for all possible pairs of reads and hence is quite slow and requires high computing requirements. Alternate method is reference-guided assembly, where in the reads are directly aligned to reference sequences. It is a faster method, uses smaller number of reads and employs reference genome which is similar to the target genome. Software that are based on both the approaches are CLC Genomics Workbench (CLC Bio, Aarhus, Denmark) and Geneious (http://geneious.com/). They are commercially available and widely used. Other freely available assembly/read mapping programs are listed in Table 4.

Table 4
List of genome assembly tools for chloroplast and mitochondrial genomes

| Programs | Algorithm | Input file | Programming language | Running platform | Required read length | Single/paired-end reads | Reference |
|---|---|---|---|---|---|---|---|
| Novoplasty | De Bruijn | FASTA/ Fastq | Perl | Unix/linux | N/A | Both | [63] |
| ALLPATHS-LG | De Bruijn | BAM/ FASTA/ Fastq | Perl | Unix/linux | ~100 bp | Both | [64] |
| SPAdes (v3.13.0) | De Bruijn | BAM/ FASTA/ Fastq | Python | Linux/Mac OS/Cygwin | N/A | Both | [65] |
| Canu (v1.8) | De Bruijn | FASTA/ Fastq | Perl | Linux/Mac OS/Cygwin | N/A | N/A | [66] |
| Trinity (v2.8.4) | De Bruijn | FASTA/ Fastq | Perl | Linux/Mac OS/Cygwin | N/A | Both | [67] |
| SOAPdenovo2 (v 2.04. r240) | De Bruijn | FASTA/ Fastq | C | Linux/Mac OS/Cygwin | N/A | Both | [68] |
| Abyss (v2.0) | De Bruijn | FASTA/ Fastq | C, C++, Perl, Shell | Linux/Mac OS/Cygwin | N/A | Both | [69] |
| Velvet (v1.2.10) | De Bruijn | FASTA/ Fastq | C | Linux/Mac OS/Cygwin | N/A | Both | [70] |
| SSAKE (v4.0) | Greedy Extension | FASTA/ Fastq/ RAW | Perl | OS with Perl Interpretater | 25–36 nt | Both | [71] |
| VCAKE | Greedy Extension | FASTA/ Fastq/ RAW | Perl | OS with Perl Interpretater | <40 nt | Sinle-end reads | [72] |

(continued)

**Table 4 (continued)**

| Programs | Algorithm | Input file | Programming language | Running platform | Required read length | Single/paired-end reads | Reference |
|---|---|---|---|---|---|---|---|
| GetOrganelle | De Bruijn | BAM/FASTA/Fastq | Python | Linux/Mac OS/Cygwin | N/A | Both | [73] |
| CLC genomic Workbench (Commercial) | N/A | BAM/FASTA/Fastq | N/A | Linux/Mac OS/Cygwin | N/A | Both | https://www.qiagenbioinformatics.com/ |
| Galaxy (Commercial) | N/A | Fastqsanger | Python, JavaScript | Linux/Mac OS/Cygwin | N/A | Both | [74] |
| Geneious (Commercial) | N/A | BAM/FASTA/Fastq | Javascript | Linux, Apple OSX and MS Windows | N/A | Both | [75] |
| Newbler (Commercial) | N/A | FASTA and . SFF | Javascript | Linux/Mac OS/Cygwin | N/A | Both | [76] |

BLASTn search software is initially used to map all the sequence reads from any sequencing platform system to the available plant chloroplast/mitochondria genome. Consequently, the reads are assembled using Amos, Minimus, and Phrap software. The detailed parameters to be used for BLASTn are: *blastn –db database_name –query input_file –out output_file –evalue 1e-5 –word_size 9 –outfmt 6*. For Amos, Minimus, and Phrap, default parameters are used. The contigs generated as a result of assembling are linked, followed by gap filling using BLASTn and PCR experiments. Chloroplast genomes are simpler and conserved and share a quadripartite structure, consisting of two copies of inverted repeats, a large-single-copy region, and a small single-copy region. On the contrary, assembly of mitochondrial genomes is far more complex due to paucity of reference genome information. Some researchers have also used Newbler software to assemble the raw reads. On the basis of read depth, the contigs can be classified as high read depth contigs, medium read depth contigs, and low read depth contigs. Read depth is directly correlated to the quantity of total DNA and its copy number. As per data from different plant genomes, chloroplast genomes and repeat sequences are present as high read depth contigs, while mitochondrial genomes are present as medium read depth contigs and low read depth contigs.

## 8.2 Free Software

(ftp://ftp.broadinstitute.org/pub/crd/ALLPATHS/Release-LG/)

### 8.2.1 ALLPATHS-LG

ALLPATHS-LG is a whole genome shotgun short reads (~100 bp) assembler that can generate high quality genome assemblies. It is not designed to assemble Sanger or 454 FLX reads, or a mix of these with short reads. It requires high sequence coverage of the genome in order to compensate for the shortness of the reads. The appropriate coverage required depends on the length and quality of the paired reads of the order 100× or above.

ALLPATHS-LG requires at least two paired-end libraries—one short and one long. The short library average separation size must be slightly less than twice the read size (for example, for 100 base reads, 180 bases should be the insert size), whereas for the long library, approximately 3000 bases should be the insert size with larger size distribution.

*Requirements*: To compile and run you will need

- 64 bit Linux and a server with at least 1 TB RAM.
- Memory Requirements: About 160 bytes per genome (i.e., 512 GB for mammals and 1 GB for bacterium).
- Time taken to complete the run: 5 Mb genome in 1 h (~8 processors) and 2500 Mb genome in 500 h (~48 processors).
- g++ compiler from GCC (v 4.7.0 or higher) http://gcc.gnu.org/

- GMP library compiled with the C++ interface https://gmplib.org/
- Picard set of Java-based command line utilities for SAM file manipulation http://picard.sourceforge.net/
- Graph command dot from the graphviz package http://www.graphviz.org/

*Input format*: .bam, .fastq, .fasta, .fa, .fq, .fastq.gz, and .fq.gz

*STEPS*:

1. ALLPATHS comprises a series of modules and each module performs assembly processes. Depending on the assembly parameters different modules may be run in varying order. A single module called *RunAllPathsLG* controls the entire pipeline and decides which module is to be run and how to run them.
2. *REFERENCE (organism) directory* (*OPTIONAL*)
   As ALLPATHS is a de novo assembler, but you can provide a reference genome, if available, to perform useful evaluation at different stages of the assembly process. However, you can create a single directory of REFERENCE for the required organism to assemble, if you do not have a reference genome.
3. *Preparing ALLPATHS input files*
   To prepare data for ALLPATHS *PrepareAllPathsInputs.pl*, Perl script is used to automatically convert a set of BAM, fasta, fastq, or fastb files to ALLPATHS input files and will also generate the necessary ploidy file (optional). Two comma-separated values (.csv) must be provided by user as input files:
   (a) *In_groups.csv*: contains files to libraries
   Each line in *in_groups.csv* provides, for each data file, the following information:

   *group_name*: a UNIQUE nickname for this particular data set.

   *library_name*: the library to which the data set belongs.

   *file_name*: the absolute path to the data file.

   Example *in_groups.csv*:

| *group_name* | *library_name* | *file_name* |
|---|---|---|
| 105 | Illumina_01 | /seq/Illumina/01/105.bam |
| 103 | Illumina_02 | /seq/Illumina/02/103.fasta |
| 500 | PacBio_07 | /seq/PacBio/07/500.*.fastq |

   (b) *In_libs.csv:* describes the information of the libraries
   Each line in *in_libs.csv* describes a library and specific fields are as follows:

library_name: the library to which the data set belongs (same as in in_groups.csv)

project_name: the project name as string.

organism_name: the organism.

type: fragment, jumping, EcoP15, and so on ( informative field).

paired: 0: Unpaired reads; 1: paired reads.

frag_size: average number of bases in the fragments (defined for FRAGMENT libraries only).

frag_stddev: expected standard deviation of the fragments sizes (defined for FRAGMENT libraries only).

insert_size: average number of bases in the inserts (defined for JUMPING libraries only; if insert size is >20 kb, the library is considered to be a LONG JUMPING library).

insert_stddev: expected standard deviation of the inserts sizes (defined for JUMPING libraries only).

read_orientation: inward or outward (Outward oriented reads will be reversed)

genomic_start: index of the FIRST genomic base in the reads. If nonzero, all the bases before genomic_start will be trimmed out.

genomic_end: index of the LAST genomic base in the reads. If nonzero, all the bases after genomic_end will be trimmed out.

Example in_libs.csv:

| library_name | Type | Paired | frag_size | frag_stddev | insert_size | insert_stddev | read_orientation | genomic_start | genomic_end |
|---|---|---|---|---|---|---|---|---|---|
| Solexa-11557 | Fragment | 1 | 180 | 10 | – | – | Inward | – | – |
| Solexa-11665 | Jumping | 1 | – | – | 4000 | 500 | Outward | 0 | 25 |

```
$ PrepareAllPathsInputs.pl
```

```
IN_GROUPS_CSV= <in groups file> /
IN_LIBS_CSV= <in libs file> /
DATA_DIR= <full path of data directory> /
PLOIDY= <ploidy, either 1 or 2> /
PICARD_TOOLS_DIR= <picard tools directory> /
HOSTS= <list of hosts to be used in parallel> /
```

where, IN_GROUPS_CSV and IN_LIBS_CSV: optional arguments to determine where the data are found with default values *./in_groups.csv* and *./in_libs.csv*. Also, DATA_DIR: imported data will be placed here.

(c) Do this:

```
$ RunAllPathsLG
PRE=<prefix path> \
REFERENCE_NAME=<reference dir> \
DATA_SUBDIR=<data dir> \
RUN=<run dir> \
```

*Output*: The assembly files are as follows:

| | |
|---|---|
| *final.contigs.fasta* | fasta contigs |
| *final.contigs.efasta* | efasta contigs |
| *final.assembly.fasta* | scaffolded fasta |
| *final.assembly.efasta* | scaffolded efasta |

### *8.2.2 SPAdes (v3.13.0): Saint Petersburg Genome Assembler*

(http://cab.spbu.ru/software/spades/)

Spades is a de novo assembler designed for prokaryotic and small eukaryotic genomes. It is multi-k and takes generally longer time and memory than other assemblers. It also supports assembly of IonTorrent and PacBio data for small genomes.

*SPAdes assembly approach*:
SPAdes uses k-mers for building the initial de Bruijn graph and on following stages it performs graph-theoretical operations which are based on graph structure, coverage and sequence lengths. Moreover, it adjusts errors iteratively. The stages of assembly in SPAdes are:

Stage 1: Assembly graph construction: It employs multisized de Bruijn graph

Stage 2: k-bimer (pairs of k-mers) adjustment. Estimation of exact distances between k-mers in the genome and edges in the assembly graph.

Stage 3: Paired assembly graph construction.

Stage 4: Contig construction. SPAdes outputs contigs and allows to map reads back to their positions in the assembly graph after graph simplification (backtracking)

*Do this*:

```
$ spades.py [options] -o <output_dir>
```

Pipeline options:
Basic options:

| | |
|---|---|
| -o- | <output_dir> directory to store all the resulting files (required) |
| --sc- | this flag is required for MDA (single-cell) data |
| --meta- | this flag is required for metagenomic sample data |
| --rna- | this flag is required for RNA-Seq data |
| --plasmid | runs plasmidSPAdes pipeline for plasmid detection |
| --iontorrent | this flag is required for IonTorrent data |
| --test- | runs SPAdes on toy dataset |
| -h/--help- | prints this usage message |
| -v/--version- | prints version |

Input data:

| |
|---|
| --12 <filename> file with forward and reverse paired-end reads |
| -1 <filename> file with forward paired-end reads |
| -2 <filename> file with reverse paired-end reads |
| -s <filename> file with unpaired reads |
| --sanger <filename> file with Sanger reads |
| --pacbio <filename> file with PacBio reads |
| --nanopore <filename> file with Nanopore reads |
| --only-error-correction runs only read error correction (without assembling) |
| --only-assembler runs only assembling (without read error correction) |
| --careful tries to reduce number of mismatches and short indels |

*Output*:

| |
|---|
| output_dir content: |
| *scaffolds.fasta*: contains resulting scaffolds (recommended for use as resulting sequences) |
| *contigs.fasta*: contains resulting contigs |
| *assembly_graph.gfa*: contains SPAdes assembly graph and scaffolds paths in GFA 1.0 format |
| *assembly_graph.fastg*: contains SPAdes assembly graph in FASTG format |
| *contigs.paths*: contains paths in the assembly graph corresponding to contigs.fasta |

(continued)

| |
|---|
| *scaffolds.paths*: contains paths in the assembly graph corresponding to scaffolds.fasta |
| *corrected*: files from read correction |
| *configs/*: configuration files for read error correction |
| *corrected.yaml*: internal configuration file |
| *param.txt*: information about SPAdes parameters in this run |
| *spades.log*: SPAdes log |
| *dataset.info*: internal configuration file |
| *input_dataset.yaml*: internal YAML data set file |
| *K<##>/*: directory containing intermediate files from the run with K=<##>. These files should not be used as assembly results; use resulting contigs/sc |

*Output format*:

Contigs/scaffolds.fasta format:

>NODE_3_length_237403_cov_243.207

where, 3: no. of contigs/scaffold

237403: sequence length in nucleotides

243.207: k-mer coverage for the largest *k* value used

### 8.2.3 Canu (v1.8)

Canu is a tool which specializes in assembling the sequences generated from PacBio or Oxford Nanopore. Canu operates in three stages: (a) correction (improve the accuracy of bases in reads), (b) trimming (trim reads to the portion that appears to be high-quality sequence), and (c) assembly (order the reads into contigs, generate consensus sequences and create graphs) [66].

*Input sequences*: FASTA or FASTQ format, uncompressed or compressed with gzip (.gz), bzip2 (.bz2), or xz (.xz). (Note: zip files (.zip) are not supported)

*Do this*:

```
$ canu [options]
```

Pipeline option:

[-correct | -trim | -assemble | -trim-assemble] \

[-s <assembly-specifications-file>] \

-p <assembly-prefix> \

-d <assembly-directory> \

genomeSize=<number>[g|m|k] \

[other-options] \
[-pacbio-raw | -pacbio-corrected | -nanopore-raw | -nanopore-corrected] *fastq

By default, all three steps (correct, trim, assemble) are computed.

To compute only a single stage, use:

| | |
|---|---|
| -correct | generate corrected reads |
| -trim | generate trimmed reads |
| -assemble | generate an assembly |
| -trim-assemble | generate trimmed reads followed by assembling them |

*Output files*:
*Reads*:

*<prefix>.correctedReads.fasta.gz*: the reads after correction.

*<prefix>.trimmedReads.fasta.gz*: the corrected reads after overlap based trimming.

*Sequences*:

*<prefix>.contigs.fasta*: contains resulting contigs

*<prefix>.unitigs.fasta*: contigs, split at alternate paths in the graph.

*<prefix>.unassembled.fasta*: reads and low-coverage contigs which could not be incorporated into the primary assembly.

*<prefix>.contigs.gfa*: unused or ambiguous edges between contig sequences. The GFA format cannot represent partial overlaps between contigs (i.e., part of contig A overlaps with part of contig B).

*<prefix>.unitigs.gfa*: since the GFA format cannot represent partial overlaps, the contigs are split at all such overlap junctions into unitigs. The unitigs capture nonbranching subsequences within the contigs and will break at any ambiguity (e.g., a haplotype switch).

*<prefix>.unitigs.bed*: the position of each unitig in a contig.

*Metadata*:

The layout provides information on where each read ended up in the final assembly, including contig and positions. It also includes the consensus sequence for each contig.

*<prefix>.contigs.layout, <prefix>.unitigs.layout* (undocumented)

*<prefix>.contigs.layout.readToTig, <prefix>.unitigs.layout.readToTig*: the position of each read in a contig (unitig).

*<prefix>.contigs.layout.tigInfo, <prefix>.unitigs.layout.tigInfo*: a list of the contigs (unitigs), lengths, coverage, number of reads, and other metadata.

#### 8.2.4 Trinity (v2.8.4)

(https://github.com/trinityrnaseq/trinityrnaseq/)

Trinity is a method which results in efficient and robust de novo reconstruction of transcriptomes from RNA-Seq data. It contains three independent software modules: *Inchworm*, *Chrysalis*, and *Butterfly*, required to process large RNA-Seq datasets. *Inchworm* and *Chrysalis* steps can prove to be memory intensive.

*Do this*:

```
$ Trinity --seqType [fa|fq] --JM <jellyfish_memory> --
left input_reads_pair_1.[fa|fq] --right input_read-
s_pair_2.[fa|fq]
```

where input_reads_pair_1.[fa|fq] and input_reads_pair_2.[fa|fq] are the input paired-end files of sequence reads in fasta/fastq format, and –seqType is the type of these input reads. The option –JM defines the number of GB of system memory required for k-mer counting by jellyfish.

*Output*:

"*Trinity.fasta*" output file in the "trinity_out_dir/" output directory (or your specified directory) will create when trinity assembly completes.

#### 8.2.5 NOVOPlasty (v2.7.2)

NOVOPlasty is a de novo assembler and variance caller for short circular mitochondrial and chloroplast genomes. It only supports whole genome Illumina paired-end reads as input [63].

*Input file*: fastq/fasta files of Illumina reads

*STEPS*:

1. Create configuration file (config.txt): structure of configuration is showing below, it is necessary to make assembler work.

```
Project:
-----------------------
Project name = Test
Type = mito
Genome Range = 12000-22000
K-mer = 39
Max memory =
Extended log = 0
Save assembled reads = no
Seed Input = Seed.fasta
Reference sequence = /path/to/reference_file/refer-
ence.fasta (optional)
Variance detection = no
Heteroplasmy =
```

```
HP exclude list =
Chloroplast sequence = /path/to/chloroplast_file/
chloroplast.fasta (only for "mito_plant" option)
Dataset 1:
-----------------------
Read Length = 151
Insert size = 300
Platform = illumina
Single/Paired = PE
Combined reads =
Forward reads = /path/to/reads/reads_1.fastq
Reverse reads = /path/to/reads/reads_2.fastq
Optional:
-----------------------
Insert size auto = yes
Insert Range = 1.8
Insert Range strict = 1.3
Use Quality Scores = no
```

2. To run NOVOPlasty:

```
$perl NOVOPlasty.pl -c config.txt
```

*Output files*:

1. *Contigs_projectname.txt*: This file contains the contigs of the assemblies.
2. *Merged_contigs_projectname.txt*: When there are multiple contigs, it will combine all contigs in to a complete circular genome along with all the different possibilities
3. *Option_nr_projectname.txt*: a separate fasta file for all possible contig combinations.
4. *contigs_tmp_projectname.txt*: you can extract some contigs from this file if the above output files are empty.

### *8.2.6 SOAPdenovo2 (v2.04-r241)*

(http://soap.genomics.org.cn/soapdenovo.html)

SOAPdenovo2 is a de novo assembler, which is designed particularly for short reads of plant and animal genomes, generated by illumina. It is an advanced version of SOAPdenovo with revised algorithm that reduces the consumption of memory, resolves more repeat regions, results in increased coverage, and optimizes the assembly for large genomes.

SOAPdenovo2 has two commands, *SOAPdenovo-63mer* and *SOAPdenovo-127mer*. The first one is suitable for assembly with k-mer values less than 63 bp, requires less memory and runs faster and the latter one works for k-mer values less than 127 bp [68].

*Input file*: FASTA, Fastq and BAM

*STEPS*:

1. configFile: In order to run SOAPdenovo2, config file needs to be created which includes the information about read length, insert size, and read location.

```
#maximal read length
max_rd_len=150
[LIB]
#average insert size of the library
avg_ins=300
#if sequences are forward-reverse of reverse-forward
reverse_seq=0
#in which part(s) the reads are used (only contigs,
only scaffolds, both contigs and scaffolds, only gap
closure)
asm_flags=3
#cut the reads to the given length
rd_len_cutoff=100
#in which order the reads are used while scaffolding
rank=1
# cutoff of pair number for a reliable connection
(at least 3 for short insert size)
pair_num_cutoff=3
#minimum aligned length to contigs for a reliable
read location (at least 32 for short insert size)
map_len=32
#paired-end fastq files, read 1 file should always be
followed by read 2 file
q1=input_reads1_pair_1.fq
q2=input_reads1_pair_2.fq
#another pair of paired-end fastq files, read 1 file
should always be followed by read 2 file
q1=input_reads2_pair_1.fq
q2=input_reads2_pair_2.fq
#paired-end fasta files, read 1 file should always be
followed by read 2 file
f1=input_reads_pair_1.fa
f2=input_reads_pair_2.fa
#fastq file for single reads
q=input_reads.fq
```

2. *Do this*:

```
$ SOAPdenovo-63mer all -s configFile -o output_directory/outputGraph -K <kmer_value>
```

where configFile is a defined configuration file, outputGraph is the prefix of the output files, and kmer_value is the value of k-mer used for building the assembly (<=*63* for SOAPdenovo-63mer and <=*127* for SOAPdenovo-127mer).

*Output*:

(a) **.contig*: contains contig sequences

(b) **.scafSeq*: contains scaffold sequences

### 8.2.7 GetOrganelle

(https://github.com/Kinggerm/GetOrganelle)

The GetOrganelle pipeline has proved to be an effective way for the assembly of chloroplast genomes of land plants, without the requirement for reference-guided scaffolding, gap filling, or start-end point closing. This pipeline can be also applied to assemble mitochondrial genomes and nuclear Ribosomal DNAs [73]. It consists of four steps:

1. recruiting plastid-like reads;
2. de novo assembly using SPAdes;
3. filtering plastid-like contigs, and
4. visualizing and editing de novo assembly graph

The first three steps can be accomplished automatically by the use of a combined command and the fourth step is to visualize and evaluate the assemblies.

*STEPS*:

1. Prepare the data: Includes both paired-end and single-end short reads (read length 75–300 bp)
2. Recruiting plastid-like reads by a Python script:

```
$ get_organelle_reads.py
```

3. De novo assembly using SPAdes
4. Filtering plastid-like contigs by script (Bowtie2)
5. Visualizing and editing de novo assembly graph by Bandage

*Output*:

*filtered_spades/assembly_graph.fastg*: produced by SPAdes consisting of assembly graph

*Utilities/slim_fastg.py*: is used to modify the *filtered_spades/assembly_graph.fastg* file and produce a new fastg file

**complete*path_sequence.fasta*: produced as the final result.

### 8.2.8 AbySS (v2.0): (Assembly By Short Sequencing)

(http://www.bcgsc.ca/downloads/abyss/abyss-2.0.tar.gz)

- It is massive parallel assembler for the assembly of the whole-genome shotgun sequencing data of the mitochondrial and plastid genomes
- ABySS efficiently identifies polymorphic sequences, insertions and deletions, and novel sequences. Moreover, coding genes and RNAs (rRNA and tRNA) were annotated from the sequence data
- It performs the assembly in two steps: first, the contigs are extended without the paired-end information; and then the paired-end information is utilized to merge contigs [69] .

### 8.2.9 SSAKE (v4.0): (Short Sequences Assembly by Progressive K-mer Search and 3′ Read Extension)

(http://www.bcgsc.ca/bioinfo/software/ssake)

- It is an algorithm of short read sequences assembly used for assembling millions of short nucleotide sequences. It is written in PERL and runs on Linux operating system.
- It progressively searches for the longest possible overlap between any two sequences and assembles them into contigs [71] .
- It works in four steps:
  (a) Sequence Overlap
  (b) Building scaffolds with SSAKE
  (c) Using a seed sequence file
  (d) Using seeds (-s) with mate pairs (-p 1)

*Do this*:

```
$ ./runSSAKE.sh read1.fq read2.fq libraryFragmentLength
basename
```

*Output*:

| | |
|---|---|
| *_contigs.fa* | All sequence contigs (fasta file) |
| *_scaffolds.fa* | All sequence scaffolds (fasta file) |
| *.log* | text file; |
| *_short.txt* | text file; Lists sequence reads shorter than a set (minimum) |
| *_singlets.fa* | All unassembled sequence reads (fasta file) |

### 8.2.10 VCAKE: (Verified Consensus Assembly by K-mer Extension)

(http://sourceforge.net/projects/vcake/)

- VCAKE, an extension of SSAKE, can partially assemble a de novo genome from short reads in a range of genomic contexts, error rates, and sequencing coverage, usually develop to efficiently handle sequencing error. We also believe that VCAKE may be a step toward the assembly of larger bacterial genomes from short reads, particularly with the development of superior base calling or paired end technology. It considers all the possible overlaps between the contigs being extended and unassembled reads rather than extending the contig based on the greedy approach or quality values [72].

### 8.2.11 Velvet (v1.2.10)

(http://www.ebi.ac.uk/~zerbino/velvet)

- It is used to perform short read assembly to eliminate errors and resolve repeats.
- Running Velvet consists of a sequence of two commands *velveth* and *velvetg*, where velveth produces a hash table of k-mers, while velvetg constructs the genome assembly. The k-mer length, also known as hash length, corresponds to the length (base pairs).

To obtain optimal assemblies, k-mer value is important as higher and lower the k-mer values increase the specificity and sensitivity respectively [70]. *Input files*: fasta, fastq, fasta.gz, fastq.gz, sam, bam, eland, gerald

*Reads*: short, shortPaired, short2, shortPaired2, long, longPaired

*Do this*:

1. The first step of running Velvet is to run velveth:

```
$ ./velveth output_directory/ hash length/k-mer length [-file_format] [-read_type] -separate input_reads_pair_1.fastq input_reads_pair_2.fastq
```

2. After running velveth, the next step is to run velvetg on the output_directory/ and files generated from velveth:

```
$ ./velvetg output_directory/ -min_contig_lgth 200
```

*Output*:

The "*contigs.fa*" is the final Velvet output fasta file that contains the assembled contigs.

### 8.3 Commercial Software

#### 8.3.1 CLC BIO

(http://www.clcbio.com/)

CLC Main Workbench is a commercial bioinformatics software (Fig. 5a) used for the analysis of biological data as it supports read mapping of Sanger and high-throughput sequencing data, de novo assembly, SNP detection on whole genomes of any size, structural variation detection, ChIP-seq analysis, RNA-Seq analysis, small RNA analysis, and BLAST analysis.

It consists of different modules such as Assembly; Cloning; Align and Phylogenetic Tree tutorial and Microarray modules (https://www.qiagenbioinformatics.com/).

Although, *CLC Genomics Workbench (v11.0.1)* is used for NGS data analysis as it comprises the following:

- Import of 454, Illumina Genome Analyzer, SOLiD, and Helicos data.
- Reference assembly of human-size genomes.
- De novo assembly.
- SNP/DIP detection.
- Graphical display of large contigs.
- Support for mixed-data assembly.
- Paired data support.
- RNA-Seq analysis.
- Expression profiling by tags.
- ChIP-Seq analysis.

#### 8.3.2 Galaxy

(https://galaxyproject.org/)

Galaxy is a freely available, open, web-based data analysis and workflow platform designed for biologist to analyze their own data with different toolsets already installed in it (Fig. 5b). It is written in Python and used for biological data analysis such as genomics, proteomics, epigenomics, transcriptomics, and also gene expression analysis and genome assembly.

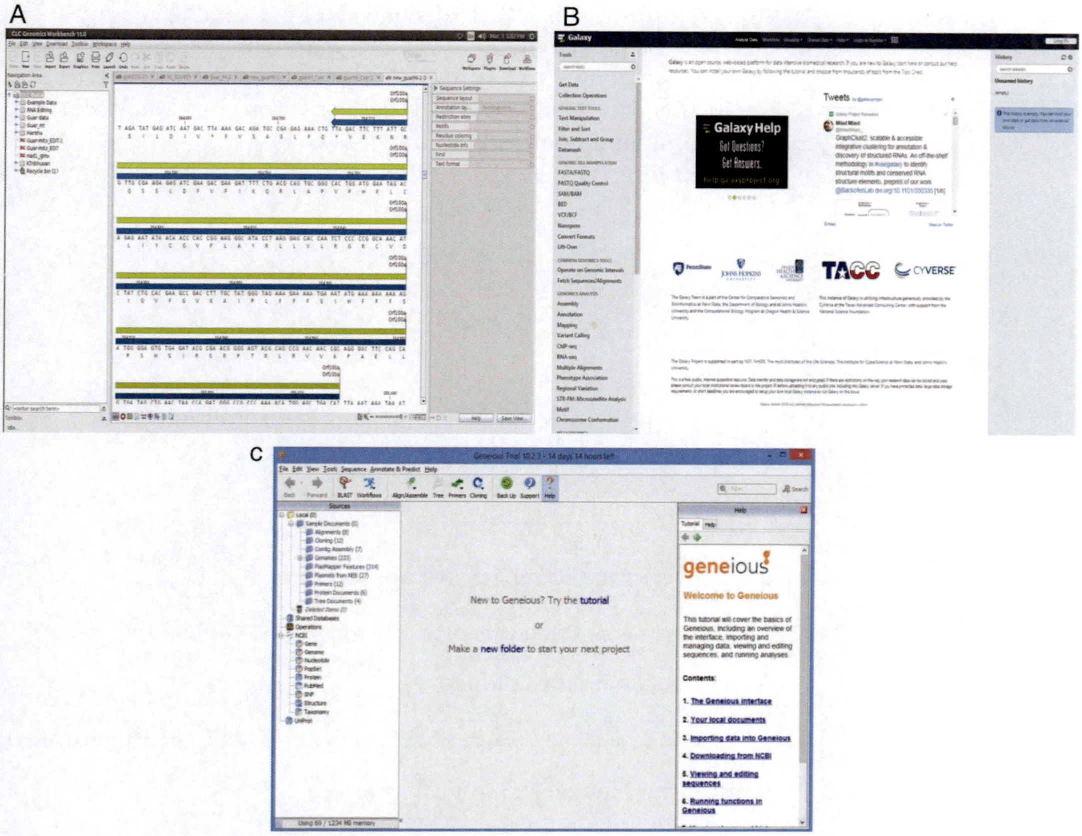


**Fig. 5** Commercial based tools available for organelle (Mitochondria and Chloroplast) genome assembly: (**a**) CLC Bio home page; (**b**) Galaxy home page; (**c**) Geneious home page

A number of different tools are implemented in Galaxy interface to perform various bioinformatics analysis, one can select from among the appropriate analysis tools to process the data [74].

### *8.3.3 Geneious*

(https://www.geneious.com/)

Geneious is easy, flexible integrated platform for molecular biology and NGS data analysis (Fig. 5c). It is graphical user interface and displays selected file(s) in a variety of different ways including sequence view (linear and circular), dot plot view, query-centric alignment view, protein domain view, 3D structure view, text view, and notes.

It is written in JAVA and applicable on any operating system (e.g., Linux, Apple OSX, and MS Windows) as a stand-alone download [75].

#### 8.3.4 Newbler

(https://github.com/etheleon/newbler)

It is a commercially available software package for assembling organelle genomes (Chloroplast and mitochondrial genomes) sequencing data which is run based on Java GUI (gsAssembler) or the command line (runAssembly) [76].

*Input files*: FASTA and .SFF

*Do this*:

```
$ runAssembly -o testproject2 -cdna 454transcriptome.sff
```

*Steps*:

The steps involved for data analysis are as follows:

1. Check the extracted data in form of SSF file.

```
$ sffinfo 454transcriptome.sff | less
```

2. Convert the sequences in fasta format.

```
$ sffinfo -seq 454transcriptome.sff > 454transcriptome.fna
```

3. Check the quality of file.

```
$ sffinfo -qual 454transcriptome.sff > 454transcriptome.qual
```

4. Run the Newbler graphical interface

```
$ gsAssembler&
```

*Output*:

1. 454transcriptome.fna
2. 454transcriptome.qual

## 9 Reference Based Assembler

After obtaining a reference genome, shorter reads from Illumina sequencing platform are mapped to reference genomes through BWA, forming a consensus sequence to determine whether there are base differences in the reference genome.

The detailed procedure describing the alignment of Illumina short reads against the reference genomes by the use of BWA is described below:

build index: bwa index –a bwtsw reference.fa

⇩

find SA coordinates: bwa aln –t 30 –f single.sai reference.fa single.fastq

⇩

convert SA coordinates and output sam: bwa samse –f single.sam reference.fa single.sai single.fastq

⇩

convert sam to bam: samtools view –bS single.sam > single.bam

⇩

extract results that can align to the reference sequence: samtools view –Bf 4 single.bam > single.F.bam

⇩

bam to fastq: bam2fastq single.F.bam –o single.fq

⇩

assembly: runAssembly –cpu 10 –het –sio –m –urt –large –o result single.fq

## 10 Gap Filling and Correction: Remaining Gap Filling

Input: related contigs with remaining gaps

Step 1: prepare the query sequence;

Step 2: "setup," specify related options and database, create a lookup table;

Step 3: "BLASTn search";

Step 4: "back-track," input the preliminary matches and locate the insertions and deletions of uncertain sequences;

Step 5: output the results in a file.

Output: related contigs

## 11 Annotation of Organelle Genome:

Once a draft genome is constructed, its constituent genes need to be identified and annotated. Before annotation, gene prediction should be done. Gene prediction is the process of identifying the regions of encoded genes that are likely to occur. This process entails translating nucleotide sequence and finding open reading frames (ORFs). Gene prediction can be performed with some computational programs such as Geneious Pro or ORF Finder. Organelle genomes are generally highly conserved, thus reference-based prediction is sufficient for organelle genome annotation. Predicted ORFs are verified using a BLAST similarity search. For coding gene annotation, BLASTx is recommended. The BLASTx program compares the six-frame conceptual translated products of a nucleotide query against a protein sequence database to provide more accurate models and to detect unknown ORF sequence. Moreover, there are some annotation tools that are available for use (e.g., DOGMA, Mitofy, and CpGAVAS), *see* more in Table 5. After annotation, two things should be checked: (1) whether the lengths of the annotated genes are similar to that of the reference, and (2) whether the proper start codon was used (*see* **Note 7**). Once all gaps and ambiguous sequences have been identified, the basic steps involved in annotation of organelle genome are as follows:

Cp or mt genome annotation by using DOGMA (Dual Organellar GenoMe Annotator)

⇩

Codon usage is predicted by using CodonW for rRNA and tRNA genes

⇩

NCBI BLASTx and BLASTn to identify genes against database using the default settings with e-value 1e−10

⇩

tRNA boundaries identification by tRNAscan-SE

⇩

Drawing gene maps using OrganellarGenome DRAW tool (OGDRAW)

⇩

Repeat sequence analysis using the REPuter web-based interface with a minimal length of 50 bp

⇩

Transposable elements and other repeated elements mapping with RepeatMasker

⇩

Cp and mt-derived sequences identification with BLASTn against reference (Identity $\geq$80%, *E*-value $\leq$ 1e−5, and Length $\geq$ 50 bp)

## 12 Bioinformatics Software Tools for Annotating Organelle Genome

There are various online freely available tools used to annotate the chloroplast and mitochondrial genome with their graphical representation (Fig. 6). These tools can be used to do the ORF prediction, predict CDS, rRNA, and tRNA, and draw the complete chloroplast and mitochondrial genome. These tools can also be helpful to further identify various structural and functional analysis including evolutionary process, genomic recombination, taxonomic markers, intron structure, and evidence of genome reduction.

**Table 5**
**List of tools and software used for annotation of organelle genome sequence data**

| Bioinformatics tool | Input file | User interface | Organalle assembly | Reference |
|---|---|---|---|---|
| MitoSuit | BAM | GUI | Human MIT | [77] |
| mtDNAprofiler | FASTA, variant | Web | Human MIT | [78] |
| MToolBOX | FastQ, BAM, SAM | CUI | Human MIT | [77] |
| MitoSeek | BAM | CUI | Animal/plant | [79] |
| MitoBamAnnotator | BAM | Web | Human MIT | [77] |
| mtDNAmanager | variant | Web | Animal/plant | [77] |
| MitoTool | FASTA, variant | Web/GUI | Human MIT | [77] |
| HAPLOFIND | FASTA | Web | Animal/plant | [77] |
| mit-o-matic | FastQ, pileup | Web/CUI | Human MIT | [80] |
| Phy-Mer | FASTA, FastQ,BAM | CUI | Animal/plant | [77] |
| MITOMASTER | FASTA | Web | Human MIT | [81] |
| Mitofy | FASTA | Web/CUI | Plant MIT | [7] |
| PLANN | FASTA and GENBANK | CUI | CHL | [82] |
| CpGAVAS | FASTA and GENBANK | Web | CHL | [82] |
| Verdant | FASTA | Web | CHL | [83] |
| GeSeq | FASTA | Web | MIT/CHL | [83] |
| AGORA | FASTA | Web | MIT/CHL | [82] |
| DOGMA | FASTA | Web | Plant CHL/animal MIT | [84] |
| MitoAnnotator | FASTA | Web | Fish MIT | [82] |

**Fig. 6** List of versatile and efficient tools for chloroplast and mitochondrial annotation

### 12.1 Mitofy

Mitofy a set of software tools used for annotations of complete mitochondrial genome sequences and also RNA editing data of different plant species and tissue sample used. It is available freely at https://omictools.com/mitofy-tool. [7]

### 12.2 ChloroSeq

ChloroSeq is among a handful of free bioinformatics software packages, dedicated to studying plastid and mitochondrial genetics [85].

### 12.3 RNAweasel and MFannot

RNAweasel and MFannot (http://megasun.bch.umontreal.ca/RNAweasel/) are online tools where one can predict and model the complex organelle RNAs and annotate introns and exons.

### 12.4 MitoSeek

Mitoseek is an open-source software tool and is designed to work with paired end sequencing reads. It can accurately extract the mitochondrial genome sequence from the whole genome sequencing data. The input sequence can be in form of a BAM file. It aids in extraction of high-throughput mitochondrial sequencing data from the existing large exome sequencing databases and also can perform mitochondrial genome assembly. It filter reads on the basis of mapping and base quality score (*see* **Note 8**). It works by evaluating alignment quality of the mitochondrial genome, estimating the mitochondrial copy number, and also works by detecting somatic mutation, structural variants, and heteroplasmy of the mitochondrial genome. It is the only known tool which reports allele counts by forward and reverse strand separately. It can be run parallel on large exome sequencing datasets. The above tool can also be used for detecting structural changes like mitochondria–nuclear genome integration (detected by discordant read pairs) and deletions in mitochondrial genome. The tool is more suitable for detecting large copy number variation rather than small copy number variation.[79]

### 12.5 Mitosuite

Mitosuite, a user friendly tool, is an all in one tool which can be used for the analysis of high throughput mitochondrial genome sequencing data. The input file can be in form of BAM format which is a binary version of SAM file (Sequence Alignment/Map). The software besides checking the mapping and sequencing quality, can also find the potential mismatches which can arise due to sequencing errors, exogenous combinations, or heteroplasmy. The summary statistics like depth of coverage, GC content, base-call quality, mapping quality, and read length are all calculated by the above tool. File conversion (like BAM > FASTA, BAM > VCF) is not required and it can directly assign mitochondrial haplogroups, based on the haplogroup-defining sites of PhyloTree [77].

### 12.6 MITOMASTER

Mitomaster is a tool used for analysis of input mitochondrial DNA sequences. The tool has been rebuilt on a new framework available at https://www.mitomap.org/foswiki/bin/view/MITOMASTER/WebHome . The present version (Beta 1) performs variant calling relative to the rCRS, haplogrouping based on Phylotree, and variant annotation based on Mitomap. The present tool is based on the interaction of three elements database (including genomic, variation catalog, seq repository, NUMTs); analysis (for variation extraction, NUMTs screening, haplogrouping and functional analysis); and interface (for sequence analysis, variant analysis and sequence builder). Full mtDNA or partial mtDNA sequences can be used as an input. While the minimum length for variant listing is 100 nucleotides, the haplogroup prediction can be done accurately only for full mitochondrial genomic sequences [81]

### 12.7 ROUSFinder

ROUSFinder (https://omictools.com/rousfinder-tool) detects nontandem repeats within mitochondrial genomes in both direct and inverted orientations by the use of BLAST. The repeats are infrequently annotated mitochondrial sequence assemblies and are not well conserved between species. As the repeats are much larger than expected, therefore, they are called repeats of unusual size (ROUS). The repeats are divided into two functional classes: (1) repeats occurring due to crossing over and (2) repeats occurring due to mutations in DNA repair genes. The script ROUSFinder.py by the use of BLASTn, performs a pairwise ungapped comparison of a sequence with itself, both strands separately, with a setting of word size 50, 1e−23, reward for a match +1, penalty for a mismatch −9 and percent identity cutoff 99%. Full length identity is deleted once the script concatenates the two output files. Alignments are then sorted and compared to identify and remove duplicate repeats, and an output file consisting of the repeats is created in fasta format. The above file can also be formatted for GenBank annotation. [86]

### 12.8 Norgal

Norgal is a pipeline (https://bitbucket.org/kosaidtu/norgal) which is used for the extraction and assembly of full or partial mitochondrial and chloroplast genomes from WGS short reads without the availability of reference sequences. A two-step procedure is followed for the plastid genome assembly. It successfully assembles a completely unknown mitochondrial genome, which cannot be assembled from any other known reference. Therefore, it has contributed to the field of discovering and assembling novel mitochondrial sequences from whole genome sequencing data. In comparison to MITObim and NOVOPlasty, which requires a small seed sequence to initiate an assembly, the Norgal tool does not require any such seed sequence to de novo assemble the mitogenome, and it relies exclusively on differential k-mer frequencies in the reads which it detects automatically. .The programming language for Norgal is Python 3. The workflow of Norgal consists of trimming and removal of adapters from NGS reads using Adapter Removal followed by a de novo assembly using MEGAHIT. The reads are then mapped back to the longest assembled sequence using *bwa mem*. The read depth for each position is calculated and kmers of size 31 are counted in all reads and those subsets having at least one 31-kmer are kept. Then a de novo assembly is done using *idba_ud*. The circularity of the longest contig is examined and potential mitochondrial and chloroplast contigs are identified. [87]

### 12.9 MITObim

MITObim is an in silico approach, which without the need of a reference genome, reconstructs complete mitochondrial genomes of nonmodel organisms directly from next-generation sequencing (NGS) data. It relies solely on mitochondrial genome information of more distantly related taxa or short mitochondrial barcoding sequences (seeds), such as the commonly used cytochrome-oxidase subunit 1 (COI), as a starting reference. MITObim is a superior tool in comparison to the existing tools in terms of accuracy, runtime, and memory requirements. With MITObim, fully automatically recovered mitochondrial genome, with accuracy exceeding 99.5% can be derived from total genomic DNA NGS data sets in less than 24 h while using a standard desktop computer. [88]

### 12.10 mitoAnalyzer

mitoAnalyzer (https://lgsun.irp.nia.nih.gov/hsgu/software/mitoAnalyzer/index.html) is a software package that provides a general approach for the analysis of mitochondrial DNA (mtDNA) in next-generation sequencing studies, using whole-genome sequencing data.

### 12.11 GeSeq: Annotation of Organellar Genomes

GeSeq is a web-based annotation tool used for annotating plant organellar sequences (Fig. 7). It can also annotate mammal mitochondrial genomes using a suitable reference. It is specially made for high quality annotation in very short time and also takes batch submissions. Due to the fact that GeSeq's functions are integrated in an easy-to-use GUI and it is highly flexible, it would be able to meet the demands of most annotation jobs. For high-quality annotation of chloroplast genomes, GeSeq is equipped with manually curated reference sequences and a corresponding profile hidden Markov model (profile HMM) database for chloroplast protein and rRNA-coding genes.

GeSeq provide options to select or upload the most appropriate reference sequences for each annotation project and provides a manually curated reference sequences set for chloroplast genomes. GeSeq relies on a BLAT-driven best-match method. Geseq recalls the hits with highest score and best nonoverlapping hits for each genes. Nucleic acid sequences provided for annotation purpose must be in (multi-)FASTA format (http://blast.ncbi.nlm.nih.gov/blastcgihelp.shtml). This annotation tool does not accept protein sequences. GeSeq is prepared with a copy of the organelle

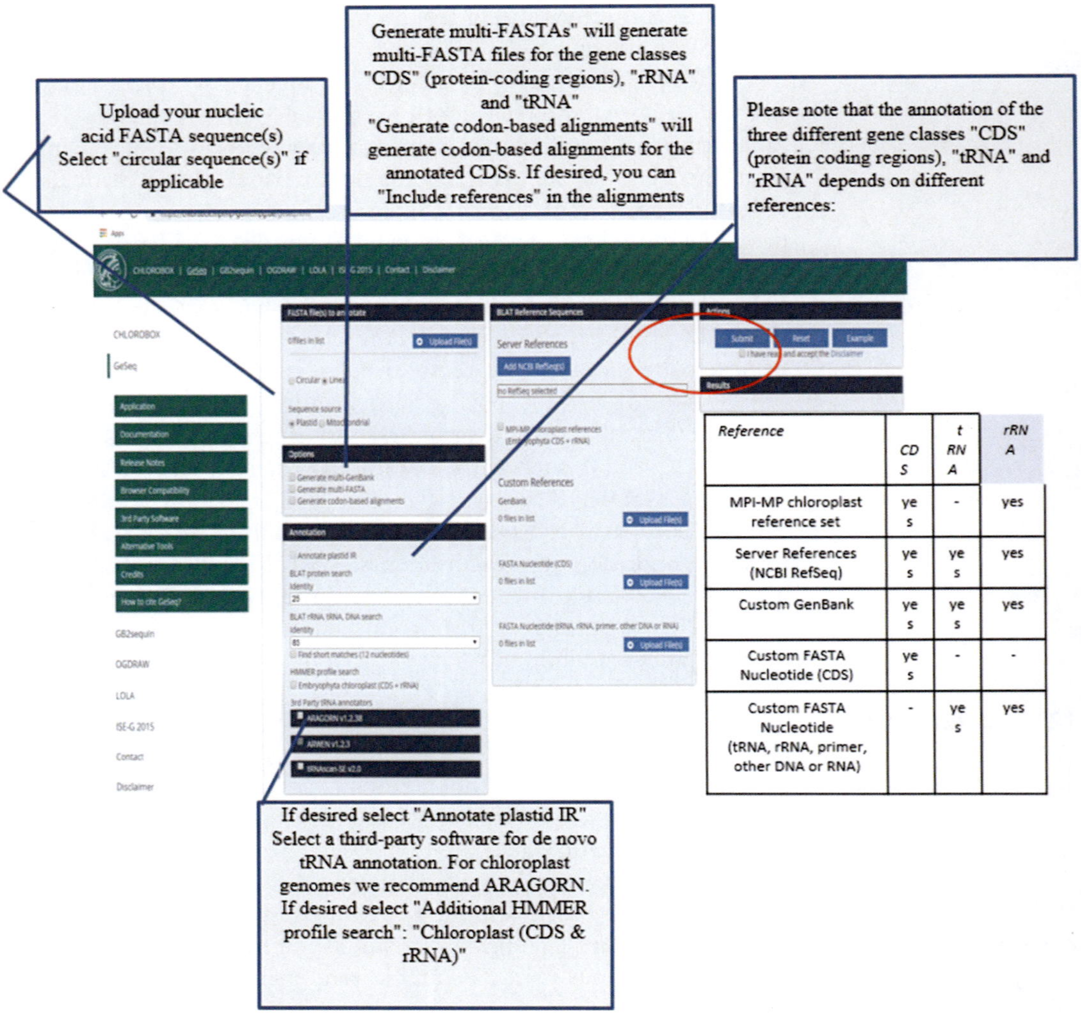


**Fig. 7** GeSeq: Home page in detail format for annotating organelle (Mitochondria and chloroplast) genome

genome records of the NCBI RefSeq project (http://www.ncbi.nlm.nih.gov/genome/organelle/). These hosted NCBI RefSeq records are updated and visualized on monthly basis. For each annotation job two database are created by GeSeq, one is protein-coding (CDS) and another is non-protein-coding (NA) database. The non-protein-coding (NA) database may contain rRNA, tRNA and other nucleic acid sequences. The program analyses all the uploaded GenBank files for CDS, tRNA, and rRNA entries. The output file of GeSeq is GenBank file which can be edited by users and visualized by OGDRAW. There is option to create multi-FASTA files containing all genes, CDSs, rRNAs, and tRNAs found in the sequence(s) submitted for annotation. It can also

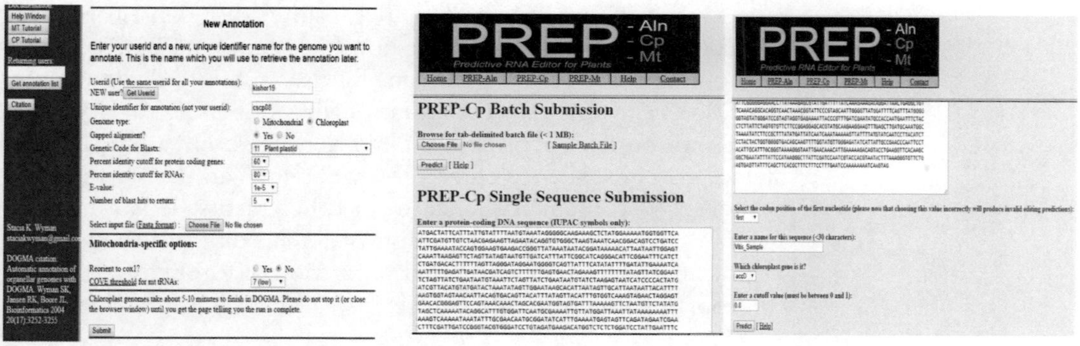


A B

**Fig. 8** Automated annotation tools used for annotating organelle genome: (**a**) Dogma homepage (plant chloroplast and animal mitochondrial genome); (**b**) PREP home page (Mitochondria and chloroplast genome)

provide the user with codon-based alignments by TranslatorX and MUSCLE for each CDS identified for phylogenetic or other analyses. [83]

### 12.12 DOGMA

DOGMA (Dual Organellar Genome Annotator) provides the automatic annotation of organellar (plant chloroplast and animal mitochondrial) genomes (Fig. 8a). It is a Web-based tool, which by the use of BLAST search against a custom database, provides the identification and annotation of the genes. The annotations provided by the DOGMA can be edited and stored in a password protected server and later the finished annotations can be extracted for submission in GenBank [83].

The following are the steps for uploading chloroplast genome sequence in DOGMA-:

1. Enter you User ID or new users can create one.
2. Enter an unique identifier for your sequence.
3. Select the genome type you want to annotate.
4. Rest of the parameters can used in default settings or change accordingly.
5. Upload the genome sequence.
6. Press the submit button.
7. After approximately 10 min a page will appear showing that the run is complete.
8. Click on the tab appeared on the page and a table of annotated genes will appear.

### 12.13 PREP Suite

RNA editing is a post transcriptional process which converts cytidines to uridines in mitochondrial and chloroplast transcripts of land plants. The above process thereby helps in amino acid residues conservation across species. The PREP suite [89]. consists of three programs PREP-Mt (for mitochondrial genes), PREP-Cp (for chloroplast genes), and PREP-Aln (for user-submitted alignments) (Fig. 8b). The software computationally predicts C to U editing sites in an organellar protein-coding gene by comparing the predicted protein sequence entered by the user to an already reported homologous protein sequence from other plants. It is a fast and accurate method for the detection of RNA editing sites. The following are the steps described for prediction of RNA editing sites in chloroplast proteins by PREP-Cp

1. Choose the fasta file of the gene sequence or paste the fasta sequence in the box given.
2. Select the codon position of the first nucleotide.
3. Enter the name of the sequence.
4. Choose the name of the gene entered.
5. Set a cutoff value between 0 and 1, default is 0.8.
6. Press the predict button.

## 13 Visualization

### 13.1 Organellar-GenomeDraw (OGDRAW)

OrganellarGenomeDraw (OGDRAW), a very simple and straightforward software tools that provide facility to the users to generate high-quality visual representations map of both circular and linear annotated genome sequences (Fig. 9a). It is used to draw organellar genome maps and provides a number of options for such as, color-coded conserved gene classes present in organellar genomes, IRs and polycistronic transcripts. It can accept all types of DNA sequences as input. If the genome sequence of an organelle to be drawn is already available in GenBank repository, users can upload GenBank files directly to OGDRAW or accession number can be used for the same. The parameters in software can be adjusted or defined by the users via a web browser interface to depict organellar genomes appropriately (http://ogdraw.mpimp-golm.mpg.de/). GenBank flat files can be read and edited. There is no size limit for the genome submitted to OGDRAW but gene-dense and/or feature-rich genomes produces relatively crowded maps. Map quality depends on the consistency and accuracy of the annotation data provided in the input file. Incorrectly or nonuniformly annotated features probably generate incorrect maps. [90]

### 13.2 Circos

Circos is a set of software for visualization of annotated data (Fig. 9b). It visualizes and presents the data in a circular map. It is easy to plot, format and layer the data with Circos. This software is flexible. It is not restricted or specific to genome sequence but it has ability to visualize any kind of data and gives circular output. It can

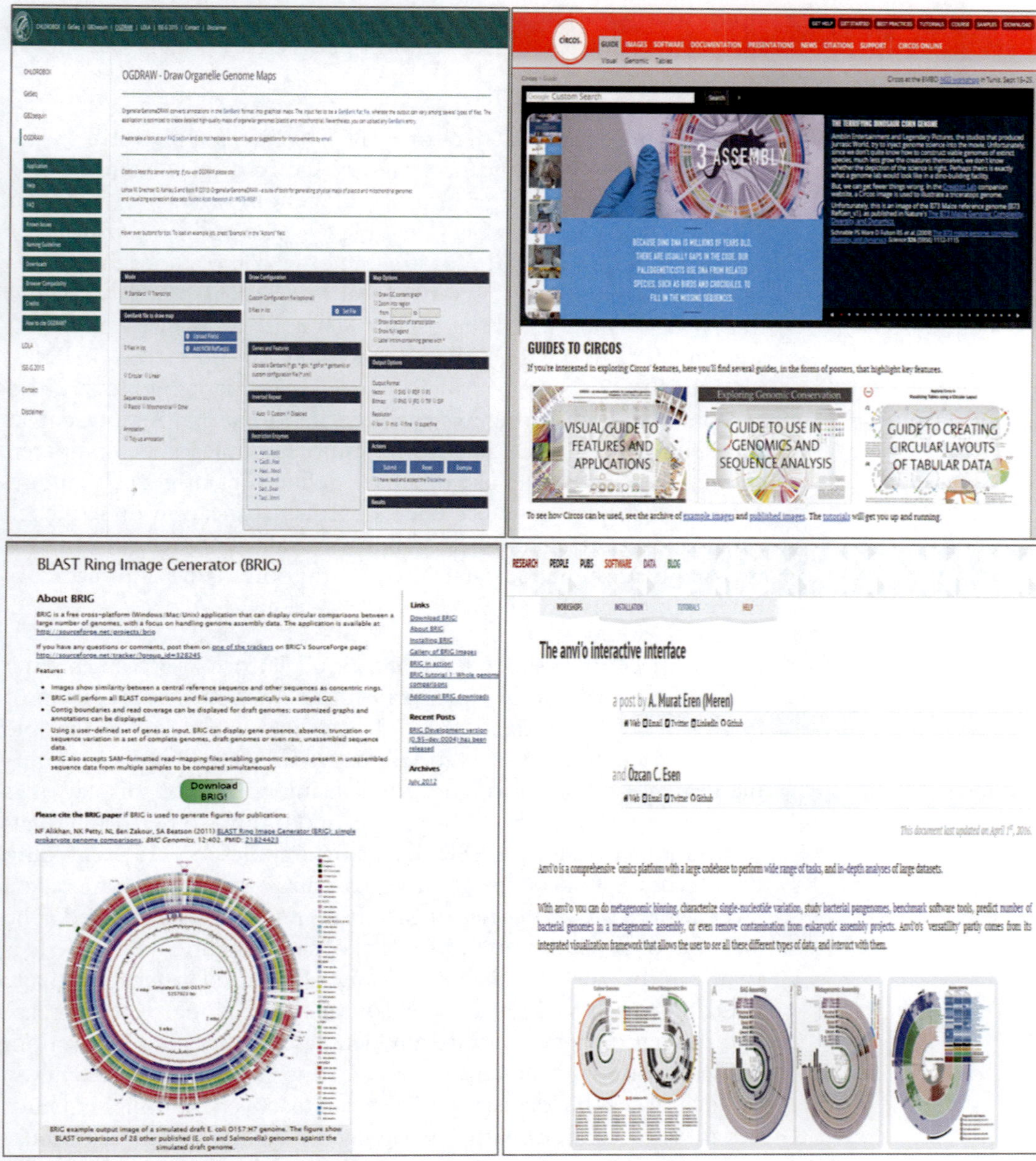


**Fig. 9** Interactive interfaces for visualizing assembled mitochondria and chloroplast genome (**a**) OGDRAW; (**b**) CIRCOS; (**c**) BRIG; and (**d**) ANVI'O home page

also visualize the data with multilayered annotations of one or more scale. Automation in Circos and measurement using plain text configuration files allows for its incorporation into different data collection and analysis pipelines. Laying down of images is controlled via a plain-text configuration file. There is no interactive user interface. The purpose of fashioning this tool was to present data in a beautiful way. The compact and crowded data in circular layout appears characteristically more attractive than a linear arrangement [90].

### 13.3 BRIG

BLAST Ring Image Generator (BRIG) is a cross-platform (Windows/Mac/Unix) application that is used to display circular comparisons between a large number of genomes, with a focus on handling genome assembly data (Fig. 9c). It utilizes CGView for image rendering and BLAST for genome comparisons automatically via a simple GUI. Its graphical interface is based on Swing framework which helps the user to generate a circular image in a step by step mode. The image generated can be in JPEG, PNG, SVG, or SVGZ format. SAM-formatted read-mapping files are also accepted by BRIG, thereby enabling genomic regions present in unassembled sequence data from multiple samples to be compared simultaneously. By the use of a user-defined set of genes as input, BRIG can display gene presence, absence, truncation or sequence variation in a set of complete genomes, draft genomes or even raw, unassembled sequence data. Also, contig boundaries and read coverage can be displayed [91].

### 13.4 Anvi'o

An advanced analysis and visualization platform for 'omics data: Anvi'o is an open source platform for analysis and visualization of large set of 'omics data (Fig. 9d). An interactive and extensible visualization interface is provided by the software that distils multiple dimensions of information into a single, intuitive display. It is written mostly in Python, JavaScript, and C, and to perform visualization job. It relies on scalable vector graphics (SVG). Regarding data layout, Anvi'o presents tree structures with data or metadata layers where the properties of each leaf on the tree are defined. The computed data is stored in the form of self-contained database files that can be further analyzed using structured query language (SQL) through SQLite, an open source transactional SQL database engine that does not require any database server or configuration. This is a command-based software. Using Anvi'o, we can reanalyze publicly accessible datasets and also can discover genomic sequence variation or changes within naturally occurring microbial populations through de novo characterization of single nucleotide variations, and linked cultivar and single-cell genomes with metagenomic and metatranscriptomic data.

## 14 Conclusion

In this chapter, we have described the recent tools and approaches for sequencing and analyzing the organelle genomes. Starting from DNA isolation from organelles to sequencing and further to bioinformatic tools and downstream analysis, we have tried to provide comprehensive information covering the mentioned topic. Due to ever improving sequencing technologies, new bioinformatics tools are being developed to analyze the data. It becomes pertinent to acquaint the reader with the updated information. Organelle genome sequencing provides great opportunity to study the detection of somatic mutation, heteroplasmy, haplogrouping, comparative genomics, and consequently the evolutionary aspect.

## 15 Notes

1. Generally, fresh tissues are used, as frozen tissues yield limited quantities and low quality of DNA.
2. Lack of good starting material/tissue will limit the choice of sequencing technology and will also affect the quality of obtained data. Besides, DNA extracts often contain carryover contaminants like polysaccharides, proteoglycans, proteins, secondary metabolites, polyphenols, humic acids, and pigments, which can impair the efficacy of library preparation particularly in PCR-free library preparation protocol like PacBio and Oxford Nanopore sequencing technology. Hence, it is very important to remove all the possible contaminants from the DNA sample before processing it for sequencing.
3. High-molecular weight DNA is fragile; therefore, gentle handling (vortexing at minimal speed, pipetting with wide-bore pipette tips, and transportation in a solid frozen stage) is advised. It is also important to keep the number of freeze–thaw cycles to a minimum, since ice crystals can mechanically damage the DNA. Therefore, one should avoid such DNA extraction protocols which involve harsh bead-beating treatment for tissue homogenization.
4. The RNA contamination in the sample will result in the overestimation of the library nucleic acid molecules concentration if we consider PacBio and 10× Chromium libraries.
5. The quality check after library preparation is a crucial step for loading onto a flow cell. If DNA quality and quantity is optimum, the sequencing reaction runs efficiently. High or low input DNA gives low-quality data possibly due to read problems resulting from flow cell saturation, or reduced coverage because of insufficient DNA respectively.
6. This is important because during PCR amplification of some sample library preparation protocols, duplicates of fragments will be generated. The consequence of duplicate fragments is that the sequencing reaction will be biased toward these fragments resulting in overrepresentation in the machine output.
7. The variation in gene length after annotation could be because of pseudogenes. Therefore, if nucleotide insertions/deletions

exist, the stop codon may occur in the middle of a gene resulting in alteration of the length, referred to as a "pseudogene."

8. Mapping and base quality score of filtered reads should be ≥20 because most of the aligners will detect alignment which are of poor quality either due to high numbers of mismatches, or the presence of high-quality secondary alignments and to remove these poor-quality reads one should use minimum threshold score

## References


1. Mower JP, Sloan DB, Alverson AJ (2012) Plant mitochondrial genome diversity: the genomics revolution. In: Leitch IJ et al (eds) Plant genome diversity, vol 1. Springer, Vienna, pp 123–144
2. Liu S, Sawada T, Lee S et al (2012) Parkinson's disease-associated kinase PINK1 regulates miro protein level and axonal transport of mitochondria. PLoS Genet 8:e1002537. https://doi.org/10.1371/journal.pgen.1002537
3. Kubo T, Newton KJ (2008) Angiosperm mitochondrial genomes and mutations. Mitochondrion 8:5–14. https://doi.org/10.1016/J.MITO.2007.10.006
4. Gualberto JM, Newton KJ (2017) Plant mitochondrial genomes: dynamics and mechanisms of mutation. Ann Rev Plant Biol 68:225–252. https://doi.org/10.1146/annurev-arplant-043015-112232
5. Sloan DB (2013) One ring to rule them all? Genome sequencing provides new insights into the 'master circle' model of plant mitochondrial DNA structure. New Phytol 200:978–985. https://doi.org/10.1111/nph.12395
6. Morley SA, Nielsen BL (2017) Plant mitochondrial DNA. Front Biosci (Landmark Ed) 22:1023–1032
7. Alverson AJ, Wei X, Rice DW et al (2010) Insights into the evolution of mitochondrial genome size from complete sequences of Citrullus lanatus and Cucurbita pepo (Cucurbitaceae). Mol Biol Evol 27:1436–1448. https://doi.org/10.1093/molbev/msq029
8. Kosa P, Valach M, Tomaska L et al (2006) Complete DNA sequences of the mitochondrial genomes of the pathogenic yeasts Candida orthopsilosis and Candida metapsilosis: insight into the evolution of linear DNA genomes from mitochondrial telomere mutants. Nucleic Acids Res 34:2472–2481. https://doi.org/10.1093/nar/gkl327
9. Schultze M, Kondorosi A (1998) Regulation of symbiotic root nodule development. Annu Rev Genet 32:33–57. https://doi.org/10.1146/annurev.genet.32.1.33
10. Schuster W, Brennicke A (1994) The Plant Mitochondrial Genome: Physical Structure, Information Content, RNA Editing, and Gene Migration to the Nucleus. Annual Review of Plant Physiology and Plant Molecular Biology 45:61–78. https://doi.org/10.1146/annurev.pp.45.060194.000425
11. Knoop V, Brennicke A (2002) Molecular biology of the plant mitochondrion. CRC Crit Rev Plant Sci 21:111–126. https://doi.org/10.1080/0735-260291044205
12. Gray MW, Cedergren R, Abel Y, Sankoff D (1989) On the evolutionary origin of the plant mitochondrion and its genome. Proc Natl Acad Sci U S A 86:2267–2271. https://doi.org/10.1073/PNAS.86.7.2267
13. Sloan DB, Alverson AJ, Chuckalovcak JP et al (2012) Rapid evolution of enormous, multichromosomal genomes in flowering plant mitochondria with exceptionally high mutation rates. PLoS Biol 10:e1001241. https://doi.org/10.1371/journal.pbio.1001241
14. Simpson CL, Stern DB (2002) The treasure trove of algal chloroplast genomes. Surprises in architecture and gene content, and their functional implications. Plant Physiol 129:957–966. https://doi.org/10.1104/pp.010908
15. Daniell H, Lee S-B, Grevich J et al (2006) Complete chloroplast genome sequences of Solanum bulbocastanum, Solanum lycopersicum and comparative analyses with other Solanaceae genomes. Theor Appl Genet 112:1503–1518. https://doi.org/10.1007/s00122-006-0254-x
16. Daniell H, Cohill PR, Kumar S, Dufourmantel N (2004) Chloroplast genetic engineering. In: Daniell H, Chase CD (eds) Molecular biology and biotechnology of plant organelles. Springer, Dordrecht, pp 443–490
17. Smith DR (2015) Mutation rates in plastid genomes: they are lower than you might

think. Genome Biol Evol 7:1227–1234. https://doi.org/10.1093/gbe/evv069

18. Boudreau E, Turmel M (1996) Extensive gene rearrangements in the chloroplast DNAs of Chlamydomonas species featuring multiple dispersed repeats. Mol Biol Evol 13:233–243. https://doi.org/10.1093/oxfordjournals.molbev.a025560
19. Chen J, Hao Z, Xu H et al (2015) The complete chloroplast genome sequence of the relict woody plant Metasequoia glyptostroboides Hu et Cheng. Front Plant Sci 6:447. https://doi.org/10.3389/fpls.2015.00447
20. Curci PL, De Paola D, Danzi D et al (2015) Complete chloroplast genome of the multifunctional crop globe artichoke and comparison with other Asteraceae. PLoS One 10: e0120589. https://doi.org/10.1371/journal.pone.0120589
21. Lin C-P, Ko C-Y, Kuo C-I et al (2015) Transcriptional slippage and RNA editing increase the diversity of transcripts in chloroplasts: insight from deep sequencing of Vigna radiata genome and transcriptome. PLoS One 10: e0129396. https://doi.org/10.1371/journal.pone.0129396
22. Redwan RM, Saidin A, Kumar SV (2015) Complete chloroplast genome sequence of MD-2 pineapple and its comparative analysis among nine other plants from the subclass Commelinidae. BMC Plant Biol 15:196. https://doi.org/10.1186/s12870-015-0587-1
23. Williams AV, Boykin LM, Howell KA et al (2015) The complete sequence of the acacia ligulata chloroplast genome reveals a highly divergent clpP1 gene. PLoS One 10: e0125768. https://doi.org/10.1371/journal.pone.0125768
24. Yan L, Lai X, Li X et al (2015) Analyses of the complete genome and gene expression of chloroplast of sweet potato [Ipomoea batata]. PLoS One 10:e0124083. https://doi.org/10.1371/journal.pone.0124083
25. Yao X, Tang P, Li Z et al (2015) The first complete chloroplast genome sequences in Actinidiaceae: genome structure and comparative analysis. PLoS One 10:e0129347. https://doi.org/10.1371/journal.pone.0129347
26. Han Z, Li W, Liu Y, Gao L (2016) The complete chloroplast genome of North American ginseng, *Panax quinquefolius*. Mitochondrial DNA A 27:3496–3497. https://doi.org/10.3109/19401736.2015.1066365
27. Choi KS, Park S (2016) The complete chloroplast genome sequence of *Euonymus japonicus* (Celastraceae). Mitochondrial DNA A 27:3577–3578. https://doi.org/10.3109/19401736.2015.1075127
28. Kaila T, Chaduvla PK, Saxena S et al (2016) Chloroplast genome sequence of Pigeonpea (Cajanus cajan (L.) Millspaugh) and Cajanus scarabaeoides (L.) Thouars: genome organization and comparison with other legumes. Front Plant Sci 7:1847. https://doi.org/10.3389/fpls.2016.01847
29. Kaila T, Chaduvla PK, Rawal HC et al (2017) Chloroplast genome sequence of clusterbean (Cyamopsis tetragonoloba L.): genome structure and comparative analysis. Genes (Basel). https://doi.org/10.3390/genes8090212
30. Keller J, Rousseau-Gueutin M, Martin GE et al (2017) The evolutionary fate of the chloroplast and nuclear rps16 genes as revealed through the sequencing and comparative analyses of four novel legume chloroplast genomes from Lupinus. DNA Res 24:343–358. https://doi.org/10.1093/dnares/dsx006
31. Cheng H, Li J, Zhang H et al (2017) The complete chloroplast genome sequence of strawberry ( *Fragaria × ananassa* Duch.) and comparison with related species of Rosaceae. Peer J 5:e3919. https://doi.org/10.7717/peerj.3919
32. Saina J, Li Z-Z, Gichira A et al (2018) The complete chloroplast genome sequence of tree of heaven (Ailanthus altissima (Mill.)) (Sapindales: Simaroubaceae), an important pantropical tree. Int J Mol Sci 19:929. https://doi.org/10.3390/ijms19040929
33. D'Agostino N, Tamburino R, Cantarella C et al (2018) The complete plastome sequences of eleven capsicum genotypes: insights into DNA variation and molecular evolution. Genes (Basel) 9:503. https://doi.org/10.3390/genes9100503
34. Liu C, Yang Z, Yang L et al (2018) The complete plastome of Panax stipuleanatus: comparative and phylogenetic analyses of the genus Panax (Araliaceae). Plant Divers 40:265–276. https://doi.org/10.1016/J.PLD.2018.11.001
35. Pinard D, Myburg AA, Mizrachi E (2019) The plastid and mitochondrial genomes of Eucalyptus grandis. BMC Genomics 20(132). https://doi.org/10.1186/s12864-019-5444-4
36. Mustafina FU, Yi D-K, Choi K et al (2019) A comparative analysis of complete plastid genomes from *Prangos fedtschenkoi* and *Prangos lipskyi* (Apiaceae). Ecol Evol 9:364–377. https://doi.org/10.1002/ece3.4753
37. Hisano H, Tsujimura M, Yoshida H et al (2016) Mitochondrial genome sequences from wild and cultivated barley (Hordeum

vulgare). BMC Genomics 17(824). https://doi.org/10.1186/s12864-016-3159-3

38. Aljohi HA, Liu W, Lin Q et al (2016) Complete sequence and analysis of coconut palm (Cocos nucifera) mitochondrial genome. PLoS One 11:e0163990. https://doi.org/10.1371/journal.pone.0163990
39. Guo W, Zhu A, Fan W, Mower JP (2017) Complete mitochondrial genomes from the ferns *Ophioglossum californicum* and *Psilotum nudum* are highly repetitive with the largest organellar introns. New Phytol 213:391–403. https://doi.org/10.1111/nph.14135
40. Bi C, Paterson AH, Wang X et al (2016) Analysis of the complete mitochondrial genome sequence of the diploid cotton *Gossypium raimondii* by comparative genomics approaches. Biomed Res Int 2016:1–18. https://doi.org/10.1155/2016/5040598
41. Kim B, Kim K, Yang T-J, Kim S (2016) Completion of the mitochondrial genome sequence of onion (Allium cepa L.) containing the CMS-S male-sterile cytoplasm and identification of an independent event of the ccmF N gene split. Curr Genet 62:873–885. https://doi.org/10.1007/s00294-016-0595-1
42. Silva SR, Alvarenga DO, Aranguren Y et al (2017) The mitochondrial genome of the terrestrial carnivorous plant Utricularia reniformis (Lentibulariaceae): structure, comparative analysis and evolutionary landmarks. PLoS One 12:e0180484. https://doi.org/10.1371/journal.pone.0180484
43. Dong S, Zhao C, Chen F et al (2018) The complete mitochondrial genome of the early flowering plant Nymphaea colorata is highly repetitive with low recombination. BMC Genomics 19(614). https://doi.org/10.1186/s12864-018-4991-4
44. Won SY, Jung J-A, Kim JS (2018) The complete mitochondrial genome sequence of *Chrysanthemum boreale* (Asteraceae). Mitochondrial DNA Part B 3:529–530. https://doi.org/10.1080/23802359.2018.1468226
45. Shi Y, Liu Y, Zhang S et al (2018) Assembly and comparative analysis of the complete mitochondrial genome sequence of Sophora japonica 'JinhuaiJ2. PLoS One 13:e0202485. https://doi.org/10.1371/journal.pone.0202485
46. Liao X, Zhao Y, Kong X et al (2018) Complete sequence of kenaf (Hibiscus cannabinus) mitochondrial genome and comparative analysis with the mitochondrial genomes of other plants. Sci Rep 8(12714). https://doi.org/10.1038/s41598-018-30297-w
47. Zheng F, Liu H, Jiang M et al (2018) The complete mitochondrial genome of the Caulerpa lentillifera (Ulvophyceae, Chlorophyta): sequence, genome content, organization structure and phylogenetic consideration. Gene 673:225–238. https://doi.org/10.1016/J.GENE.2018.06.050
48. Shi C, Hu N, Huang H et al (2012) An improved chloroplast DNA extraction procedure for whole plastid genome sequencing. PLoS One 7:e31468. https://doi.org/10.1371/journal.pone.0031468
49. Keech O, Dizengremel P, Gardeström P (2005) Preparation of leaf mitochondria from *Arabidopsis thaliana*. Physiol Plant 124:403–409. https://doi.org/10.1111/j.1399-3054.2005.00521.x
50. Sweetlove LJ, Taylor NL, Leaver CJ (2007) Isolation of intact, functional mitochondria from the model plant Arabidopsis thaliana. Humana, New York, NY, pp 125–136
51. Murcha MW, Whelan J (2015) Isolation of intact mitochondria from the model plant species Arabidopsis thaliana and Oryza sativa. Humana, New York, NY, pp 1–12
52. Taylor NL, Ströher E, Millar AH (2014) Arabidopsis organelle isolation and characterization. Humana, Totowa, NJ, pp 551–572
53. Benamar A, Tallon C, Macherel D (2003) Membrane integrity and oxidative properties of mitochondria isolated from imbibing pea seeds after priming or accelerated ageing. Seed Sci Res 13:35–45. https://doi.org/10.1079/SSR2002122
54. Chaudhary A, Chaudhary S, Ghosh A et al (2015) A rapid, low cost, and efficient method for isolation of high quality mitochondrial DNA from Oryza sativa. J Crop Sci Biotechnol 18:155–160. https://doi.org/10.1007/s12892-014-0095-0
55. Meyer EH, Millar AH (2008) Isolation of mitochondria from plant cell culture. Humana, Totowa, NJ, pp 163–169
56. Hirai A, Ishibashi T, Morikami A et al (1985) Rice chloroplast DNA: a physical map and the location of the genes for the large subunit of ribulose 1,5-bisphosphate carboxylase and the 32 KD photosystem II reaction center protein. Theor Appl Genet 70:117–122. https://doi.org/10.1007/BF00275309
57. Bookjans G, Stummann BM, Henningsen KW (1984) Preparation of chloroplast DNA from pea plastids isolated in a medium of high ionic strength. Anal Biochem 141:244–247. https://doi.org/10.1016/0003-2697(84)90452-4
58. Kaneko K, Takamatsu T, Inomata T et al (2016) *N*-Glycomic and microscopic subcellular localization analyses of NPP1, 2 and 6 strongly indicate that *trans*-Golgi

compartments participate in the Golgi to plastid traffic of nucleotide pyrophosphatase/phosphodiesterases in rice. Plant Cell Physiol 57:1610–1628. https://doi.org/10.1093/pcp/pcw089

59. Jansen RK, Raubeson LA, Boore JL et al (2005) Methods for obtaining and analyzing whole chloroplast genome sequences. Methods Enzymol 395:348–384. https://doi.org/10.1016/S0076-6879(05)95020-9
60. Millar AH, Liddell A, Leaver CJ (2001) Isolation and subfractionation of mitochondria from plants. Methods Cell Biol 65:53–74. https://doi.org/10.1016/S0091-679X(01)65004-0
61. Pomeroy MK (1974) Studies on the respiratory properties of mitochondria isolated from developing winter wheat seedlings. Plant Physiol 53:653–657. https://doi.org/10.1104/PP.53.4.653
62. Ahmed Z, Fu Y-B (2015) An improved method with a wider applicability to isolate plant mitochondria for mtDNA extraction. Plant Methods 11:56. https://doi.org/10.1186/s13007-015-0099-x
63. Dierckxsens N, Mardulyn P, Smits G (2016) NOVOPlasty: *de novo* assembly of organelle genomes from whole genome data. Nucleic Acids Res 45:gkw955. https://doi.org/10.1093/nar/gkw955
64. Butler J, MacCallum I, Kleber M et al (2008) ALLPATHS: de novo assembly of whole-genome shotgun microreads. Genome Res 18:810–820. https://doi.org/10.1101/gr.7337908
65. Bankevich A, Nurk S, Antipov D et al (2012) SPAdes: a new genome assembly algorithm and its applications to single-cell sequencing. J Comput Biol 19:455–477. https://doi.org/10.1089/cmb.2012.0021
66. Koren S, Walenz BP, Berlin K et al (2017) Canu: scalable and accurate long-read assembly via adaptive k-mer weighting and repeat separation. Genome Res 27:722–736. https://doi.org/10.1101/gr.215087.116
67. Grabherr MG, Haas BJ, Yassour M et al (2011) Full-length transcriptome assembly from RNA-Seq data without a reference genome. Nat Biotechnol 29:644–652. https://doi.org/10.1038/nbt.1883
68. Luo R, Liu B, Xie Y et al (2012) SOAPdenovo2: an empirically improved memory-efficient short-read de novo assembler. Gigascience 1(18). https://doi.org/10.1186/2047-217X-1-18
69. Simpson JT, Wong K, Jackman SD et al (2009) ABySS: a parallel assembler for short read sequence data. Genome Res 19:1117–1123. https://doi.org/10.1101/gr.089532.108
70. Zerbino DR, Birney E (2008) Velvet: algorithms for de novo short read assembly using de Bruijn graphs. Genome Res 18:821–829. https://doi.org/10.1101/gr.074492.107
71. Warren RL, Sutton GG, Jones SJM, Holt RA (2007) Assembling millions of short DNA sequences using SSAKE. Bioinformatics 23:500–501. https://doi.org/10.1093/bioinformatics/btl629
72. Jeck WR, Reinhardt JA, Baltrus DA et al (2007) Extending assembly of short DNA sequences to handle error. Bioinformatics 23:2942–2944. https://doi.org/10.1093/bioinformatics/btm451
73. Jin J-J, Yu W-B, Yang J-B et al (2018) GetOrganelle: a simple and fast pipeline for de novo assembly of a complete circular chloroplast genome using genome skimming data. bioRxiv:256479. https://doi.org/10.1101/256479
74. Giardine B, Riemer C, Hardison RC et al (2005) Galaxy: a platform for interactive large-scale genome analysis. Genome Res 15:1451–1455. https://doi.org/10.1101/gr.4086505
75. Kearse M, Moir R, Wilson A et al (2012) Geneious basic: an integrated and extendable desktop software platform for the organization and analysis of sequence data. Bioinformatics 28:1647–1649. https://doi.org/10.1093/bioinformatics/bts199
76. Chaisson MJ, Pevzner PA (2008) Short read fragment assembly of bacterial genomes. Genome Res 18:324–330. https://doi.org/10.1101/gr.7088808
77. Ishiya K, Ueda S (2017) MitoSuite: a graphical tool for human mitochondrial genome profiling in massive parallel sequencing. PeerJ 5:e3406. https://doi.org/10.7717/peerj.3406
78. Yang IS, Lee HY, Yang WI, Shin K-J (2013) mtDNAprofiler: a Web application for the nomenclature and comparison of human mitochondrial DNA sequences. J Forensic Sci 58:972–980. https://doi.org/10.1111/1556-4029.12139
79. Guo Y, Li J, Li C-I et al (2013) MitoSeek: extracting mitochondria information and performing high-throughput mitochondria sequencing analysis. Bioinformatics 29:1210–1211. https://doi.org/10.1093/bioinformatics/btt118
80. Vellarikkal SK, Dhiman H, Joshi K et al (2015) mit-o-matic: a comprehensive computational pipeline for clinical evaluation of mitochondrial

variations from next-generation sequencing datasets. Hum Mutat 36:419–424. https://doi.org/10.1002/humu.22767

81. Brandon MC, Ruiz-Pesini E, Mishmar D et al (2009) MITOMASTER: a bioinformatics tool for the analysis of mitochondrial DNA sequences. Hum Mutat 30:1–6. https://doi.org/10.1002/humu.20801
82. Huang DI, Cronk QCB (2015) Plann: a command-line application for annotating plastome sequences. Appl Plant Sci 3. https://doi.org/10.3732/apps.1500026
83. Tillich M, Lehwark P, Pellizzer T et al (2017) GeSeq – versatile and accurate annotation of organelle genomes. Nucleic Acids Res 45: W6–W11. https://doi.org/10.1093/nar/gkx391
84. Wyman SK, Jansen RK, Boore JL (2004) Automatic annotation of organellar genomes with DOGMA. Bioinformatics 20:3252–3255. https://doi.org/10.1093/bioinformatics/bth352
85. Castandet B, Hotto AM, Strickler SR, Stern DB (2016) ChloroSeq, an optimized chloroplast RNA-seq bioinformatic pipeline, reveals remodeling of the organellar transcriptome under heat stress. G3 (Bethesda) 6:2817–2827. https://doi.org/10.1534/g3.116.030783
86. Wynn EL, Christensen AC (2019) Repeats of unusual size in plant mitochondrial genomes: identification, incidence and evolution. G3 (Bethesda) 9:549–559. https://doi.org/10.1534/g3.118.200948
87. Al-Nakeeb K, Petersen TN, Sicheritz-Pontén T (2017) Norgal: extraction and de novo assembly of mitochondrial DNA from whole-genome sequencing data. BMC Bioinformatics 18:510. https://doi.org/10.1186/s12859-017-1927-y
88. Hahn C, Bachmann L, Chevreux B (2013) Reconstructing mitochondrial genomes directly from genomic next-generation sequencing reads—a baiting and iterative mapping approach. Nucleic Acids Res 41: e129–e129. https://doi.org/10.1093/nar/gkt371
89. Mower JP (2009) The PREP suite: predictive RNA editors for plant mitochondrial genes, chloroplast genes and user-defined alignments. Nucleic Acids Res 37:W253–W259. https://doi.org/10.1093/nar/gkp337
90. Lohse M, Drechsel O, Bock R (2007) OrganellarGenomeDRAW (OGDRAW): a tool for the easy generation of high-quality custom graphical maps of plastid and mitochondrial genomes. Curr Genet 52:267–274. https://doi.org/10.1007/s00294-007-0161-y
91. Alikhan N-F, Petty NK, Ben Zakour NL, Beatson SA (2011) BLAST ring image generator (BRIG): simple prokaryote genome comparisons. BMC Genomics 12:402

# Chapter 5

# Methods of Gene Expression Profiling to Understand Abiotic Stress Perception and Response in Legume Crops

**Meenu Bala, Ragini Sinha, M. A. Mallick, Tilak Raj Sharma, and Anil Kumar Singh**


## Abstract

Legume crops offer a wide genetic diversity that can be exploited to raise improved crop varieties with higher tolerance against adverse climatic conditions. In order to achieve food and nutritional security, legume breeding programs should also incorporate advanced genomics tools. Genomes of many model and nonmodel legume crops have been sequenced, which provide opportunities to identify and characterize candidate genes to develop abiotic stress tolerant crops. Gene expression profiling is a powerful tool to identify candidate genes and understand their function. The present chapter describes two such strategies, that is, candidate gene expression profiling approach and global transcriptome profiling approach. The methods like RT-PCR and qRT-PCR that are being traditionally used to study expression of target genes under defined experimental conditions are discussed. In addition, global transcriptome analysis approach and its advancements are discussed. Details of next-generation sequencing (NGS) based RNA-sequencing (RNA-seq) and associated advanced bioinformatics tools to identify differentially expressing genes at a global level are also described.


**Key words** Abiotic stress, Bioinformatics, Differential gene expression (DGE), Legume crops, Next-generation sequencing (NGS), qRT-PCR, RNA-seq

## 1 Introduction

The legume crops, owing to their nutritional seeds and potential to fix atmospheric nitrogen and vast biodiversity can be an effective mean to ensure global food and nutritional security under changing climatic conditions. The family leguminosae (or fabaceae), has nearly 18,000 species, with approximately 700 genera in many species, for example, *Glycine max* (soybean), *Cicer arietinum* (chickpea), *Arachis hypogaea* (groundnut), *Phaseolus vulgaris* (common bean), *Pisum sativum* (garden pea), *Cajanus cajan* (pigeon pea), *Lens culinaris* (lentil), and *Medicago sativa* (alfalfa) [1–6]. However, legume crop production is drastically affected by environmental stresses, like drought, salinity, cold, heat, and soil

Mukesh Jain and Rohini Garg (eds.), *Legume Genomics: Methods and Protocols*, Methods in Molecular Biology, vol. 2107, https://doi.org/10.1007/978-1-0716-0235-5_5, 

nutrient deficiencies [7–11]. Considering the erratic climate change, efforts are required to maximize efficient use of water and natural resources, by developing improved crop varieties. Thus, major breakthrough is required in legume breeding programs, by employing advanced genetic and genomics approaches. During the last decade, genomes of many legume species have been sequenced, like *Lotus japonicus*, soybean, *M. truncatula*, pigeon pea, and chickpea [12–17]. In addition, transcriptomic studies in various legume crops have been carried out for important traits, thus identifying many genes and genic markers [10, 18–21]. The genomic information provided by genome and transcriptome level studies must be utilized to develop stress resilient legume crops.

There may be two approaches for studying transcript level. First is the candidate gene approach that may be performed for selected genes, by employing techniques, like semiquantitative reverse transcription polymerase chain reaction (RT-PCR) and quantitative RT-PCR (qRT-PCR). These are the targeted methods, where genes expected to show altered expression under given experimental conditions are selected. Thus, it is presumed to have some prior information of plants genome and nucleotide sequence. Second approach is to study global gene expression through next-generation sequencing (NGS), which may then be used to identify differentially expressed genes (DEGs). Besides, it may also generate other valuable data regarding, single nucleotide polymorphism (SNPs), simple sequence repeats (SSRs) and small RNAs with the help of bioinformatics. This chapter describes the protocols for both the approaches with maximum possible information and techniques.

### 1.1 Candidate Gene Approach

Plant adaptation/response to any environmental stress is triggered by activation of a molecular network cascade, involved in signal transduction and expression of specific stress-related genes and metabolites, interacting at many points [22]. The stress perception and subsequent molecular signaling is the first step in any stress response and involves a plethora of genes, whose expression determines stress response. Abscisic acid (ABA) has been shown to play important role as signaling molecule in response to most of the abiotic stresses. A drastic increase in ABA level has been observed at the onset of drought, which then triggers downstream signal transduction cascade [23]. This initiates ABA signaling process, by activating ABA receptors and enzymes; protein phosphatase 2C family ($PP_2C$) and SNF-related kinase ($SnRK_2$). In the absence of ABA, $PP_2C$ inhibits, $SnRK_2$, however, during drought stress, ABA binds $PP_2C$ and $SnRK_2$ is activated, which phosphorylates downstream ABFs (ABA responsive element binding factors). ABF binds to the promoter regions of its target genes and activates their expression, which help plant to cope with the stress [24]. Besides ABA, other signaling molecules may be triggered in different responses like cytosolic Calcium ($Ca^{2+}$) that is triggered by cold, salinity and drought. The $Ca^{2+}$ ions bind to calcium binding protein, calmodulin (CaM), calcium-dependent protein kinases (CDPKs), CBL-interacting protein kinases (CIPKs), and mitogen activated protein kinase (MAPKs), which transduce the signals to switch on transcriptional cascades [25, 26]. Interestingly, a cross talk also exists between different signaling molecules triggered by different abiotic stress [22]. ABA biosynthesis is also activated during salinity stress [27, 28], where it provokes a specific SOS (salt overlay sensitive) pathway. It plays an important role in regulating $Na^+/H^+$ homeostasis and salt tolerance. The SOS pathway consists of various components like SOS1 ($Na^+/H^+$ antiporter), SOS2 (serine/threonine-specific protein kinase), and SOS3 (myristoylated $Ca^{2+}$ binding protein) that mediate ion flux at different tissues [29–33]. The $Ca^{2+}$-activated SOS pathway directly or indirectly affects $K^+$ homeostasis. Besides, the SOS pathway also controls many other activities, such as endocytosis, shape and function of vacuole, and intracellular pH [34]. The SOS pathway genes have been found to be conserved in many plant species [30, 35–37].

Most of the stress responsive ABA-dependent/independent pathways involve various transcription factors (TFs). The major TFs involved in ABA-dependent pathway are ABA responsive element binding proteins (AREBs), MYB, Drought responsive element binding 2 (DREB2), and NAC (NAM;ATAF1, 2; and CUC) TFs, while cold stress triggers ABA-independent pathway through *DREB1* TFs [38–40]. These TFs are primarily responsible for regulating expression of majority of genes linked directly or indirectly to stress tolerance [41–43].

Thus, after receiving signals of environmental stress, the TFs trigger the induction or repression of stress-responsive genes or pathways. These may include genes/proteins related to osmotic adjustment, osmoprotectant accumulation, oxidative stress management, induction of stress proteins (late embryogenesis abundant proteins, chaperonin, etc.) and other physiological adaptations such as modifications in root and shoot growth and transpiration. However, the mode of adaptation may vary in different plant species. Where, some plant species prefer use of ions for osmotic adjustment, others may prefer biosynthesis of organic osmolyte, to tolerate osmotic stresses. These osmolytes include proline, glycine betaine, polyols, sugar alcohols, and soluble sugars. Genes involved in osmoprotectant biosynthesis are also upregulated under salt stress, and concentrations of accumulated osmoprotectants correlate with osmotic stress tolerance [44]. The enzymes, like superoxide dismutase (SOD), catalase (CAT), ascorbate peroxidase (APX), glutathione reductase (GR), glutathione peroxidase (GPX), ascorbate peroxidase (APX), dehydro-ascorbate reductase (DHAR), monodehydro-ascorbate reductase (MDHAR) scavenge ROS and protect plant against oxidative stress and determine the stress tolerance potential of cell. Stress proteins, like late embryogenesis abundant (LEA) proteins, osmotins, antifreeze proteins, chaperones, and mRNA binding proteins are also important for protection of macromolecules. Analysis of differential expression of genes related to above discussed mechanism may be helpful in determining the degree of stress perception and response in legumes.

### 1.2 Global Transcriptome Profiling

With the advent of next generation sequencing (NGS) technologies, RNA sequencing (RNA-seq) has emerged as a standard tool for global transcriptome profiling outperforming other hybridization- and sequence-based gene expression methods with regards to reproducibility, precision, low background noise and nondependency on a priori knowledge of the gene sequence [45]. The dropping cost of NGS assays and availability of web-based genome and transcriptome data resources have enabled researchers to extensively employ RNA-seq approach to study plant response to differential treatment conditions. Detailed analysis of RNA-seq data using bioinformatics tools yields a wealth of information crucial for understanding plant response to stress condition.

Most of the transcriptome profiling studies reported in legumes have been aimed to understand the underlying mechanism of change in gene expression resulting from abiotic and biotic stress, various developmental processes in different stages of the plant's life cycle and comparisons of such transcriptome between closely related species or varieties [19, 46–65]. Such RNA-seq based transcriptome profiling of legume crops range from having simple experimental design involving single genotype and one abiotic stress factor with control to more complex experimental design with many experimental factors.

Transcriptome analysis for DEGs can be divided into some basic steps such as generation of reads, quality assessment and preprocessing of reads, alignment and assembly of reads, expression quantification, normalization, and identification of differentially expressed genes. In each step, there are many alternatives available depending on the objectives of the experiment and many pipelines have been developed with combination of such alternatives. Though, some benchmark studies have been done to compare different protocols, there is no standard pipeline of RNA-seq data analysis for a specific experimental condition or a genotype [66–72].

Here, we present the basic protocols of the most common applications of RNA-seq experiments, that is, differential gene expression (DGE) analysis with paired-end NGS reads as overviewed in Fig. 1.

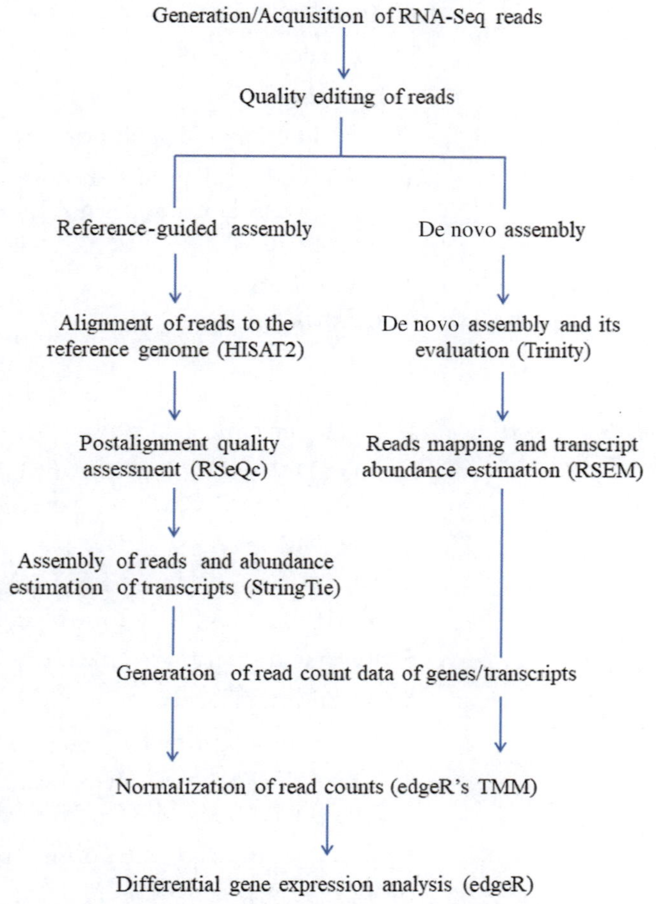


**Fig. 1** An overview of NGS-based differential gene expression profiling

## 2 Materials

Prepare all solutions using ultrapure water (e.g., milli Q Type I water) to attain a sensitivity of 18 MΩ cm at 25 °C) and analytical grade reagents. For preparation of all standard buffers, please refer Green and Sambrook [73]. Prepare and store all reagents at room temperature (unless indicated otherwise). Carefully follow all waste disposal regulations, when disposing of waste materials.

### 2.1 Candidate Gene Approach

#### 2.1.1 RNA Isolation

1. DEPC-treated water: Add 1% DEPC to milli Q type I water, stir overnight at room temperature. Autoclave twice (*see* **Note 1**).
2. Iris solution: 0.1% SDS (w/v), 0.01 M EDTA (pH 8.0) (w/v), 0.32 M Sodium acetate, Tris-equilibrated phenol (pH 8.0–10.0).
3. 10× MOPS buffer: for 100 ml, dissolve 4.18 g MOPS in DEPC-treated water, adjust pH 7.0 with NaOH; add 2 ml sodium acetate (1 M) and 2 ml EDTA (0.5 M, pH 8); adjust final volume to 100 ml with DEPC-treated water.

4. Agarose.
5. Formaldehyde.
6. Formamide.
7. Ethidium bromide solution.
8. RNA loading dye: 50% glycerol (v/v), 10 mM EDTA (pH 8) (v/v), 0.25% bromophenol blue (w/v), 0.25% xylene cyanol (w/v).
9. Chloroform.
10. Isopropanol.
11. Ethanol.

#### 2.1.2 RT-PCR Analysis

1. DNase enzyme 1 U.
2. 10× DNase buffer: 100 mM Tris–HCl (pH 7.5), 25 mM $MgCl_2$, 1 mM $CaCl_2$, prepared in DEPC-treated water.
3. 50 mM EDTA.
4. 10× Reverse Transcriptase buffer: 20–80 mM Tris pH 8.4, 50–100 mM KCl, 1–2 mM $MgCl_2$.
5. Reverse transcriptase enzyme 1 U.
6. 10 mM dNTP mix.
7. Oligo $(dT)_{18}$ primers.
8. 10× Taq DNA polymerase buffer: 200 mM Tris–HCl (pH 8.4), 500 mM KCl, 15 mM $MgCl_2$.
9. Gene of interest (GOI) primer pairs (10 pM) (*see* **Note 2**).
10. Taq DNA polymerases (1 U).
11. Agarose.
12. 50× TAE: dissolve 24.7 g of Tris base in milli Q Type I water, add 5.7 ml of glacial acetic acid and 10 ml of 0.5 M EDTA (pH 8).

#### 2.1.3 qRT-PCR Analysis

1. Commercially available SYBR green master mix.
2. GOI primer pairs (200 nM) (*see* **Note 2**).
3. 1× ROX dye.

### 2.2 Global Transcriptome Profiling

#### 2.2.1 Hardware Requirements

Depending on the size of the data to be analyzed and the memory footprints of the tools, hardware requirements for transcriptome analysis might be chosen accordingly. The RNA-seq analysis can also be done using cloud computing [74] and/or online platforms like Galaxy (http://usegalaxy.org) or CyVerse (http://www.cyverse.org), but to deal regularly with substantial amount of data, generally high performance computing with 64-bits architecture running Unix-based operating systems are required and connected with high speed internet. Most of the protocols use commands and codes, which run on Unix shell prompt and R statistical computing environment, while some user-friendly commercial software are also available with graphical user interfaces.

#### 2.2.2 Data Analysis Tools

1. FastQC (http://www.bioinformatics.babraham.ac.uk/projects/fastqc/).
2. Trimmomatic (http://www.usadellab.org/cms/?page=trimmomatic).
3. HISAT2 (https://ccb.jhu.edu/software/hisat2/index.shtml).
4. RSeQC (http://rseqc.sourceforge.net/).
5. Samtools (http://samtools.sourceforge.net/).
6. GffCompare (https://github.com/gpertea/gffcompare).
7. Stringtie (https://ccb.jhu.edu/software/stringtie/).
8. Trinity (https://github.com/alyssafrazee/ballgown).
9. CD-HIT (http://weizhongli-lab.org/cd-hit/).
10. RSEM (https://github.com/alyssafrazee/ballgown).
11. BLAST (ftp://ftp.ncbi.nlm.nih.gov/blast/executables/blast+/LATEST/).
12. edgeR (https://bioconductor.org/packages/release/bioc/html/edgeR.html).

## 3 Methods

### 3.1 Candidate Gene Approach

#### 3.1.1 RNA Isolation and Gel Electrophoresis

1. RNA isolation is carried out following Ghawana et al. [75].
2. Prepare Iris solution, by adding 0.1% SDS, 0.32 M sodium acetate, 0.01 M EDTA (pH 8.0), to tris-equilibrated phenol (pH 8.0–10.0).
3. Grind 100 mg of leaf sample in liquid nitrogen, add 2 ml Iris solution, homogenize properly (*see* **Note 3**).
4. Add 800 μl of DEPC-treated water, mix properly, and transfer to two 2 ml tubes. Leave at room temperature for 5 min.
5. Add 200 μl chloroform and mix vigorously for 10 s. Leave at room temperature for 10 min.
6. Centrifuge at 13,000 rpm (17000 × *g*) for 10 min at 4 °C.
7. Transfer upper aqueous phase to fresh tube, add 0.6 ml isopropanol, and mix properly.
8. Leave at room temperature for 10 min. Centrifuge as above.

9. Discard supernatant, wash the pellet with prechilled, 70% ethanol (prepared with DEPC-treated water).
10. Dry the pellet properly and dissolve in 20–50 μl, DEPC-treated water.
11. RNA can be quantified by measuring absorbance at 260 and 280 nm using UV-Vis spectrophotometer.
12. Quality of the RNA can also be assessed by running formaldehyde-agarose gel.
13. 1.5% formaldehyde-agarose gel. Dissolve 0.6 g of agarose in DEPC-treated water, allow cooling to nearly 55 °C and add 10× MOPS (final concentration 1×) and 7.2 ml formaldehyde.
14. Pour the gel in casting trays in fume hood.
15. Perform electrophoresis with 1× MOPS at 5 V/cm, till the bromophenol tacking dye reached two-thirds of the gel.
16. Take image of the gel using gel documentation system.
17. RNA quality is assessed by intensity of bands and minimum smearing.
18. The RNA quality may also be assessed by UV-Vis spectroscopy.
19. Ratio of absorbance at 260 and 280 nm determines, protein contamination (preferable value ranges from 1.8 to 2.0) and ratio of absorbance at 260 and 230 nm determines the purity of nucleic acid (preferable value ranges from 2.0 to 2.2).
20. The purity of RNA may also be checked through Agilent Bioanalyzer.

#### *3.1.2 cDNA Synthesis*

1. Perform the DNase treatment of total RNA before preparing cDNA. Add 1× DNase buffer, to 1 μg of RNA, 1 U of DNase enzyme and adjust the volume to10 μl with DEPC-treated water.
2. Incubate at 37 °C, for 30 min.
3. Stop the reaction by adding 1 μl of 50 mM EDTA. Proceed for cDNA synthesis.
4. Prepare a reaction mixture by adding, 10× Reverse transcriptase synthesis buffer (final concentration 1×), 10 mM dNTP mix (final concentration 500 μM), Oligo $(dT)_{18}$ primer (final concentration 1×), reverse transcriptase enzyme (1 U). Add the DNase treated RNA template, adjust total volume to 20 μl.
5. Incubate at 42 °C, for 30 min followed by 95 °C for 2 min.
6. Dilute this cDNA ten times and use for RT-PCR and/or qRT-PCR.

### 3.1.3 RT-PCR

RT-PCR is routinely used to check the performance of cDNA and for primer screening. However, semiquantitative RT-PCR may also be used for analyzing gene expression. This is generally done by taking small aliquots (5 μl) of cDNA after partial amplifications like 20, 25, 30, and 35 cycles. A difference in band intensity may be calculated through densitometric method to analyze relative gene expression.

1. Use the diluted cDNA as templates.
2. Use any commercially available Taq DNA Polymerases, and prepare the reaction mixture (use 100 μl PCR tubes), by mixing 10× Taq DNA polymerase buffer (final concentration 1×), 10 mM dNTP mix (final concentration 200 μM), forward and reverse primers (final concentration 10 pM), template (2.5 μl), and Taq DNA polymerases (1 U). Adjust the volume to 50 μl with milli Q type I water.
3. Set the thermal cycler: Step I—initial denaturation at 94 °C, 10 min; Step II (35 cycles)—denaturation 94 °C, 30 s, annealing at Tm °C, 30 s, extension at 72 °C 1 min; Step III—final extension at 72 °C for 10 min.
4. After the reaction is over, check for amplifications by running agarose gel.
5. Generally, small aliquots of amplification products are checked initially for optimum band intensity, the PCR cycles may be increased (maximum 40 cycles) accordingly, depending on the band intensity.
6. Prepare 1% agarose gel by dissolving 1 g agarose in water. Add 50× TAE (1× final concentration). Allow to cool, and then add 2 μl of ethidium bromide solution (10 mg/ml). Mix and pour on gel casting tray.
7. Perform electrophoresis in 1× TAE at 5 V/cm, till bromophenol blue tracking dye come toward the end.

### 3.1.4 qRT-PCR

1. For qRT-PCR, reactions are prepared in qRT-PCR compatible PCR tubes or plates, with recommended commercial qRT-PCR reagents as suggested by particular brand of machine.
2. Next step is to perform standard curve assay of primers, to determine primer efficiency. Always check the proper amplification of primers by RT-PCR.
3. For standard curve generation, a reaction mixture is prepared having, 10× SYBR green master mix (final concentration 1×), 1× ROX dye, forward and reverses primers (200 nM), 2.5 μl templates is added in a10× series of dilution (0–10,000), and volume is adjusted to 10 μl (*see* **Note 4**).

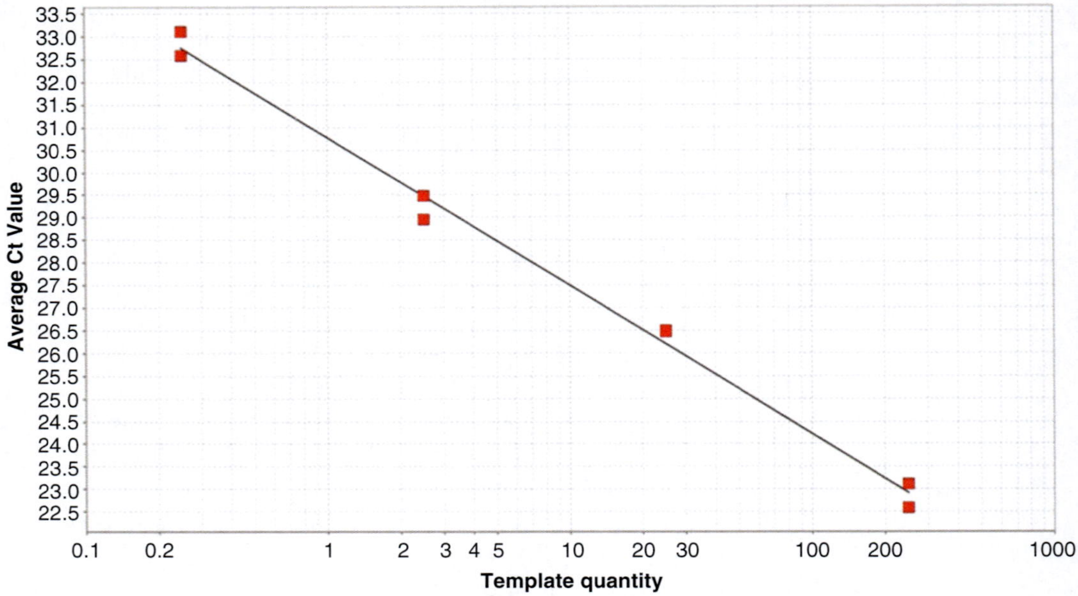


**Fig. 2** Standard curve generated to determine primer efficiency. The curve indicates slope, *Y* intercept, $R^2$, and PCR efficiency

4. Set the reaction as per the default program. Plot a standard curve and determine correlation coefficient, slope, and primer efficiency (*see* Fig. 2).
5. The primers having efficiency range between 80% and 120%, may be accepted.
6. Also perform melting curve analysis of primer pair to screen primer specificity. Single peak in melting curve indicates specific amplification (*see* Fig. 3). If required, the amplified products may be verified by agarose gel electrophoresis (as mentioned before).
7. To compare differential expression of genes, ΔCt value is determined for each gene at given set of conditions.
8. To compare ΔCt value of genes, prepare the reactions as mentioned before in triplicates. Every time there should be a no-template control (NTC) in each plate.
9. ΔCt value is the difference in average Ct value of GOI with respect to the average Ct value of HKG (i.e., $\Delta Ct_{\text{control}} = Ct_{\text{GOI}} - Ct_{\text{HKG}}$).

   Taking the ΔCt value, ΔΔCt value can be calculated as $\Delta\Delta Ct = \Delta Ct_{\text{stress}} - \Delta Ct_{\text{control}}$.

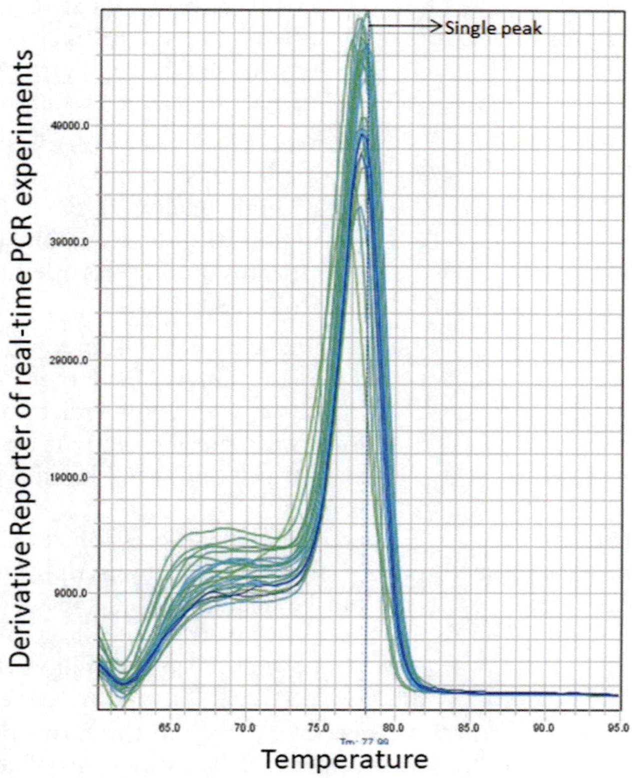


**Fig. 3** Typical melt curve for depicting amplification of a specific amplicon using a primer pair generated by the real time PCR instrument

10. The average fold change of the gene under any given situation can be calculated as $2^{-\Delta\Delta Ct}$.
11. Thus relative expression of the gene can be compared between control and stressed samples [76].

### 3.2 Global Transcriptome Profiling

#### 3.2.1 Experimental Design

The first step to start studying DEG profiling through RNA-seq analysis is to determine the settings for each step of the experiments appropriate for available data and data to be generated. Before deployment of the RNA-seq approach for transcriptome study, one needs to plan experimental design carefully, with consideration of the methods of sample preparations, selection of an NGS platform, different parameters of sequencing such as sequencing depth or coverage, sequencing length, numbers of technical and biological replicates, randomization of replicates, single-end or paired-end sequencing, and other methods of data analysis according to the objectives of the studies. Such considerations on the experimental design of DEG profiling are widely covered in the literature [66–72].

*3.2.2 Generation of Sequencing Reads*

1. Extract total RNA from samples (as described in Subheading 3.1 or with other commercial kit suited to the experiment) and check the quality and integrity of RNA isolate before proceeding further.
2. Construct cDNA libraries using commercially available library preparation kits according to the manufacturer's protocol. Ensure high quality of RNA (mRNA enriched) and cDNA preparation with degenerating agarose gel electrophoresis and other spectrophotometric methods before it using the standard protocols [73].
3. Illumina sequencing—After validating the insert size and quantifying the library, load them onto the flow cell for cluster generation with commercially available cluster generation kits along with specific sequencing primers. Following this, load the flow cell onto the Illumina sequencing machine.

*3.2.3 Acquisition of Genome, Transcriptome, and Annotation Data*

To determine the expression level of the genes of interest in a previously performed transcriptome sequencing, the sequencing data can be acquired from the sequence repositories like NCBI's Sequence Read Archive (SRA, https://www.ncbi.nlm.nih.gov/sra), Gene Expression Omnibus(GEO, https://www.ncbi.nlm.nih.gov/geo) and European Nucleotide Archive (ENA, https://www.ebi.ac.uk/ena) in the form of SRA and/or FASTQ file format, respectively with the help of accession number of the project of the RNA-seq experiment. SRA files can be downloaded from NCBI by HTTP and FTP or with Aspera/ascp client software for faster speed. NCBI also provides a tool named SRA toolkit to download SRA files and derive fastq files from them.

For reference-based RNA-seq analysis of legume plants, the genome and/or transcriptome data and their annotation can also be retrieved from the several plants sequence databases. One such major database is Phytozome (https://phytozome.jgi.doe.gov/pz/portal.html) that currently (v12.1.6) hosts 93 assembled and annotated plant genomes, including economically important legume plants. Several other databases, exclusive for legume group (e.g., https://legumeinfo.org/, https://www.coolseasonfoodlegume.org/ and https://knowpulse.usask.ca/) or specific for a model legume species, also provide the required data, for example, the databases for *G. max* (SoyBase, https://soybase.org/), *M. trucatula* (*Medicago truncatula* genome database, http://www.medicagogenome.org/), and *C. arietinum* (The Chickpea Portal, http://www.cicer.info/databases.php) etc.

The genome sequence data are generally provided in the FASTA format while the transcriptome data are available in SRA or FASTQ format. For annotation, the data are commonly stored as GTF or GFF3 formats.

#### *3.2.4 Quality Assessment and Preprocessing of Reads*

The initial step in any RNA-seq analysis involves general assessment of the quality of raw sequence data generated by sequencing machine (*see* Fig. 1). There are several quality control metrics which provides the information about the quality of data and detects the possible technical problems occurred during sample preparation and sequencing process in the NGS machine. The most widely used tool for it, is FastQC which requires java runtime environment (JVM) and can be run on command line interface as well as graphical user interface (GUI). It provides a graphical overview of quality of NGS reads in the form of HTML based report for quick assessment. The report contains information about basic statistics, per base sequence quality based on *Q*-score (*see* **Note 5**), per tile sequence quality, per base "N" content ("N" represents the base that cannot be called by NGS machine), GC content, sequence length distribution, adapter content, overrepresented sequences and sequence duplication. For details, a comprehensive documentation on FastQC report has been made available by the developers. The interpretation of these metrics also required the consideration for the nature of the samples, the sample and library preparation methods, and the sequencing platform used.

Based on the interpretation of quality reports, the reads sequences are subjected to the preprocessing, which generally involves removal of low quality reads, adapter sequences, filtering of contaminants (sequences derived from nontarget species), and trimming the fraction of sequences with low quality scores. To conduct these steps, the common publicly available tools include FASTX-Toolkit (http://hannonlab.cshl.edu/fastx_toolkit/)), PRINSEQ [77], NGSQC-Toolkit [78], Trimmomatic [79], and Cutadapt [80]. Among them, Trimmomatic has been used most widely for Illumina paired-end and single-ended read data, which is a command line tool written in Java and contains a database of Illumina adapter sequences for the available sequencing kits. The additional sequences with their complement sequences can also be added to customize the adapter database of Trimmomatic, if needed.

Using FastQC and Trimmomatic programs, the prealignment quality control method requires following steps:

1. Evaluate the quality of the sequenced reads with FastQC, which provides different graphical reports on reads statistics and quality control metrics.
2. Using Trimmomatic, apply following preprocessing methods according to the need interpreted from FastQC report (good quality sequences or preprocessed sequences downloaded from databases might not require any editing; on the other hand, the editing might not be enough to make poor quality reads qualify for further analysis):

3. Perform ILLUMINACLIP step to remove adapters or other Illumina-specific sequences.
4. Filter the low quality reads on the basis of average quality lower than a specified level with AVGQUAL.
5. Perform quality trimming using CROP/HEADCROP/LEADING/TRALING. Generally, it is done by removing low-quality bases from 3′ end based on per base quality report with CROP. Trimmomatic also allows use of sliding window trimming starting from 5′ end with SLIDINGWINDOW and cuts the read when average quality drops below a threshold within the window. The typical values for window size and the threshold average quality score are set 4 and 20 respectively.
6. Filter the retained reads on the basis of read length and drop the reads below a specified threshold length (typically <36) with MINLEN.
7. Check again to ensure the enhanced quality of the edited sequences using FastQC to confirm good quality reads for further analysis.

#### *3.2.5 Reference-Guided RNA-Seq Assembly*

When the annotated genome of the plant of interest is available, the processed reads can be mapped to it by reference-based alignment (*see* Fig. 1). BWA (http://bio-bwa.sourceforge.net/), Bowtie2 (http://bowtie-bio.sourceforge.net/bowtie2/index.shtml), STAR (https://code.google.com/archive/p/rna-star/), and HISAT2 are the tools of common choice for mapping of reads to the reference genome. Contrary to the traditional alignment tools, these aligners employ Burrow–Wheeler Transformation algorithm based on suffix trees for faster alignment and lower memory consumption convenient for computationally demanding NGS data and fulfilling the need to address additional problem encountered with RNA-seq data like mapping of the reads spanning multiple exons having varying length of introns in-between [81]. Some tools, like Bowtie and HISAT2, also use Ferragina–Manzini (FM) Index on the transformation of reference genomes to further reduce the memory footprint [82, 83].

The quality of the reference-based alignment can be evaluated on the basis of mapping position of reads with respect to the annotation of known gene coding regions. Some modules and scripts provided by samtools, Picard Toolkit [84] and RSeQC [85] allow for assessment and evaluation of alignment using various parameters like (a) proportion of reads mapped and unmapped to the reference, (b) distribution of reads over different genomic regions viz. Coding regions, introns, untranslated regions (UTRs) and intergenic regions, (c) 5’-3’ bias estimates and (d) read duplication based on similar mapping patterns, etc.

After a successful alignment, the next step is to quantify the gene expression, in terms of total number of counts mapped to gene, exon, transcript or any other feature of the reference. Gaining advantage from the previous annotation, the most widely used assembly tool is Cufflinks, which assembles reads into transcript with different isoforms according to the observed splice junctions using directed acyclic graph of the read alignments. Later, for abundance estimation at transcript isoform level, reads are assigned to those transcripts [86]. The successor of Cufflinks, StringTie performs an assembly using a network flow algorithm that computes transcript abundance and exon–intron structure simultaneously making the assembly process faster [82, 87].

The following protocol describes reference-guided RNA-seq alignment and its evaluation using HISAT2 and RSeQC tools, followed by assembly and qualification using Stringtie:

#### Alignment of Reads to the Reference Genome

1. Download the reference genome (in the form of fasta file) available in a database and build indices of the genome sequence using "hisat2-build" command of the HISAT2 suite. It creates the indices files with .ht2 extension. These indices files for a specific organism can be used for mapping of any RNA-seq sample, which is derived from the same organism or any close relative of the organism.
2. Align the fastq files of each sample to the indexed genome using "hisat2" command of HISAT2 suite.
3. The above step generates the SAM files which contain a list of read alignments for each sample. Convert them into BAM files and sort those files using Samtools. To get basic metrics for post-alignment quality assessment, use the sorted BAM files as an input to run the scripts of RSeQC python package:
4. bam_stat.py to get basic statistics of alignment, including numbers of unmapped reads, reads mapped to "+" and "−" strands, reads from forward or reverse sequencing, reads mapped in proper pairs, and so on.
5. read_duplication.py to infer about mapping based duplication, that is, reads mapped to identical gene mapping location.
6. geneBody_coverage.py to check 5′ or 3′ bias of the mapping. It requires sorted and indexed BAM files and also reference gene model in BED file format.

#### Gene/Transcript Assembly and Abundance Estimation

1. Using the sorted bam files as an input, assemble the transcripts for each sample with "stringtie" command of the String Tie assembly tool.
2. Create a single merged transcriptome annotation (.gtf file) with "stringtie –merge" command using the assembly files (.gtf files) of all the samples.

3. Download the reference genome annotation (.gtf or .gff3 file) and compare the merged transcriptome annotation with it using gffcompare to get the novel transcripts. To annotate the novel transcripts, BLAST can be performed against nonredundant protein databases.
4. Run "stingtie" command again to estimate transcript abundance with "-e" parameter using the merged annotation file. This command generates GTF file for each sample containing mapped genomic features and their raw count data and within-sample normalized expression values in FPKM and TPM (*see* **Note 6**).
5. Using a Python script "preDE.py" (http://ccb.jhu.edu/software/stringtie/dl/prepDE.py), extract read count data for genes and transcripts from the GTF files generated in the previous step.

### *3.2.6 De Novo RNA-Seq Assembly*

De novo transcriptome assembly of RNA-seq data enables to study the gene expression in differential experimental conditions in a nonmodel organism, when a well-annotated genome is not available [88]. It can also be used to discover novel transcripts for already annotated genomes of model organisms enriching the previous annotations [89] and to develop biomarkers in plants [60, 90]. Because of comparatively complex algorithm in comparison to reference-guided transcriptome assembly, de novo assembly is slower and computationally more demanding.

The most cited tools for de novo transcriptome assembly are Velvet, Oases, Trinity, SOAPdenovo, and TransABySS, which employ de Bruijn graph, a mathematical graph theory to derive contigs and scaffolds from a hash table of short k-mers of RNA-seq library [91]. With the fixed setting of k-mer size (equal to 25), assembly with Trinity involves three modules namely, Ichworm, Chrysalis, and Butterfly. First all k-mers are extracted from reads, ranked and overlapped to generate contigs using Ichworm which are clustered together into de Bruijn graph components by Chrysalis. Finally, Butterfly processes the graph and reconstructs isoforms by tracing the reads through the graph [92].

To correctly assign reads to the assembled transcripts for transcript abundance estimation, Trinity supports several qualification methods including RSEM (RNA-Seq by Expectation Maximization) though it needs to be downloaded separately. RSEM is an alignment-based quantification method that fractionally allocates reads to each transcript iteratively on the basis of probabilities of the reads being derived from each transcript [93].

#### De Novo Assembly Construction

1. To construct a single reference assembly, concatenate the RNA-seq data of all the samples using "cat" command in terminal. For paired-end data, the forward end reads across all the samples are combined into a single Fastq file and the reverse-end reads into another one.

2. Assemble the concatenated forward and reverse end reads into transcripts using de novo assembly tool Trinity. The k-mer length is fixed as 25 in the normalization step of trinity. Run the "trinity.pl" script with parameters for name and directory path of input files, input file type, RAM usage, and number of processors.
3. To check the success of assembly process, run the script "TrinityStates.pl" to generate assembly statistics report which includes length and number of transcripts generated and the contig N50 value (*see* **Note 7**).
4. Remove the redundancy from the assembled transcripts using CD-HIT. To predict protein coding regions, perform homology search of assembly transcripts to the nonredundant protein database using BLASTx alignment tool.

#### Reads Mapping and Transcript Abundance Estimation

1. Map the reads of each sample separately to the assembled transcript using the perl script "align_n_estimate.pl" with parameter "--est_method RSEM". This step generates two output files from each mapping process containing abundance estimates for transcripts and Trinity component (a proxy of gene).
2. Build a matrix of count data of transcripts of each sample in a tab-delimited text file using the files produced in the previous step by running the perl script "abundance_estimate_to_matrix.pl".

### *3.2.7 Normalization of Raw Count Data and Differential Gene Expression Analysis*

In RNA-seq analysis, the level of expression is measured from the read count data of total number of mapped fragmented transcripts which is expected to be directly proportional to their abundance in RNA samples. To compare the expression levels of genes in different samples, it is mandatory to normalize the read count data to remove nonuniformities across the samples generated from the technical sources of variation like library size, sequencing depth, and nucleotide compositions [66]. Some of the general approaches implemented in the software packages for DGE analysis to normalize the RNA-seq data are as follows:

#### RPKM (Reads per Kilo Base per Million) and FPKM (Fragments per Kilobase per Million)

These two simplest normalization methods perform normalization by dividing the gene/transcript count by the total number of reads in each library. RPKM is used for single-end reads while FPKM is for paired end reads. Both of these methods are the most widely used within sample as well as between-sample normalization methods and employed in most of the RNA-seq analysis tools, including RSeQC and edgeR mentioned in this protocol, however, these methods can be biased against short reads, and in case, if very few number of genes are highly expressed [94].

TPM (Transcripts per Million)

TPM in contrary to total read counts used by RPKM or FPKM uses total transcript count and performs more effective normalization for the differences in composition of the transcripts [68].

TMM (Trimmed Means of *M* Values)

The default between-sample normalization method of edgeR, TMM is based on the hypothesis that the number of differentially expressed genes is very low and computes a scaling factor for the library sizes for a set of genes in which differentially expressed genes with high fold change across treatments and genes with large average expression are excluded [95].

DESeq

Similar to TMM, the normalization method included in DESeq package, it also assumes that most of the genes are not differentially expressed. DESeq estimates a scaling factor for a given sample by calculating the median of the ratio of read count of a gene and its geometric mean across all samples. Such median values give a correction factor for each sample which is applied to the read counts in that sample [96].

Upper Quantile

Some DGE analysis tools like baySeq [97] use upper quantile normalization with the assumption that majority of genes have low number of reads. This method divides all expression level by the gene counts at 75th percentile in each sample.

Among the commonly used normalization methods, DESeq and/or TMM (edgeR) have been shown to give best results with simulated and real data [66, 94, 98].

Again, there are many different approaches to find significant DEGs based on various statistical model and assumptions which generally use log fold change and list of adjusted $p$-value or $q$-value to identify them (*See* **Note 8**). Most of them use one of the two discrete distributions, either Poisson or Negative binomial distribution to model gene/transcripts counts and parametric methods to estimate variance [69]. Including edgeR, some of the widely tools for DGE analysis are RSEM [93], DESeq2 [99], and Cuffdiff2 [100].

Here, we describe DGE analysis using edgeR which can be used for both simple as well as complex experimental designs with two or more experimental factors. For multifactor experimental designs, more flexible General Linearized Model (GLM) approach of edgeR can be used.

*3.2.8 Differential Gene Expression Analysis Using edgeR*

1. Load the Bioconductor package "edgeR" and its dependencies in R console.
2. Import the count data in R and create an object holding the count data and other metadata like sample names and gene names in a single R variable using "DGElist()" function.

3. Filter the data to remove genes with zero or a few counts, which can dominate the distribution of gene expression values. Filtering the genes with zero or very few counts reduces the false negatives and thus enhances the sensitivity of detecting differentially expressed genes.
4. Normalize them using "CalcNormFactors()" function, which calculates TMM scaling factors for normalization.
5. Explore the Inter-sample relationship using "plotMDS()" function for multidimensional scaling plot (MDS) which shows the relationships among the samples. In MDS plot, similar samples with similar gene expression are located close to each other than the dissimilar ones.
6. To estimate dispersion, first setup a model and design a matrix with "model.matrix()" function and then run "estimateGLM-CommonDisp()", "estimateGLMTrendedDisp()", "estimateGLMTaggedDisp()" functions.
7. Assess the quality of dispersion estimate by plotting the biological coefficient of variation (BCV) against gene abundance in log2 counts per million ($\log_2$CPM) using "plotBCV" function.
8. Fit GLM with "glmfit()" function and use "glmlrt()" to perform differential expression test using likelihood ration test.
9. Extract the top differentially expressed genes sorted *p*-value using "topTag()" function. For log-fold change, firstly the function glmTreat() should be called on the object produced by glmFit() function before using the "topTag()" function.

### 3.3 Data Visualization

The most commonly used visualization tool to view RNA-seq data is Integrative Genomics Viewer (IGV) [101], which may be downloaded from http://software.broadindtitute.org/software/igv/home. Briefly, IGV is a java-based application which can take input in various file formats including fasta, fastq, SAM, BAM, and VCF. The annotation for the reference guided alignment and assembly can also be loaded in form of BED or GFF file format.

### 3.4 Downstream Analysis

Apart from the gene expression analysis, merging with other available metadata, RNA-seq analysis can add dimensions to the study with other downstream analyses, like gene set enrichment analysis and other network analysis, biomarkers identification etc. Once, the list of assembled transcripts is obtained, they can be statistically correlated and grouped into gene sets on the basis of various known relationships, like cellular localization, biological functions or enzymatic functions for better annotation. Similarly, the DEGs can also be mapped to known pathways and statistically overrepresented pathways can be examined by pathway enrichment analysis [81, 102]. Further, RNA-seq data can also be integrated with other sequencing data, like genomic data and ChIP-seq data to study allele-specific expression and epigenetics of plant stress responses [81].

### 3.5 Data Validation

The genes identified as differentially expressed using above computational tools can be further validated by protocols discussed in candidate gene approach, like qRT-PCR.

### 3.6 Conclusions

#### 3.6.1 Candidate Gene Approach

While performing gene expression profiling experiments, the major point of concern is selection of appropriate genes. The researchers should plan to incorporate genes related to all the major events starting from stress perception to ultimate stress responsive effector genes.

##### Stress Sensing and Signal Transduction

Genes related to histidine kinase (HK) pathway and receptor-like kinase (RLK) play important roles in stress sensing, thus, their expression during early time points of stress exposure may be examined. ABA signaling is a major determinant of the stress responses, therefore, expression analysis of genes related to ABA biosynthesis and signaling may also be performed [28]. Likewise, SOS pathway genes, like *SOS1* ($Na^+/H^+$ antiporter), *SOS2* (serine/threonine-specific protein kinase) and *SOS3* (myristoylated $Ca^{2+}$ binding protein), may be selected specifically during salinity stress. Genes related to calcium-binding proteins and calcium signaling, like calmodulin (CaM), calcineurin (CaN), and $Ca^{2+}$-dependent protein kinases (CDPKs), and genes involved in mitogen-activated protein kinase (MAPK) signaling cascade may also be targeted for carrying out expression analysis.

##### Transcription Factors

Many TFs are considered as markers of stress response (e.g., bZIP motifs, APETALA2 (AP2) domain-containing TFs or ethylene responsive element binding proteins (EREBP), NAC TFs, MYC and MYB-like TFs, RDs, ERDs, and DREB isoforms, zinc finger protein (ZFP), WRKY, and heat shock factor TFs (HSFs), as reviewed by Libault et al. [103]. Differential expression of these genes may be analyzed under stress conditions, to understand stress response mechanisms.

##### Effector Proteins

Following stress stimulus and receiving stress signals, the cell starts showing response at different organellar levels, which ultimately decide the stress tolerance of the plant.

Increased activity of SOD, Cat, and APX in legumes has been reported under stresses, like salinity [104], drought [105], and UV radiations [106]. Transgenic lines with altered level of these enzymes have shown tolerance in many crop species [107]. Thus, studying genes related to these enzymes will be helpful in determining their level of synthesis and mode of action. Induction of enzymes related to proline biosynthesis under various abiotic stresses has been reported in *M. truncatula* [108].

Other important effector proteins are chaperonins/heat shock proteins (HSPs), which contribute in cellular homeostasis, by regulating protein folding, assembly, translocation and degradation [109]. Thus, expression profiling of HSPs may also be performed to understand the plant response against abiotic stresses.

Overexpression of genes related to ubiquitination protein system (UPS) also confers tolerance to salinity and drought [110]. Upregulation of transcripts of legume UPS enzymes also showed upregulation in response to drought stress [111]. Therefore, genes related to UPS are also potential targets to understand stress response of plants.

The authors have tried to incorporate maximum possible genes expected to respond to abiotic stress conditions. However, stress response is a complex phenomenon, involving a plethora of genes; thus, it is recommended that the researchers select the genes, wisely. However, to overcome this limitation of candidate gene approach, it is advisable to perform global gene expression profile by employing NGS as discussed below.

#### 3.6.2 Global Transcriptome Profiling

Transcriptome profiling of many major legume plants have revealed the changes in expression pattern of genes involved in signaling, transcriptional regulation, cellular protection, and metabolic pathways in response to the experimental stress conditions (e.g., *Macrotyloma uniflorum* [46], *P. vulgaris* [48], *M. falcata* [49], *L. japonicas* [52], *V. radiata* [53], *G. max* [57], *L. culinaris* [58], *C. arietinum* [60], and *A. hypogaea* [63]). Comparative transcriptome profiling of cultivars of a species with contrasting stress tolerance also provides valuable insights in to the mechanism of stress tolerance.

With advancing techniques of sequencing, evolving methods of data analysis and accumulating sequence data and metadata, it is anticipated that by using transcriptome profiling, the accuracy and resolution of subtle details of the adaptive mechanisms in legumes and other plants will be enriched.

## 4 Notes

1. DEPC water treatments: prepare 1% DEPC water, stir overnight at room temperature, Autoclave at least twice to degrade DEPC. Treat all plasticware, glassware, pestle and mortar with DEPC water overnight, autoclave, and then use for RNA isolation.
2. GOI primer pair: Design primers for selected genes of interest (GOI), for expression analysis prefer to make primers from CDS. For qRT-PCR, small amplicon size is preferred like 80–120 bp. Another important point to be considered is the

choice of internal control gene(s), which is mandatory for data normalization [76].

3. RNA isolation should be preferably done in cold conditions. Grind tissue to fine powder, then add Iris Solution, it immediately gets frozen, continue grinding to fine powder as it thaws into a homogeneous mixture.
4. For generating qPCR standard curve, make serial dilutions of cDNA template as 0, 10, 100, 1000, and 10,000. Prepare reaction mixture in triplicate, also include one NTC. Plot a graph with average Ct value vs. cDNA dilutions. The straight line from origin will give slope, correlation coefficient, and PCR efficiency.
5. *Q*-score represents the estimated log transformed error probability along the sequence reads and is calculated by the following formula:

$$Q = -\log_{10} P$$

where $P$ is the estimated error probability. Thus, $Q$ score of 30 (Q30) tells that there is the probability of finding 1 incorrect base call in 1000 base calling cycle (99% accuracy) while Q20 represents error probability of 1/100 (99.9% accuracy) and so on.

6. In addition to "-e" parameter in stringtie command during transcripts abundance estimation "–B" parameter generates an R object to be used by "Ballgown" package for differential gene expression and isoforms identification.
7. The N50 of an assembly is the maximum length by which at least half of the total assembled sequences are present in contigs of at least that length.
8. In order to adjust statistical confidence measures, *q*-value is used for multiple testing correction, which is a statistical method to estimate False Discovery Rate (FDR) calculated from p-value using Benjamini–Hochberg procedure [112].

## Acknowledgments

A.K.S. acknowledges funding support by Indian Council of Agricultural Research, New Delhi in the form of projects IXX12585 and IXX12644. R.S. acknowledges DST-SERB for the National-Postdoctoral Fellowship (PDF/2016/000924), and DST, Govt. of India for funding under the Women Scientists Scheme-A (SR/WOS-A/LS-160/2018). M.B. acknowledges University Grant Commission (UGC), New Delhi for JRF.

## References


1. Foyer CH, Lam H, Nguyen HT, Siddique KHM, Varshney RK, Colmer TD et al (2016) Neglecting legumes have compromised health and sustainable food production. Nat Plants 2:1–10. https://doi.org/10.1038/nplants.2016.112
2. Song L, Prince S, Valliyodan B, Joshi T, Maldonado JV, Wang J et al (2016) Genome-wide transcriptome analysis of soybean primary root under varying water- deficit conditions. BMC Genomics 17:57. https://doi.org/10.1186/s12864-016-2378-y
3. Valliyodan B, Qiu D, Patil G, Zeng P, Huang J, Dia L et al (2016) Landscape of genomic diversity and trait discovery in soybean. Sci Rep 6:23598. https://doi.org/10.1038/srep23598
4. Cao Y, Halance MK, Gassmann W, Stacey G (2017) The role of plant innate immunity in the legume-Rhizobium symbiosis. Annu Rev Plant Biol 68:535–561. https://doi.org/10.1146/annurev-arplant-042916-041030
5. Considine MJ, Siddique KHM, Foyer CH (2017) Nature's pulse power: legumes, food security and climate change. J Exp Bot 68:1815–1818. https://doi.org/10.1093/jxb/erx099
6. Sosa-Valencia G, Palomar M, Covarrubias AA, Reyes JL (2017) The legume miR1514a modulates a NAC transcription factor transcript to trigger phasiRNA formation in response to drought. J Exp Bot 68:2013–2026. https://doi.org/10.1093/jxb/erw380
7. Araujo SS, Beebe S, Crespi M, Delbreil B, Gonzalez EM, Gruber V et al (2015) Abiotic stress responses in legumes: strategies used to cope with environmental challenges. Crit Rev Plant Sci 34:237–280. https://doi.org/10.1080/07352689.2014.898450
8. Kunert KJ, Vorster BJ, Fenta BA, Kibido T, Davis TS (2016) Drought stress responses in soybean roots and nodules. Front Plant Sci 7:1015. https://doi.org/10.3389/fpls.2016.01015
9. Valdés-lópez O, Batek J, Gomez-hernandez-N, Stacey G, Sayre R (2016) Soybean roots grown under heat stress show global changes in their transcriptional and proteomic profiles. Front Plant Sci 7:517. https://doi.org/10.3389/fpls.2016.00517
10. Nasr-Esfahani M, Inoue K, Chu HD, Nguyen KH, Ha CV, Watanabe Y et al (2017) Comparative transcriptome analysis of nodules of two *Mesorhizobium*–chickpea associations with differential symbiotic efficiency under phosphate deficiency. Plant J 91:911–926. https://doi.org/10.1111/tpj.13616.
11. Maqbool MA, Aslam M, Ali H (2017) Breeding for improved drought tolerance in Chickpea (*Cicer arietinum* L.). Plant Breed 136:300–318. https://doi.org/10.1111/pbr.12477
12. Sato S, Nakamura Y, Kaneko T, Asamizu E, Kato T, Nakao M et al (2008) Genome structure of the legume lotus japonicus. DNA Res 15:227–239. https://doi.org/10.1093/dnares/dsn008
13. Schmutz J, Cannon SB, Schlueter J, Ma J, Mitros T, Nelson W et al (2010) Genome sequence of the palaeopolyploid soybean. Nature 463:178–183. https://doi.org/10.1038/nature08670
14. Young ND, Debellé F, Oldroyd GE, Geurts R, Cannon SB, Udvardi MK et al (2011) The *Medicago* genome provides insight into the evolution of rhizobial symbioses. Nature 480:520–524. https://doi.org/10.1038/nature10625
15. Varshney RK, Chen W, Li Y, Bharti AK, Saxena RK, Schlueter JA et al (2011) Draft genome sequence of pigeonpea (*Cajanus cajan*), an orphan legume crop of resource-poor farmers. Nat Biotechnol 30:83–89. https://doi.org/10.1038/nbt.2022
16. Varshney RK, Song C, Saxena RK, Azam S, Yu S, Sharpe AG (2013) Draft genome sequence of chickpea (*Cicer arietinum*) provides a resource for trait improvement. Nat Biotechnol 31:240–246. https://doi.org/10.1038/nbt.2491
17. Jain M, Misra G, Patel RK, Priya P, Jhanwar S, Khan AW et al (2013) A Draft genome sequence of the pulse crop chickpea (Cicer arietinum L.). Plant J 74:715–729. https://doi.org/10.1111/tpj.12173
18. Ha CV, Watanabe Y, Tran UT, Le DT, Tanaka M, Nguyen KH et al (2015) Comparative analysis of root transcriptomes from two contrasting drought-responsive Williams 82 and DT2008 soybean cultivars under normal and dehydration conditions. Front Plant Sci 6:551. https://doi.org/10.3389/fpls.2015.00551.
19. Garg R, Shankar R, Thakkar B, Kudapa H (2016) Transcriptome analyses reveal genotype- and developmental stage-specific molecular responses to drought and salinity stresses in chickpea. Sci Rep 6:19288. https://doi.org/10.1038/srep19228

20. Garg R, Singh VK, Rajkumar MS, Kumar V, Jain M (2017) Global transcriptome and co-expression network analyses reveal cultivar-specific molecular signatures associated with seed development and seed size/weight determination in chickpea. Plant J 91:1088–1107. https://doi.org/10.1111/tpj.13621
21. Patil G, Chaudhary J, Vuong TD, Jenkins B, Qiu D, Kadam S et al (2017) Development of SNP genotyping assays for seed composition traits in soybean. Int J Plant Genome 6:572969. https://doi.org/10.1155/2017/6572969.
22. Chinnusamy V, Schumaker K, Zhu JK (2004) Molecular genetics perspectives on cross-talk and specificity in abiotic stress signalling in plants. J Exp Bot 55:225–236. https://doi.org/10.1093/jxb/erh005
23. Xiong L, Zhu JK (2002) Molecular and genetic aspects of plant responses to osmotic stress. Plant Cell Environ 25:131–139. https://doi.org/10.1046/j.1365-3040.2002.00782.x
24. Park SY, Fung P, Nishimura N, Jensen DR, Fujii H, Zhao Y et al (2009) Abscisic acid inhibits type 2C protein phosphatases via the PYR/PYL family of START proteins. Science 324:1068–1071. https://doi.org/10.1126/science.1173041
25. Solanke AU, Sharma AK (2008) Signal transduction during cold stress in plants. Physiol Mol Biol Plants 14:69–79. https://doi.org/10.1007/s12298-008-0006-2
26. Huang GT, Ma SL, Bai LP, Zhang L, Ma H, Jia P et al (2012) Signal transduction during cold, salt, and drought stresses in plants. Mol Biol Rep 39:969–987. https://doi.org/10.1007/s11033-011-0823-1
27. Jia W, Wang Y, Zhang S, Zhang J (2002) Salt-stress-induced ABA accumulation is more sensitively triggered in roots than in shoots. J Exp Bot 53:2201–2206. https://doi.org/10.1093/jxb/erf079
28. Xiong L, Zhu JK (2003) Regulation of abscisic acid biosynthesis. Plant Physiol 133:29–36. https://doi.org/10.1104/pp.103.025395
29. Munns R (2002) Comparative physiology of salt and water stress. Plant Cell Environ 25:239–250. https://doi.org/10.1046/j.0016-8025.2001.00808.x
30. Olías R, Eljakaoui Z, Li J, De Morales PA, Marin-Manzano MC, Pardo JM et al (2009) The plasma membrane Na+/H+ antiporter SOS1 is essential for salt tolerance in tomato and affects the partitioning of $Na^+$ between plant organs. Plant Cell Environ 32:904–916. https://doi.org/10.1111/j.1365-3040.2009.01971.x
31. Shi H, Ishitani M, Kim C, Zhu JK (2000) The *Arabidopsis thaliana* salt tolerance gene *SOS1* encodes a putative $Na^+/H^+$ antiporter. Proc Natl Acad Sci USA 97:6896–6901. https://doi.org/10.1073/pnas.120170197
32. Shi H, Quintero FJ, Pardo JM, Zhu JK (2002) The putative plasma membrane $Na^+/H^+$ antiporter SOS1 controls long-distance $Na^+$ transport in plants. Plant Cell 14:465–477. https://doi.org/10.1105/tpc.010371
33. Zhu JK (2003) Regulation of ion homeostasis under salt stress. Curr Opin Plant Biol 6:441–445. https://doi.org/10.1016/S1369-5266(03)00085-2
34. Oh DH, Lee SY, Bressan RA, Yun DJ, Bohnert HJ (2010) Intracellular consequences of SOS1 deficiency during salt stress. J Exp Bot 61:1205–1213. https://doi.org/10.1093/jxb/erp391.
35. Martínez-Atienza J, Jiang X, Garciadeblas B, Mendoza I, Zhu JK, Pardo JM et al (2007) Conservation of the salt overly sensitive pathway in rice. Plant Physiol 143:1001–1012. https://doi.org/10.1104/pp.106.092635
36. Tang RJ, Liu H, Bao Y, Lv QD, Yang L, Zhang HX (2010) The woody plant poplar has a functionally conserved salt overly sensitive pathway in response to salinity stress. Plant Mol Biol 74:367–380. https://doi.org/10.1007/s11103-010-9680-x
37. Ji H, Pardo JM, Batelli G, Van Oosten MJ, Bressan RA, Li X (2013) The salt overly sensitive (SOS) pathway: established and emerging roles. Mol Plant 6:275–286. https://doi.org/10.1093/mp/sst017
38. Hoang XLT, Nhi DNH, Thu NBA, Thao NP, Tran LSP (2017) Transcription factors and their roles in signal transduction in plants under abiotic stresses. Curr Genomics 18:483–497. https://doi.org/10.2174/1389202918666170227150057
39. Zhu JK (2016) Abiotic stress signaling and responses in plants. Cell 167:313–324. https://doi.org/10.1016/j.cell.2016.08.029
40. Lata C, Prasad M (2011) Role of DREBs in regulation of abiotic stress responses in plants. J Exp Bot 62:4731–4748. https://doi.org/10.1093/jxb/err210
41. Agarwal P, Agarwal PK, Joshi AJ, Sopory SK, Reddy MK (2010) Overexpression of pgDREB2A transcription factors enhances abiotic stress tolerance and activates

downstream responsive genes. Mol Biol Rep 37:1125–1135. https://doi.org/10.1007/s11033-009-9885-8.

42. Nakashima K, Yamaguchi-Shinozaki K, Shinozaki K (2014) The transcriptional regulatory network in the drought responses and its crosstalk in abiotic stress responses including drought, cold and heat. Front Plant Sci 5:25. https://doi.org/10.3389/fpls.2014.00170
43. Joshi R, Wani SH, Singh B, Bohra A, Dar ZH, Lone AA et al (2016) Transcription factors and plants response to drought stress: current understanding and future directions. Front Plant Sci 7:1029. https://doi.org/10.3389/fpls.2016.01029
44. Zhu JK (2002) Salt and drought stress signal transduction in plants. Annu Rev Plant Biol 53:247–273. https://doi.org/10.1146/annurev.arplant.53.091401.143329
45. Wang Z, Gerstein M, Snyder M (2009) RNA-seq: a revolutionary tool for transcriptomics. Nat Rev Genet 10:57–63. https://doi.org/10.1038/nrg2484
46. Bhardwaj J, Chauhan R, Swarnkar MK, Chahota RK, Singh AK, Shankar R et al (2013) Comprehensive transcriptomic study on horse gram (*Macrotyloma uniflorum*): de novo assembly, functional characterization and comparative analysis in relation to drought stress. BMC Genomics 14:647. https://doi.org/10.1186/1471-2164-14-647
47. Yates SA, Swain MT, Hegarty MJ, Chernukin I, Lowe M, Allison GG et al (2014) De novo assembly of red clover transcriptome based on RNA-Seq data provides insight into drought response, gene discovery and marker identification. BMC Genomics 15:453. https://doi.org/10.1186/1471-2164-15-453.
48. Hiz MC, Canher B, Niron H, Turet M (2014) Transcriptome analysis of salt tolerant common bean (*Phaseolus vulgaris* L.) under saline conditions. PLoS One 9:e92598. https://doi.org/10.1371/journal.pone.0092598.
49. Miao Z, Xu W, Li D, Hu X, Liu J, Zhang RT (2015) De novo transcriptome analysis of *Medicago falcata* reveals novel insights about the mechanisms underlying abiotic stress-responsive pathway. BMC Genomics 16:818. https://doi.org/10.1186/s12864-015-2019-x
50. Prince SJ, Joshi T, Mutava RN, Syed N, Vitor MSJ, Patil G et al (2015) Comparative analysis of the drought-responsive transcriptome in soybean lines contrasting for canopy wilting. Plant Sci 240:65–78. https://doi.org/10.1016/j.plantsci.2015.08.017
51. Li S, Fan C, Li Y, Zhang J, Sun J, Chen Y et al (2016) Effects of drought and salt-stresses on gene expression in *Caragana korshinskii* seedlings revealed by RNA-seq. BMC Genomics 17:200. https://doi.org/10.1186/s12864-016-2562-0
52. Calzadilla PI, Maiale SJ, Ruiz OA, Escaray FJ (2016) Transcriptome response mediated by cold stress in *Lotus japonicus*. Front Plant Sci 7:374. https://doi.org/10.3389/fpls.2016.00374
53. Tian X, Li S, Liu Y, Liu X (2016) Transcriptomic profiling reveals metabolic and regulatory pathways in the desiccation tolerance of Mungbean (*Vigna radiata* [L.] R. Wilczek) front. Plant Sci 7:1921. https://doi.org/10.3389/fpls.2016.01921
54. Tan H, Huang H, Tie M, Tang Y, Lai Y, Li H (2016) Transcriptome profiling of two asparagus bean (*Vigna unguiculata* subsp. *sesquipedalis*) cultivars differing in chilling tolerance under cold stress. PLoS One 11:e0151105. https://doi.org/10.1371/journal.pone.0151105
55. Nakayama TJ, Rodrigues FA, Neumaier N, Marcolino-Gomes J, Molinari HBC, Santiago TR et al (2017) Insights into soybean transcriptome reconfiguration under hypoxic stress: functional, regulatory, structural, and compositional characterization. PLoS ONE 12:e0187920. https://doi.org/10.1371/journal.pone.0187920
56. Waldeck N, Burkey K, Carter T, Dickey D, Song Q, Taliercio E (2017) RNA-Seq study reveals genetic responses of diverse wild soybean accessions to increased ozone levels. BMC Genomics 18:498. https://doi.org/10.1186/s12864-017-3876-2
57. Leisner CP, Yendrel CR, Ainsworth EA (2017) Physiological and transcriptomic responses in the seed coat of field-grown soybean (*Glycine max* L. Merr.) to abiotic stress. BMC Plant Biol 17:242. https://doi.org/10.1186/s12870-017-1188-y
58. Singh D, Singh CK, Taunk J, Sewak R, Tomar S, Chaturvedi AK (2017) Transcriptome analysis of lentil (*Lens culinaris* Medikus) in response to seedling drought stress. BMC Genomics 18:206. https://doi.org/10.1186/s12864-017-3596-7.
59. Zhang H, Zuelsdorf C, Penneys D, Fan S, Kofsky J, Song B-H (2018) Transcriptome profiling of a beach-adapted wild legume for dissecting novel mechanisms of salinity tolerance. Sci Data 5:180290. https://doi.org/10.1038/sdata.2018.290.

60. Kudapa H, Garg V, Chitikineni A, Varshney RK (2018) The RNA-Seq-based high resolution gene expression atlas of chickpea (*Cicer arietinum* L.) reveals dynamic spatio-temporal changes associated with growth and development. Plant Cell Environ 41:2209–2225. https://doi.org/10.1111/pce.13210
61. Vinson CC, Mota APZ, Oliveira TN, Guimaraes LA, Leal-Bertioli SCM, Williams TCR et al (2018) Early responses to dehydration in contrasting wild *Arachis* species. PLoS One 13:e0198191. https://doi.org/10.1371/journal.pone.0198191
62. Mashaki KM, Garg V, Nasrollahnezhad GAA, Kudapa H, Chitikineni A, Nezhad KZ et al (2018) RNA-Seq analysis revealed genes associated with drought stress response in kabuli chickpea (*Cicer arietinum* L.). PLoS ONE 13:e0199774. https://doi.org/10.1371/journal.pone.0199774
63. Cui F, Sui N, Duan G, Liu Y, Han Y, Liu S, Wan S, Li G (2018) Identification of metabolites and transcripts involved in salt stress and recovery in peanut. Front Plant Sci 9:217. https://doi.org/10.3389/fpls.2018.00217
64. Silva D, Tsai SM, Chiorato AF, da Silva ASC, Esteves JAF, Recchia GH et al (2019) Analysis of the common bean (*Phaseolus vulgaris* L.) transcriptome regarding efficiency of phosphorus use. PLoS One 14:e0210428. https://doi.org/10.1371/journal.pone.0210428
65. Henriet C, Aimé D, Térézol M, Kilandamoko A, Rossin N, Combes-Soia L et al (2019) Water stress combined with sulfur deficiency in pea affects yield components but mitigates the effect of deficiency on seed globulin composition. J Exp Bot 70(16):4287–4304. https://doi.org/10.1093/jxb/erz114
66. Dillies MA, Rau A, Aubert J, Hennequet-Antier C, Jeanmougin M, Servant N et al (2012) A comprehensive evaluation of normalization methods for Illumina high-throughput RNA sequencing data analysis. Brief Bioinform 14:671–683. https://doi.org/10.1093/bib/bbs046.
67. Williams AG, Thomas S, Wyman SK, Holloway AK (2014) RNA-seq data: challenges in and recommendations for experimental design and analysis. Curr Protoc Hum Genet 83:11.13.1–11.13.20. https://doi.org/10.1002/0471142905.hg1113s83
68. Conesa A, Madrigal P, Tarazona S, Gomez-Cabrero D, Cervera A, McPherson A et al (2016) A survey of best practices for RNA-seq data analysis. Genome Biol 17:13. https://doi.org/10.1186/s13059-016-0881-8
69. Seyednasrollah F, Laiho A, Elo LL (2015) Comparison of software packages for detecting differential expression in RNA-seq studies. Brief Bioinform 16:59–70. https://doi.org/10.1093/bib/bbt086
70. Baruzzo G, Hayer KE, Kim EJ, Di Camillo B, FitzGerald GA, Grant GR (2017) Simulation-based comprehensive benchmarking of RNA-seq aligners. Nat Methods 14:135–139. https://doi.org/10.1038/nmeth.4106
71. Costa-Silva J, Domingues D, Lopes FM (2017) RNA-Seq differential expression analysis: an extended review and a software tool. PLoS One 12:e0190152. https://doi.org/10.1371/journal.pone.0190152
72. Corley SM, MacKenzie KL, Beverdam A, Roddam LF, Wilkins MR (2017) Differentially expressed genes from RNA-Seq and functional enrichment results are affected by the choice of single-end versus paired-end reads and stranded versus non-stranded protocols. BMC Genomics 18:399. https://doi.org/10.1186/s12864-017-3797-0
73. Green MR, Sambrook J (2014) Molecular cloning, a laboratory manual. Cold Spring Harbor Laboratory Press, Cold Spring Harbor, NY
74. Griffith M, Walker JR, Spies NC, Ainscough BJ, Griffith OL (2015) Informatics for RNA sequencing: a web resource for analysis on the cloud. PLoS Comput Biol 11:e1004393. https://doi.org/10.1371/journal.pcbi.1004393
75. Ghawana S, Paul A, Kumar H, Kumar A, Singh H, Bhardwaj PK et al (2011) An RNA isolation system for plant tissues rich in secondary metabolites. BMC Res Notes 4:85–89. https://doi.org/10.1186/1756-0500-4-85
76. Sinha R, Sharma TR, Singh AK (2019) Validation of reference genes for qRT-PCR data normalisation in lentil (*Lens culinaris*) under leaf developmental stages and abiotic stresses. Physiol Mol Biol Plants 25:123–134. https://doi.org/10.1007/s12298-018-0609-1
77. Schmieder R, Edwards R (2011) Quality control and preprocessing of metagenomic datasets. Bioinformatics 27:863–864. https://doi.org/10.1093/bioinformatics/btr026
78. Patel RK, Jain M (2012) NGS QC Toolkit: a toolkit for quality control of next generation sequencing data. PLoS One 7:e30619. https://doi.org/10.1371/journal.pone.0030619

79. Bolger AM, Lohse M, Usadel B (2014) Trimmomatic: a flexible trimmer for Illumina sequence data. Bioinformatics 30:2114–2120. https://doi.org/10.1093/bioinformatics/btu170
80. Martin M (2011) Cutadapt removes adapter sequences from high-throughput sequencing reads. EMB Net J 17:10–12. https://doi.org/10.14806/ej.17.1.200
81. Brown SM (2015) Next-generation DNA sequencing informatics, 2nd edn. Cold Spring Harbor Laboratory Press, New York, NY
82. Pertea M, Kim D, Pertea GM, Leek JT, Salzberg SL (2016) Transcript-level expression analysis of RNA-seq experiments with HISAT, StringTie and Ballgown. Nat Protoc 11:1650–1667. https://doi.org/10.1038/nprot.2016.095
83. Kim D, Langmead B, Salzberg SL (2015) HISAT: a fast spliced aligner with low memory requirements. Nat Methods 12:357–360. https://doi.org/10.1038/nmeth.3317
84. DeLuca DS, Levin JZ, Sivachenko A, Fennell T, Nazaire MD, Williams C et al (2012) RNA-SeQC: RNA-seq metrics for quality control and process optimization. Bioinformatics 28:1530–1532. https://doi.org/10.1093/bioinformatics/bts196
85. Wang L, Wang S, Li W (2012) RSeQC: quality control of RNA-seq experiments. Bioinformatics 28:2184–2185. https://doi.org/10.1093/bioinformatics/bts356
86. Trapnell C, Williams BA, Pertea G, Mortazavi A, Kwan G, van Baren MJ et al (2010) Transcript assembly and quantification by RNA-Seq reveals unannotated transcripts and isoform switching during cell differentiation. Nat Biotechnol 28:511–515. https://doi.org/10.1038/nbt.1621
87. Pertea M, Pertea GM, Antonescu CM, Chang TC, Mendell JT, Salzberg SL (2015) StringTie enables improved reconstruction of a transcriptome from RNA-seq reads. Nat Biotechnol 33:290–295. https://doi.org/10.1038/nbt.3122
88. Garg R, Jain M (2013) RNA-Seq for transcriptome analysis in non-model plants. Methods Mol Biol 1069:43–58. https://doi.org/10.1007/978-1-62703-613-9_4
89. Chen G, Li R, Shi L, Qi J, Hu P, Luo J et al (2011) Revealing the missing expressed genes beyond the human reference genome by RNA-Seq. BMC Genomics 12:590. https://doi.org/10.1186/1471-2164-12-590
90. Garg R, Patel RK, Tyagi AK, Jain M (2011) De novo assembly of chickpea transcriptome using short reads for gene discovery and marker identification. DNA Res 18:53–63. https://doi.org/10.1093/dnares/dsq028
91. Geniza M, Jaiswal P (2017) Tools for building de novo transcriptome assembly. Curr Plant Biol 11–12:41–45. https://doi.org/10.1016/j.cpb.2017.12.004
92. Hass BJ, Papanicolaou A, Yassour M, Grabherr M, Blood PD, Bowden J et al (2013) *De novo* transcript sequence reconstruction from RNA-seq using the Trinity platform for reference generation and analysis. Nat Protoc 8:1494–1512. https://doi.org/10.1038/nprot.2013.084
93. Li B, Dewey CN (2011) RSEM: accurate transcript quantification from RNA-Seq data with or without a reference genome. BMC Bioinformatics 12:323. https://doi.org/10.1186/1471-2105-12-323
94. Li X, Brock GN, Rouchka EC, Cooper N, Wu D, O'Toole TE et al (2017) A comparison of per sample global scaling and per gene normalization methods for differential expression analysis of RNA-seq data. PLoS One 12:e0176185. https://doi.org/10.1371/journal.pone.0176185
95. Robinson MD, Oshlack A (2010) A scaling normalization method for differential expression analysis of RNA-seq data. Genome Biol 11:R25. https://doi.org/10.1186/gb-2010-11-3-r25
96. Anders S, Huber W (2010) Differential expression analysis for sequence count data. Genome Biol 11:R106. https://doi.org/10.1186/gb-2010-11-10-r106
97. Hardcastle TJ, Kelly KA (2010) baySeq: empirical Bayesian methods for identifying differential expression in sequence count data. BMC Bioinformatics 11:422. https://doi.org/10.1186/1471-2105-11-422
98. Evans C, Hardin J, Stoebel DM (2018) Selecting between sample RNA-seq normalisation methods from the perspective of their assumptions. Brief Bioinform 19:776–792. https://doi.org/10.1093/bib/bbx008.
99. Love MI, Huber W, Anders S (2014) Moderated estimation of fold change and dispersion for RNA-seq data with DESeq2. Genome Biol 15:550. https://doi.org/10.1186/s13059-014-0550-8
100. Trapnell C, Hendrickson DG, Sauvageau M, Goff L, Rinn JL, Pachter L (2013) Differential analysis of gene regulation at transcript resolution with RNA-seq. Nat Biotechnol 31:46–53. https://doi.org/10.1038/nbt.2450
101. Robinson JT, Thorvaldsdóttir H, Winckler W, Guttman M, Lander ES, Getz

G et al (2011) Integrative genomics viewer. Nat Biotechnol 29:24–26. https://doi.org/10.1038/nbt.1754

102. Reimand J, Isserlin R, Voisin V, Kucera M, Tannus-Lopes C, Rostamianfar A et al (2019) Pathway enrichment analysis and visualization of omics data using g:Profiler, GSEA, Cytoscape and EnrichmentMap. Nat Protoc 14:482–517. https://doi.org/10.1038/s41596-018-0103-9
103. Libault M, Joshi T, Benedito VA, Xu D, Udvardi MK, Stacey G (2009) Legume transcription factor genes: what makes legumes so special? Plant Physiol 151:991–1001. https://doi.org/10.1104/pp.109.144105
104. Eyidigan F, Oz T (2007) Effect of salinity on antioxidant responses of chickpea seedlings. Act Physiol Plant 29:485–493. https://doi.org/10.1007/s11738-007-0059-9.
105. Kukreja S, Nandwal AS, Kumar N, Sharma SK, Sharma SK, Unvi V et al (2005) Plant water status, $H_2O_2$ scavenging enzymes, ethylene evolution and membrane integrity of *Cicer arietinum* roots as affected by salinity. Biol Plant 49:305–308. https://doi.org/10.1007/s10535-005-5308-4
106. Agarwal SB, Singh S, Agarwal M (2009) Ultraviolet-B induced changes in gene expression and antioxidants in plants. In: Jacquot J-P (ed) Advances in botanical research, vol 52. Academic, Burlington, VT, pp 47–86
107. Hasanuzzaman M, Nahar K, Alam M, Roychowdhury R, Fujita M (2013) Physiological, biochemical, and molecular mechanisms of heat stress tolerance in plants. Int J Mol Sci 14:9643–9684. https://doi.org/10.3390/ijms14059643
108. Kim G-B, Nam Y-W (2013) A novel 1-pyrroline-5-carboxylate synthetase gene of *Medicago truncatula* plays a predominant role in stress-induced proline accumulation during symbiotic nitrogen fixation. J Plant Physiol 170:291–302. https://doi.org/10.1016/j.jplph.2012.10.004
109. Wang W, Vinocur B, Shoseyov O, Altman A (2004) Role of plant heat-shock proteins and molecular chaperones in the abiotic stress response. Trends Plant Sci 9:244–252. https://doi.org/10.1016/j.tplants.2004.03.006
110. Guo Q, Zhang J, Gao Q, Xing S, Li F, Wang W (2008) Drought tolerance through overexpression of monoubiquitin in transgenic tobacco. Plant Physiol 165:1745–1755. https://doi.org/10.1016/j.jplph.2007.10.002
111. Chung E, Cho CW, So HA, Kang JS, Chung YS, Lee JH (2013) Overexpression of VrUBC1, a mung bean E2 ubiquitin-conjugating enzyme, enhances osmotic stress tolerance in *Arabidopsis*. PLoS One 8: e66056. https://doi.org/10.1371/journal.pone.0066056
112. Noble WS (2009) How does multiple testing correction work? Nat Biotechnol 27:1135–1137. https://doi.org/10.1038/nbt1209-1135

# Chapter 6

# Method for Bisulfite Sequencing Data Analysis for Whole-Genome Level DNA Methylation Detection in Legumes


## Khushboo Gupta and Rohini Garg



## Abstract

Methylation of cytosines in DNA is the most stable type of epigenetic modification that is established and maintained by different enzymes. In plants, DNA methylation is inherited from one generation to another leaving an epigenetic mark as a memory of previous state, which may include encounter with stress or pathogen. Advancement in the next generation sequencing technologies has enabled the profiling of methylation marks. Whole-genome bisulfite sequencing (WGBS) has the potential to unravel the patterns of DNA methylation at single-base resolution. Though the sequencing technologies have evolved drastically, analysis of WGBS data still remains challenging. Here, we provide a methodology for performing WGBS data analysis along with critical steps for identification of methylation marks in plant genomes including legumes.



**Key words** Whole-genome bisulfite sequencing, Mapping, Differentially methylated regions, Methylated cytosine, Bisulfite conversion, Error rate, Annotation


## 1 Introduction

DNA methylation occurs at cytosine residues in DNA in three different sequence contexts: CG, CHG, and CHH (where H refers to A, C, or T) in plants [1]. Over the last decade, remarkable progress has been made to develop methodologies to understand DNA methylation. DNA methylation is one of the epigenetic modifications that have a crucial role in gene regulation, genome stability and silencing transposable elements (TEs) in response to environmental stresses, and growth and development in plants [2–5]. Recent advances in sequencing technologies have made it feasible to profile the whole genome DNA methylation via methods such as affinity enrichment (MeDIP-seq, MBD), enzyme digestion (MRE-seq) and treatment of DNA with sodium bisulfite [6, 7]. Sodium bisulfite treatment-based techniques include, reduced representation of bisulfite sequencing (RRBS) and


Mukesh Jain and Rohini Garg (eds.), *Legume Genomics: Methods and Protocols*, Methods in Molecular Biology, vol. 2107, https://doi.org/10.1007/978-1-0716-0235-5_6, © Springer Science+Business Media, LLC, part of Springer Nature 2020

whole-genome bisulfite sequencing (WGBS). WGBS technique is the most widely used method for DNA methylation studies. In WGBS, firstly, high-quality DNA is extracted from the samples followed by bisulfite conversion (for conversion of unmethylated cytosines), this is followed by library preparation and read sequencing [8]. Once the sequencing data has been obtained, it needs to be processed through various steps as given below for the identification of methylated cytosines and/or differentially methylated regions using different tools (Table 1).

### 1.1 Data Preprocessing

Analysis for methylation detection begins by quality control step of the reads. Assessing the quality of raw reads is required before proceeding for downstream analysis for removal of low-quality sequences, adapter or primer sequence contamination. Several software tools like NGS QC toolkit [9], FASTQC [10], and HTQC [11] have been developed that can detect the quality issues in the sequencing reads data.

### 1.2 Mapping of High-Quality Reads to the Reference Genome

After preprocessing of the bisulfite sequencing data, the next step is alignment on to the reference genome. Mapping bisulfite reads to the genome requires alignment tools that can map C residue to C, and T residue to C or T rather than a mismatch resulting in a computationally challenging step. This complex step involves analysis of four DNA strands and the possibility that each read can theoretically exist in all possible methylation states [6]. For mapping, the bisulfite reads to the reference genome, different approaches can be used. For example, tools like BSMAP [12] and BRAT [13] maps the masked bisulfite reads to the indexed reference genome in the position-specific manner [29]. However, tools like Bismark [14], MethylCoder [15] and BS-seeker [16] are based on reconversion of the reference genome and the sequence reads from C to T, followed by mapping the sequences using alignment tools like bowtie2 [17] or BWA [18]. Bismark first transforms bisulfite reads into C to T and G to A, then each of them is aligned to preconverted reference genome parallelly using bowtie [30] and all the four alignment outputs are read simultaneously to determine if the sequence can be mapped uniquely.

After mapping the bisulfite converted reads, another recommended step for WGBS is the removal of duplicated reads called as deduplication using tools like SAMtools [26] and Bismark. The next step involves calculation of methylation coverage and depth on all the chromosomes, to estimate coverage of genome by the bisulfite converted reads. After deduplication, methylKit can be used to identify methylated cytosines from the deduplicated files based on their context (CG, CHG, and CHH). Another very crucial measure for the quality of WGBS experiment is the estimation of bisulfite conversion efficiency and error rate calculation [31]. If the genomic

**Table 1**
**Various tools for DNA methylation data analysis**

| | Tool | URL | Reference | Suitable for | based on |
|---|---|---|---|---|---|
| Quality control | NGS QC toolkit | http://www.nipgr.res.in/ngsqctoolkit.html | [9] | Preprocessing of high throughput data | Perl |
| | FASTQC | http://www.bioinformatics.bbsrc.ac.uk/projects/fastqc | [10] | Quality control analysis of high throughput data | Java |
| | HTQC | https://omictools.com/htqc-tool | [11] | Quality control analysis of Illumina data | C++ |
| Mapping | BSMAP | https://code.google.com/p/bsmap/ | [12] | Aligns bisulfite (BS) converted reads including mapping asymmetry of WGBS and RRBS data | C++ |
| | BRAT | http://compbio.cs.ucr.edu/brat/ | [13] | Aligns BS converted reads including mapping asymmetry of only WGBS data and methylation calling | C++ |
| | Bismark | http://www.bioinformatics.babraham.ac.uk/projects/bismark/ | [14] | Aligns BS converted reads by reconversion of the reference genome in WGBS and RRBS | Perl |
| | MethylCoder | https://github.com/brentp/methylcode | [15] | Maps BS converted reads of WGBS and RRBS data | Python |
| | BS-Seeker | http://pellegrini.mcdb.ucla.edu/BS_Seeker2/ | [16] | Maps BS converted reads of WGBS data | Python |
| | Bowtie2 | https://sourceforge.net/projects/bowtie-bio/files/bowtie2/2.3.5/ | [17] | Mapping BS converted reads to the BS converted genome (dependency for Bismark and BS-Seeker) | C++ |
| | BWA | https://sourceforge.net/projects/bio-bwa/files/ | [18] | Mapping BS converted reads to the BS converted genome (dependency for Bismark) | C & Javascript |

(continued)

**Table 1 (continued)**

| | Tool | URL | Reference | Suitable for | based on |
|---|---|---|---|---|---|
| Differentially methylated regions (DMRs) | bsseq | https://bioconductor.org/packages/release/bioc/html/bsseq.html | [19] | DMRs identification in CG context using smoothing algorithm | R |
| | RnReads | http://bioconductor.org/packages/RnBeads/ | [20] | DMRs identification in CG context using hierarchical linear models followed by FDR | R |
| | DMRCaller | http://bioconductor.org/packages/DMRcaller/ | [21] | DMRs identification in CG, CHG and CHH context using Fisher's exact test, Score test and Beta regression test | R |
| | MethylKit | https://github.com/al2na/methylKit | [22] | DMRs identification in CG, CHG and CHH context using Fisher's exact test or Logistic regression followed by SLIM method for significance calculation | R |
| | methylSig | https://github.com/sartorlab/methylSig | [23] | DMCs or DMRs test using beta-binomial model and FDR | R |
| | MethylPipe | http://bioconductor.org/packages/methylPipe/ | [24] | DMRs identification using Wilcoxon-rank test or Kruskal–Wallis nonparametric statistical test | R |
| Annotation | BEDTools | https://github.com/arq5x/bedtools2 | [25] | Identifying overlaps between two files | C++ |
| | Samtools | https://sourceforge.net/projects/samtools/files/ | [26] | Format conversion, depth calculation and sorting bam file | C |
| Visualization | IGV | https://www.broadinstitute.org/igv/ | [27] | visualization of methylation data using bed files | Java |
| | Circos | http://circos.ca/software/download/circos/ | [28] | visualization of methylation data | Perl |

DNA is not converted completely, C rather than a T will be sequenced, this will be read as a methylated base [32].

### 1.3 Identification of Methylated Cytosines (mCs)

Methylation level of a site or region is calculated based on the count of aligned reads giving methylation and unmethylation calls. Statistical analysis is performed using binomial distribution to calculate statistical significance or *p*-value to identify true methylated cytosines [32, 33]. Other than binomial distribution, nearest-neighbor method can also be used with tools like DMRcaller [21]. Since in plants non-CG methylation also occurs, it is important to implement method or use tools that can identify both CG and non-CG methylation. Tools like bsseq [19] and RnBeads [20] do not identify non-CG methylation, whereas tools like DMRCaller (Fisher's exact test, the Score test, and Beta regression test) [21], methylkit (Fisher's exact test, Logistic regression followed by SLIM method) [22], methylSig (Beta-Binomial model and false discovery rate) [23], and methylPipe (Wilcoxon-rank test or Kruskal–Wallis non-parametric statistical test) [24] can call CG and non-CG methylation.

### 1.4 Identification of Differentially Methylated Regions (DMRs)

The next step is to identify regions with differences in methylation levels between different groups/conditions/samples (e.g., control vs. stress). Differential methylation pattern studies are important to identify unique/differential DNA methylation profiles under different conditions with respect to control conditions. Differential methylation can reveal important locations of divergent changes between the comparisons. Packages like methylSig [23], methylPipe [24], and methylKit [22] can be used to identify DMRs in CG, CHG and CHH contexts. A detailed description of different software that could be used for various in silico analysis steps is given in Table 1. For DMRs calculation, methylKit selects statistical tests depending upon the number of samples. If treatment and control have multiple samples, logistic regression is applied; else Fisher's exact test is applied [22].

Advancement in next generation sequencing and availability of sequenced and annotated genome has revolutionized plant epigenetic studies. Recently, epigenomic profiling of organ specific genome-wide DNA methylation patterns in chickpea has been published [34] which unraveled DNA methylation patterns to understand the regulation of gene expression in different organs. Legumes like chickpea are the third most widely consumed crop because of its high nutritional value and agricultural importance. Crop performance, yield and quality are sensitive to interactions between the plant genotype and environment through epigenetic changes [35]. For deeper understanding of these epigenetic changes in response to environment or developmental processes, genome-wide DNA methylation studies are required [36, 37]. This chapter describes a general methodology to perform whole genome

DNA methylation analysis with important considerations, which will be helpful in methylation profiling in legumes and other plants.

## 2 Materials

### 2.1 Computational Requirements

Linux based computer, preferably with 16 core processor with 128 GB RAM (depends on data size).

### 2.2 Software

NGS QC toolkit (v2.3.3) [9] http://www.nipgr.res.in/ngsqctoolkit.html

Bismark (v0.14.3) [14] https://www.bioinformatics.babraham.ac.uk/projects/bismark/

SAMtools (v1.9) [26] http://www.htslib.org/download/

Bowtie2 (v2–2.2.4) [17] https://sourceforge.net/projects/bowtie-bio/files/bowtie2/2.2.4/

R (v3.4.4) https://www.r-project.org/

MethylKit (v1.5.3) [22] https://bioconductor.org/packages/release/bioc/html/methylKit.html

BEDTools (v2.28.0) [25] https://bedtools.readthedocs.io/en/latest/content/installation.html

IGV [27] https://software.broadinstitute.org/software/igv/download

## 3 Methods

After successful preparation of WGBS library and sequencing by high throughput technology, data analysis for identification of genome-wide methylation profiles involves many steps. These steps are discussed below with general framework as outlined in Fig. 1.

### 3.1 Data Preprocessing

WGBS Illumina reads in both paired-end or single-end reads can be used for DNA methylation analysis. Preprocessing is done by performing quality control of the sequencing output by NGS QC toolkit [10]. This toolkit provides various parameters for optimal QC and filtering to obtain high-quality reads from raw data. Here, we outline major steps for doing QC using NGS QC toolkit (v2.3.3).

1. Download the source code of the toolkit from the link given in Materials section and extract the compressed folder.
2. Download and install the dependencies (perl modules) required for the usage of the tool as mentioned in the user manual.

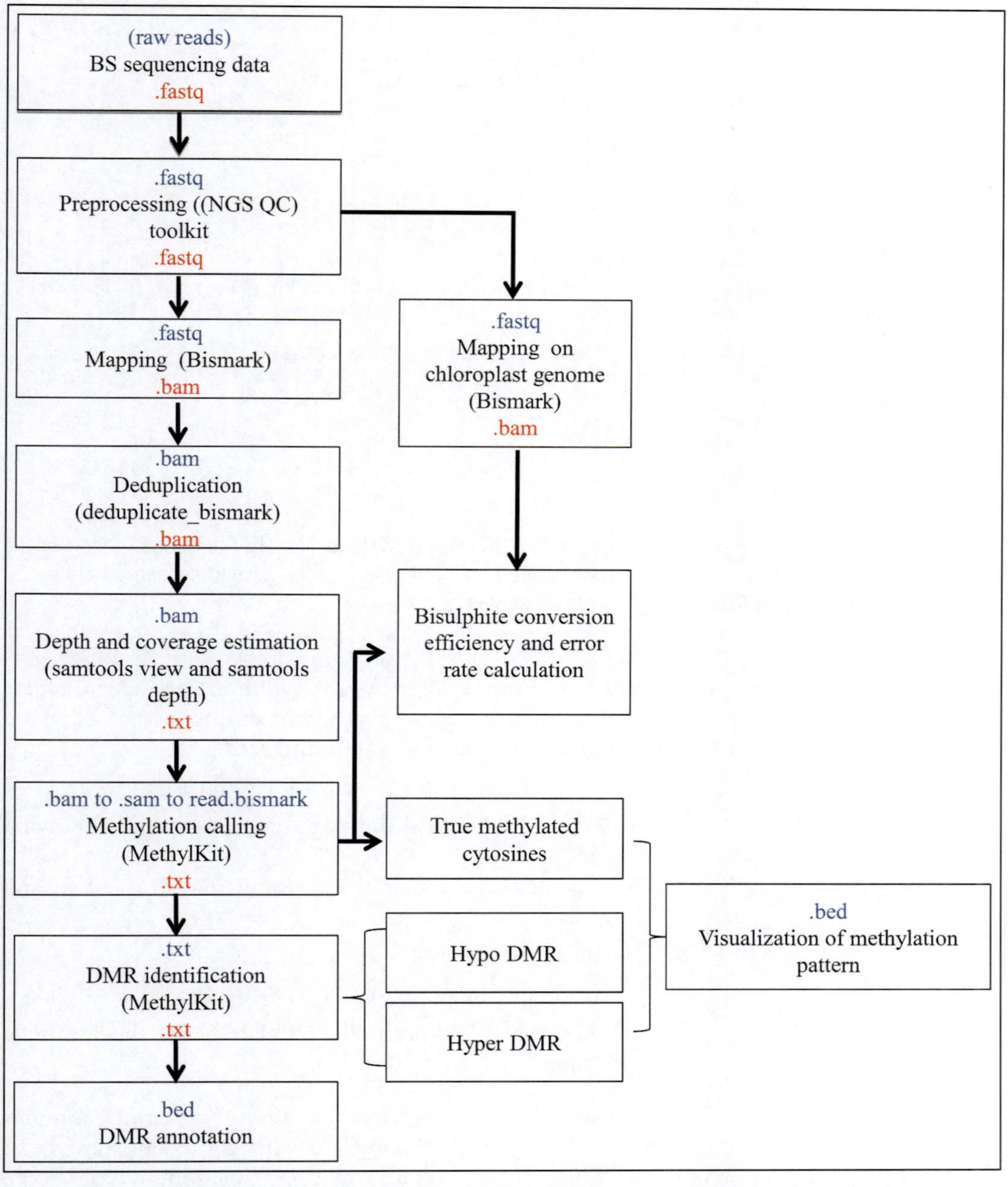


**Fig. 1** An outline of the steps involved in WGBS data analysis: Flowchart depicting steps involved in WGBS data analysis, blue color text represents input file format, red color text represents output file format, and the text in brackets represents tools used

```
Parameters
  Library type                    Paired-end
  Input files                     Input1.fq  Input2.fq
  Primer/Adaptor library          Paired End DNA Library
  Cut-off read length for HQ      70%
  Cut-off quality score           20
  Only statistics                 Off
  Number of CPUs                  8


QC statistics
  File name                                           Input1.fq       Input2.fq
  Total number of reads                               150723895       150723895
  Total number of HQ reads                            150723895       150723895
  Percentage of HQ reads                              100.00%         100.00%
  Total number of bases                               15072389500     15072389500
  Total number of bases in HQ reads                   15072389500     15072389500
  Total number of HQ bases in HQ reads                15072389500     15072389500
  Percentage of HQ bases in HQ reads                  100.00%         100.00%
  Number of Primer/Adaptor contaminated HQ reads      53              55
  Total number of HQ filtered reads                   150723787       150723787
  Percentage of HQ filtered reads                     100.00%         100.00%
```

**Fig. 2** Output statistics file generated after performing QC using NGS QC toolkit: The first section describes the parameters used for quality control analysis of paired-end input file and the second section describes QC statistics of the input fastq files and high-quality fastq file

3. User can perform QC analysis of Illumina sequencing data depending on the type of reads. For example, the following command can be used for paired-end reads:

   - `perl IllQC_PRLL.pl -pe -1 Input1.fq -2 Input2.fq 2 A -c 4`

   where *IllQC_PRLL.pl* is the perl script with the following parameters:

   "-pe" = names of the paired end input files (Input1.fq and Input2.fq).

   "2" = Paired end library.

   "A" = Automatic detection of FASTQ variant.

   "-c" = Number of multiple CPUs (4 in the example command).

The output files *Input1_filtered.fq*, *Input2_filtered.fq* contains high-quality filtered reads. Alongside with these statistics file is generated which contains information, including base score (Phred score), base composition, percentage of contaminated reads; low-quality reads and high-quality reads (*see* **Note 1**). An example QC statistics output file generated after this step is shown in Fig. 2.

### 3.2 Mapping of High-Quality Reads to the Reference Genome

After quality control analysis, high-quality reads obtained in fastq format are mapped to the genome using Bismark tool using the following steps:

1. Download and install Bismark (v0.14.3), bowtie (2-2.2.4), SAMtools (v1.9), and other dependencies using the instruction manual provided with the packages.
2. Download the fasta format of genome sequence and chloroplast genome sequence from the respective plant genome database.
3. Convert the genome using "bismark_genome_preparation" tool from the Bismark package using the command:
   - *bismark_genome_preparation --path_to_bowtie /usr/local/bowtie2--2.2.4/ /genomes/Chickpea*

   where "/usr/local/bowtie2–2.2.4/" is the path to the bowtie and "/genomes/Chickpea" is the path to the genome folder containing the fasta file of the genome to be converted.

   The output will be created inside the genome folder and has two folders, namely (1) CT conversion and (2) GA conversion, containing converted genome indices. Repeat the same for the chloroplast genome as well, which will be helpful in calculating the error rate (*see* **Note 2**).
4. Map the high-quality reads to the converted genome using "Bismark" tool with command:
   - *bismark --bowtie2 --path_to_bowtie /usr/local/bowtie2-2.2.4/ --prefix genome Chickpea -1 Input1_filtered.fq -2 Input2_filtered.fq*

   where, "--bowtie2" option will allow Bismark to use bowtie2, "/usr/local/bowtie2-2.2.4/" is the path of the bowtie, "--prefix genome" is the tag that will be added before the name of the output files, "Chickpea" is the name of the genome folder and "Input1_filtered.fq -2 Input2_filtered.fq" are the input files.

   Other parameters that can be used in the above command are:

   "-N", number of mismatch (is usually set as one, which increases the alignment sensitivity).

   "-X", maximum insert size for valid paired end alignment.

   "multicore" option for running the process on multiple cores.

   The same step is repeated for mapping the reads to the chloroplast genome (*see* **Notes 3** and **4**). The output files will include mapped reads in BAM format and a report file (e.g., genome.Input1_filtered.fq _bismark_bt2_pe.bam and genome.Input1_filtered.fq_bismark_bt2_PE_report.txt files will be created by the above given command). The report file

```
Final Alignment report
======================
Sequence pairs analysed in total:       108194029
Number of paired-end alignments with a unique best hit: 775496
Mapping efficiency:     0.7%
Sequence pairs with no alignments under any condition:  107418514
Sequence pairs did not map uniquely:    19
Sequence pairs which were discarded because genomic sequence could not be extracted:    14

Number of sequence pairs with unique best (first) alignment came from the bowtie output:
CT/GA/CT:       447958  ((converted) top strand)
GA/CT/CT:       0       (complementary to (converted) top strand)
GA/CT/GA:       0       (complementary to (converted) bottom strand)
CT/GA/GA:       327524  ((converted) bottom strand)

Number of alignments to (merely theoretical) complementary strands being rejected in total:     0

Final Cytosine Methylation Report
=================================
Total number of C's analysed:   24999571

Total methylated C's in CpG context:    70877
Total methylated C's in CHG context:    57024
Total methylated C's in CHH context:    74278
Total methylated C's in Unknown context:        0


Total unmethylated C's in CpG context:  4704221
Total unmethylated C's in CHG context:  4037697
Total unmethylated C's in CHH context:  16055474
Total unmethylated C's in Unknown context:      0
```

**Fig. 3** Output statistics file generated after mapping reads using Bismark: The figure represents output file obtained after the mapping of bisulfite converted reads to the genome, consisting of mapping statistics along with cytosine methylation report in CG, CHG, and CHH context

contains mapping statistics like total sequences analyzed; uniquely mapped reads; mapping efficiency; number of unaligned sequences; total C's analyzed; total methylated cytosines in CG, CHG, and CHH context; and percentage of cytosines in three contexts as shown in Fig. 3.

5. Optional step for the removal of deduplicated reads, "deduplicate_bismark" tool of "Bismark" package can be used using the command:

   - *deduplicate_bismark -p genome.Input1_filtered.fq_bismark_bt2_pe.bam --bam --samtools_path /usr/local/samtools/*

   where, "-p" is for paired-end input files in BAM format, "--bam" to write output files in BAM format, "/usr/local/samtools/" represents path to SAMtools (if, SAMtools is not in the path already).

   Output of this step will generate deduplicated BAM file (for example, genome.Input1_filtered.fq_bismark_bt2_pe.deduplicated.bam) and a statistics report file (genome.Input1_filtered.fq_bismark_bt2_pe.deduplication_report.txt) containing the information about the number of alignments

```
----------- Genome deduplication report ---------

Total number of alignments analysed in Genome INPUT.bam:        56332836
Total number duplicated alignments removed:     1621274 (2.88%)
Duplicated alignments were found at:     1060924 different position(s)

Total count of deduplicated leftover sequences: 54711562 (97.12% of total)
```

**Fig. 4** An output statistics file generated after performing deduplication: The genome deduplication report describes statistics for the deduplicated reads

analyzed, total number of duplicated alignments removed and the number of total deduplicated leftover sequences, as shown in Fig. 4 (*see* **Note 5**).

6. The next step involves calculating genome coverage and sequencing depth using SAMtools package. Firstly, sort the deduplicated BAM file as follows:

   - *samtools sort genome.Input1_filtered.fq_bismark_bt2_pe.deduplicated.bam genome.Input1_filtered.fq_bismark_bt2_pe.deduplicated.bam_sort.bam*

   where, "sort" is the option for the SAMtools to sort the file, "genome.Input1_filtered.fq_bismark_bt2_pe.deduplicated.bam" is the input file and "genome.Input1_filtered.fq_bismark_bt2_pe.deduplicated.bam_sort.bam" is the output file.

   For depth calculation use the following command:

   - *samtools depth genome.Input1_filtered.fq_bismark_bt2_pe.deduplicated.bam_sort.bam -r chromosome_size > genome.Input1_filtered.fq_bismark_bt2_pe.deduplicated.bam_sort.bam.depth*

   where, "depth" is the SAMtools option for depth calculation, "genome.Input1_filtered.fq_bismark_bt2_pe.deduplicated.bam_sort.bam" is the input file, "-r" is for the name of the file containing information about chromosome numbers and chromosome size in two columns as (column 1 = Chr1 and column 2 = Chromosome size), for calculation of genome coverage and "genome.Input1_filtered.fq_bismark_bt2_pe.deduplicated.bam_sort.bam.depth" is the output file (*see* **Notes 6** and 7).

7. In the next step, methylated cytosines are called context wise and stored in three different files as CG, CHG, and CHH context. For this, "methylKit" an R package can be used. This requires conversion of BAM file format to SAM file format as mentioned below:

- *samtools view -h genome.Input1_filtered.fq_bismark_bt2_pe.deduplicated.bam_sort.bam >genome.Input1_filtered.fq_bismark_bt2_pe.deduplicated.bam_sort.bam.sam*

where "view" is the option for SAMtools to convert BAM to SAM file format, "h" is to print the header for the SAM output, "genome.Input1_filtered.fq_bismark_bt2_pe.deduplicated.bam" is the input file and "genome.Input1_filtered.fq_bismark_bt2_pe.deduplicated.bam.sam" is the output file.

Launch R from the terminal and load library (methylKit) using the command:

*library(methylKit)*

For calling CG methylated cytosines (CpG) following command can be used:

- *sample.id <-genome.Input1_filtered.fq_bismark_bt2_pe.deduplicated.bam_sort.bam.sam*
- *CpG =processBismarkAln (genome.Input1_filtered.fq_bismark_bt2_pe.deduplicated.bam_sort.bam.sam, sample.id, assembly = "Chickpea", save.context = c("CpG"), read.context = "CpG")*
- *write.table (CpG, file = "test1_CpG.txt", sep='\t')*

where "genome.Input1_filtered.fq_bismark_bt2_pe.deduplicated.bam_sort.bam.sam" is the input file for methylation calling, "sample.id" is the sample name, "Chickpea" is the genome assembly name, "CpG" represents the context of file being created. An example of output file (test1_CpG.txt) generated after methylated cytosine calling is shown in Fig. 5.

| chrBase | chr | base | strand | coverage | freqC | freqT |
|---|---|---|---|---|---|---|
| Ca1.160 | Ca1 | 160 | F | 8 | 87.50 | 12.50 |
| Ca1.20 | Ca1 | 20 | F | 1 | 0.00 | 100.00 |
| Ca1.261 | Ca1 | 261 | F | 16 | 93.75 | 6.25 |
| Ca1.307 | Ca1 | 307 | F | 31 | 90.32 | 9.68 |
| Ca1.318 | Ca1 | 318 | F | 36 | 88.89 | 11.11 |
| Ca1.392 | Ca1 | 392 | F | 28 | 71.43 | 28.57 |
| Ca1.43 | Ca1 | 43 | F | 2 | 100.00 | 0.00 |
| Ca1.465 | Ca1 | 465 | F | 30 | 93.33 | 6.67 |
| Ca1.618 | Ca1 | 618 | F | 10 | 20.00 | 80.00 |
| Ca1.876 | Ca1 | 876 | F | 4 | 75.00 | 25.00 |
| Ca1.892 | Ca1 | 892 | F | 5 | 0.00 | 100.00 |
| Ca1.966 | Ca1 | 966 | F | 4 | 50.00 | 50.00 |
| Ca1.161 | Ca1 | 161 | R | 3 | 66.67 | 33.33 |
| Ca1.262 | Ca1 | 262 | R | 92 | 90.22 | 9.78 |

**Fig. 5** Output file for CG context methylated cytosine generated by methylkit: The image of the file contains information about position and frequency of reads mapped to that position for each methylated cytosine in CG context in the genome

Repeat the same step for CHG and CHH context files.

Based on the mapping of reads to the chloroplast genome (as described in Subheading 3.4) the bisulfite conversion efficiency and error rate can be calculated. Error rate can be defined as the ratio of total number of methylated cytosines analyzed in different context (CG, CHG, and CHH) to the number of cytosines analyzed in the chloroplast genome. This error rate can further be used for the estimation of bisulfite conversion efficiency.

### 3.3 Identification of Differentially Methylated Regions (DMRs)

For identification of DMRs using methylKit package for a test and control comparison (two replicates each), these steps can be followed:

1. Launch R in terminal to load the library "methylKit" and read the individual context methylated cytosines file generated from the above step for control and test using "methRead" function using the following commands (*see* **Notes 8** and **9**):

   - `library(methylKit)`
   - `sample1= "test1_CpG.txt"`
   - `sample2= "test2_CpG.txt"`
   - `sample3= "control1_CpG.txt"`
   - `sample4= "control2_CpG.txt"`
   - `file.list <- list(sample1,sample2,sample3,sample4)`

   These commands will load the test and control files for both the replicates into "file.list" function.

   - `myobj = methRead (file.list, sample.id=list("test1","-test2","control1","control2"), assembly="Chickpea", treatment=c(1,1,0,0), context= "CpG" )`

2. The next step involves filtering of samples based on read coverage; this step is useful to discard very high read coverage bases that may have occurred due to PCR bias along with the removal of low-coverage reads.

   - `filtered.myobj = filterByCoverage (myobj, lo.count = 10, lo.perc = NULL, hi.count = NULL, hi.perc = 99.9)`

   User can decide the "low.count" and "hi.perc" depending upon the minimum and maximum read coverage obtained (*see* **Note 10**).

3. In the next step, tiling window method is applied, to tile the genome with a window size of different length. For example, 100 bp length and 100 bp step size and outlines the methylation information on those tiles (*see* **Note 11**).

- `tiles <- tileMethylCounts (filtered.myobj, win.size = 100, step.size = 100)`

4. Merge all the samples using the "unite" function. This function will merge all samples into one object for base pair locations that are present throughout the samples and will also return an object that can be used for all comparative analysis (*see* **Note 12**).

- `meth= unite (tiles, destrand=FALSE)`

5. For identification of differentially methylated regions "calculateDiffMeth" function is used. This function compares the methylation in each tile/window in control and treated samples. Depending on the sample, Fisher's exact test or logistic regression will be applied to calculate *p*-value and SLIM to calculate the corrected *p*-value or *q*-value.

- `myDiff = calculateDiffMeth(meth)`

6. To categorize identified DMRs into hyper and hypo (hypo, loss of methylation, whereas hyper, gain in methylation as compared to control). One needs to specify the cut-off value for methylation difference and *p*-value or *q*-value along with specifying the type as "hyper" or "hypo." For example, to identify hyper- or hypomethylated DMRs, with a cutoff percentage of methylation difference > =25% and *q*-value = 0.01, below mentioned command can be used:

- `myDiff_25.hyper = getMethylDiff(myDiff,difference=25, qvalue=0.01, type="hyper")`
- `myDiff_25.hypo = getMethylDiff(myDiff,difference=25, qvalue=0.01, type="hypo")`
- `myDiff_25.all = getMethylDiff(myDiff,difference=25, qvalue=0.01)`
- `write.table (myDiff_25.hyper, file="DMR-CpG-hyper.txt", sep="\t")`
- `write.table (myDiff_25.hypo, file ="DMR-CpG-hypo.txt", sep="\t")`
- `write.table (myDiff_25.all, file ="DMR-CpG-all.txt", sep="\t")`

Repeat all the steps from Subheading 3.3 to obtain DMRs in CHG and CHH context.

This step will generate a text file containing information of hyper, hypo, and all DMRs in the specified context (i.e., CG, CHG or CHH). A hypomethylated DMRs file generated using this command is given in Fig. 6 as an example.

| chr | start | end | strand | pvalue | qvalue | meth.diff |
|---|---|---|---|---|---|---|
| Ca1 | 49051 | 49150 | * | 9.32920929946265e-18 | 3.1858600111165e-15 | -30.7904411764706 |
| Ca1 | 314201 | 314300 | * | 3.115582689346e-15 | 7.93006056758371e-13 | -32.7345117885496 |
| Ca1 | 356751 | 356850 | * | 1.78955288235751e-07 | 1.32504916044821e-05 | -30.6007751937984 |
| Ca1 | 421301 | 421400 | * | 6.37306805977132e-16 | 1.76244971733565e-13 | -30.1191151446398 |
| Ca1 | 635601 | 635700 | * | 1.57386980994597e-08 | 1.44845766852457e-06 | -27.9493076283212 |
| Ca1 | 737651 | 737750 | * | 1.91164273032621e-10 | 2.48189152739328e-08 | -27.1516046033867 |
| Ca1 | 1106951 | 1107050 | * | 2.05879276625818e-14 | 4.734916729169e-12 | -29.7856150334149 |
| Ca1 | 1127251 | 1127350 | * | 8.41284312989287e-14 | 1.79557959087069e-11 | -27.0833333333333 |
| Ca1 | 1455651 | 1455750 | * | 1.20605417565498e-11 | 1.89097048817123e-09 | -29.8415416062475 |
| Ca1 | 1555151 | 1555250 | * | 5.67969242569563e-17 | 1.7705667665154e-14 | -37.3328839456241 |
| Ca1 | 1555201 | 1555300 | * | 5.67969242569563e-17 | 1.7705667665154e-14 | -37.3328839456241 |
| Ca1 | 1891801 | 1891900 | * | 1.39831819730781e-30 | 1.48660618810033e-27 | -37.2484146677695 |

**Fig. 6** Differentially methylated regions (DMRs) output file generated by methylkit: The figure describes DMRs identified in CG context along with their statistical significance and methylation difference

### *3.4 Annotation of DMRs*

After DMRs identification, annotation of these methylated regions is required to understand the possible function of methylation with respect to their location in the genome [38, 39]. One can locate DMRs in different genomic regions such as genic or their flanking regions and correlations with their expression. For annotation, tools like "BEDTools" can be used to find the overlaps between the two files [25]. Since the preliminary requirement for WGBS experiment is that the genome should be sequenced and annotated, these genome annotation bed files can be created from the genome annotation file (gtf or gff file format) and DMR.bed file can be created from the DMR.txt file obtained in the previous step. The basic format for a bed file used for BEDTools is chromosome, start position, and end position for each feature file (genic, promoter, exon, intron, etc.). To find the DMRs present in the genic regions using BEDTools, use the below mentioned command:

- *bedtools intersect -a genic.bed -b DMR.bed -wo > genic_DMR_out.bed*

Where "-a" represents first input file (genic.bed), "-b" is the second input file (DMR.bed), "-wo" is for writing all the columns of both the files in the output along with overlap bases information and "genic_DMR_out.bed" is the output file created at default parameters.

### *3.5 Visualization of DNA Methylation Data*

For visualization of DNA methylation patterns, Integrative Genome Viewer (IGV) [27] or circos [28] can be used. IGV is a freely available java-based application in which user can load BED files using a simple drop-down option. It is very user friendly tool and can be used by researchers without any bioinformatics expertise because of its highly interactive interface. Circos, on the other hand, requires good understanding of the in-built scripts that are interrelated to one another and a specific format of input file which contains the count of methylated cytosines in a bin that requires making additional script for creating such files.

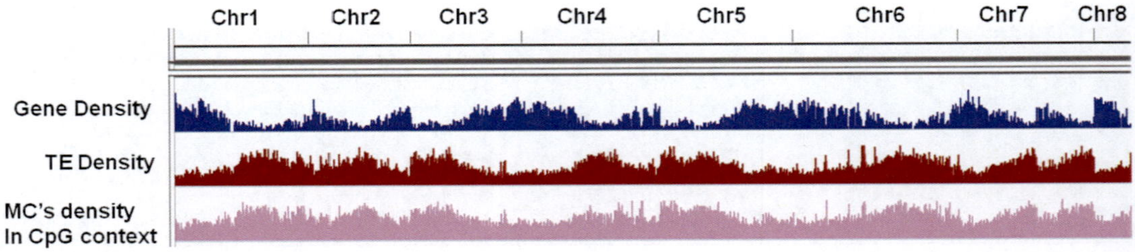


**Fig. 7** IGV view of methylation data: The example figure represents distribution of methylated cytosines throughout the chickpea genome along with the gene and TE density in IGV

To visualize methylated cytosines density using IGV follow these steps:

- Download and install using the link provided in the material section.

Launch IGV in the terminal using:

- *sh igv.sh*

Graphical user interface for IGV will open.

To load the genome using these steps:

- Click on the **Genome tab → load from file/url**
- Locate your genome fasta file and press **open**

To load the BED files for genomic features and metylated cytosines or DMRs:

- Click on the drop-down option from **file tab→load from file**
- Locate your file and press **open**

An example IGV view of the methylation data along with the genomic features is shown in Fig. 7.

## 4 Notes

1. Quality control analysis of the raw reads is crucial for downstream analysis. All the low-quality sequences and adapter or primer sequence contamination should be trimmed.
2. Error rate can be calculated by determining the ratio of total number of methylated C's in different context to the number of C's analyzed in the chloroplast genome (since it is expected that there is no methylation in the chloroplast genome), which can give an estimate of the conversion efficiency.
3. Bowtie2 should be installed before mapping bisulfite converted reads to the converted genome.

4. Both the input files along with the converted genome for mapping should be in the Bismark folder or else their path should be on the user's path in order for Bismark to run successfully.
5. SAMtools being a dependency tool for deduplication using Bismark, should be installed prior to the execution of the command.
6. For calculation of genome coverage, the deduplicated BAM file needs to be sorted first using SAMtools sort otherwise unsorted BAM file will give error during the SAMtools depth command.
7. Calculate the percentage of reads covering each chromosome and then average it to calculate the genome coverage.
8. Use the same version of methylKit for the analysis of all samples in a particular experiment, or for comparison of different experiments.
9. All the input files of methylated cytosines (two replicates of test and control) should be in the working directory. If not in the directory, provide the complete path of the files.
10. The function "getMethylationStats" and "getCoverageStats" can be used to decide the parameters.
11. The function "tileMethylCounts," tiles the genome into specified windows sizes and sums up C and T counts for each covered cytosine and returns the total C and T count for each tile.
12. Setting "destrand = TRUE" will merge reads on both strands of CG methylation. It is advisable when interested in studying only CG methylation.

## Acknowledgments

K.G. acknowledges Shiv Nadar University for Research Fellowship. R.G. acknowledges Women Excellence Award from the Science and Engineering Research Board, Department of Science and Technology, Government of India, and Young Scientist Award scheme from Indian National Science Academy, New Delhi, for funding support.

### References

1. Bartels A, Ha Q, Nair P, Stacey L, Gaynier H, Mosley M, Xiao W (2018) Dynamic DNA methylation in plant growth and development. Int J Mol Sci 19:2144
2. Candaele J, Demuynck K, Mosoti D, Beemster GT, Inzé D, Nelissen H (2014) Differential methylation during maize leaf growth targets developmentally regulated genes. Plant Physiol 164:1350–1364
3. Wada Y, Miyamoto K, Kusano T, Sano H (2004) Association between up-regulation of stress-responsive genes and hypomethylation

of genomic DNA in tobacco plants. Mol Gen Genomics 271:658–666

4. Kawakatsu T, Nery JR, Castanon R, Ecker JR (2017) Dynamic DNA methylation reconfiguration during seed development and germination. Genome Biol 18:171
5. Xing MQ, Zhang YJ, Zhou SR, Hu WY, Wu XT, Ye YJ, Wu XX, Xiao YP, Li X, Xue HW (2015) Global analysis reveals the crucial roles of DNA methylation during rice seed development. Plant Physiol 168:1417–1432
6. Laird PW (2010) Principles and challenges of genome-wide DNA methylation analysis. Nat Rev Genet 11:191–203
7. Cokus SJ, Feng S, Zhang X, Chen Z, Merriman B, Haudenschild CD, Pradhan S, Nelson SF, Pellegrini M, Jacobsen SE (2008) Shotgun bisulfite sequencing of the Arabidopsis genome reveals DNA methylation patterning. Nature 13:215–224
8. Trivedi UH, Cézard T, Bridgett S, Montazam A, Nichols J, Blaxter M, Gharbi K (2014) Quality control of next-generation sequencing data without a reference. Front Genet 5:111
9. Patel RK, Jain M (2012) NGS QC toolkit: a toolkit for quality control of next generation sequencing data. PLoS One 7:e30619
10. FastQC: a quality control tool for high throughput sequence data Babraham Institute Bioinformatics. https://www.bioinformatics.babraham.ac.uk/projects/fastqc/
11. Yang X, Liu D, Liu F, Wu J, Zou J, Xiao X, Zhao F, Zhu B (2013) HTQC: a fast quality control toolkit for Illumina sequencing data. BMC Bioinformatics 14:33
12. Xi Y, Li W (2009) BSMAP: whole genome bisulfite sequence MAPping program. BMC Bioinformatics 10:232
13. Harris EY, Ponts N, Levchuk A, Roch KL, Lonardi S (2010) BRAT: bisulfite-treated reads analysis tool. Bioinformatics 26:572–573
14. Krueger F, Andrews SR (2011) Bismark: a flexible aligner and methylation caller for bisulfite-Seq applications. Bioinformatics 27:1571–1573
15. Pedersen B, Hsieh T-F, Ibarra C, Fischer RL (2011) MethylCoder: software pipeline for bisulfite-treated sequences. Bioinformatics 27:2435–2441
16. Chen P-Y, Cokus SJ, Pellegrini M (2010) BS Seeker: precise mapping for bisulfite sequencing. BMC Bioinformatics 11:203
17. Langmead B, Salzberg SL (2012) Fast gapped-read alignment with Bowtie 2. Nat Methods 9:357–359
18. Li H, Durbin R (2009) Fast and accurate short read alignment with Burrows-Wheeler transform. Bioinformatics 25:1754–1760
19. Hansen KD, Langmead B, Irizarry RA (2012) BSmooth: from whole genome bisulfite sequencing reads to differentially methylated regions. Genome Biol 13:R83
20. Assenov Y, Mueller F, Lutsik P, Walter J, Lengauer T, Bock C (2014) Comprehensive analysis of DNA methylation data with RnBeads. Nat Methods 11:1138–1140
21. Catoni M, Tsang JM, Greco AP, Zabet NR (2018) DMRcaller: a versatile R/Bioconductor package for detection and visualization of differentially methylated regions in CpG and non-CpG contexts. Nucleic Acids Res 46:e114
22. Akalin A, Kormaksson M, Li S, Garrett-Bakelman FE, Figueroa ME, Melnick A, Mason CE (2012) methylKit: a comprehensive R package for the analysis of genome-wide DNA methylation profiles. Genome Biol 13:R87
23. Park Y, Figueroa ME, Rozek LS, Sartor MA (2014) MethylSig: a whole genome DNA methylation analysis pipeline. Bioinformatics 30:2414–2422
24. Kishore K, de Pretis S, Lister R, Morelli MJ, Bianchi V, Amati B, Pelizzola M (2015) methylPipe and compEpiTools: a suite of R packages for the integrative analysis of epigenomics data. BMC bioinformatics 16:313
25. Quinlan AR, Hall IM (2010) BEDTools: a flexible suite of utilities for comparing genomic features. Bioinformatics 26:841–842
26. Li H, Handsaker B, Wysoker A, Fennell T, Ruan J, Homer N, Marth G, Abecasis G, Durbin R, 1000 Genome Project Data Processing Subgroup (2009) The sequence alignment/map (SAM) format and SAMtools. Bioinformatics 25:2078–2087
27. Robinson JT, Thorvaldsdóttir H, Winckler W, Guttman M, Lander ES, Getz G, Mesirov JP (2011) Integrative genomics viewer. Nat Biotechnol 29:24–26
28. Krzywinski M, Schein J, Birol I, Connors J, Gascoyne R, Horsman D, Jones SJ, Marra MA (2009) Circos: an information aesthetic for comparative genomics. Genome Res 9:1639–1684
29. Tran H, Porter J, Sun MA, Xie H, Zhang L (2014) Objective and comprehensive evaluation of bisulfite short read mapping tools. Adv Bioinforma 2014:472045
30. Langmead B, Trapnell C, Pop M, Salzberg SL (2009) Ultrafast and memory-efficient alignment of short DNA sequences to the human genome. Genome Biol 10:R25

31. Grehl C, Kuhlmann M, Becker C, Glaser B, Grosse I (2018) How to design a whole-genome bisulfite sequencing experiment. Epigenomes 2:21
32. Greaves IK, Groszmann M, Ying H, Taylor JM, Peacock WJ, Dennis ES (2012) Trans chromosomal methylation in Arabidopsis hybrids. Proc Natl Acad Sci U S A 109:3570–3575
33. Becker C, Hagmann J, Müller J, Koenig D, Stegle O, Borgwardt K, Weigel D (2011) Spontaneous epigenetic variation in the *Arabidopsis thaliana* methylome. Nature 480:245–254
34. Bhatia H, Khemka N, Jain M, Garg R (2018) Genome-wide bisulfite-sequencing reveals organ-specific methylation patterns in chickpea. Sci Rep 8:9704
35. King GJ (2015) Crop epigenetics and the molecular hardware of genotype × environment interactions. Front Plant 6:968
36. He Y, Li Z (2018) Epigenetic environmental memories in plants: establishment, maintenance, and reprogramming. Trends Genet 34:856–866
37. Garg R, Narayana Chevala V, Shankar R, Jain M (2015) Divergent DNA methylation patterns associated with gene expression in rice cultivars with contrasting drought and salinity stress response. Sci Rep 5:14922
38. Boyko A, Kathiria P, Zemp FJ, Yao Y, Pogribny I, Kovalchuk I (2007) Transgenerational changes in the genome stability and methylation in pathogen-infected plants: (virus-induced plant genome instability). Nucleic Acids Res 5:1714–1725
39. Miura A, Yonebayashi S, Watanabe K, Toyama T, Shimada H, Kakutani T (2001) Mobilization of transposons by a mutation abolishing full DNA methylation in Arabidopsis. Nature 411:212–214

Chapter 7

# Method for Genome-Wide Association Study: A Soybean Example

**Robyn Anderson, Cassandria Tay Fernandez, Yuxuan Yuan, Agnieszka A. Golicz, David Edwards, and Philipp E. Bayer**


## Abstract

Genome-wide association studies (GWAS) are a valuable approach to identify single nucleotide polymorphisms (SNPs) associated with a phenotype of interest. There are now a variety of R-packages and command line tools available to perform GWAS. Here, we provide an example downloading and filtering SNP data, followed by GWAS analysis using the R-package rMVP.


**Key words** Genome Wide Association Study, Soybean, vcf, rMVP, R, bcftools

## 1 Introduction

Genome-wide association studies (GWAS) are a common approach to identify genomic variants (usually single nucleotide polymorphisms (SNPs)) associated with phenotypes of interest. GWAS involve the comparison of SNPs either between two groups of individuals (a case/control study) or individuals differing in a quantitative trait. Agronomic traits such as seed number and weight, oil composition, and flowering time are quantitative and these traits are often influenced by multiple genes. GWAS has the power to identify SNPs/genomic regions associated with traits which makes it a useful tool for understanding the genetic basis of complex traits. In many cases, the associated markers cosegregate with or are located within candidate causal genes, which enhances our understanding of the molecular basis of trait variation.

GWAS has become a popular approach for exploring the genetic architecture of complex traits including yield and stress tolerance. For example, a GWAS conducted using 211 chickpea individuals identified 22 loci associated with pod and seed number [1], and a GWAS in *Medicago truncatula* using 132 accessions identified 12 genomic regions associated with salinity responsive

Mukesh Jain and Rohini Garg (eds.), *Legume Genomics: Methods and Protocols*, Methods in Molecular Biology, vol. 2107, https://doi.org/10.1007/978-1-0716-0235-5_7, 

[2]. A GWAS of 809 soybean accessions identified 249 loci significantly associated with traits including yield, seed color and composition, growth period, and plant architecture [3]. Additionally, evidence of epistasis between two genes influencing plant height was detected, and several loci were found to influence more than one trait.

There are several approaches and packages to carry out GWAS, including PLINK2 [4], TASSEL [5], GAPIT 3 [6], and rMVP (https://github.com/XiaoleiLiuBio/rMVP). PLINK/PLINK2 is a command-line tool which focuses on SNP data management and filtering, but also offers relatively simple approaches to SNP–phenotype associations. TASSEL is especially suited to programming beginners as it can be run using a graphical user interface with many functions, from calculating kinship matrices or SNP-based phylogenies, to genome-wide association studies using MLM or GLM. GAPIT and rMVP are both R packages which, in one command, perform an entire genome-wide association study using a filtered set of SNPs, including plotting of all relevant plots such as Manhattan or Q-Q-plots.

These programs implement different statistical approaches to determine genomic regions and phenotype associations, which most commonly include generalized linear models (GLMs), mixed linear models (MLMs), and FarmCPU [7], however other technologies such as SUPER [8] are also in use (*see* Table 1).

In GWAS, uncontrolled population structure can lead to false-positive SNP associations, as SNPs that are more abundant in sub-populations due to population history may appear to be linked to a phenotype associated with that sub-population. To control for population structure, it is necessary to use special covariates such as principal components (PCs) and a kinship matrix as covariates in the GWAS. PCs and the kinship matrix can be calculated using many different software packages and are part of the GWAS pipelines TASSEL, GAPIT, and rMVP presented here.

## 2 Materials

### 2.1 File Retrieval

These steps are run on the command line (Linux or MacOS). This step will download the publicly available vcf data from https://soybase.org/snps/ (*see* **Note 1**). The vcf data contains SNP information for 20,087 soybean accessions.

1. `wget https://soybase.org/snps/soysnp50k_wm82.a2_41317.vcf.gz`

The tool gunzip is used to extract the gzip-compressed vcf.

2. `gunzip soysnp50k_wm82.a2_41317.vcf.gz`

**Table 1**
**Comparison of commonly used GWAS software packages**

| Package name | Statistical approaches implemented | User interface | Data management functionality | Plotting support |
|---|---|---|---|---|
| PLINK2 | MLM | Command line | Yes | No |
| TASSEL | MLM, GLM | Graphical user interface | Yes | Yes |
| GAPIT 3 | MLM, GLM, CMLM, ECMLM, FaST-LMM, SUPER, MLMM, FarmCPU | R package | No | Yes |
| rMVP | MLM, GLM, FarmCPU | R package | No | Yes |

### 2.2 Bioinformatics Tools and Scripts

1. bcftools (http://www.htslib.org/doc/bcftools.html) is a flexible data management program to handle SNPs and Indels called in VCF or BCF format [9].
2. rMVP (https://github.com/XiaoleiLiuBio/rMVP) is a R library which performs genome-wide association studies and plots the results.
3. bedtools (https://bedtools.readthedocs.io/en/latest/) is a versatile tool to compare different ranges of genomic features [10].
4. SnpEff (http://snpeff.sourceforge.net/) is a Java tool which predicts changes in protein sequences based on SNPs in coding regions [11].

This pipeline can be run on a single workstation with moderate memory requirements (at least 10GB).

## 3 Methods

This chapter presents the workflow from SNP filtering, to carrying out a GWAS using rMVP, to search for candidate genes based on proximity, and to predict effects of SNPs in coding regions (Fig. 1).

### 3.1 Quality Control

It is common to filter SNPs by Minor Allele Frequency (MAF), the number of reads supporting the SNP (Depth or DP), and SNP quality as reported by the SNP calling program (QUAL). SNPs with quality scores lower than 20, a number of reads aligning (depth) lower than 10, and a minor allele frequency below 1% are removed. The –q flag is used for quality control, in this case, each SNP's minor allele frequency has to be more than 0.05 (5%). The –i flag denotes SNPs that should be included in the output, in this case, SNPs with a quality above 20 and a depth above 10 (*see*

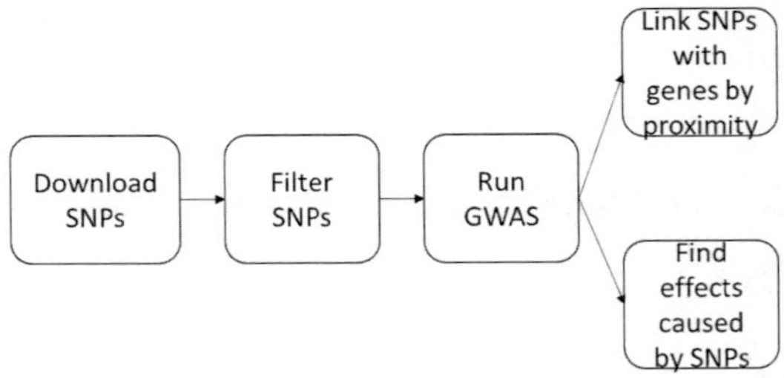


**Fig. 1** Workflow of GWAS and candidate gene search presented in this chapter

**Notes 2** and **3**). The output of bcftools view is redirected to a new file called *soysnp50k_wm82.a2_41317.filtered.vcf.*

```
1. bcftools view -q 0.05:minor -i 'QUAL>20 && INFO/DP>10' soysnp50k_wm82.a2_41317.vcf > soysnp50k_wm82.a2_41317.filtered.vcf
```

If there are any outlier individuals, it is possible to use *bcftools view* to remove these outliers from the original vcf file. Bcftools is run on the command line, and for example, to remove the individual AC-Bravor, perhaps because we miss phenotyping data for this individual, or because it is an outlier (*see* **Note 4**). The –s flag is used to include or exclude individuals, excluded individuals are prefixed with a "^".

```
2. bcftools view -s ^AC-Bravor soysnp50k_wm82.a2_41317.filtered.vcf > soysnp50k_wm82.a2_41317.filtered.noAC-Bravor.vcf
```

Similarly, it is possible to use bcftools to isolate SNPs located within specific regions. The –t flag is used to extract SNPs from a given region, in this case, 1–10,000 bp on pseudomolecule Gm01, with the extracted SNPs moved to the file *soysnp50k_wm82.a2_41317.filtered.Gm01_1_10000.vcf.*

```
3. bcftools view -t Gm01:1-10000 soysnp50k_wm82.a2_41317.filtered.vcf > soysnp50k_wm82.a2_41317.filtered.Gm01_1_10000.vcf
```

### 3.2 GWAS Using rMVP

rMVP takes as input a vcf file and a tab-delimited text file containing quantitative or qualitative phenotypic data of the accessions of interest. The following commands are run in R. First, the vcf file has to be transformed into the rMVP format. The parameter *filePhe* describes the path of the file of phenotypes, *fileKin* describes the path of an optional file with the kinship matrix, and *filePC* describes the path to the file with optional principal components. If *fileKin* and *filePC* are set to FALSE as in this case, both are calculated by rMVP. This step assumes that we have a tab-delimited file of phenotypes of interest, have filtered SNPs in the file "soysnp50k_wm82.a2_41317.filtered", have no file of kinship or principal components, and that the output files should be prefixed with "mvp.vcf".

```
1. MVP.Data(fileVCF=" soysnp50k_wm82.a2_41317.filtered.vcf",
 filePhe="Phenotype.txt",
 fileKin=FALSE,
 filePC=FALSE,
 out="mvp.vcf"
 )
```

This step generates several output files starting with "mvp.vcf". These files have to be loaded into R. The following three commands will load the genotype and phenotype table into memory for subsequent steps.

```
2. genotype <- attach.big.matrix("mvp.vcf.geno.desc")
phenotype <- read.table("mvp.vcf.phe",head=TRUE)
map <- read.table("mvp.vcf.map" , head = TRUE)
```

Then, to perform the GWAS for one phenotype. rMVP takes three files as input which were prepared above: *phe* is the name of the phenotype table, *geno* is the name of the genotyping table, and *map* is the name of the table of SNP positions. These three files are generated by rMVP from the input data. This command will perform a PC analysis, a kinship analysis, and three GWAS methods using the following statistical models as set in the *method* parameter: Generalized Linear Model (GLM), a Mixed Linear Model (MLM), and FarmCPU. For the GLM it will use five PC as covariates (*nPC.GLM*), and for the MLM and FarmCPU it will use three PCs as covariates (*nPC.MLM*, *nPC.FarmCPU*). The statistical significance cut off is set as the *p*-value threshold of 5% (0.05) divided by the number of SNPs (parameter *threshold*). rMVP will use 100% (parameter *perc*) of SNPs for the PCA analysis. The parameters *maxLoop* and *method.bin* control the false discovery rate of FarmCPU. The parameter *priority* can be set to "memory" or "speed," where "speed" consumes more memory but is faster, and "memory" consumes less memory at the cost of speed, which is suitable for work stations with small amounts of memory. The parameter *ncpus* sets the number of CPUs used for the analysis, by default all CPUs are used.

```
3. imMVP <- MVP(
 phe=phenotype,
 geno=genotype,
 map=map,
 nPC.GLM=5,
 nPC.MLM=3,
 nPC.FarmCPU=3,
 perc=1,
 priority="speed",
 ncpus=10,
 vc.method="EMMA",
 maxLoop=10,
 method.bin="FaST-LMM",#"FaST-LMM","EMMA", "static"
 threshold=0.05,
 method=c("GLM", "MLM", "FarmCPU")
)
```

This step will run for several hours or days, after which rMVP will have generated all output files necessary (*see* **Notes 5–7**).

### 3.3 Searching for Gene Overlaps

For each phenotype, rMVP produces one table of output for each statistical model (GLM, MLM, and FarmCPU) containing *p*-values, and an effect size of each SNP. Bedtools [10] is used to search for overlap between genes and the candidate SNPs as reported by rMVP. Figure 2 shows an example of the table of *p*-values as reported by rMVP. First, quotation marks in the rMVP output need to be replaced. The flag –i means that the replacement happens in-place (no new file is generated), s stands for “substitute,” and substitute quotation marks (“) are substituted with nothing (*see* **Note 8**).

```
1. sed -i 's/"//g' MVP.seedsize.FarmCPU.csv
```

Then, sort by *p*-value and keep only the ten SNPs with the lowest *p*-value. The –k flag denotes the column on which the table should be sorted—in this case, the fifth column is to be sorted numerically (*n*). The –t flag tells *sort* how the input table is delimited, in this case, the table is comma-delimited. The output of the sorting is piped into *head*, which takes only the first ten rows (*-10*).

```
2. sort -k 5n -t ',' MVP.seedsize.FarmCPU.csv | head -10 >
top10_seedsize_hits.csv
```

```
"SNP","CHROM","POS","effect","seedweight.FarmCPU"
"Gm01_1","Gm01","        1","0.00426297061517689","0.744828773484174"
"Gm01_499","Gm01","      499","-0.0105687113729187","0.471805273694811"
"Gm01_873","Gm01","      873","-0.0284574927834433","0.525282832695838"
"Gm01_950","Gm01","      950","-0.00679152832562124","0.713460549135325"
"Gm01_1212","Gm01","     1212","0.0277090590493785","0.203677060100663"
"Gm01_1384","Gm01","     1384","-0.00117330979335106","0.949382628645025"
"Gm01_1424","Gm01","     1424","0.0059999976459828","0.739673958389251"
"Gm01_1452","Gm01","     1452","0.0181166105774593","0.515728652115829"
```

**Fig. 2** Output example of *p*-values and effect as reported by rMVP using FarmCP

```
SNP,CHROM,POS,effect,seedweight.FarmCPU
Gm01_14027,Gm01,     14027,0.0400685760383093,0.00565200017282935
Gm01_14031,Gm01,     14031,-0.0370355893566465,0.00740391089922914
Gm01_8287,Gm01,      8287,-0.0502699094286143,0.0134431895877045
Gm01_8201,Gm01,      8201,0.0293514531621071,0.0352831595716529
Gm01_8742,Gm01,      8742,0.0201535645197577,0.0596190948588376
Gm01_13336,Gm01,     13336,-0.0305847414808458,0.0708720857190618
Gm01_8286,Gm01,      8286,-0.0337330762191179,0.0750195057619282
Gm01_13629,Gm01,     13629,-0.0259942635011505,0.0956786280424518
Gm01_8193,Gm01,      8193,-0.0232367281397141,0.102322334333685
```

**Fig. 3** The ten SNPs with the lowest *p*-values

```
Gm01    14026   14028
Gm01    14030   14032
Gm01    8286    8288
Gm01    8200    8202
Gm01    8741    8743
Gm01    13335   13337
Gm01    8285    8287
Gm01    13628   13630
Gm01    8192    8194
```

**Fig. 4** Output example of the bed file with the positions of the ten SNPs with the lowest *p*-values

Figure 3 shows the content of *top10_seedsize_hits.csv*.

Then this table is reformatted into the bed format so that it can be used by bedtools. This will print the chromosome name (second column, \$2) and the position of the SNP into a new file. One base pair is subtracted from the start position (third column \$3) as the bed format starts at 0, while the MVP results start at 1 (*see* **Note 9**).

One base pair is added to the end position as in the bed format, the end position is excluded. The header line including CHROM is removed by using *grep –v* (remove all lines with hits).

```
3. awk -F ',' '{print $2"\t"$3-1"\t"$3+1}' top10_seedsize_-
hits.cv | grep -v CHROM > top10_seedsize_hits.bed
```

Figure 4 shows the resulting bed file.

The annotation of *Glycine max* Wm82.a2, the assembly on which the above vcf-file's positions are based, is available at SoyBase or at JGI. This step will generate a new file with all positions where the top 10 GWAS hits overlap with any feature reported in the gff3 file.

```
4. bedtools intersect -a top10_seedsize_hits.bed -b Wm82_anno-
tation.gff3 > Hits.bed
```

It is also possible to search for a window around the features reported in the gff3 file, by default 1000 bp upstream ("-l 1000") and 1000 bp downstream ("-r 1000"):

```
5. bedtools window -a top10_seedsize_hits.bed -b Wm82_annota-
tion.gff3 -l 1000 -r 1000 > Hits.bed
```

### 3.4 Identifying Changes Caused by Associated SNPs

Many SNPs located within genes and associated with phenotypes have been shown to cause synonymous or nonsynonymous amino acid changes in the protein the gene encodes. SnpEff [11] is one of the toolkits which can be used for this purpose. When working with plant genomes it is often necessary to first format the plant genome to the SnpEff format.

First, a new genome entry needs to be created in the *SnpEff.config* file in the SnpEff directory, for example this line for *Glycine max* Wm82.a2:

```
1. gm.wm82.a2.genome : Williams82
```

This new line in *snpEff.config* tells SnpEff what the prefix of all files is (*gm.wm82.a2*) to be added in a new *genome*, and that the long name of the new genome is *Williams82*.

Next, a new folder is created in the SnpEff data folder. The current working directory is changed to the data directory of SnpEff.

```
2. cd /path/to/snpEff/data/
```

A new folder is created which will contain the genome assembly and annotation—the name of the folder has to be identical to the identifier that was added to *snpEff.config*.

```
3. mkdir gm.wm82.a2
```

The current working directory is changed to the folder where the annotation of this assembly has to be stored using the name *genes.gtf*. This file has to be stored here by the user, either by using a file manager or *wget* as above.

```
4. cd gm.wm82.a2
```

This annotation has to be compressed using gzip.

```
5. gzip genes.gtf
```

Now the reference assembly itself needs to be stored in the data/genomes folder, so the current working directory is moved to the *genomes* folder in the *data* folder:

```
6. cd /path/to/snpEff/data/genomes
```

In this example, the assembly needs to be named *gm.wm82.a2.fa* as this is the name that was used in *snpEff.config*, and again compressed with gzip:

```
7. gzip gm.wm82.a2.fa
```

Lastly, the new SnpEff database has to be built from the assembly and annotation:

```
8. cd /path/to/snpEff
```

The current working directory is changed to snpEff's main directory, so that snpEff can be run. The flag –jar tells Java where the main program is located, in this case, a file called *snpEff.jar*. The –gtf22 flag tells snpEff that the annotation is in *gtf* format. The –v flag tells snpEff to report more details of the database building process. Lastly, SnpEff needs to know the name of the genome that was just prepared (*gm.wm82.a2*).

```
9. java -jar snpEff.jar build -gtf22 -v gm.wm82.a2
```

This step will take several minutes. It is advisable to watch SnpEff's output to scan for any errors.

Now SnpEff can be run with the above filtered vcf file. Again, Java has to know which .jar file is run using the –jar flag. The first argument is the name of the genome build and annotation with which the SNPs will be compared (*gm.wm82.a2*). The second argument is the name of the vcf file containing all SNPs (*soysnp50k_wm82.a2_41317.filtered.vcf*). This will create a new vcf file with a new "ANN" field in the eighth column. This field contains information as to whether the SNP in question is a silent, missense, nonsense, or no change at all, with more details as to the changed protein and the exact change.

```
10. java -jar snpEff.jar gm.wm82.a2 soysnp50k_wm82.a2_41317.
filtered.vcf > soysnp50k_wm82.a2_41317.filtered.annotated.vcf
```

This step will also generate a summary html and txt file—the txt file contains information as to how many changes of which type occurred in each gene, while the summary html file contains graphs summarizing the changes in all SNPs and genes.

## 4 Notes

1. On MacOS, *wget* is not installed by default, so it needs to be installed by the user using for example homebrew: *brew install wget*.
2. Quality and SNP frequency filtering helps remove variants which may be due to sequencing/mapping errors and/or are in very low abundance in the population. The parameters for SNP quality and read depth chosen here are relatively arbitrary and depend on many factors including depth of sequencing, quality of the sequencing reads, and more.
3. The way bcftools is used has changed several times over the last years. If the commands printed here were tried with bcftools v1.8, readers are advised to upgrade their bcftools version on encountering a "unrecognized command" error when running bcftools.
4. Outliers may be detected using principal components analysis (PCA). Outliers will be represented by points grouping away from the other individuals.
5. The parameters presented here are the default parameters taken from rMVP's manual. From these parameters, the number of principal components (PCs via the parameter *nPC*) has the largest effect on the outcome of the GWAS. If the user chooses too few PCs, residual population structure may be unaccounted for and false-positive SNP associations appear in the results. If the user chooses too many PCs, overcorrection will

happen and no SNPs with *p*-values below the significance cutoff may appear in the results. In most GWAS-publications, three PCs are chosen as covariates, but it is up to the user to try different PCs for their GWAS. The R-package *GAPIT* for performing GWAS contains a parameter *model.selection=-TRUE* which will evaluate different numbers of PCs and chooses the optimal number of PCs.

6. The GWAS software programs differ not only in the statistical methods applied but also in computational time required and dataset scalability. Often the statistical analysis to be performed, the size of the dataset, and the computational resources available all need to be considered while choosing the most appropriate GWAS software.
7. For long-running commands, it is advisable to learn how to use *screen* or *tmux* so that the user can disconnect from the server the command is running on.
8. When run on a Mac-based system *sed* behaves slightly different from a Linux-based system. On MacOS, an extension in the –i flag is mandatory: *sed –i.bak* will store the unchanged file in a new copy ending in *.bak*.
9. Conversion of SNP position to bed coordinate system is required before bedtools can be used. Bed coordinate system is 0 based, open ended.

## References


1. Kujur A, Bajaj D, Upadhyaya HD, Das S, Ranjan R, Shree T, Saxena MS, Badoni S, Kumar V, Tripathi S, Gowda CL, Sharma S, Singh S, Tyagi AK, Parida SK (2015) A genome-wide SNP scan accelerates trait-regulatory genomic loci identification in chickpea. Sci Rep 5:11166. https://doi.org/10.1038/srep11166
2. Kang Y, Torres-Jerez I, An Z, Greve V, Huhman D, Krom N, Cui Y, Udvardi M (2018) Genome-wide association analysis of salinity responsive traits in Medicago truncatula. Plant Cell Environ. https://doi.org/10.1111/pce.13508
3. Fang C, Ma Y, Wu S, Liu Z, Wang Z, Yang R, Hu G, Zhou Z, Yu H, Zhang M, Pan Y, Zhou G, Ren H, Du W, Yan H, Wang Y, Han D, Shen Y, Liu S, Liu T, Zhang J, Qin H, Yuan J, Yuan X, Kong F, Liu B, Li J, Zhang Z, Wang G, Zhu B, Tian Z (2017) Genome-wide association studies dissect the genetic networks underlying agronomical traits in soybean. Genome Biol 18(1):161. https://doi.org/10.1186/s13059-017-1289-9
4. Purcell S, Neale B, Todd-Brown K, Thomas L, Ferreira MA, Bender D, Maller J, Sklar P, de Bakker PI, Daly MJ, Sham PC (2007) PLINK: a tool set for whole-genome association and population-based linkage analyses. Am J Hum Genet 81(3):559–575. https://doi.org/10.1086/519795
5. Bradbury PJ, Zhang Z, Kroon DE, Casstevens TM, Ramdoss Y, Buckler ES (2007) TASSEL: software for association mapping of complex traits in diverse samples. Bioinformatics 23 (19):2633–2635. https://doi.org/10.1093/bioinformatics/btm308
6. Tang Y, Liu X, Wang J, Li M, Wang Q, Tian F, Su Z, Pan Y, Liu D, Lipka AE, Buckler ES, Zhang Z (2016) GAPIT version 2: an enhanced integrated tool for genomic association and prediction. Plant Genome 9(2). https://doi.org/10.3835/plantgenome2015.11.0120
7. Kusmec A, Schnable PS (2018) Farm CPU pp: efficient large-scale genomewide association studies. Plant Direct 2(4):e00053
8. Wang Q, Tian F, Pan Y, Buckler ES, Zhang Z (2014) A SUPER powerful method for

genome wide association study. PLoS One 9 (9):e107684. https://doi.org/10.1371/journal.pone.0107684
9. Li H (2011) A statistical framework for SNP calling, mutation discovery, association mapping and population genetical parameter estimation from sequencing data. Bioinformatics 27(21):2987–2993. https://doi.org/10.1093/bioinformatics/btr509
10. Quinlan AR, Hall IM (2010) BEDTools: a flexible suite of utilities for comparing genomic features. Bioinformatics 26(6):841–842. https://doi.org/10.1093/bioinformatics/btq033
11. Cingolani P, Platts A, Wang le L, Coon M, Nguyen T, Wang L, Land SJ, Lu X, Ruden DM (2012) A program for annotating and predicting the effects of single nucleotide polymorphisms, SnpEff: SNPs in the genome of Drosophila melanogaster strain w1118; iso-2; iso-3. Fly (Austin) 6(2):80–92. https://doi.org/10.4161/fly.19695

# Chapter 8

# Genotyping for Species Identification and Diversity Assessment Using Double-Digest Restriction Site-Associated DNA Sequencing (ddRAD-Seq)

**Anita A. Severn-Ellis, Armin Scheben, Ting Xiang Neik, Nur Shuhadah Mohd Saad, Aneeta Pradhan, and Jacqueline Batley**

## Abstract

Genotyping-by-sequencing (GBS) is a powerful approach for studying the genetic diversity of legume species. By using restriction enzymes or other methods to generate a reduced representation of the genome for sequencing, GBS can provide genome-wide single nucleotide polymorphisms (SNP) for diversity analysis at high throughput and low cost. Here we describe a novel double-digest restriction site-associated DNA sequencing (ddRAD-seq) approach. We also describe the downstream bioinformatic analysis of the sequencing data, including alignment to a reference genome, de novo assembly, SNP calling, phylogenetic analysis, and structure analysis.

**Key words** Genotyping-by-sequencing, GBS, Double-digest restriction site-associated DNA sequencing, ddRAD-seq, Whole-genome sequencing, Restriction enzyme, Single nucleotide polymorphism

## 1 Introduction

In the past two decades the molecular marker tools for legume species identification have evolved rapidly to meet the rising demand of crop improvement for global food security using next generation sequencing technologies. Current genomic advancement has contributed to the genome assemblies of at least ten legume crops [1] including soybean (*Glycine max*) [2], cultivated and wild peanuts (*Arachis hypogaea*, *Arachis monticola*, *Arachis duranensis*, *Arachis ipaensis*) [3–5], barrel medic (*Medicago truncatula*) [6–8], lotus (*Lotus japonicus*) [9], mung bean (*Vigna radiata*) [10], common bean (*Phaseolus vulgaris*) [11], pigeon pea (*Cajanus cajan*) [12–14], chickpea (*Cicer arietinum*) [15, 16], cowpea (*Vigna unguiculata*) [17], and alfalfa (*Medicago sativa*) [18].

Mukesh Jain and Rohini Garg (eds.), *Legume Genomics: Methods and Protocols*, Methods in Molecular Biology, vol. 2107, https://doi.org/10.1007/978-1-0716-0235-5_8, 

The availability and accessibility of these genome assemblies have brought the study of the legume species to the next level where the evolutionary history, candidate gene identification, marker–trait association, genetic diversity can all be studied within and between species at a whole-genome level [19]. When a reference genome is available, a straightforward approach for genotyping legume diversity panels or mapping populations would be whole genome resequencing, where each individual is sequenced at moderate coverage (5–15×). Single nucleotide polymorphisms (SNPs) are then called genome-wide by aligning reads to a reference genome. However, the cost of resequencing large panels of individuals at moderate coverage is high, particularly when the genome size is large, and reference genomes are not available for many legumes. An alternative to whole-genome resequencing is a suite of reduced representation genotyping approaches known under the umbrella term genotyping-by-sequencing (GBS), with one of the earliest specific methods used being restriction site-associated DNA sequencing (RAD-seq) [20].

In GBS, only part of the plant genome is sequenced, which is achieved through digestion of the genomic DNA into large amounts of DNA fragments using one or more restriction enzymes [21, 22]. GBS can offer the best of both worlds enabling high-resolution evaluation of genetic diversity, while at the same providing a more economical alternative to whole-genome resequencing [22]. This is particularly valid in the processing and genotyping of large populations [23] and has a further advantage that the SNP discovery using this approach do not necessarily require the availability of a sequenced reference genome.

In soybean, GBS was used to develop 10,120 high-quality SNPs [24]. The GBS method for genome-wide SNP discovery was also applied in 466 cultivated and 16 wild mung bean accessions by Noble et al. [25], who identified more than 22,000 polymorphic SNPs and five genomic regions associated with seed coat color [25]. In chickpea, 44,844 SNPs were discovered using GBS in 211 accessions, including 92 diverse cultivated desi and kabuli accessions, representing 21 countries [26]. The first SNP-based linkage map was generated from the cultivated peanut (*Arachis hypogaea*) using GBS [27]. Furthermore, GBS was also used to distinguish 11 peanut accessions covering seven cultivars and four wild accessions [28]. The study successfully uncovered 15,678 polymorphic SNPs that genetically clustered the wild accessions into two groups while 3248 polymorphic SNPs grouped the cultivated peanuts into three genetic clusters.

The GBS method described in this chapter is based on the "double-digest restriction site-associated DNA sequencing" or ddRAD-seq method adapted from the GBS methods described by Clark et al. [29], Elshire et al. [21], Fu et al. [30], and Peterson et al. [31, 32]. ddRAD-seq is a versatile SNP discovery and genotyping approach that does not require a reference genome sequence but utilizes two restriction enzymes simultaneously followed by size selection during library preparation [31].

## 2 Materials

### 2.1 DNA Starting Material

1. EB buffer (10 mM Tris–hydrochloride, pH 8.5).
2. Nuclease-free water (Qiagen, 129114).
3. Qubit dsDNA BR Assay Kit, 500 (Thermo Fisher Scientific/ Life Technologies, Q32853).
4. HT DNA gDNA reagents (PerkinElmer, CLS760685).
5. Qubit® 3.0 Fluorometer (Thermo Fisher Scientific, Q33216).
6. LabChip GX Touch Nucleic Acid Analyzer (PerkinElmer, CLS138162).

### 2.2 Adapter Preparation

1. TRIS hydrochloride, pH 8 (Sigma-Aldrich, 93363).
2. Ethylenediaminetetraacetic acid (EDTA) (Sigma-Aldrich, 03677).
3. Nuclease-free water (Qiagen, 129114).
4. 10× Annealing buffer: 1 mL 1 M Tris–HCl, pH 8 (100 mM final concentration), 1 mL 5 M NaCl (500 mM final concentration), 0.2 mL 0.5 M EDTA (10 mM final concentration), 7.8 mL nuclease-free water. Aliquot and autoclave the stock solution before use. The remainder of the stock solution can be stored at −20 °C for later use.
5. Adapter oligonucleotides (*see* Table 3, **Note 1**)
6. 8-SnapStrip 0.2 mL PCR tubes with caps (SSI, 3245-00).
7. Thermal cycler.
8. Benchtop microcentrifuge.

### 2.3 Preparation of Adapter Working Solution

1. 8-SnapStrip 0.2 mL PCR tubes with caps (SSI, 3245-00).
2. Nuclease-free water (Qiagen, 129114).
3. 1× Annealing buffer: 1 mL 10× Annealing buffer (prepared above), 9 mL Nuclease-free water.

### 2.4 Restriction Digestion

1. HpyCH4IV Restriction enzyme (New England Biotechnologies (NEB), R619L).
2. HinfI Restriction enzyme (NEB, R0155L).
3. Nuclease-free water (Qiagen, 129114).
4. 8-SnapStrip 0.2 mL PCR tubes with caps (SSI, 3245-00) or 96 well Semi-Skirted PCR Plates (SSI, 3421-00).
5. Benchtop microcentrifuge or tabletop centrifuge to spin plates.
6. Thermal cycler.

**Table 3**
**Adapter and primer oligonucleotides**

| Restriction enzyme | Restriction site |
|---|---|
| HpyCH4VI | A/CGT |
| HinfI | G/ANTC |
| *Indexed primers* | |
| PCR1 | AATGATACGGCGACCACCGAGATCTACACTCTTTCCCTACACGACG |
| PCR2_Idx_1_ATCACG | CAAGCAGAAGACGGCATACGAGAT**CGTGAT**GTGACTGGAGTTCAGACGTGTGC |
| PCR2_Idx_2_CGATGT | CAAGCAGAAGACGGCATACGAGAT**ACATCG**GTGACTGGAGTTCAGACGTGTGC |
| PCR2_Idx_3_TTAGGC | CAAGCAGAAGACGGCATACGAGAT**GCCTAA**GTGACTGGAGTTCAGACGTGTGC |
| PCR2_Idx_4_TGACCA | CAAGCAGAAGACGGCATACGAGAT**TGGTCA**GTGACTGGAGTTCAGACGTGTGC |
| PCR2_Idx_5_ACAGTG | CAAGCAGAAGACGGCATACGAGAT**CACTGT**GTGACTGGAGTTCAGACGTGTGC |
| PCR2_Idx_6_GCCAAT | CAAGCAGAAGACGGCATACGAGAT**ATTGGC**GTGACTGGAGTTCAGACGTGTGC |
| PCR2_Idx_7_CAGATC | CAAGCAGAAGACGGCATACGAGAT**GATCTG**GTGACTGGAGTTCAGACGTGTGC |
| PCR2_Idx_8_ACTTGA | CAAGCAGAAGACGGCATACGAGAT**TCAAGT**GTGACTGGAGTTCAGACGTGTGC |
| PCR2_Idx_9_GATCAG | CAAGCAGAAGACGGCATACGAGAT**CTGATC**GTGACTGGAGTTCAGACGTGTGC |
| PCR2_Idx_10_TAGCTT | CAAGCAGAAGACGGCATACGAGAT**AAGCTA**GTGACTGGAGTTCAGACGTGTGC |
| PCR2_Idx_11_GGCTAC | CAAGCAGAAGACGGCATACGAGAT**GTAGCC**GTGACTGGAGTTCAGACGTGTGC |
| PCR2_Idx_12_CTTGTA | CAAGCAGAAGACGGCATACGAGAT**TACAAG**GTGACTGGAGTTCAGACGTGTGC |
| *Common adapter* | |
| HinfI_P2.1 (common adapter) | GTGACTGGAGTTCAGACGTGTGCTCTTCCGATCT |
| HinfI_P2.2 (common adapter) | [Phos]ANTAGATCGGAAGAGCGAGAACAA |
| *Barcoded adapters* | |

| | |
|---|---|
| 1-GCATG_ HpyCHIV_P1.1 | ACACTCTTTCCCTACACGACGCTCTTCCGATCTGCATG |
| 2-AACCA_HpyCHIV_P1.1 | ACACTCTTTCCCTACACGACGCTCTTCCGATCTAACCA |
| 3-CGATC_HpyCHIV_P1.1 | ACACTCTTTCCCTACACGACGCTCTTCCGATCTCGATC |
| 4-TCGAT_HpyCHIV_P1.1 | ACACTCTTTCCCTACACGACGCTCTTCCGATCTTCGAT |
| 5-TGCAT_HpyCHIV_P1.1 | ACACTCTTTCCCTACACGACGCTCTTCCGATCTTGCAT |
| 6-CAACC_HpyCHIV_P1.1 | ACACTCTTTCCCTACACGACGCTCTTCCGATCTCAACC |
| 7-GGTTG_HpyCHIV_P1.1 | ACACTCTTTCCCTACACGACGCTCTTCCGATCTGGTTG |
| 8-AAGGA_HpyCHIV_P1.1 | ACACTCTTTCCCTACACGACGCTCTTCCGATCTAAGGA |
| 9-AGCTA_HpyCHIV_P1.1 | ACACTCTTTCCCTACACGACGCTCTTCCGATCTAGCTA |
| 10-ACACA_HpyCHIV_P1.1 | ACACTCTTTCCCTACACGACGCTCTTCCGATCTACACA |
| 11-AATTA_HpyCHIV_P1.1 | ACACTCTTTCCCTACACGACGCTCTTCCGATCTAATTA |
| 12-ACGGT_HpyCHIV_P1.1 | ACACTCTTTCCCTACACGACGCTCTTCCGATCTACGGT |
| 13-ACTGG_HpyCHIV_P1.1 | ACACTCTTTCCCTACACGACGCTCTTCCGATCTACTGG |
| 14-ACTTC_HpyCHIV_P1.1 | ACACTCTTTCCCTACACGACGCTCTTCCGATCTACTTC |
| 15-ATACG_HpyCHIV_P1.1 | ACACTCTTTCCCTACACGACGCTCTTCCGATCTATACG |
| 16-ATGAG_HpyCHIV_P1.1 | ACACTCTTTCCCTACACGACGCTCTTCCGATCTATGAG |
| 17-ATTAC_HpyCHIV_P1.1 | ACACTCTTTCCCTACACGACGCTCTTCCGATCTATTAC |
| 18-CATAT_HpyCHIV_P1.1 | ACACTCTTTCCCTACACGACGCTCTTCCGATCTCATAT |
| 19-CGAAT_HpyCHIV_P1.1 | ACACTCTTTCCCTACACGACGCTCTTCCGATCTCGAAT |
| 20-CGGCT_HpyCHIV_P1.1 | ACACTCTTTCCCTACACGACGCTCTTCCGATCTCGGCT |
| 21-CGGTA_HpyCHIV_P1.1 | ACACTCTTTCCCTACACGACGCTCTTCCGATCTCGGTA |
| 22-CGTAC_HpyCHIV_P1.1 | ACACTCTTTCCCTACACGACGCTCTTCCGATCTCGTAC |
| 23-CGTCG_HpyCHIV_P1.1 | ACACTCTTTCCCTACACGACGCTCTTCCGATCTCGTCG |

(continued)

**Table 3 (continued)**

| Restriction enzyme | Restriction site |
|---|---|
| 24-CTGAT_HpyCHIV_P1.1 | ACACTCTTTCCCTACACGACGCTCTTCCGATCTCTGAT |
| 25-CTGCG_HpyCHIV_P1.1 | ACACTCTTTCCCTACACGACGCTCTTCCGATCTCTGCG |
| 26-CTGTC_HpyCHIV_P1.1 | ACACTCTTTCCCTACACGACGCTCTTCCGATCTCTGTC |
| 27-CTTGG_HpyCHIV_P1.1 | ACACTCTTTCCCTACACGACGCTCTTCCGATCTCTTGG |
| 28-GACAC_HpyCHIV_P1.1 | ACACTCTTTCCCTACACGACGCTCTTCCGATCTGACAC |
| 29-GAGAT_HpyCHIV_P1.1 | ACACTCTTTCCCTACACGACGCTCTTCCGATCTGAGAT |
| 30-GAGTC_HpyCHIV_P1.1 | ACACTCTTTCCCTACACGACGCTCTTCCGATCTGAGTC |
| 31-GCCGT_HpyCHIV_P1.1 | ACACTCTTTCCCTACACGACGCTCTTCCGATCTGCCGT |
| 32-GCTGA_HpyCHIV_P1.1 | ACACTCTTTCCCTACACGACGCTCTTCCGATCTGCTGA |
| 33-GGATA_HpyCHIV_P1.1 | ACACTCTTTCCCTACACGACGCTCTTCCGATCTGGATA |
| 34-GGCCA_HpyCHIV_P1.1 | ACACTCTTTCCCTACACGACGCTCTTCCGATCTGGCCA |
| 35-GGCTC_HpyCHIV_P1.1 | ACACTCTTTCCCTACACGACGCTCTTCCGATCTGGCTC |
| 36-GTAGT_HpyCHIV_P1.1 | ACACTCTTTCCCTACACGACGCTCTTCCGATCTGTAGT |
| 37-GTCCG_HpyCHIV_P1.1 | ACACTCTTTCCCTACACGACGCTCTTCCGATCTGTCCG |
| 38-GTCGA_HpyCHIV_P1.1 | ACACTCTTTCCCTACACGACGCTCTTCCGATCTGTCGA |
| 39-TACCG_HpyCHIV_P1.1 | ACACTCTTTCCCTACACGACGCTCTTCCGATCTTACCG |
| 40-TACGT_HpyCHIV_P1.1 | ACACTCTTTCCCTACACGACGCTCTTCCGATCTTACGT |
| 41-TAGTA_HpyCHIV_P1.1 | ACACTCTTTCCCTACACGACGCTCTTCCGATCTTAGTA |
| 42-TATAC_HpyCHIV_P1.1 | ACACTCTTTCCCTACACGACGCTCTTCCGATCTTATAC |
| 43-TCACG_HpyCHIV_P1.1 | ACACTCTTTCCCTACACGACGCTCTTCCGATCTTCACG |

| | |
|---|---|
| 44-TCAGT_HpyCHIV_P1.1 | ACACTCTTTCCCTACACGACGCTCTTCCGATCTTCAGT |
| 45-TCCGG_HpyCHIV_P1.1 | ACACTCTTTCCCTACACGACGCTCTTCCGATCTTCCGG |
| 46-TCTGC_HpyCHIV_P1.1 | ACACTCTTTCCCTACACGACGCTCTTCCGATCTTCTGC |
| 47-TGGAA_HpyCHIV_P1.1 | ACACTCTTTCCCTACACGACGCTCTTCCGATCTTGGAA |
| 48-TTACC_HpyCHIV_P1.1 | ACACTCTTTCCCTACACGACGCTCTTCCGATCTTTACC |
| 1-GCATG_HpyCHIV_P1.2 | [Phos]CGCATGCAGATCGGAAGAGCGTCGTGTAGGGAAAGAGTGT |
| 2-AACCA_HpyCHIV_P1.2 | [Phos]CGTGGTTAGATCGGAAGAGCGTCGTGTAGGGAAAGAGTGT |
| 3-CGATC_HpyCHIV_P1.2 | [Phos]CGGATCGAGATCGGAAGAGCGTCGTGTAGGGAAAGAGTGT |
| 4-TCGAT_HpyCHIV_P1.2 | [Phos]CGATCGAAGATCGGAAGAGCGTCGTGTAGGGAAAGAGTGT |
| 5-TGCAT_HpyCHIV_P1.2 | [Phos]CGATGCAAGATCGGAAGAGCGTCGTGTAGGGAAAGAGTGT |
| 6-CAACC_HpyCHIV_P1.2 | [Phos]CGGGTTGAGATCGGAAGAGCGTCGTGTAGGGAAAGAGTGT |
| 7-GGTTG_HpyCHIV_P1.2 | [Phos]CGCAACCAGATCGGAAGAGCGTCGTGTAGGGAAAGAGTGT |
| 8-AAGGA_HpyCHIV_P1.2 | [Phos]CGTCCTTAGATCGGAAGAGCGTCGTGTAGGGAAAGAGTGT |
| 9-AGCTA_HpyCHIV_P1.2 | [Phos]CGTAGCTAGATCGGAAGAGCGTCGTGTAGGGAAAGAGTGT |
| 10-ACACA_HpyCHIV_P1.2 | [Phos]CGTGTGTAGATCGGAAGAGCGTCGTGTAGGGAAAGAGTGT |
| 11-AATTA_HpyCHIV_P1.2 | [Phos]CGTAATTAGATCGGAAGAGCGTCGTGTAGGGAAAGAGTGT |
| 12-ACGGT_HpyCHIV_P1.2 | [Phos]CGACCGTAGATCGGAAGAGCGTCGTGTAGGGAAAGAGTGT |
| 13-ACTGG_HpyCHIV_P1.2 | [Phos]CGCCAGTAGATCGGAAGAGCGTCGTGTAGGGAAAGAGTGT |
| 14-ACTTC_HpyCHIV_P1.2 | [Phos]CGGAAGTAGATCGGAAGAGCGTCGTGTAGGGAAAGAGTGT |
| 15-ATACG_HpyCHIV_P1.2 | [Phos]CGCGTATAGATCGGAAGAGCGTCGTGTAGGGAAAGAGTGT |
| 16-ATGAG_HpyCHIV_P1.2 | [Phos]CGCTCATAGATCGGAAGAGCGTCGTGTAGGGAAAGAGTGT |
| 17-ATTAC_HpyCHIV_P1.2 | [Phos]CGGTAATAGATCGGAAGAGCGTCGTGTAGGGAAAGAGTGT |
| 18-CATAT_HpyCHIV_P1.2 | [Phos]CGATATGAGATCGGAAGAGCGTCGTGTAGGGAAAGAGTGT |

(continued)

**Table 3**
**(continued)**

| Restriction enzyme | Restriction site |
|---|---|
| 19-CGAAT_HpyCHIV_P1.2 | [Phos]CGATTCGAGATCGGAAGAGCGTCGTGTAGGGAAAGAGTGT |
| 20-CGGCT_HpyCHIV_P1.2 | [Phos]CGAGCCGAGATCGGAAGAGCGTCGTGTAGGGAAAGAGTGT |
| 21-CGGTA_HpyCHIV_P1.2 | [Phos]CGTACCGAGATCGGAAGAGCGTCGTGTAGGGAAAGAGTGT |
| 22-CGTAC_HpyCHIV_P1.2 | [Phos]CGGTACGAGATCGGAAGAGCGTCGTGTAGGGAAAGAGTGT |
| 23-CGTCG_HpyCHIV_P1.2 | [Phos]CGCGACGAGATCGGAAGAGCGTCGTGTAGGGAAAGAGTGT |
| 24-CTGAT_HpyCHIV_P1.2 | [Phos]CGATCAGAGATCGGAAGAGCGTCGTGTAGGGAAAGAGTGT |
| 25-CTGCG_HpyCHIV_P1.2 | [Phos]CGCGCAGAGATCGGAAGAGCGTCGTGTAGGGAAAGAGTGT |
| 26-CTGTC_HpyCHIV_P1.2 | [Phos]CGGACAGAGATCGGAAGAGCGTCGTGTAGGGAAAGAGTGT |
| 27-CTTGG_HpyCHIV_P1.2 | [Phos]CGCCAAGAGATCGGAAGAGCGTCGTGTAGGGAAAGAGTGT |
| 28-GACAC_HpyCHIV_P1.2 | [Phos]CGGTGTCAGATCGGAAGAGCGTCGTGTAGGGAAAGAGTGT |
| 29-GAGAT_HpyCHIV_P1.2 | [Phos]CGATCTCAGATCGGAAGAGCGTCGTGTAGGGAAAGAGTGT |
| 30-GAGTC_HpyCHIV_P1.2 | [Phos]CGGACTCAGATCGGAAGAGCGTCGTGTAGGGAAAGAGTGT |
| 31-GCCGT_HpyCHIV_P1.2 | [Phos]CGACGGCAGATCGGAAGAGCGTCGTGTAGGGAAAGAGTGT |
| 32-GCTGA_HpyCHIV_P1.2 | [Phos]CGTCAGCAGATCGGAAGAGCGTCGTGTAGGGAAAGAGTGT |
| 33-GGATA_HpyCHIV_P1.2 | [Phos]CGTATCCAGATCGGAAGAGCGTCGTGTAGGGAAAGAGTGT |
| 34-GGCCA_HpyCHIV_P1.2 | [Phos]CGTGGCCAGATCGGAAGAGCGTCGTGTAGGGAAAGAGTGT |
| 35-GGCTC_HpyCHIV_P1.2 | [Phos]CGGAGCCAGATCGGAAGAGCGTCGTGTAGGGAAAGAGTGT |
| 36-GTAGT_HpyCHIV_P1.2 | [Phos]CGACTACAGATCGGAAGAGCGTCGTGTAGGGAAAGAGTGT |
| 37-GTCCG_HpyCHIV_P1.2 | [Phos]CGCGGACAGATCGGAAGAGCGTCGTGTAGGGAAAGAGTGT |
| 38-GTCGA_HpyCHIV_P1.2 | [Phos]CGTCGACAGATCGGAAGAGCGTCGTGTAGGGAAAGAGTGT |

| | |
|---|---|
| 39-TACCG_HpyCHIV_P1.2 | [Phos]CGCGGTAAGATCGGAAGAGCGTCGTGTAGGGAAAGAGTGT |
| 40-TACGT_HpyCHIV_P1.2 | [Phos]CGACGTAAGATCGGAAGAGCGTCGTGTAGGGAAAGAGTGT |
| 41-TAGTA_HpyCHIV_P1.2 | [Phos]CGTACTAAGATCGGAAGAGCGTCGTGTAGGGAAAGAGTGT |
| 42-TATAC_HpyCHIV_P1.2 | [Phos]CGGTATAAGATCGGAAGAGCGTCGTGTAGGGAAAGAGTGT |
| 43-TCACG_HpyCHIV_P1.2 | [Phos]CGCGTGAAGATCGGAAGAGCGTCGTGTAGGGAAAGAGTGT |
| 44-TCAGT_HpyCHIV_P1.2 | [Phos]CGACTGAAGATCGGAAGAGCGTCGTGTAGGGAAAGAGTGT |
| 45-TCCGG_HpyCHIV_P1.2 | [Phos]CGCCGGAAGATCGGAAGAGCGTCGTGTAGGGAAAGAGTGT |
| 46-TCTGC_HpyCHIV_P1.2 | [Phos]CGGCAGAAGATCGGAAGAGCGTCGTGTAGGGAAAGAGTGT |
| 47-TGGAA_HpyCHIV_P1.2 | [Phos]CGTTCCAAGATCGGAAGAGCGTCGTGTAGGGAAAGAGTGT |
| 48-TTACC_HpyCHIV_P1.2 | [Phos]CGGGTAAAGATCGGAAGAGCGTCGTGTAGGGAAAGAGTGT |

### 2.5 Ligation of Adapters

1. T4 DNA ligase 1 U/μL (Thermo Fisher Scientific, 15224090).
2. Nuclease-free water (Qiagen, 129114).
3. VistaLab Technologies Reagent reservoirs (Fischer Scientific, 21-381-27E).
4. Multichannel pipette.
5. Benchtop microcentrifuge or tabletop centrifuge to spin plates.
6. Thermal cycler.

### 2.6 Sample Cleanup and Double Size Selection

1. Ethanol 200 proof, for molecular biology (Sigma-Aldrich, E7023-1L).
2. Nuclease-free water (Qiagen, 129114).
3. Polyethylene glycol (PEG) 6000 (Sigma-Aldrich, 81189).
4. Sodium Chloride (NaCl) (Sigma-Aldrich, S3014).
5. Agencourt AMPure XP Beads (Beckman Coulter, A63881).
6. PEG buffer: (20% PEG w/v, 2.5 M NaCl): 10 g PEG 6000, 7.3 g NaCl, top up to 50 mL with nuclease-free water. Aliquot into smaller volumes and sterilize the PEG buffer before use.
7. 80% Ethanol: 10 mL nuclease-free water, 40 mL ethanol.
8. Semi-Skirted PCR Plates (SSI, 3421-00).
9. Adhesive Microseal® 'B' PCR Plate Sealing Film (Bio-Rad, MSB1001).
10. 8-SnapStrip 0.2 mL PCR tubes with caps (SSI, 3245-00).
11. Multichannel pipette.
12. Magnetic Stand-96 (Thermo Fisher Scientific/Invitrogen, AM10027).

### 2.7 PCR Amplification of Libraries

1. Indexed forward and reverse PCR primers (*see* Table 3, **Note 1**).
2. Phusion Hot Start II High-Fidelity (Thermo Fisher Scientific, F565L).
3. Nuclease-free water (Qiagen, 129114).
4. VistaLab Technologies Reagent reservoirs (Fischer Scientific, 21-381-27E).
5. Adhesive Microseal® 'B' PCR Plate Sealing Film (Bio-Rad, MSB1001).
6. Multichannel pipette.
7. Benchtop microcentrifuge or tabletop centrifuge to spin plates.
8. Thermal cycler.

### 2.8 Ampure XP Bead Cleaning of Post-PCR Samples

1. Ethanol 200 proof, for molecular biology (Sigma-Aldrich, E7023-1L).
2. Nuclease-free water (Qiagen, 129114).
3. Agencourt AMPure XP Beads (Beckman Coulter, A63881).
4. 80% Ethanol: 10 mL nuclease-free water, 40 mL ethanol.
5. Semi-Skirted PCR Plates (SSI, 3421-00).
6. Adhesive Microseal® 'B' PCR Plate Sealing Film (Bio-Rad, MSB1001).
7. 8-SnapStrip 0.2 mL PCR tubes with caps (SSI, 3245-00).
8. Multichannel pipette.
9. Magnetic Stand-96 (Thermo Fisher Scientific/Invitrogen, AM10027).

### 2.9 Quantification, Visualization, and Pooling of Individual Libraries for Sequencing

1. TRIS–hydrochloride, pH 8 (Sigma-Aldrich, 93363).
2. Tris buffer/EB buffer: 10 mM Tris, pH 8.5.
3. 1.5 mL low-bind Eppendorf tubes (Eppendorf, 0030122348).
4. Semi-Skirted PCR Plates (SSI, 3421-00).
5. Adhesive Microseal® 'B' PCR Plate Sealing Film (Bio-Rad, MSB1001).
6. 8-SnapStrip 0.2 mL PCR tubes with caps (SSI, 3245-00).
7. Qubit dsDNA HS Assay Kit, 500 (Thermo Fisher Scientific/ Life Technologies, Q32854).
8. HT DNA HiSens Dual Protocol Reagents (PerkinElmer, CLS760672).
9. HT DNA HiSens Dual Protocol Reagents (PerkinElmer, CLS760672).
10. 24 DNA Extended Range Chip (PerkinElmer, CLS138948).
11. Qubit® 3.0 Fluorometer (Thermo Fisher Scientific, Q33216).
12. LabChip GX Touch Nucleic Acid Analyzer (PerkinElmer, CLS138162).

### 2.10 Final Library Clean and Quality Check

1. Ethanol 200 proof, for molecular biology (Sigma-Aldrich, E7023-1L).
2. Nuclease-free water (Qiagen, 129114).
3. TWEEN® 20 for molecular biology (Sigma-Aldrich, P9416).
4. Agencourt AMPure XP Beads (Beckman Coulter, A63881).
5. 80% Ethanol: 10 mL nuclease-free water, 40 mL ethanol.
6. 1.5 mL low-bind Eppendorf tubes (Eppendorf, 0030122348).
7. Normalization buffer: 10mM Tris, pH 8.5, 0.1% Tween 20. Aliquot and sterilize before use.

8. 16-Tube SureBeads™ Magnetic Rack (Bio-Rad, 1614916).
9. HT DNA HiSens Protocol Reagents (PerkinElmer, CLS760672).
10. 24 DNA Extended Range Chip (PerkinElmer, CLS138948).
11. Qubit® 3.0 Fluorometer (Thermo Fisher Scientific, Q33216).
12. LabChip GX Touch Nucleic Acid Analyzer (PerkinElmer, CLS138162).

## 3 Methods

### *3.1 DNA Starting Material*

1. Determine the sample DNA concentration using the Qubit® 3.0 Fluorometer.
2. It is recommended that the quality of the sample DNA is verified using the LabChip GX Touch with HT DNA gDNA reagents. The DNA should preferably contain no RNA and be of high molecular weight.
3. The sample DNA should be normalized to 100 ng/μL using EB buffer or nuclease-free water. A total of 200 ng of DNA will be used to create the double digest GBS sequencing library.

### *3.2 Adapter Preparation*

1. Prepare the 10× annealing buffer and set the annealing program on PCR machine: 95 °C for 5 min, slowly reduce the temperature by ramping down 1 °C every 20 s or 3 °C every minute until 25 °C is reached, hold at 25 °C for 30 min and finally hold at 4 °C.
2. Prepare the 10 μM adapter stock solution by adding the following into a 0.2 mL PCR tube for each of the barcoded adapters: 20 μL HpyCH4IV P1/Barcoded (Forward/Top strand) 100 μM, 20 μL HpyCH4IV P1/Barcoded—(Reverse complement/Bottom strand) 100 μM, 20 μL 10× annealing buffer, 140 μL nuclease-free water.
3. Mix well by pipetting up and down and transfer 100 μL into a second labeled 0.2 mL PCR tube.
4. Repeat the steps with the HinfI/Common P2 adapters. Prepare several tubes of the P2 Common Adapter.
5. Place the tubes on a PCR machine and run the annealing program.

### *3.3 Preparation of Adapter Working Solution*

1. Add 50 μL of the annealed Barcoded Adapter (P1) prepared in the previous step to 150 μL of 1× annealing buffer in labeled 8-Snapstrip PCR tubes to create the Barcoded Adapter Working stock.
2. Add 101 μL annealed Common Adapter (P2) Stock to 99 μL of 1× annealing buffer to create the Common Adapter

(HinfI_P2) working stock in a separate tube. Prepare several tubes of working stock for the Common Adapter at the same time as a larger amount is required during the ligation step. Alternatively, the dilution of the stock solutions can be calculated using the Molarity Adapter table described by Petersen et al. [31] (*see* **Note 2**).

3. Store the annealed Adapter stock and working solution at −20 °C.

### 3.4 Restriction Digestion

1. Prepare the Restriction Master Mix as follows (volumes required per sample): 2 μL NEB CutSmart Buffer (10×), HpyCH4IV 5 Units (0.5 μL), HinfI 5 Units (0.5 μL), nuclease-free water to a total of 18 μL (*see* **Note 3**). The amount of restriction enzyme to be added depends on the restriction enzyme concentration (i.e., Units/μL).
2. Transfer 18 μL of the Master Mix to 0.2 μL PCR tubes or PCR plate.
3. Add 2 μL of DNA (100 ng/μL) to a final volume of 20 μL.
4. Tap to mix.
5. Spin down briefly using a tabletop centrifuge.
6. Incubate at 37 °C for at least 4 h.
7. Continue directly to the ligation step. Alternatively, the reactions can be stored at −20 °C overnight.

### 3.5 Ligation of Adapters

*See* **Notes 2**, **4**, and **5** before starting with the ligation step.

1. Thaw T4 DNA ligase buffer. The buffer is sensitive to freeze/thaw cycles and it is recommended to aliquot the buffer if repeated use is intended. The T4 DNA ligase enzyme is unstable on ice for long periods. Keep the enzyme at −20 °C until just before use and return immediately to −20 °C after use.
2. Remove P1 adapters from fridge/freezer and allow to come to room temperature.
3. Vortex and pulse spin tubes before opening.
4. Set the ligation program on a Thermocycler as follows: Incubate at 22 °C for 2 h, deactivate the ligase by incubating at 65 °C for 20 min, cool to 4 °C at a rate of 2 °C per 90 s.
5. Prepare the ligation master mix as follows (volumes as required per sample): 8 μL 5× T4 ligation buffer, 2 μL P2 Common Adapter working solution, 1 μL T4 DNA ligase, and 7 μL nuclease-free water $dH_2O$ to a final volume of 18 μL.
6. Using a multichannel pipette add 2 μL (0.1 μM) of the P1 Barcoded Adapter working stock to the 20 μL of the digested samples from the previous step.

7. Now add 18 μL of ligation master mix to each of the digested samples in the PCR tubes or PCR plate. A multichannel pipette can be used in this step to distribute the ligation master mix evenly. Pipet up and down to mix.
8. The sample volume is now 40 μL after adding the ligation master mix and barcoded adapters.
9. Spin down briefly using a tabletop centrifuge.
10. Place the samples on the thermocycler and run the ligation program.
11. Continue directly with the next step. Alternatively, the ligated samples can be stored at −20 °C.

### 3.6 Sample Cleanup and Double Size Selection

This step is aimed at removing unligated adapters while simultaneously selecting and enriching fragments within the correct size range of 250–800 bp before PCR amplification. These steps are valuable in the normalization of samples that may differ in quality. Size selection of the libraries also reduces the number of unique reads before sequencing. Accurate pipetting is critical when using Ampure XP beads. This is particularly important during the size selection steps. *See* **Note 6** for more information on the size selection.

1. Prepare the Ampure XP Mix 1 by adding together 10 μL Ampure XP beads and 40 μL prepared PEG buffer (volumes listed are for one sample).
2. Prepare Ampure XP Mix 2 1 by adding together 10 μL Ampure XP beads and 10 μL PEG buffer (volumes required per sample).
3. Prepare fresh 80% ethanol. A total of 400 μL is required per sample.
4. Add 60 μL nuclease-free water to each ligated sample to bring the total volume to 100 μL.
5. Pipet the Ampure Mix1 up and down to ensure it is well mixed. Now add 50 μL of Ampure Mix 1 to each well and pipet up and down 10× to mix. Inspect the plate for even distribution of the beads within wells after pipetting.
6. Incubate for at least 30 min on the bench.
7. Place the plate on magnet for 3–5 min. It is important to make sure that the liquid is clear before moving on to the next step.
8. In a NEW PLATE (labeled PLATE2), add 20 μL of well mixed Ampure XP Mix 2 to each well ready for sample transfer.
9. While the original plate is on the magnet, transfer 147 μL of the SUPERNATANT to the corresponding wells on PLATE2 containing the 20 μL Ampure Mix 2. Pipet up and down 10× to mix. It is important not to disturb the beads in the original

plate or to transfer any beads of the original plate to the next plate in this step.

10. Incubate PLATE2 off the magnet for at least 30 min.
11. Transfer plate to magnet and incubate for 5 min (or at least until the solution clears).
12. Remove and discard the supernatant.
13. With the plate remaining on the magnet wash the samples by adding 200 μL of 80% ethanol to each sample. Take care not to disturb the beads.
14. After 30 s remove the 80% ethanol and discard.
15. Repeat washing **steps 15** and **16** once more without disturbing the beads.
16. Without removing the plate from magnet and allow the beads to air dry for 2–3 min. Make sure that all the ethanol has evaporated without over drying the beads. *See* **Note 7**.
17. Once the ethanol has evaporated, remove the plate from the magnet and add 30 μL of nuclease-free water.
18. Pipet up and down 10× to mix, making sure all the beads have been resuspended.
19. Allow to incubate for at least 15 min.
20. Place the plate back on the magnet for 5 min (or at least until the solution clears).
21. Transfer 10 μL aliquots to two different plates or sets of 8-SnapStrip PCR tubes. One set to be used for PCR and one backup set if required. Seal the plates.
22. The ligated and Ampure bead cleaned samples can be stored at −20 °C.

### 3.7 PCR Amplification of Libraries

The PCR amplification in this step is performed to add the sequencing adapter and additional PCR-based indexes onto the end of the barcode ligated DNA fragments. When preparing 48 samples or less for sequencing it is not necessary to make use of more than one of the Indexed primers. However, if more than 48 samples are to be sequenced it is recommended that at least 6 of the indexed primers are used to ensure that there is sufficient diversity among indexes which will help in the de-multiplexing of samples after sequencing. The barcode selection tool provided should be used to select the best combination of indexed primer barcodes (https://github.com/ascheben/RAD_analysis_workflow).

1. Prepare the PCR Master Mix as follows, using Phusion hot-start high-fidelity master mix polymerase (*see* **Note 8**). 25.0 μL high-fidelity buffer, 10 μL nuclease-free water, 2.5 μL indexed forward primer (10 μM), 2.5 μL reverse primer (10 μM), to a final volume of 40 μL.

2. Set the PCR program on a Thermocycler as follows: 98 °C for 2 min, 12–18 cycles of 98 °C for 15 s, 62 °C for 30 s, 72 °C for 30 s, final extension for 5 min at 72 °C and hold at 4 °C.
3. Add 40 μL of the prepared PCR master mix to the aliquoted 10 μL of ligated Ampure bead cleaned DNA from the previous step.
4. Centrifuge tubes or sealed plate briefly to collect the reaction mixture at the bottom and place on the Thermocycler.
5. Run the PCR program.
6. Proceed to the next step or alternatively, the samples can be stored at −20 °C.

### 3.8 Ampure XP Bead Cleaning of Post-PCR Samples

1. Remove the Ampure XP beads from the fridge and allow to warm to room temperature.
2. Prepare 80% ethanol, 400 μL is required per sample.
3. Add 1.5× of Ampure beads (75 μL) to sample volume to the samples in a PCR plate.
4. Incubate for at least 15 min on bench.
5. Place the plate on the magnet for 5 min or until the liquid has cleared completely.
6. Remove and discard the supernatant without disturbing the beads.
7. Wash samples by adding 200 μL of 80% ethanol to each sample and wait 30 s.
8. Remove and discard ethanol without disturbing the beads using a pipette.
9. Repeat **steps** 7 and **8** once more.
10. Keeping the plate on the magnet allow to air dry for 2–3 min. Make sure that all the ethanol has been removed and evaporated without over drying the beads.
11. Once the ethanol has evaporated, remove the plate from the magnet and add 30 μL of nuclease-free water.
12. Pipet up and down 10× to mix, making sure all the beads have been resuspended.
13. Allow to incubate for at least 10 min.
14. Place the plate back on magnet for 5 min (or at least until the solution has cleared completely).
15. Remove 25–28 μL of the supernatant containing the prepared individual ddRAD libraries without disturbing the beads and transfer to a new labeled tube/plate.

### 3.9 Quantification, Visualization, and Pooling of Individual Libraries for Sequencing

This step is aimed at pooling equimolar amounts of each individual sample library to ensure that equal amounts of the individual samples are sequenced.

1. Determine the concentration of each of the individual libraries prepared using the Qubit BR/HS DNA Assay Kit.
2. The library concentration should be around 20 ng/μL.
3. Verify the quality of the individual libraries on the LabChip GX Touch using the HT DNA HiSens Dual Protocol Reagents and determine the average fragment size.
4. The libraries should resemble a smooth curve or smear from 200 to maximum 700 bp, usually peaking between 320 and 360 bp. The presence of sharp peaks may indicate over amplification or as suggested by [29] that the enzymes were cutting in a repetitive region of the genome, in which case it is best to choose different enzymes.
5. To prepare the final pooled ddRAD library for sequencing, equal nM amounts of each individual sample should be combined in a single tube. It is recommended that at least 20–30 nM per individual library is added to the pool to create the final pooled library for sequencing.
6. To convert the library quantity from ng/μL to nM use the following the formula Concentration in ng/μL $\div$ (660 g/mol $\times$ average library fragment size in bp) $\times 10^6 =$ concentration in nM.
7. First create 20–30 nM equimolar dilutions of each individual library in a new plate or 8-SnapStrip PCR tube using Tris buffer (EB). (The volume per individual library should preferably not exceed 10 μL.)
8. Combine all the normalized individual libraries into a single low bind 1.5 mL Eppendorf tube.

### 3.10 Final Library Clean and Quality Check Before Sequencing

A final bead clean, as well as quality and quantity check, of the pooled library is recommended.

1. Remove the Ampure XP beads from the fridge and allow to warm to room temperature.
2. Prepare 80% ethanol and Normalization buffer.
3. Quantify the pooled library using the Qubit BR/HS DNA Assay Kit.
4. Run 2 μL of the pooled library on the LabChip GX Touch using the DNA High Sensitivity Reagents and protocol.
5. Determine the mean fragment size and size distribution of the library. Inspect the electropherogram or gel for traces of primer dimers or fragments smaller than 200 bp. To remove possible

remaining traces of primer dimers or fragments smaller than 200 bp a 0.8× Ampure bead clean step is recommended. If the library size distribution is optimal a 1.2× Ampure bead cleanup is recommended to concentrate the library.

6. Add either 0.8× or 1.2× Ampure XP beads: pooled sequencing library volume, to the pooled library.
7. Incubate for at least 10–15 min.
8. Place the tube on the magnetic stand and wait for the liquid to clear in 2–5 min.
9. Remove the supernatant without disturbing the beads.
10. Wash the beads while keeping the tube on the magnet by adding 200–500 μL of 80% ethanol for without disturbing the beads. Wait for 30 s and remove the ethanol.
11. Repeat **step 11** once more.
12. Remove all the ethanol and leave the beads to dry for 2–3 min while on the magnet. Do not let the beads dry out completely.
13. Remove the tube from the magnet and resuspend the beads using nuclease-free water or EB buffer. Depending on the initial library concentration between 0.6× and 1× of the original library volume should be used to resuspend the beads.
14. Quantify library concentration again using the Qubit BR/HS DNA Assay Kit and run the final bead cleaned sequencing library on LabChip GX Touch using the DNA High Sensitivity Reagents and protocol to verify library quality and determine the average fragment size.
15. Calculate the molarity of the final library using the formula as described before.
16. Prepare 50–100 μL of the ddRAD bead cleaned library by diluting it to a final concentration of 10–20 nM using Normalization or EB buffer.
17. The library is now ready for sequencing. *See* **Note 9**.

### *3.11 Bioinformatic Analysis of Reduced Representation Sequencing Data*

The open-source bioinformatics tools available for SNP calling and diversity analyses using ddRAD-seq data are mature and can be integrated into a simple workflow. A basic understanding of the Linux environment and a shell programming language, such as bash, are required to execute the workflow. As data volumes for most current analyses can exceed the capacity of personal computers, access to a high-performance computing environment is recommended. Software can be easily installed using the package manager conda (https://conda.io/en/latest/). The workflow and supplementary scripts have been documented online (https://github.com/ascheben/RAD_analysis_workflow).

#### 3.11.1 SNP Calling Workflow

SNPs can be called from ddRAD-seq using a reference genome or de novo. Both approaches are described in the workflow below, although reference-based SNP calling provides more robust results when a reference genome of the species or a closely related species is available [33]. There are nine software packages that should be installed to carry out the workflow (Table 1).

1. Demultiplex pooled paired-end ddRAD-seq reads in a fastq file using barcodes (*see* **Note 10**), and discard low-quality reads and reads missing restriction sites for the two enzymes.

```
process_radtags -i gzfastq -P -1 sample_R1.fq.gz -2 sam-
ple_R2.fq.gz -b ./barcodes.txt -o /path/to/output/directory
-r -c -q --renz_1 <enzyme1> --renz_2 <enzyme2>
```

2. Trim reads with adapter contamination and discard reads when read length is below 120 bp (*see* **Note 11**), using a maximum mismatch score of 2 for the adapter sequence, a palindrome clip score threshold of 30 and a simple clip score threshold of 10. Unpaired reads after filtering are written to separate fastq files.

```
trimmomatic sample_01.1.fq sample_01.2.fq sample_01_pe_R1.fq
sample_01_se_R1.fq sample_01_pe_R2.fq sample_01_se_R2.fq IL-
LUMINACLIP:TruSeq3-PE-2.fa:2:30:10 MINLEN:120
```

3. Generate individual quality reports for all uncompressed fastq files in the working directory.

```
for sample in *.fq; do
fastqc ${sample}
done
```

4. Generate multi-sample quality report based on individual fastqc quality reports.

```
multiqc /path/to/fastqc/outputs/
```

5. Sketch all R1 fastq files to create reduced sequence representations with at least four copies per k-mer and 1000 nonredundant min-hashes.

```
for sample in *.1.fq; do
mash sketch -r -m 4 -s 1000 ${sample}
done
```

6. Estimate pairwise distances between all sequence sketches and write to output file. The genetic distance can be used for preliminary species identification and diversity assessment.

```
for sample_x in *.msh; do
for sample_y in *.msh; do
mash dist ${sample_x} ${sample_y}
done
done > All_Distances.txt
```

**Table 1**
**SNP calling software requirements**

| Tool | Link | Reference |
|---|---|---|
| stacks | http://catchenlab.life.illinois.edu/stacks/ | [34] |
| trimmomatic | http://www.usadellab.org/cms/?page=trimmomatic | [35] |
| FastQC | https://www.bioinformatics.babraham.ac.uk/projects/download.html | – |
| MultiQC | https://multiqc.info/ | [36] |
| mash | https://github.com/marbl/Mash | [37] |
| BWA | https://sourceforge.net/projects/bio-bwa/files/ | [38] |
| SAMtools | http://www.htslib.org/download/ | [39] |
| BCFtools | http://www.htslib.org/download/ | [39] |
| VCFtools | https://github.com/vcftools/vcftools | [40] |

7. If there is no reference genome available, execute de novo assembly of restriction site-associated loci, genotype individuals and output SNPs in vcf format, allowing for a maximum distance of three nucleotides between stacks (−M) in an individual and a maximum three mismatches in a locus between individuals (*see* **Note 12**). Skip **steps 8–12**. If there is a reference genome available, skip only this step (**step 7**).

```
denovo_map.pl -M 3 -n 3 -T 16 --paired -o ./stacks --popmap ./
treestudy_popmap --samples ./samples -X "populations:--vcf"
```

8. Index the reference genome.

```
samtools faidx reference.fa
```

9. Align reads to a reference genome with default settings and output a bam file for indexing.

```
bwa mem reference.fa -t 1 -M -R '@RG\tID:sample_01\tPL:
illumina\tPU:sample_01\tSM:sample_01' sample_01.1.fq sam-
ple_01.2.fq | samtools view -bS -h | samtools sort > sam-
ple_01.bam
```

10. Merge aligned samples into a single file and index the file.

```
samtools merge merged.bam *.bam
samtools index merged.bam
```

11. Generate a pileup file using all reads of all samples (*see* **Note 13**), with a maximum read depth of 1000 and outputting genotype depth information.

```
samtools mpileup -d 1000 -I -go merged.bcf -ugf reference.fa
-t DP merged.bam
```

12. Call SNPs and output an uncompressed vcf file.

```
bcftools call -mv -O v -o merged.vcf merged.bcf
```

13. Calculate genotype missingness per individual.

```
vcftools --vcf merged.vcf --missing-indv --out merged
```

14. Identify individuals with >90% genotypes missing (*see* **Note 14**).

```
tail -n +2 merged.imiss | awk '$5>0.9' | cut -f1 > remove.txt
```

15. Apply stringent filters to retain only biallelic, high-quality SNPs.

```
vcftools --vcf merged.vcf --remove remove.txt --maf 0.05 --
max-missing 0.8 --remove-indels --max-alleles 2 --min-alleles
2 --minDP 5 --recode --stdout > merged_filtered.vcf
```

16. Retain only a single random SNP from each ddRAD locus (*see* **Note 15**).

```
vcftools --vcf merged_filtered.vcf -thin 500 --recode
--stdout > merged_filtered_thin.vcf
```

*3.11.2 Diversity Analysis*

To identify unknown species and investigate genetic diversity within the sampled population, the detected SNPs can be used to carry out principal component analysis (PCA), estimate population structure and infer a phylogenetic tree. Here, five software packages will be used to analyze and visualize genetic diversity using SNPs (Table 2).

1. Calculate summary population statistics and corrected Fst statistics using population information.

```
populations -V merged_filtered.vcf -M popmap.txt --fstats --
fst_correction p_value -O.
```

2. Carry out PCA and output plots with individuals colored by population (*see* **Note 16**).

```
Rscript pca.R merged_filtered_thin.vcf popmap.txt merged_out
```

**Table 2**
**Diversity analysis software requirements**

| Tool | Link | Reference |
|---|---|---|
| SNPRelate | https://github.com/zhengxwen/SNPRelate | [41] |
| plink | http://zzz.bwh.harvard.edu/plink/download.shtml | [42] |
| fastStructure | https://rajanil.github.io/fastStructure/ | [43] |
| RAxML | https://cme.h-its.org/exelixis/software.html | [44] |
| ggtree | https://github.com/GuangchuangYu/ggtree | [45] |

3. Convert vcf to plink format (*see* **Note 17**).

```
plink --vcf merged_filtered_thin.vcf --double-id --allow-
extra-chr --recode --out merged_filtered_thin -make-bed
```

4. Infer population structure for population sizes (−K) of 1–10 (*see* **Note 18**).

```
for 1 in {1..10};do
structure.py -K $1 --input=merged_filtered_thin -- output=-
merged_structure
done
```

5. Calculate the number of populations with the highest likelihood in the directory with the outputs of the above structure. py command.

```
chooseK.py -input=merged_structure
```

6. Convert vcf to phylip alignment format (*see* **Note 19**).

```
vcf2phylip.py -i merged_filtered.vcf
```

7. Infer maximum likelihood phylogeny (*see* **Note 20**).

```
raxmlHPC-PTHREADS-SSE3 -f a -V -T 12 -m ASC_GTRCAT --asc-corr
lewis -p 12345 -x 12345 -# 100 -s merged_filtered.phy -n
merged_filtered -o outgroup
```

8. Plot dendrogram and colour branches by population (*see* **Note 21**).

```
Rscript ggtree.R RAxML_bipartitions.merged_filtered popmap.
txt
```

## 4 Notes

1. Barcoded adapters are designed to accommodate the restriction sites created by the restriction enzymes to be used. Care must be taken that the restriction site is not recreated by the adapters. The adapters are ordered as oligonucleotides and then annealed to create the adapters. Additional important considerations when designing adapters have been previously documented [30, 31]. The adapter oligos and PCR primers used in this protocol are provided in Table 3 and are also available at https://github.com/ascheben/RAD_analysis_workflow.

2. Ligation efficiency depends on the total number of adapters and the number of fragment ends in a digested sample. It is therefore recommended to use a two- to tenfold excess of adapters to the complementary sticky ends produced to ensure efficient ligation. A lack of adapters may promote chimera

formation among sample DNA molecules, at the same time an excess of adapters may result in the formation of adapter dimers. Adapter dimers are, however, easily removed after ligation during the Ampure XP bead cleanup and size selection step.

3. If alternative restriction enzymes are used the experimental conditions should be adjusted. Adapters and experimental conditions described in this protocol are suited to the ddRAD library preparations using the restriction enzymes HpyCH4IV and HinfI only. Important considerations in the choice of restriction enzymes have been documented by Fu et al. [30] and Peterson et al. [31].
4. Forty eight barcoded adapters are used in this protocol. If less than 48 samples are prepared for sequencing it would be necessary to check the compatibility of the barcodes to enable creating a balanced library for sequencing. Use the "Barcode diversity calculator" before starting this step to assign the correct combination of barcodes. A barcode selection tool is available at https://github.com/ascheben/RAD_analysis_workflow.
5. It is recommended to create a sample table before starting. Note sample information, unique barcode and index information to be used for each sample. An example of the interactive table is provided at https://github.com/ascheben/RAD_analysis_workflow.
6. In the first size selection step the beads are used to bind to the high molecular weight DNA (fragments > ~800 bp). The supernatant containing fragments shorter than 800 bp is transferred after which the second bead bind is used to collect >200 bp fragments. The fragments bound to the beads are kept and used in the preparation of the ddRAD library.
7. If the beads are allowed to dry for too long they will start to crack and become hard to resuspend. This will be evident in the precipitation of the beads within 2–3 min after resuspension. Additional pipetting may be required to resuspend the beads. Always check the plates for the even distribution of the beads within each well after resuspension.
8. Alternative high-fidelity polymerases can be used in this step. This may require adjustments to the amount of the reagents, as well as amplification conditions specific to the polymerase to be used. In addition, adjustments to the number of PCR cycles may be required depending on the starting amount of DNA, genome size and amplification yield. Care should be taken when considering increasing the number of amplifications cycles as this may result in an increase in base mis-incorporation, as well as size and composition bias in final libraries.

9. A trial sequencing run might initially be required as the coverage, number of reads and number of expected loci to be obtained are unknown.
10. Barcodes are unique sequences (usually 4–8 bp) included in the sequencing reads (usually at the start of the Rl read after adapter removal); barcodes are provided to process_radtags in a two-column tab-separated text file with barcode sequences in the first column and corresponding sample names in the second column. The demultiplexing command must be adjusted if only a single restriction enzyme or a different barcoding approach was used. All available enzymes recognized by process_radtags are listed in the manual (http://catchenlab.life.illinois.edu/stacks/comp/process_radtags.php). If enzymes are not available, the RAD site check can be skipped using the parameter "--disable_rad_check".
11. Illumina adapter sequences are provided with the trimmomatic installation, and other adapters specific to a protocol used can be added manually. Discarding reads with adapters is only necessary when using the stacks de novo SNP calling approach, which requires reads of uniform length; in this case the value for MINLEN should be the total read length. The MINLEN option can be omitted when using a reference-based approach.
12. The stacks de novo parameters "-M" and "-n" are key to optimizing assembly and SNP calling performance, and a thorough exploration of the effects of altering these parameters should be conducted for each dataset [46]. The popmap file require for the analysis is a two-column tab-separated text file with sample names in the first column and corresponding populations names (e.g., pop1, pop2) in the second column.
13. The mpileup and SNP calling steps are not multi-threaded, therefore when the volume of data is high, the merged bam file should be split by chromosome and each chromosome analyzed in parallel to speed up the analysis.
14. The threshold for genotype missingness should be adjusted as appropriate for each dataset to exclude outliers.
15. By retaining only a single SNP of groups of SNPs that are within 500 bp of one another, linkage between markers is reduced. This is important because PCA and structure analysis assume SNPs are independent. The length in bp for thinning depends on the combined length covered by the paired reads. As an alternative to thinning, SNPs can be filtered based on linkage using plink or SNPRelate.
16. The pca.R script is available online (https://github.com/ascheben/RAD_analysis_workflow/).

17. Plink version > =1.9 should be used and chromosome or scaffold names may need to be adjusted to the "chr<#>" format (e.g., "chr1", "chr2", and "chr3"). The bed, bim, and fam output files are required by fastStructure.
18. The number of population sizes tested should be determined roughly by expectation of population diversity. The analyses can be carried out in parallel.
19. The vcf2phylip script can be found online (https://github.com/edgardomortiz/vcf2phylip).
20. The model and correction method are selected to handle the SNP input rather than the more common gene sequences. These setting require all sites in the input alignment to be variable and the sites that are heterozygous in all samples must be removed.
21. The ggtree.R script is available online (https://github.com/ascheben/RAD_analysis_workflow/).

## References


1. Wang J, Sun P, Li Y, Liu Y, Yu J, Ma X, Sun S, Yang N, Xia R, Lei T, Liu X, Jiao B, Xing Y, Ge W, Wang L, Wang Z, Song X, Yuan M, Guo D, Zhang L, Zhang J, Jin D, Chen W, Pan Y, Liu T, Jin L, Sun J, Yu J, Cheng R, Duan X, Shen S, Qin J, Zhang M-C, Paterson AH, Wang X (2017) Hierarchically aligning 10 legume genomes establishes a family-level genomics platform. Plant Physiol 174 (1):284–300. https://doi.org/10.1104/pp.16.01981
2. Schmutz J, Cannon SB, Schlueter J, Ma J, Mitros T, Nelson W, Hyten DL, Song Q, Thelen JJ, Cheng J, Xu D, Hellsten U, May GD, Yu Y, Sakurai T, Umezawa T, Bhattacharyya MK, Sandhu D, Valliyodan B, Lindquist E, Peto M, Grant D, Shu S, Goodstein D, Barry K, Futrell-Griggs M, Abernathy B, Du J, Tian Z, Zhu L, Gill N, Joshi T, Libault M, Sethuraman A, Zhang X-C, Shinozaki K, Nguyen HT, Wing RA, Cregan P, Specht J, Grimwood J, Rokhsar D, Stacey G, Shoemaker RC, Jackson SA (2010) Genome sequence of the palaeopolyploid soybean. Nature 463:178. https://doi.org/10.1038/nature08670
3. Yin D, Ji C, Ma X, Li H, Zhang W, Li S, Liu F, Zhao K, Li F, Li K, Ning L, He J, Wang Y, Zhao F, Xie Y, Zheng H, Zhang X, Zhang Y, Zhang J (2018) Genome of an allotetraploid wild peanut *Arachis monticola*: a *de novo* assembly. GigaScience 7(6):giy066. https://doi.org/10.1093/gigascience/giy066
4. Bertioli DJ, Cannon SB, Froenicke L, Huang G, Farmer AD, Cannon EKS, Liu X, Gao D, Clevenger J, Dash S, Ren L, Moretzsohn MC, Shirasawa K, Huang W, Vidigal B, Abernathy B, Chu Y, Niederhuth CE, Umale P, Araújo ACG, Kozik A, Do Kim K, Burow MD, Varshney RK, Wang X, Zhang X, Barkley N, Guimarães PM, Isobe S, Guo B, Liao B, Stalker HT, Schmitz RJ, Scheffler BE, Leal-Bertioli SCM, Xun X, Jackson SA, Michelmore R, Ozias-Akins P (2016) The genome sequences of *Arachis duranensis* and *Arachis ipaensis*, the diploid ancestors of cultivated peanut. Nat Genet 48:438. https://doi.org/10.1038/ng.3517
5. PeanutBase (2017) International peanut genome initiative. Accessed 23 March 2019
6. Young ND, Debellé F, Oldroyd GED, Geurts R, Cannon SB, Udvardi MK, Benedito VA, Mayer KFX, Gouzy J, Schoof H, Van de Peer Y, Proost S, Cook DR, Meyers BC, Spannagl M, Cheung F, De Mita S, Krishnakumar V, Gundlach H, Zhou S,

Mudge J, Bharti AK, Murray JD, Naoumkina MA, Rosen B, Silverstein KAT, Tang H, Rombauts S, Zhao PX, Zhou P, Barbe V, Bardou P, Bechner M, Bellec A, Berger A, Bergès H, Bidwell S, Bisseling T, Choisne N, Couloux A, Denny R, Deshpande S, Dai X, Doyle JJ, Dudez A-M, Farmer AD, Fouteau S, Franken C, Gibelin C, Gish J, Goldstein S, González AJ, Green PJ, Hallab A, Hartog M, Hua A, Humphray SJ, Jeong D-H, Jing Y, Jöcker A, Kenton SM, Kim D-J, Klee K, Lai H, Lang C, Lin S, Macmil SL, Magdelenat G, Matthews L, McCorrison J, Monaghan EL, Mun J-H, Najar FZ, Nicholson C, Noirot C, O'Bleness M, Paule CR, Poulain J, Prion F, Qin B, Qu C, Retzel EF, Riddle C, Sallet E, Samain S, Samson N, Sanders I, Saurat O, Scarpelli C, Schiex T, Segurens B, Severin AJ, Sherrier DJ, Shi R, Sims S, Singer SR, Sinharoy S, Sterck L, Viollet A, Wang B-B, Wang K, Wang M, Wang X, Warfsmann J, Weissenbach J, White DD, White JD, Wiley GB, Wincker P, Xing Y, Yang L, Yao Z, Ying F, Zhai J, Zhou L, Zuber A, Dénarié J, Dixon RA, May GD, Schwartz DC, Rogers J, Quétier F, Town CD, Roe BA (2011) The *Medicago* genome provides insight into the evolution of rhizobial symbioses. Nature 480:520. https://doi.org/10.1038/nature10625

7. Tang H, Krishnakumar V, Bidwell S, Rosen B, Chan A, Zhou S, Gentzbittel L, Childs KL, Yandell M, Gundlach H, Mayer KFX, Schwartz DC, Town CD (2014) An improved genome release (version Mt4.0) for the model legume *Medicago truncatula*. BMC Genomics 15 (1):312. https://doi.org/10.1186/1471-2164-15-312
8. Zhou P, Silverstein KAT, Ramaraj T, Guhlin J, Denny R, Liu J, Farmer AD, Steele KP, Stupar RM, Miller JR, Tiffin P, Mudge J, Young ND (2017) Exploring structural variation and gene family architecture with *de novo* assemblies of 15 *Medicago* genomes. BMC Genomics 18 (1):261. https://doi.org/10.1186/s12864-017-3654-1
9. Sato S, Nakamura Y, Kaneko T, Asamizu E, Kato T, Nakao M, Sasamoto S, Watanabe A, Ono A, Kawashima K, Fujishiro T, Katoh M, Kohara M, Kishida Y, Minami C, Nakayama S, Nakazaki N, Shimizu Y, Shinpo S, Takahashi C, Wada T, Yamada M, Ohmido N, Hayashi M, Fukui K, Baba T, Nakamichi T, Mori H, Tabata S (2008) Genome structure of the legume, *Lotus japonicus*. DNA Res 15(4):227–239. https://doi.org/10.1093/dnares/dsn008
10. Kang YJ, Kim SK, Kim MY, Lestari P, Kim KH, Ha B-K, Jun TH, Hwang WJ, Lee T, Lee J, Shim S, Yoon MY, Jang YE, Han KS, Taeprayoon P, Yoon N, Somta P, Tanya P, Kim KS, Gwag J-G, Moon J-K, Lee Y-H, Park B-S, Bombarely A, Doyle JJ, Jackson SA, Schafleitner R, Srinives P, Varshney RK, Lee S-H (2014) Genome sequence of mungbean and insights into evolution within *Vigna* species. Nat Commun 5:5443. https://doi.org/10.1038/ncomms6443
11. Schmutz J, McClean PE, Mamidi S, Wu GA, Cannon SB, Grimwood J, Jenkins J, Shu S, Song Q, Chavarro C, Torres-Torres M, Geffroy V, Moghaddam SM, Gao D, Abernathy B, Barry K, Blair M, Brick MA, Chovatia M, Gepts P, Goodstein DM, Gonzales M, Hellsten U, Hyten DL, Jia G, Kelly JD, Kudrna D, Lee R, Richard MMS, Miklas PN, Osorno JM, Rodrigues J, Thareau V, Urrea CA, Wang M, Yu Y, Zhang M, Wing RA, Cregan PB, Rokhsar DS, Jackson SA (2014) A reference genome for common bean and genome-wide analysis of dual domestications. Nat Genet 46:707. https://doi.org/10.1038/ng.3008. https://www.nature.com/articles/ng.3008#supplementary-information
12. Varshney RK, Chen W, Li Y, Bharti AK, Saxena RK, Schlueter JA, Donoghue MTA, Azam S, Fan G, Whaley AM, Farmer AD, Sheridan J, Iwata A, Tuteja R, Penmetsa RV, Wu W, Upadhyaya HD, Yang S-P, Shah T, Saxena KB, Michael T, McCombie WR, Yang B, Zhang G, Yang H, Wang J, Spillane C, Cook DR, May GD, Xu X, Jackson SA (2011) Draft genome sequence of pigeonpea (*Cajanus cajan*), an orphan legume crop of resource-poor farmers. Nat Biotechnol 30:83. https://doi.org/10.1038/nbt.2022. https://www.nature.com/articles/nbt.2022#supplementary-information
13. Varshney RK, Saxena RK, Upadhyaya HD, Khan AW, Yu Y, Kim C, Rathore A, Kim D, Kim J, An S, Kumar V, Anuradha G, Yamini KN, Zhang W, Muniswamy S, Kim J-S, Penmetsa RV, von Wettberg E, Datta SK (2017) Whole-genome resequencing of 292 pigeonpea

accessions identifies genomic regions associated with domestication and agronomic traits. Nat Genet 49:1082. https://doi.org/10.1038/ng.3872

14. Singh NK, Gupta DK, Jayaswal PK, Mahato AK, Dutta S, Singh S, Bhutani S, Dogra V, Singh BP, Kumawat G, Pal JK, Pandit A, Singh A, Rawal H, Kumar A, Rama Prashat G, Khare A, Yadav R, Raje RS, Singh MN, Datta S, Fakrudin B, Wanjari KB, Kansal R, Dash PK, Jain PK, Bhattacharya R, Gaikwad K, Mohapatra T, Srinivasan R, Sharma TR (2012) The first draft of the pigeonpea genome sequence. J Plant Biochem Biotechnol 21 (1):98–112. https://doi.org/10.1007/s13562-011-0088-8
15. Varshney RK, Song C, Saxena RK, Azam S, Yu S, Sharpe AG, Cannon S, Baek J, Rosen BD, Tar'an B, Millan T, Zhang X, Ramsay LD, Iwata A, Wang Y, Nelson W, Farmer AD, Gaur PM, Soderlund C, Penmetsa RV, Xu C, Bharti AK, He W, Winter P, Zhao S, Hane JK, Carrasquilla-Garcia N, Condie JA, Upadhyaya HD, Luo M-C, Thudi M, Gowda CLL, Singh NP, Lichtenzveig J, Gali KK, Rubio J, Nadarajan N, Dolezel J, Bansal KC, Xu X, Edwards D, Zhang G, Kahl G, Gil J, Singh KB, Datta SK, Jackson SA, Wang J, Cook DR (2013) Draft genome sequence of chickpea (*Cicer arietinum*) provides a resource for trait improvement. Nat Biotechnol 31:240. https://doi.org/10.1038/nbt.2491
16. Jain M, Misra G, Patel RK, Priya P, Jhanwar S, Khan AW, Shah N, Singh VK, Garg R, Jeena G, Yadav M, Kant C, Sharma P, Yadav G, Bhatia S, Tyagi AK, Chattopadhyay D (2013) A draft genome sequence of the pulse crop chickpea (*Cicer arietinum* L.). Plant J 74(5):715–729. https://doi.org/10.1111/tpj.12173
17. Spriggs A, Henderson ST, Hand ML, Johnson SD, Taylor JM, Koltunow A (2018) Assembled genomic and tissue-specific transcriptomic data resources for two genetically distinct lines of Cowpea (*Vigna unguiculata* (L.) Walp). Gates Open Res 2:7–7. https://doi.org/10.12688/gatesopenres.12777.1
18. Han Y, Kang Y, Torres-Jerez I, Cheung F, Town CD, Zhao PX, Udvardi MK, Monteros MJ (2011) Genome-wide SNP discovery in tetraploid alfalfa using 454 sequencing and high resolution melting analysis. BMC Genomics 12:350. https://doi.org/10.1186/1471-2164-12-350
19. Pandey MK, Roorkiwal M, Singh VK, Ramalingam A, Kudapa H, Thudi M, Chitikineni A, Rathore A, Varshney RK (2016) Emerging genomic tools for legume breeding: current status and future prospects. Front Plant Sci 7:455–455. https://doi.org/10.3389/fpls.2016.00455
20. Andrews KR, Good JM, Miller MR, Luikart G, Hohenlohe PA (2016) Harnessing the power of RADseq for ecological and evolutionary genomics. Nat Rev Genet 17:81–92. https://doi.org/10.1038/nrg.2015.28
21. Elshire RJ, Glaubitz JC, Sun Q, Poland JA, Kawamoto K, Buckler ES, Mitchell SE (2011) A robust, simple genotyping-by-sequencing (GBS) approach for high diversity species. PLoS One 6:e19379. https://doi.org/10.1371/journal.pone.0019379
22. Poland JA, Rife TW (2012) Genotyping-by-sequencing for plant breeding and genetics. Plant Genome 5:92–102. https://doi.org/10.3835/plantgenome2012.05.0005
23. Scheben A, Batley J, Edwards D (2017) Genotyping-by-sequencing approaches to characterize crop genomes: choosing the right tool for the right application. Plant Biotechnol J 15:149–161. https://doi.org/10.1111/pbi.12645
24. Sonah H, Bastien M, Iquira E, Tardivel A, Légaré G, Boyle B, Normandeau É, Laroche J, Larose S, Jean M, Belzile F (2013) An improved genotyping by sequencing (GBS) approach offering increased versatility and efficiency of SNP discovery and genotyping. PLoS One 8:e54603. https://doi.org/10.1371/journal.pone.0054603
25. Noble TJ, Tao Y, Mace ES, Williams B, Jordan DR, Douglas CA, Mundree SG (2018) Characterization of linkage disequilibrium and population structure in a mungbean diversity panel. Front Plant Sci 8:2102. https://doi.org/10.3389/fpls.2017.02102
26. Kujur A, Bajaj D, Upadhyaya HD, Das S, Ranjan R, Shree T, Saxena MS, Badoni S, Kumar V, Tripathi S, Gowda CLL, Sharma S, Singh S, Tyagi AK, Parida SK (2015) A genome-wide SNP scan accelerates trait-regulatory genomic loci identification in

chickpea. Sci Rep 5:11166. https://doi.org/10.1038/srep11166

27. Zhou X, Xia Y, Ren X, Chen Y, Huang L, Huang S, Liao B, Lei Y, Yan L, Jiang H (2014) Construction of a SNP-based genetic linkage map in cultivated peanut based on large scale marker development using next-generation double-digest restriction-site-associated DNA sequencing (ddRADseq). BMC Genomics 15:351. https://doi.org/10.1186/1471-2164-15-351
28. Gupta SK, Baek J, Carrasquilla-Garcia N, Penmetsa RV (2015) Genome-wide polymorphism detection in peanut using next-generation restriction-site-associated DNA (RAD) sequencing. Mol Breed 35:145. https://doi.org/10.1007/s11032-015-0343-0
29. Clark LV, Brummer JE, Głowacka K, Hall MC, Heo K, Peng J, Yamada T, Yoo JH, Yu CY, Zhao H, Long SP, Sacks EJ (2014) A footprint of past climate change on the diversity and population structure of *Miscanthus sinensis*. Ann Bot 114:97–107. https://doi.org/10.1093/aob/mcu084
30. Fu YB, Peterson GW, Dong Y (2016) Increasing genome sampling and improving SNP genotyping for genotyping-by-sequencing with new combinations of restriction enzymes. G3 Genes Genomes Genet 6:845. https://doi.org/10.1534/g3.115.025775
31. Peterson BK, Weber JN, Kay EH, Fisher HS, Hoekstra HE (2012) Double digest RADseq: an inexpensive method for *de novo* SNP discovery and genotyping in model and non-model species. PLoS One 7:e37135. https://doi.org/10.1371/journal.pone.0037135
32. Peterson G, Dong Y, Horbach C, Fu Y-B (2014) Genotyping-by-sequencing for plant genetic diversity analysis: a lab guide for SNP genotyping. Diversity 6:665–680
33. Shafer ABA, Peart CR, Tusso S, Maayan I, Brelsford A, Wheat CW, Wolf JBW (2017) Bioinformatic processing of RAD-seq data dramatically impacts downstream population genetic inference. Methods Ecol Evol 8:907–917. https://doi.org/10.1111/2041-210X.12700
34. Catchen J, Hohenlohe PA, Bassham S, Amores A, Cresko WA (2013) Stacks: an analysis tool set for population genomics. Mol Ecol 22:3124–3140. https://doi.org/10.1111/mec.12354
35. Bolger AM, Lohse M, Usadel B (2014) Trimmomatic: a flexible trimmer for Illumina sequence data. Bioinformatics 30:2114–2120. https://doi.org/10.1093/bioinformatics/btu170
36. Ewels P, Magnusson M, Käller M, Lundin S (2016) MultiQC: summarize analysis results for multiple tools and samples in a single report. Bioinformatics 32:3047–3048. https://doi.org/10.1093/bioinformatics/btw354
37. Ondov BD, Treangen TJ, Melsted P, Mallonee AB, Bergman NH, Koren S, Phillippy AM (2016) Mash: fast genome and metagenome distance estimation using MinHash. Genome Biol 17:132. https://doi.org/10.1186/s13059-016-0997-x
38. Li H (2013) Aligning sequence reads, clone sequences and assembly contigs with BWA-MEM. arXiv 2013:1303.3997
39. Li H, Handsaker B, Wysoker A, Fennell T, Ruan J, Homer N, Marth G, Abecasis G, Durbin R, Genome Project Data Processing S (2009) The sequence alignment/map format and SAMtools. Bioinformatics 25:2078–2079. https://doi.org/10.1093/bioinformatics/btp352
40. Danecek P, Auton A, Abecasis G, Albers CA, Banks E, DePristo MA, Handsaker RE, Lunter G, Marth GT, Sherry ST, McVean G, Durbin R, Genomes Project Analysis G (2011) The variant call format and VCFtools. Bioinformatics 27:2156–2158. https://doi.org/10.1093/bioinformatics/btr330
41. Zheng X, Levine D, Shen J, Gogarten SM, Laurie C, Bruce SW (2012) A high-performance computing toolset for relatedness and principal component analysis of SNP data. Bioinformatics 28:3326–3328. https://doi.org/10.1093/bioinformatics/bts606
42. Purcell S, Neale B, Todd-Brown K, Thomas L, Ferreira MA, Bender D, Maller J, Sklar P, de Bakker PI, Daly MJ (2007) PLINK: a tool set for whole-genome association and population-based linkage analyses. Am J Hum Genet 81:559–575. https://doi.org/10.1086/519795
43. Raj A, Stephens M, Pritchard JK (2014) fastSTRUCTURE: variational inference of

population structure in large SNP data sets. Genetics 197:573

44. Stamatakis A (2014) RAxML version 8: a tool for phylogenetic analysis and post-analysis of large phylogenies. Bioinformatics 30:1312–1313. https://doi.org/10.1093/bioinformatics/btu033
45. Yu G, Smith DK, Zhu H, Guan Y, Lam TT-Y (2017) ggtree: an r package for visualization and annotation of phylogenetic trees with their covariates and other associated data. Methods Ecol Evol 8:28–36. https://doi.org/10.1111/2041-210X.12628
46. Paris JR, Stevens JR, Catchen JM (2017) Lost in parameter space: a road map for stacks. Methods Ecol Evol 8:1360–1373. https://doi.org/10.1111/2041-210X.12775

Chapter 9

# Functional Genomic Selection in Crop Breeding

Yongle Li


## Abstract

Genomic selection (GS) is rapidly being adopted by many plant and animal breeding programs. New statistical methods that increase prediction accuracy are needed to enable effective GS. This chapter will introduce a prediction method that makes use of the biological functions of DNA markers.


**Key words** Genomic selection, GWAS, Molecular markers, Prediction accuracy, Plant breeding

## 1 Introduction

Conventional plant and animal breeding programs rely primarily on the use of phenotypic and pedigree information to inform selection decisions. Complementing this approach, marker-assisted selection (MAS) has been increasingly adopted over the last three decades by many private and public plant breeding programs with impressive impacts upon genetic gain [1]. However, the major limitation of MAS is that it can only use a few DNA markers with large effect to facilitate selection and is thus generally not suitable for selecting traits that are genetically complex such as yield. Genomic selection (GS), on the contrary, uses the collective effects of all DNA markers simultaneously from across the whole genome and thus is more suitable for selecting complex traits [2, 3]. Another major benefit of GS is the possibility to shorten breeding cycle time using early-generation intercrossing schemes in inbreeding crops parental selection can be made in the early generation, thus skipping the lengthy inbreeding stage [4].

Many statistical models for GS have been developed and implemented in various software/R packages [5–7]. One of the most common models is gBLUP/RR-BLUP, originated from the animal breeding context, where the pedigree matrix is replaced by a genomic relationship/kinship matrix obtained from whole-genome markers [3]. Another class of model, Bayesian alphabet models (Bayesian A, B, C), was developed to increase prediction accuracy

Mukesh Jain and Rohini Garg (eds.), *Legume Genomics: Methods and Protocols*, Methods in Molecular Biology, vol. 2107, https://doi.org/10.1007/978-1-0716-0235-5_9, 

by making use of markers/QTL with large effect during a variables (DNA markers) selection process [8, 9]. The Bayesian alphabet models should perform better than the gBLUP model theoretically because it can remove markers with little or no effect on the traits. However, several empirical studies failed to demonstrate the power of the Bayesian model [9]. Recently, a two-step prediction approach demonstrated increased prediction accuracy, by up to twofold, based on a subset of markers discovered through genome-wide association study (GWAS) in the first step [10]. Another similar method called sBLUP also demonstrated its merit in terms of prediction accuracy by deriving the kinship matrix using associated markers identified through the SUPER GWAS method (*see* **Note 1**) [11]. This method is implemented in an R package GAPIT (Genome Association and Prediction Integrated Tool) and will be used here to demonstrate its superiority over the traditional gBLUP method.

## 2 Software Setup

1. *R Setup*

   R is a statistical computing software and can be downloaded from https://www.r-project.org/.

2. *RStudio Setup*

   RStudio is an integrated development environment for R and can be downloaded from https://www.rstudio.com/.

3. *Loading GAPIT Functions*

   GAPIT contains a set of R functions that efficiently performs association mapping and genomic prediction with large genotypic datasets. The user manual can be downloaded from http://www.zzlab.net/GAPIT/. If users have further questions, there is an active Google group for GAPIT (https://groups.google.com/forum/#!forum/gapit-forum).

   The required libraries can be loaded by typing the following command in R:

   ```
   source("http://www.zzlab.net/GAPIT/GAPIT.library.R")
   ```

   The R functions can be loaded by typing (Fig. 1):

   ```
   source("http://www.zzlab.net/GAPIT/gapit_functions.txt")
   ```

## 3 Input Data

1. *Input data includes three files (DOI: 10.13140/RG.2.2.25095.85921) which can be downloaded from* https://www.researchgate.net/publication/330555072_Data_set_for_a_book_chapter_in_MiMB.

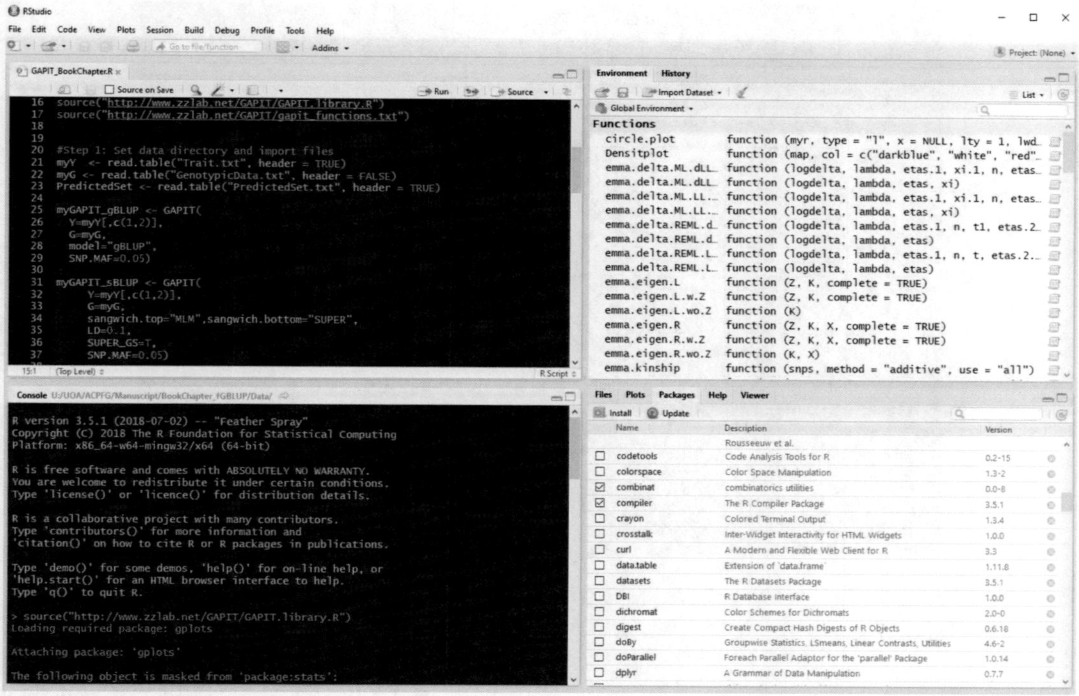

**Fig. 1** Screenshot of the RStudio environment. The top left panel shows the code editor, bottom left panel the R console, top right panel the console's workspace and history, and bottom right panel the packages, files, plots, and help

2. *Setting the working directory.*

   The working directory is the path where input and output data are stored on the local computer. The readers can set the working directory where the downloaded input data is stored by typing:

```
setwd("C:/")
```

3. *Phenotypic data.*

   The first phenotypic data is stored in a file named "Trait.txt" with the first column containing chickpea genotype names. The second column contains trait scores for the reaction of each genotype for phytophthora root rot (PRR) resistance. This score is based on the area under the disease progress stairs (AUDPS), being similar to the area under the disease progress curve (AUDPC), which combines multiple observations of disease into a single value [12]. AUDPS was calculated based on the percentage of dead chickpea plants caused by PPR measured in an irrigated field environment at the Hermitage Research Station, Australia 2015 (trial details can be found in

```
> myY  <- read.table("Trait.txt", header = TRUE)
> dim(myY)
[1] 408   2
> head(myY)
                  ID     AUDPS
1 D09001A.F6RIL.001 32493.41
2 D09001B.F6RIL.018 44102.96
3 D09001B.F6RIL.029 45750.90
4 D09001A.F6RIL.014 33896.43
5 D09001B.F6RIL.026 32378.45
6 D09001D.F6RIL.041 23491.63
> PredictedSet <- read.table("PredictedSet.txt", header = TRUE)
> dim(PredictedSet)
[1] 227   2
> head(PredictedSet)
                               ID    AUDPS
1 07H042K.07HG1002.F4TMWR2AB001 117852.5
2          07H088K.F4TMWR2AB001 121900.9
3          07H135K.F4TMWR2AB013 119565.1
4 08H074K.08HG1001.F4TMWR2AB002 125225.0
5 08H074K.08HG1001.F4TMWR2AB013 118825.8
6 08H075K.08HG1002.F4TMWR2AB004 126649.2
```

**Fig. 2** Screenshot of the R console showing the dimension and partial content of the phenotypic data

[13]). There are 408 genotypes in this phenotypic data. The first 181 genotypes with trait values are RIL lines of chickpea parents Yorker and Genesis114. They will be used as a training population to estimate the marker effect. The remaining genotypes, with missing trait values indicated as "NA," are diverse Australian chickpea lines. They will be used as the predicted set. This file can be read into R by typing (Fig. 2):

```
myY <- read.table("Trait.txt", header = TRUE)
```

In order to estimate prediction accuracy, the predicted set has been evaluated for PRR resistance at the Hermitage Research Station in 2016. AUDPS was calculated in the same way as for the training set. This file can be read into R by typing:

```
PredictedSet <- read.table("PredictedSet.txt", header = TRUE)
```

4. *Genotypic data.*

GAPIT can read genotypic data either in numeric format or HapMap format (*see* **Note 2**). The chickpea genotypic data in this example is obtained using a DArTseq genotyping method and is stored in HapMap format. The first 11 columns are the attributes of the SNP and the remaining columns show the

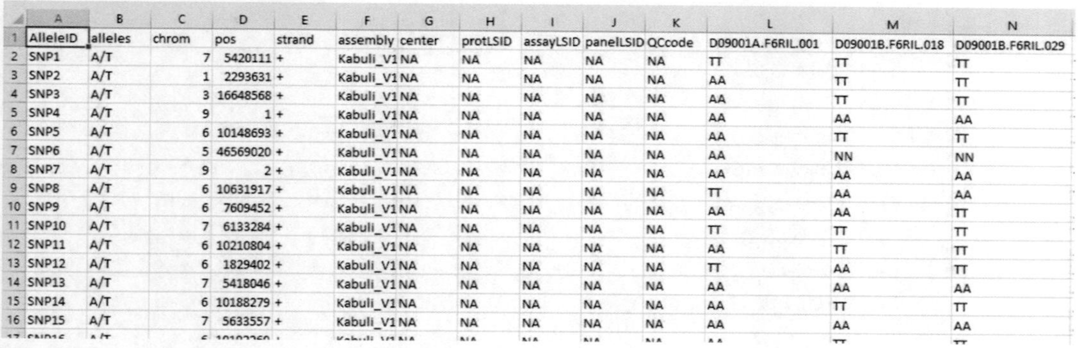

| AlleleID | alleles | chrom | pos | strand | assembly | center | protLSID | assayLSID | panelLSID | QCcode | D09001A.F6RIL.001 | D09001B.F6RIL.018 | D09001B.F6RIL.029 |
|---|---|---|---|---|---|---|---|---|---|---|---|---|---|
| SNP1 | A/T | 7 | 5420111 | + | Kabuli_V1 | NA | NA | NA | NA | NA | TT | TT | TT |
| SNP2 | A/T | 1 | 2293631 | + | Kabuli_V1 | NA | NA | NA | NA | NA | AA | TT | TT |
| SNP3 | A/T | 3 | 16648568 | + | Kabuli_V1 | NA | NA | NA | NA | NA | AA | TT | TT |
| SNP4 | A/T | 9 | 1 | + | Kabuli_V1 | NA | NA | NA | NA | NA | AA | AA | AA |
| SNP5 | A/T | 6 | 10148693 | + | Kabuli_V1 | NA | NA | NA | NA | NA | AA | TT | TT |
| SNP6 | A/T | 5 | 46569020 | + | Kabuli_V1 | NA | NA | NA | NA | NA | AA | NN | NN |
| SNP7 | A/T | 9 | 2 | + | Kabuli_V1 | NA | NA | NA | NA | NA | AA | AA | AA |
| SNP8 | A/T | 6 | 10631917 | + | Kabuli_V1 | NA | NA | NA | NA | NA | TT | AA | AA |
| SNP9 | A/T | 6 | 7609452 | + | Kabuli_V1 | NA | NA | NA | NA | NA | AA | AA | TT |
| SNP10 | A/T | 7 | 6133284 | + | Kabuli_V1 | NA | NA | NA | NA | NA | TT | TT | TT |
| SNP11 | A/T | 6 | 10210804 | + | Kabuli_V1 | NA | NA | NA | NA | NA | AA | TT | TT |
| SNP12 | A/T | 6 | 1829402 | + | Kabuli_V1 | NA | NA | NA | NA | NA | TT | AA | TT |
| SNP13 | A/T | 7 | 5418046 | + | Kabuli_V1 | NA | NA | NA | NA | NA | AA | AA | AA |
| SNP14 | A/T | 6 | 10188279 | + | Kabuli_V1 | NA | NA | NA | NA | NA | AA | TT | TT |
| SNP15 | A/T | 7 | 5633557 | + | Kabuli_V1 | NA | NA | NA | NA | NA | AA | AA | AA |

**Fig. 3** Screenshot showing partial content of the genotypic data

allele observed at each SNP for the training and prediction sets. This file can be read into R by typing (Fig. 3):

```
myG <- read.table("GenotypicData.txt", header = FALSE)
```

## 4 Genomic Prediction

GS generally includes two sets of plant populations: the training set and the predicted set. The former includes genotypic and phenotypic data and is used to estimate marker effect. The latter only includes lines that have genotypic data (same marker set as the training set) but without phenotypic data. The marker effect, estimated using the training set, is used to predict the genomic estimated breeding value (GEBV) for the predicted set (*see* **Note 3**).

### 4.1 Genomic Prediction Using gBLUP

The following code is used to execute gBLUP:

```
myGAPIT_gBLUP <- GAPIT(Y=myY[,c(1,2)], G=myG, model="gBLUP",
SNP.MAF=0.05)
```

In the code, Y is the AUDPS in the phenotypic data, G is the genotypic data, model is gBLUP, SNP.MAF is a marker filtering function based on minor allele frequency. There are a few output files saved in the working directory including the GEBV for the predicted set at the last column of the Excel file name "GAPIT.MLM.Pred.result.csv" (Fig. 4). The column with name "RefInf" indicates the genotype is in the training set (1) or the predicted set (2).

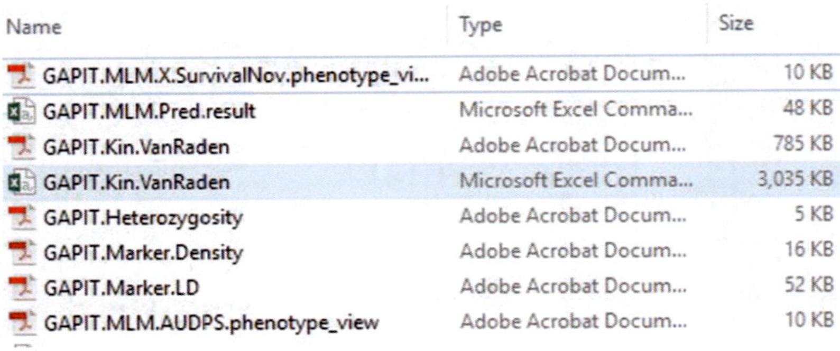


**Fig. 4** Screenshot showing the output files after running the gBLUP model

### 4.2 Genomic Prediction Using sBLUP

The following code is used to execute sBLUP:

```
myGAPIT_sBLUP <- GAPIT(Y=myY[,c(1,2)], G=myG, sangwich.
top="MLM", sangwich.bottom="SUPER", LD=0.1, SUPER_GS=T, SNP.
MAF=0.05)
```

In the code, sangwich.top gives an option of selecting different GWAS models (GLM or MLM) to identify pseudo QTNs. Sangwich.bottom is a prediction model that only uses the markers significantly associated with the trait to derive the kinship matrix. In contrast, gBLUP uses all markers to derive the kinship matrix. LD defines the r2 threshold to exclude markers with high linkage disequilibrium (LD). SUPER_GS gives an option to turn prediction on (=T) or off (=F).

### 4.3 Kinship Heat Map and Kinship Matrix

GAPIT produces a kinship heat map (GAPIT.Kin.VanRaden.pdf) and a kinship matrix (GAPIT.Kin.VanRaden.csv) in the working directory by default. The kinship heat map is a useful way to visualize the relationship between the training set and the predicted set (Fig. 5). Generally speaking, the prediction accuracy increases when the strength of the genotypic relationship between the two sets increases.

### 4.4 Prediction Accuracy

The GEBVs are also stored in the object myGAPIT_gBLUP$Pred and can be viewed using the command (Fig. 6):

```
head(myGAPIT_gBLUP$Pred)
head(myGAPIT_sBLUP$Pred)
```

These tables include both the training set and the predicted set. To calculate the prediction accuracy for the predicted set, the subset function of R can be used to only retain the prediction set using the following command:

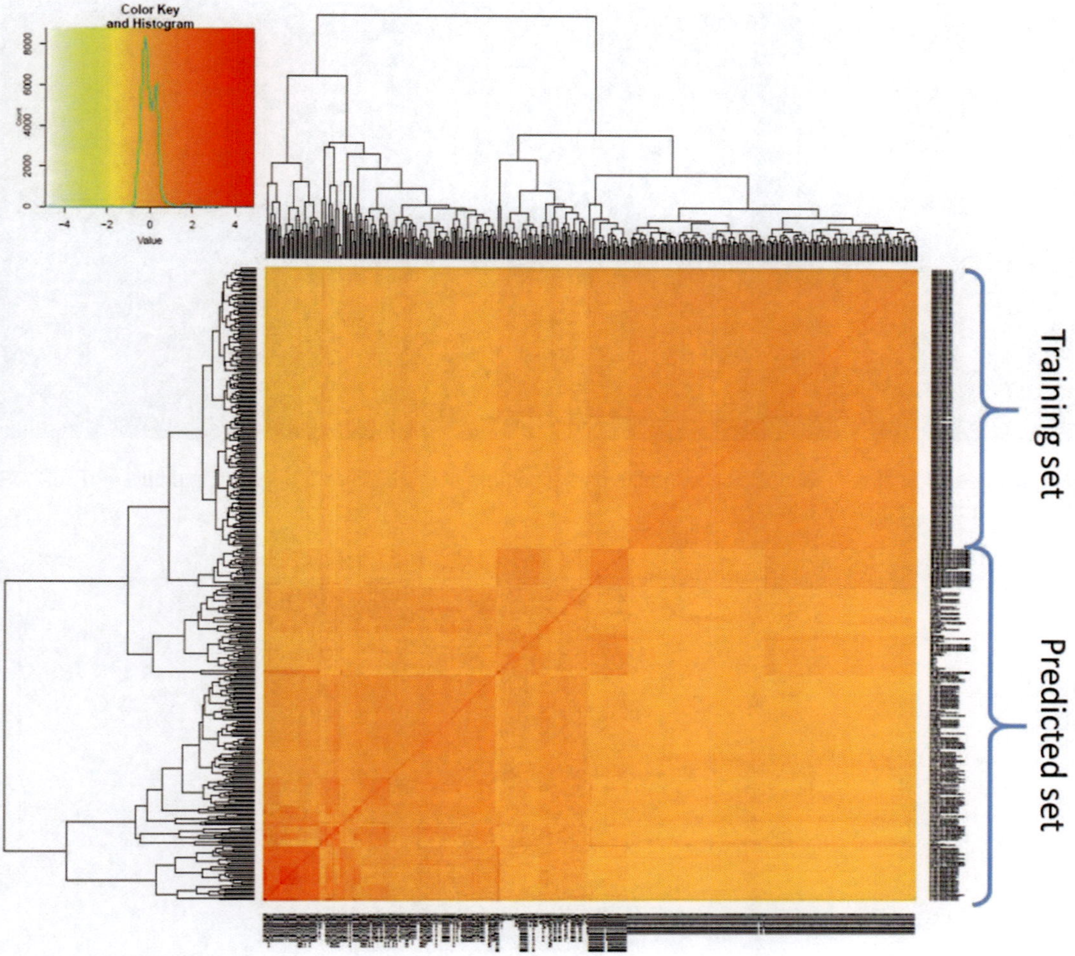


**Fig. 5** Kinship heat map showing the relationship of each genotype. The larger the kinship value, the closer the relationship between the two genotypes

```
GEBV_gBLUP <- subset(myGAPIT_gBLUP$Pred, RefInf=="2")
GEBV_sBLUP <- subset(myGAPIT_sBLUP$Pred, RefInf=="2")
```

Prediction accuracy can be calculated by correlating the GEBVs and the "true" AUDPS of the predicted set evaluated in the field:

```
cor(GEBV_gBLUP$Prediction,PredictedSet$AUDPS)
cor(GEBV_sBLUP$Prediction,PredictedSet$AUDPS)
```

The correlation coefficient of the "true" AUDPS measured in the field and the GEBVs estimated using the sBLUP model is 0.30 which is 43% larger than the correlation estimated using the

```
> head(myGAPIT_gBLUP$Pred)
                         Taxa Group RefInf ID       BLUP        PEV             BLUE Prediction
1 07H042K.07HG1002.F4TMWR2AB001     1      2  1  184.7643  2299079.9 32623.695594188   32808.46
2          07H088K.F4TMWR2AB001     2      2  2 1170.4183  -804965.8 32623.695594188   33794.11
3          07H135K.F4TMWR2AB013     3      2  3 -184.4320 -5003494.2 32623.695594188   32439.26
4 08H074K.08HG1001.F4TMWR2AB002     4      2  4 2274.7811 -1322021.7 32623.695594188   34898.48
5 08H074K.08HG1001.F4TMWR2AB013     5      2  5  572.2845  -575823.8 32623.695594188   33195.98
6 08H075K.08HG1002.F4TMWR2AB004     6      2  6 1790.7667 -1065724.7 32623.695594188   34414.46
> head(myGAPIT_sBLUP$Pred)
                         Taxa Group RefInf ID      BLUP       PEV             BLUE Prediction
1 07H042K.07HG1002.F4TMWR2AB001     1      2  1 -2191.580 -246670.9 30615.4110179834   28423.83
2          07H088K.F4TMWR2AB001     2      2  2 -4467.324 4056236.1 30615.4110179834   26148.09
3          07H135K.F4TMWR2AB013     3      2  3 -2191.580 -246670.9 30615.4110179834   28423.83
4 08H074K.08HG1001.F4TMWR2AB002     4      2  4  1623.220  323238.3 30615.4110179834   32238.63
5 08H074K.08HG1001.F4TMWR2AB013     5      2  5 -2191.580 -246670.9 30615.4110179834   28423.83
6 08H075K.08HG1002.F4TMWR2AB004     6      2  6 -2191.580 -246670.9 30615.4110179834   28423.83
> cor(GEBV_gBLUP$Prediction,PredictedSet$AUDPS)
[1] 0.208393
> cor(GEBV_sBLUP$Prediction,PredictedSet$AUDPS)
[1] 0.3035352
```

**Fig. 6** Screenshot of the R console showing the partial content of myGAPIT_gBLUP$Pred and myGAPIT_gBLUP $Pred and the prediction accuracy

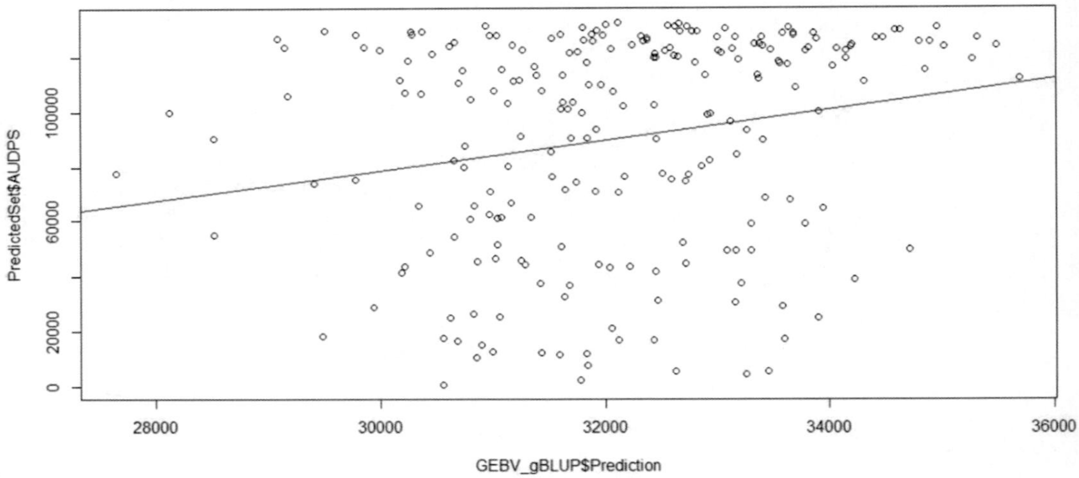


**Fig. 7** Scatter plot of the "true" AUDPS and GEBV (based on gBLUP) of the predicted set

gBLUP model ($r = 0.21$). The correlation can be visualized using simple box plots with regression fitted lines (Figs. 7 and 8):

```
plot(GEBV_gBLUP$Prediction,PredictedSet$AUDPS)
 abline(lm(PredictedSet$AUDPS~GEBV_gBLUP$Prediction),
col="red")
plot(GEBV_sBLUP$Prediction,PredictedSet$AUDPS)
 abline(lm(PredictedSet$AUDPS~GEBV_sBLUP$Prediction),
col="red")
```

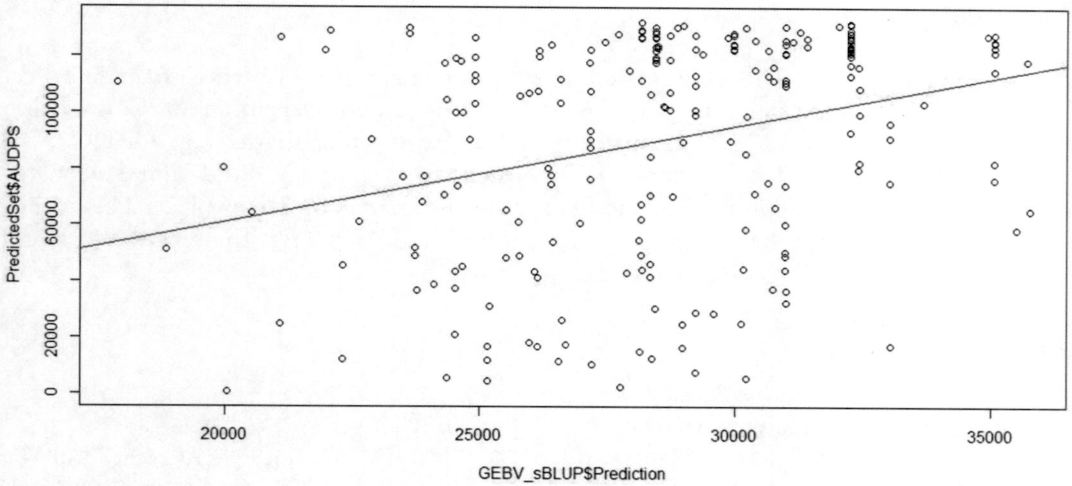


**Fig. 8** Scatter plot of the "true" AUDPS and GEBV (based on sBLUP) of the predicted set

## 5 Notes

1. Besides genomic prediction, GAPIT can efficiently perform GWAS with many model options such as SUPER-GWAS, Farm-CPU.
2. In some cases, the readers may need to manipulate their genotypic data. The author recommends to use the software TASSEL which accepts HapMap, VCF, PLINK input formats and has many easy and powerful functions to manipulate genotypic data [14].
3. GAPIT can also perform cross-validation to evaluate prediction accuracy. The readers can find the necessary R code in the user manual of GAPIT.

## 6 Conclusion

This chapter describes a step-by-step guide for performing genomic prediction using GAPIT. Compared to the conventional gBLUP, prediction model sBLUP incorporating QTL-linked markers increased prediction accuracy. It is worth testing other ways to improve prediction accuracy through using the biological function of SNP such as SNP annotation in the genome.

## Acknowledgments

The dataset used in this book chapter was obtained from a project funded by the Grains Research and Development Corporation (GRDC), Australia. Y.L.'s position is funded by GRDC. The author thanks Sean Bithell, Kristy Hobson, and Tim Sutton for conducting the field trials and managing the project. The author thanks John Harris and Judith Atieno for their comments and testing the R code.

### References

1. Bernardo R (2008) Molecular markers and selection for complex traits in plants: Learning from the last 20 years. Crop Sci 48 (5):1649–1664. https://doi.org/10.2135/cropsci2008.03.0131
2. Crossa J, Perez P, Hickey J, Burgueno J, Ornella L, Ceron-Rojas J, Zhang X, Dreisigacker S, Babu R, Li Y, Bonnett D, Mathews K (2014) Genomic prediction in CIMMYT maize and wheat breeding programs. Heredity (Edinb) 112(1):48–60. https://doi.org/10.1038/hdy.2013.16
3. Meuwissen THE, Hayes BJ, Goddard ME (2001) Prediction of total genetic value using genome-wide dense marker maps. Genetics 157(4):1819–1829
4. Asoro FG, Newell MA, Beavis WD, Scott MP, Tinker NA, Jannink JL (2013) Genomic, marker-assisted, and pedigree-BLUP selection methods for beta-glucan concentration in elite oat. Crop Sci 53(5):1894–1906. https://doi.org/10.2135/cropsci2012.09.0526
5. Wang J, Zhou Z, Zhang Z, Li H, Liu D, Zhang Q, Bradbury PJ, Buckler ES, Zhang Z (2018) Expanding the BLUP alphabet for genomic prediction adaptable to the genetic architectures of complex traits. Heredity (Edinb) 121(6):648–662. https://doi.org/10.1038/s41437-018-0075-0
6. Endelman JB (2011) Ridge regression and other kernels for genomic selection with R package rrBLUP. Plant Genome 4 (3):250–255. https://doi.org/10.3835/plantgenome2011.08.0024
7. Perez P, de los Campos G (2014) Genome-wide regression and prediction with the BGLR statistical package. Genetics 198 (2):483–495. https://doi.org/10.1534/genetics.114.164442
8. Gianola D (2013) Priors in whole-genome regression: the bayesian alphabet returns. Genetics 194(3):573–596. https://doi.org/10.1534/genetics.113.151753
9. Wimmer V, Lehermeier C, Albrecht T, Auinger HJ, Wang Y, Schon CC (2013) Genome-wide prediction of traits with different genetic architecture through efficient variable selection. Genetics 195(2):573–587. https://doi.org/10.1534/genetics.113.150078
10. Li YL, Ruperao P, Batley J, Edwards D, Khan T, Colmer TD, Pang JY, Siddique KHM, Sutton T (2018) Investigating drought tolerance in chickpea using genome-wide association mapping and genomic selection based on whole-genome resequencing data. Front Plant Sci 9:ARTN 190. https://doi.org/10.3389/fpls.2018.00190
11. Wang Q, Tian F, Pan Y, Buckler ES, Zhang Z (2014) A SUPER powerful method for genome wide association study. PLoS One 9 (9):e107684. https://doi.org/10.1371/journal.pone.0107684
12. Simko I, Piepho HP (2012) The area under the disease progress stairs: calculation, advantage, and application. Phytopathology 102 (4):381–389. https://doi.org/10.1094/PHYTO-07-11-0216
13. Amalraj A, Taylor J, Bithell S, Li Y, Moore K, Hobson K, Sutton T (2018) Mapping resistance to Phytophthora root rot identifies independent loci from cultivated (*Cicer arietinum* L.) and wild (*Cicer echinospermum* P.H. Davis) chickpea. Theor Appl Genet. https://doi.org/10.1007/s00122-018-3256-6
14. Bradbury PJ, Zhang Z, Kroon DE, Casstevens TM, Ramdoss Y, Buckler ES (2007) TASSEL: software for association mapping of complex traits in diverse samples. Bioinformatics 23 (19):2633–2635. https://doi.org/10.1093/bioinformatics/btm308

# Chapter 10

# Methods for Target Enrichment Sequencing via Probe Capture in Legumes

**Ze Peng, Dev Paudel, Liping Wang, Ziliang Luo, Qian You, and Jianping Wang**

## Abstract

Target enrichment sequencing (TES) is a powerful approach to deep-sequencing the exome or genomic regions of interest with great depth. Although successfully and widely adopted in many plant species, TES is currently applied for genotyping of only a couple legume species. Here we describe an in-solution probe capture based method for application of TES in legumes. The topics cover probe design, library preparation, probe hybridization, as well as bioinformatic analysis for evaluation of target capture efficiency and identifying single nucleotide polymorphisms using generated sequencing data.

**Key words** Legume, Next generation sequencing, Probe design, Probe hybridization, Single nucleotide polymorphism, Target enrichment sequencing

## 1 Introduction

The next generation sequencing (NGS) has revolutionized the field of biological study in the last decade. It is a more economical, efficient and high-throughput approach in retrieving the DNA sequences compared to Sanger sequencing because of its ability in handling millions of DNA/cDNA fragments in parallel. This game-changing technology has greatly facilitated genome sequencing [1, 2] and transcriptome profiling [3–5] of any living organisms. As the price decreases, NGS has become a routine tool in various biological studies.

With its capacity, NGS has extensively been utilized for high throughput genotyping and single nucleotide polymorphism (SNP) or other genetic variants calling [6]. Despite of the exponential reduction of NGS cost in recent years, sequencing the whole genome of a big population is still not practical especially for most plants that have large genome size with high ploidy level or large amounts of repetitive sequences. Considering that many researches

Mukesh Jain and Rohini Garg (eds.), *Legume Genomics: Methods and Protocols*, Methods in Molecular Biology, vol. 2107, https://doi.org/10.1007/978-1-0716-0235-5_10, © Springer Science+Business Media, LLC, part of Springer Nature 2020

involved in genotyping does not require information of the whole genome, reduced representation sequence strategies were developed for NGS-enabled high-throughput genotyping [7, 8]. The restriction enzyme (RE)-based strategy takes advantage of the versatility of the enzyme to sample specific sequence regions or fragments for sequencing [8]. This idea was firstly implemented through Complexity Reduction of Polymorphic Sequences (CRoPS) [9]. In this method, fragments generated by RE were ligated to adaptors and then sequenced. Van Tassell et al. [10] sequenced reduced-representation libraries (RRLs) for genome-wide SNP identification, in which enzymes with frequent cut sites were used for genomic DNA digestion and fragments with suitable size were selected for sequencing. Restriction site-associated DNA sequencing (RAD-seq) is another similar approach that sequence all cutting site by shearing fragments to specific size for sequencing without length selection [11]. Elshire et al. [12] demonstrated a robust and simplified procedure called genotyping by sequencing (GBS) which utilized REs to cut DNA into small pieces and implemented simultaneous SNP discovery and genotyping. Further improved approaches such as Double Digest Restriction Associated DNA Sequencing (ddRAD-seq) and Modified GBS were recently developed to reduce the cost, increase the coverage, and improve marker identification accuracy [13, 14]. Targeted enrichment sequencing (TES) is another solution for genome complexity reduction. In TES, genomic regions of interest are captured by PCR amplification, molecular inversion probes (MIPs), or homology hybridization to consequent enrichment and sequencing [15, 16]. TES allows identification of markers outside of RE cutting regions and can be customized to analyze specific genomic regions.

Target enrichment can be achieved by the following approaches for deep sequencing:

1. Uniplex PCR: In the uniplex PCR [17] approach, single amplicons are generated in each reaction for sequencing.
2. Multiplex PCR: In the multiple PCR approach, multiple amplicons are generated by using several pairs of primers in a single reaction for sequencing. However, often the reaction can generate nonspecific amplification due to interactions between primers and sometimes fail to amplify [18, 19].
3. Microdroplet PCR: In this approach, two libraries of lipid encapsulated water droplets are prepared, which contain test DNA samples and distinct pairs of primer respectively. These two libraries are then fused to generate highly multiplexed total emulsion PCR that can utilize up to 20,000 primer pairs in a single tube [20].

4. Molecular inversion probes (MIPs): In the MIP approach, single-stranded oligonucleotides that consist of universal spacer regions flanked by target-specific sequences are designed for each amplicon [21, 22]. These probes anneal to their target regions and are circularized by filling the gap to produce closed circles by a DNA polymerase and a ligase. Uncircularized DNA is digested to reduce background. The circularized targeted DNA are PCR-amplified and sequenced. Protocols for direct sequencing of MIP amplicons have also been developed that enable simultaneous amplification and accurate shotgun sequencing, thus making massively parallel exon capture and resequencing possible [23].
5. Hybrid capture: In this approach, hybridization of target regions could occur either on a microarray surface or in-solution. In the microarray process, DNA library is hybridized to a single microarray consisting of thousands to millions of isothermal probes that were identified from reference genome sequence. Background DNA is washed away, and the fragments bound to the probes are eluted. Then the selected fragments are amplified and sequenced [24]. On the other hand, in-solution based capture approach uses excess of probes over template using smaller amount of library [25].

The enrichment method to use varies based on field of application. The approaches 1–4 are basically PCR-based, thus are highly relying on primer design from the target region sequences. They are preferred when small target regions need to be examined. Microdroplet based approach is suitable for amplification of up to 4000 target sequences and requires specialist equipment [26]. In contrast, hybrid capture approach can handle large amounts of target regions. Therefore, it can be used when many Mb of regions need to be analyzed. The microarray surface hybrid capture requires a large amount of input DNA (3–10 μg) and high-tech equipment, but it can be multiplexed and ready to use when kits are available [16]. In comparison, solution-based hybrid capture is relatively easier to use and requires less amount of input DNA (<1–3 μg). Currently, the hybrid capture approach, especially the in-solution hybrid capture, is becoming popular in research society.

TES was first utilized to sequence the complete "exome" or exon regions of the genome, or smaller sets of genes that are potentially implicated in complex diseases [16]. TES was able to identify rare and common variants in more than 300 Mb of human coding sequence [27]. Exome sequencing (exome-seq) captures highly interpretable coding regions of the genome in a more affordable way and allows a large number of samples to be analyzed [28]. In this approach, probes are designed to capture the target genomic regions based on sequence homology hybridization. Then the captured fragments are retrieved and sequenced. TES has been used in several crops such as barley to underpin comparison of modern and ancient samples (6,000 year old barley grains) [29, 30], switchgrass to identify SNP association with heading date [31], and to identify SNPs in peanuts [32]. TES has also been used in deciphering novel prokaryotic taxa in metagenomic samples where the hybridization capture revealed greater microbial diversity than 16S rDNA amplicon sequencing and shotgun sequencing [33]. TES has also been shown to be useful for phylogenetics to understand the evolutionary relationships among birds [34].

TES is a powerful approach to capturing and deeply sequencing a subset of the genome of interest, such as exome-seq, a major application of TES [35]. Compared with restriction enzyme associated methods such as GBS, which randomly surveys the genome based on restriction sites, TES retrieve sequences of specific customized interests, thus has greater control of the sampled sequences. Meanwhile, TES confers a couple of benefits compared to whole genome sequencing (WGS). (1) By focusing on a small portion of the genome, it largely reduces the complexity of whole genome and avoids repetitive sequences. (2) TES could greatly increase the sequence depth and reduce the sequencing cost by allowing multiplexing more samples for sequencing. For certain cases (e.g., high-throughput genotyping), sequencing the whole genome is not necessary. Targeting coding regions, which only accounts for approximately 1–2% of the genome [36] will be much more efficient. For example, the coding regions of wheat is 170–340 Mb [37] in contrast to the large hexaploid genome size of 17 Gb, which would be intractable for WGS. Sequencing depth is very important for the reliability of sequence variation calling specifically for the highly polyploidy species. On the other hand, TES also has several limitations. TES requires a sequence source for probe design. Thus, some available sequence information is a prerequisite for its application. For species with limited sequence resources and without a genome reference even in its close relatives, TES may not be preferably considered due to the hurdles of probe design. Additionally, TES usually allows only several Mb of DNA regions to be targeted by probes for subsequent sequencing [35]. Thus, even for exome-seq, not all genes in the genome are necessarily covered by this method.

The various applications of TES in crops can be summarized as following. (1) TES has been widely utilized in phylogenetic studies. Since phylogenetic analysis does not require to know every single base in the genome, TES is a good method to sequence the homologous genomic regions across different species or genotypes. Weitemier et al. [38] performed TES in milkweed (*Asclepias syriaca*) by designing probes from genome and transcriptome assemblies, which was proved to be efficient in revolving the phylogenetic relationship of 12 milkweed genotypes, including *Asclepias* species and related genera. Comer et al [39] applied both hybrid gene capture and long-range PCR to construct phylogeny for 31 samples in 14 Arecoideae tribes, and further compared their efficiency. (2) TES is efficient for studying natural variations and genetic diversities. In maize, a subset of genomic DNA (29 Mb) was captured and enriched for identification of sequence polymorphisms among 21 diverse inbred lines, which demonstrated that TES is powerful for variant discovery [40]. In addition, several studies in wheat applied exome capture to identify natural variations and genetic diversities between different wheat accessions or varieties [41, 42]. (3) TES can serve as a tool to monitor alien genome introgressions. Wendler et al. [43] investigated the genetic diversity of introgression lines of barley using exome-seq. Large amounts of informative SNPs were identified revealing the genome introgressions from *Hordeum bulbosum*, a wild relative of barley. (4) TES can be used for genetic map construction. High density genetic maps consisting of 7434 sequence variants segregating for 3787 genes were generated from exome-seq in pine [44]. Synteny based mapping-by-sequencing was enabled by targeted enrichment in *Arabidopsis thaliana* gene models ordered by synteny with genome of distant relative *Brassica rapa* [45]. Additionally, TES using probes from close related species has proven to be effective for less resourced species. For example, before the availability of sugarcane genome reference, exome-seq was applied by using probes designed from sorghum sequences to characterize natural allelic variations [46] and to perform population genomic studies [47] in *Saccharum* species.

Currently, two studies using TES in legumes were reported. In soybean, exome-seq was applied to investigate the intracultivar variations of different Williams 82 genotypes, which revealed the unwanted heterogeneity within this cultivar [48]. Recently, we made an attempt of TES in peanut by designing probes from transcript sequences [32]. This study demonstrated that TES is effective to identify SNP markers among peanut genotypes. For species without a genome, expressed sequence tag (EST) sequences can also serve as a sequence source for probe design. With numerous successful applications of this powerful sequencing approach, an expansion of TES applications is expected in legumes and other plant species. In this chapter, we describe the process of TES using probe capture, including designing probes from a sequence source, DNA library preparation, hybridization, capture, and enrichment for NGS (Fig. 1a). In addition, we present a bioinformatics pipeline for evaluating the efficiency of target capture and identifying SNPs using generated sequencing data.

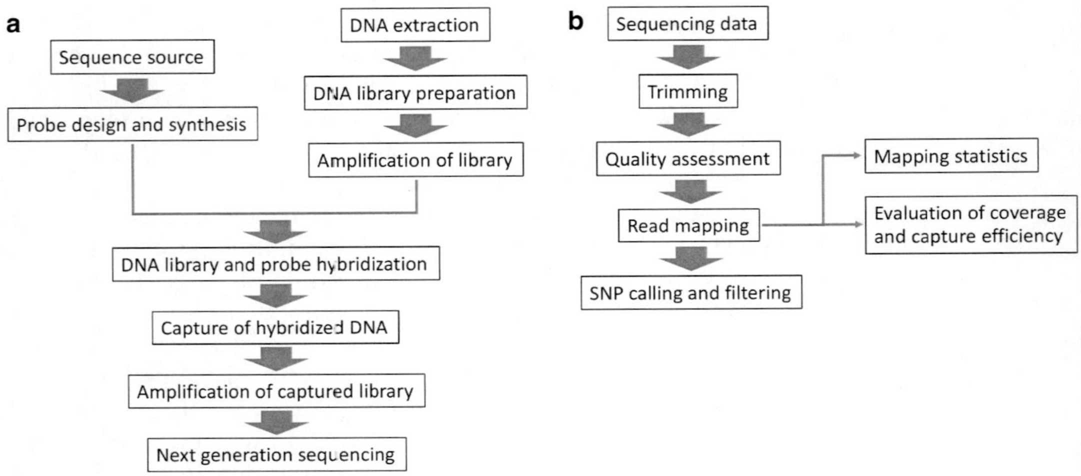


**Fig. 1** Overview of target enrichment sequencing. (**a**) General process of target enrichment sequencing via probe capture. (**b**) Bioinformatics analysis for evaluation of target capture and single nucleotide polymorphism calling

## 2 Materials

In this chapter, we mainly refer to and recommend to use Agilent SureSelect$^{XT}$ Target Enrichment System for Illumina Multiplexed Sequencing for supplies, reagents, and equipment required for this method (https://www.agilent.com/cs/library/usermanuals/public/G9681-90000.pdf). All consumables such as tubes, tips, PCR plates and water are nuclease-free.

### 2.1 Genomic DNA Extraction

1. Nuclease-free water.
2. Liquid nitrogen and appropriate containers.
3. Mortar and pestle.
4. Water bath.
5. CTAB buffer: 2% cetyltrimethylammonium bromide (CTAB), 100 mM Tris–HCl, 1.4 M NaCl, 2% polyvinylpyrrolidone-40 (PVP-40), 20 mM EDTA (pH 8.0).
6. 10× low TE buffer: 10 mM Tris–HCl and 0.1 mM EDTA (pH 8.0).
7. 1× low TE buffer: 10× low TE buffer diluted ten times.
8. 70% and 95% ethanol.
9. 0.2% β mercaptoethanol.
10. Chloroform.
11. Isopropanol.
12. RNAse A (Life Technologies).

13. NanoDrop™ 2000 Spectrophotometers (Thermo Fisher Scientific).
14. Quanti-iT PicoGreen ds DNA kit (Life Technologies).

#### 2.2 DNA Shearing and Purification

1. Covaris E220 sonicator (Covaris).
2. microTUBE AFA Fiber Pre-Slit Cap vials (Covaris).
3. PCR plates.
4. 2100 Bioanalyzer (Agilent).
5. Agencourt AMPure XP Kit (Beckman Coulter Genomics).
6. Magnetic separator (Thermo Fisher Scientific).
7. 70% ethanol.
8. Nuclease-free water.

#### 2.3 End Repair, 3′ Adenylation, and Ligation to Paired-End Adaptors

1. PCR plates.
2. Nuclease-free water.
3. 10× End Repair Buffer.
4. dNTP Mix.
5. T4 DNA Polymerase.
6. Klenow DNA Polymerase.
7. T4 Polynucleotide Kinase.
8. 10× Klenow Polymerase Buffer.
9. dATP.
10. SureSelect Adaptor Oligo Mix.
11. 5× T4 DNA Ligase Buffer.
12. T4 DNA Ligase.

#### 2.4 Amplification of Adaptor-Ligated Library and Assessment of Quality and Quantity

1. Nuclease-free water.
2. SureSelect Primer.
3. SureSelect ILM Indexing Pre-Capture PCR Reverse Primer.
4. 5× Herculase II Reaction Buffer.
5. dNTP Mix.
6. Herculase II Fusion DNA Polymerase.
7. 2100 Bioanalyzer (Agilent).
8. Quanti-iT PicoGreen ds DNA kit (Life Technologies).

#### 2.5 DNA-Probe Hybridization

1. Nuclease-free water.
2. PCR plates.
3. Hybridization buffers: SureSelect Hyb 1, SureSelect Hyb 2, SureSelect Hyb 3, and SureSelect Hyb 4.

4. SureSelect Block Mix: SureSelect Indexing Block 1, SureSelect Block 2, SureSelect ILM Indexing Block 3.
5. SureSelect RNase Block.
6. SureSelect Wash Buffer 2.
7. Streptavidin T1 beads.
8. SureSelect Binding Buffer.
9. SureSelect Wash Buffer 1.

#### 2.6 Amplification of Captured Libraries with Indexing Primers and Pooling

1. Nuclease-free water.
2. 5× Herculase II Reaction Buffer.
3. Herculase II Fusion DNA Polymerase.
4. 100 mM dNTP Mix.
5. SureSelect ILM Indexing Post-Capture Forward PCR Primer.
6. 2100 Bioanalyzer (Agilent).
7. KAPA Library Quantification Kits (KAPA Biosystems).

#### 2.7 Tools for Bioinformatics Analysis

1. Unix-capable computer: 64-bit computer with at least 1 Tb hard disk and 16 Gb memory. Alternatively, we suggest performing analysis on a supercomputer platform if available.
2. Installation of R program (https://www.r-project.org).
3. CD-HIT [49].
4. BLAT [50].
5. EMBOSS [51].
6. Trimmomatic v0.36 [52].
7. Fastqc (http://www.bioinformatics.babraham.ac.uk/projects/fastqc/).
8. BWA [53].
9. Samtools [54].
10. Bedtools [55].

## 3 Methods

#### 3.1 Probe Design and Selection for TES

Probe design is a critical step of TES study, since it influences the efficiency of probe hybridization and the effectiveness of TES. The probe design also determines what genomic regions to target for enrichment and sequencing.

*3.1.1 Sequence Source (Transcript and Genomic Sequences)*

Probes can be designed from transcribed sequences such as ESTs, reference genomes, or assemblies from genomic sequences (whole genome sequencing, reduced representation library, genotyping by sequencing) of the species of interest or closely related species. Typically, a TES study could focus on genic regions or exons. For this case, the EST sequences or gene sequences from reference genome can be used for probe design. Alternatively, a TES study may have some genes of interest that can be used for probe design. A pipeline is described below for identifying genes or regions with specific functions or associated with specific traits, which will be used for probe design. A reference genome of species of interest or closely related species is assumed to be publicly available.

*3.1.2 Target Genes or Regions for Probe Design*

Target genes or regions of interest need to be identified in order to design the probes. This can be done by following this procedure below:

1. Search for genes related to a certain trait (e.g., nodulation) from the literatures.
2. Extract sequences for related genes using the Refseq database [56]. In this example, we search for nodulation related genes using the keyword "nodulation" in the Refseq database available at https://www.ncbi.nlm.nih.gov/refseq/.
3. Filter the sequences using the "Taxonomic Groups" tab to retrieve only the sequences that are present in "plants." This is done to avoid searching hits in other organisms like bacteria.
4. In order to select particular genes, click on the "check box" symbol on the left-hand side of the name of the gene. Not checking any of the boxes will select all the sequences of the search by default. The sequences are then downloaded by clicking "Send to:" drop-down button and selecting "Complete Record > Choose Destination: File > Download Format: FASTA > Create File". This downloads a Fasta file that contains the DNA sequences of the genes that are in the Refseq database corresponding to the keyword provided. Save this file as "refseq_cds.fasta".
5. Merge this dataset with other gene sequences that are downloaded from literature review (saved in file "literature_cds.fasta"). This can be done using following commands in UNIX/LINUX system:

```
cat literature_cds.fasta refseq_cds.fasta > merged_cds.fasta
```

6. Since there can be redundant genes in the merged database, we use CD-HIT to cluster the genes and remove redundancy. This can be done using following commands:

```
cd-hit-est -i merged_cds.fasta -o nr_merged_cds.fasta
```

   -i represents input file.

-o represents output file.

This gives us a nonredundant set of genes that will be targeted for probe design.

7. Now these nonredundant set of genes need to be compared to the set of genes in the reference genome, in order to identify the trait related genes in the reference genome. For this, the coding DNA sequences of the reference genome needs to be downloaded (e.g., from Phytozome: https://phytozome.jgi.doe.gov/pz/portal.html):

   Select the reference species. Click on Bulk Data. Click on the appropriate version of the reference genome. Click "annotation." Finally, click the file that ends with ".cds.fa.gz" in order to download the coding DNA sequence of the reference genome. Save this file as "reference_cds.fasta".

8. Align the gene set that was downloaded earlier (nr_merged_cds.fasta) to the gene set from the reference genome. We can use BLAT tool for this using following commands:

```
blat reference_cds.fasta nr_merged_cds.fasta blat_output
-out=blast8 -noHead
```

   Here, the output of the BLAT results will be saved in "blat_output" file. The -out=blast8 refers to NCBI blast tabular format and -noHead suppresses ".psl" header (so it is a tab-separated file).

9. Extract gene names in reference genome that match >80% identity and length >400 bp with CDS of nonredundant gene set using following commands:

```
awk '{if($3>80 && $4>400) {print}}' blat_output| awk '{print
$2}' | sort | uniq > unique_cds_name_reference
```

   This command produces the names of the unique genes (unique_cds_name_reference) that are present in the reference genome and have >80% identity to the selected genes of interest and are longer than 400 bp.

10. Extract the CDS of these nonredundant genes in the reference genome using following commands from Kent software (http://hgdownload.soe.ucsc.edu/admin/exe/):

```
faSomeRecords reference_cds.fasta unique_cds_name_reference
cds_uniq_reference.fasta
```

#### 3.1.3 Probe design

1. Design probes with this set of CDS in the reference genome using EMBOSS software:

```
splitter -sequence cds_uniq_reference.fasta -outseq pro-
bes_cds -size 120 -overlap 60
```

-sequence refers to the CDS sequences from which probes will be designed.

-outseq refers to the filename of output file that will contain the probes.

-size refers to the length of each probe.

-overlap refers to the amount of overlap that is permitted between two probes.

2. Remove probes that are shorter than 120 bp using the following perl script:

```
perl remove_small.pl 120 probes_cds > probes_cds_120bp
```

Here 120 refers to the minimum length of the probes. The perl script "remove_small.pl" is provided (*see* **Note 1**).

#### 3.1.4 Probe Selection

In order to identify effective probes, that is, probes that do not have multiple hits on the reference genome, use following commands:

1. Align the designed probes to the reference genome using BLAT:

```
blat reference_genome.fa probes_cds_120bp output_blat_probes
-out=blast8 -noHead
```

2. Filter good hits (defined under cutoff: *e*-value $\leq 1e{-}05$; alignment length $\times$ percentage of identity $\geq$ 96 (i.e., 120 bp $\times$ 80% = 96 bp)).

```
awk '$3*$4/120>=80' output_blat_probes | awk '{print $1}' >
output_blat_goodProbes_name
```

3. Find probes that are unique:

```
cat output_blat_goodProbes_name | sort | awk '{a[$0]++}END
{for (i in a)if(a[i]<2)print i}' > output_blat_goodProbes_u-
niq
```

4. Extract the sequences of the good probes.

```
faSomeRecords probes_cds_120bp output_blat_goodProbes_uniq
probes_good.fasta
```

#### *3.1.5 Check for Repeat Regions Within Probes*

1. Probes that contain repeat regions need to be removed. For this, we will use Repeatmasker program (http://www.repeat-masker.org/):

```
Repeatmasker probes_good.fasta -species Medicago -dir . -pa 8
```

-species refers to the species or clade of the input sequence. The species name must be a valid NCBI Taxonomy Database species name and be contained in the RepeatMasker repeat database.

-dir refers to the current working directory (.)

-pa refers to the number of processors to use in parallel (8 in this example)

The output file will be named probes_good.fasta.masked.

2. Remove probes having repeats.

```
fasta_formatter -i probes_good.fasta.masked -o probes_goo-
d_repeatMasked0 -w 0
sed -s 'N;/N/!P;D' probes_good_repeatMasked0 > probes_final.
fasta
```

Here, the first command formats the fasta file to have unlimited width. The second command then removes any probes that are masked or having repeat regions in them (represented by N).

3. This gives us the final probe set: probes_final.fasta that can be now used for downstream processing.

### ***3.2 Genomic DNA Extraction***

1. Grind frozen peanut leaf samples in liquid nitrogen with mortar and pestle. Keep the powder frozen until extraction; 100–150 mg of powder aliquot can be used for one extraction.
2. Turn on water bath to 65 °C.
3. Warm up proteinase K (Sigma) at room temperature.

4. Make CTAB Extraction Buffer by adding 0.2% β mercaptoethanol and 0.1 mg/mL proteinase K to the CTAB buffer.
5. Warm CTAB Extraction Buffer at 65 °C for 15 min.
6. Add 800 μL CTAB extraction buffer to each tube while frozen and quickly mix samples powder with the buffer by vortex. To avoid DNA degradation, do one sample a time.
7. Incubate the samples at 65 °C for 45 min and vortex three times during the incubation.
8. Let the solution cool and add 500 μL of chloroform to each sample and vortex.
9. Centrifuge at 12,000 × *g* for 5 min at room temperature.
10. Transfer 500 μL upper aqueous layer to new tubes.
11. Repeat **steps 9** and **10**.
12. Remove 400 μL aqueous layer to new tubes.
13. Add an equal volume of isopropanol to each sample, invert the tubes a few times and incubate for 20 min at room temperature.
14. Centrifuge at 12,000 × *g* for 15 min.
15. Discard the supernatant and wash pellet with 300 μL of 70% ethanol.
16. Spin down pellet at 12,000 × *g* for 1 min and remove ethanol.
17. Wash pellet with 300 μL of 95% ethanol, spin down for 1 min and discard ethanol.
18. Quick spin and remove residual ethanol by pipette tip.
19. Dry the pellet for 20 min at room temperature.
20. Add 100 μL of 1× low TE and add 10 μL/mL of RNAse A (Life Technologies) to the tube.
21. Put the tube at 4 °C overnight.
22. Gently pipet to mix DNA.
23. Read NanoDrop for DNA concentration or perform PicoGreen DNA quantitation assay using Quanti-iT PicoGreen ds DNA kit (*see* **Note 2**). PicoGreen Assay is more accurate for dsDNA measurement.
24. Run 100 ng of DNA on 1% agarose gel to check quality.
25. Store DNA at −20 °C until use.

### 3.3 Shear DNA by Covaris Sonicator (Covaris)

1. Turn on Covaris E220, set temperature at 6 °C and degas for 1 h.
2. Set sonication parameters: Duty cycle 10%, Intensity 4, Cycles per burst 200, Time 360 s (*see* **Note 3**).

3. Prepare 3 μg of DNA samples in 130 μL low TE in Low DNA binding tubes.
4. Transfer samples to microTUBE AFA Fiber Pre-Slit Cap vials (Covaris) and keep the vials on ice.
5. Load vials to sample rack and start sonication.
6. Transfer the samples (130 μL/sample) to a PCR plate and put the plate on ice.
7. Save 1 μL of sheared each DNA sample to assess shearing quality on 2100 Bioanalyzer using DNA 1000 chip and move on to the next section.

#### 3.4 Purify the Sample Using AMPure XP Beads (Refer to SureSelect$^{XT}$ Manual, Page 24)

1. Add 180 μL of beads to each sheared DNA sample well.
2. Seal the plate and mix on a plate mixer.
3. Incubate the samples for 5 min at room temperature.
4. Put the plate in a 96-well magnetic separation device and wait for solutions to be cleared.
5. Centrifuge briefly and put the plate back on the magnetic device and wait until the solution becomes clear (2–4 min).
6. Carefully open the cap and remove and discard the supernatant without disturbing the beads.
7. While the plate is on the magnetic device, add 200 μL of fresh 70% ethanol to each well and wait for 1 min.
8. Remove the ethanol carefully.
9. Repeat **steps** 7 and **8**.
10. Briefly Spin the plate and put it back to the magnetic device.
11. Wait for 1 min and remove the residual ethanol without disturbing the beads.
12. Put the plate without caps into a PCR machine and run 37 °C for 3–5 min to dry the beads.
13. Take the plate out of the PCR machine and add 50 μL nuclease-free water to each well.
14. Seal the wells with caps/film and mix well on a vortex mixer.
15. Briefly spin the plate and let it sit on bench for 2 min.
16. Put the plate in the magnetic device and wait until the solution is cleared.
17. Transfer the solution containing DNA (about 48 μL) to a new plate.
18. Run 1 μL of each sample on 2100 Bioanalyzer with DNA 1000 chip following Bioanalyzer protocol and a peak of 150–200 bp should be present.

    We suggest stopping here and store samples at −20 °C.

19. Assess DNA quality using 2100 Bioanalyzer and DNA 1000 chip (refer to SureSelect$^{XT}$ manual page 26). The peak should be around 175 bp.

### 3.5 Repair the Ends (Refer to SureSelect$^{XT}$ Manual, Page 28 and Table 10)

1. Thaw reagents on ice and mix well.
2. Thaw PCR plate at room temperature.
3. Prepare end-repair master mix. For each reaction, combining the following:
   (a) 35.2 μL of nuclease-free water.
   (b) 10 μL of 10× End Repair Buffer.
   (c) 1.6 μL of dNTP Mix.
   (d) 1 μL of T4 DNA Polymerase.
   (e) 2 μL of Klenow DNA Polymerase.
   (f) 2.2 μL of T4 Polynucleotide Kinase.
4. Add the 52 μL master mix to each well containing 48 μL sheared and purified DNA.
5. Mix by pipetting and run 20 °C for 30 min and hold at 4 °C for 2 min in a PCR machine.

### 3.6 Purify End-Repaired DNA with AMPure XP Beads (Refer to SureSelect$^{XT}$ Manual Page 29)

1. Warm the AMPure XP beads at room temperature for at least 30 min.
2. Mix the beads to homogeneity.
3. Add 180 μL of beads to each sheared DNA sample well.
4. Seal the plate and mix on a plate mixer.
5. Incubate the samples for 5 min at room temperature.
6. Put the plate in a 96-well magnetic separation device and wait for solutions to be cleared.
7. Centrifuge briefly and put the plate back on the magnetic device and wait until the solution becomes clear (2–4 min).
8. Carefully open the cap and remove and discard the supernatant without disturbing the beads.
9. While the plate is on the magnetic device, add 200 μL of fresh 70% ethanol to each well and wait for 1 min.
10. Remove the ethanol carefully.
11. Repeat **steps 9** and **10**.
12. Briefly Spin the plate and put it back to the magnetic device.
13. Wait for 1 min and remove the residual ethanol without disturbing the beads.
14. Put the plate without caps into a PCR machine and run 37 °C for 3–5 min to dry the beads.

15. Take the plate out of the PCR machine and add 32 μL nuclease-free water to each well.
16. Seal the wells with caps/film and mix well on a vortex mixer.
17. Briefly spin the plate and let it sit on bench for 2 min.
18. Put the plate in the magnetic device and wait until the solution is cleared.
19. Transfer the solution containing DNA (about 30 μL) to a new plate.
20. Run 1 μL of each sample on 2100 Bioanalyzer with DNA 1000 chip following Bioanalyzer protocol and a peak of 150–200 bp should be present. You can stop here and store samples at −20 °C overnight.

### 3.7 Adenylate the 3′ End of the Repaired DNA (Refer to SureSelect$^{XT}$ Manual, Page 30 and Tables 12 and 13)

1. Thaw reagents on ice and mix well.
2. Prepare Adenylation master mix for each reaction:
   (a) 11 μL of Nuclease-free water.
   (b) 5 μL of 10× Klenow Polymerase Buffer.
   (c) 1 μL dATP.
   (d) 3 μL of Exo (−) Klenow for one reaction.
3. Add the 20 μL master mix to each sample (about 30 μL) in the PCR plate.
4. Mix well by pipetting.
5. Incubate the plate at 37 °C for 30 min and then hold at 4 °C in a PCR machine.

### 3.8 Purify DNA with AMPure Beads (Refer to SureSelect$^{XT}$ Manual Page 31)

1. Warm the AMPure XP beads at room temperature for at least 30 min.
2. Mix the beads to homogeneity.
3. Add 90 μL of beads to each sheared DNA sample well.
4. Seal the plate and mix on a plate mixer.
5. Incubate the samples for 5 min at room temperature.
6. Put the plate in a 96-well magnetic separation device and wait for solutions to be cleared.
7. Centrifuge briefly and put the plate back on the magnetic device and wait until the solution becomes clear (2–4 min).
8. Carefully open the cap and remove and discard the supernatant without disturbing the beads.
9. While the plate is on the magnetic device, add 200 μL of fresh 70% ethanol to each well and wait for 1 min.
10. Remove the ethanol carefully.
11. Repeat **steps 9** and **10**.

12. Briefly Spin the plate and put it back to the magnetic device.
13. Wait for 1 min and remove the residual ethanol without disturbing the beads.
14. Put the plate without caps into a PCR machine and run 37 °C for 2 min to dry the beads.
15. Take the plate out of the PCR machine and add 15 μL nuclease-free water to each well.
16. Seal the wells with caps/film and mix well on a vortex mixer.
17. Briefly spin the plate and let it sit on bench for 2 min.
18. Put the plate in the magnetic device and wait until the solution is cleared.
19. Transfer the solution containing DNA (about 13 μL) to a new plate.

Proceed to the next step immediately.

### 3.9 Ligate dA-Tailed DNA to Paired-End Adaptor (Refer to SureSelect$^{XT}$ Manual Page 32 and Table 14)

1. Thaw reagents on ice and mix well.
2. Prepare master mix for each ligation reaction:
   (a) 15.5 μL Nuclease-free water.
   (b) 10 μL 5× T4 DNA Ligase Buffer.
   (c) 10 μL SureSelect Adaptor Oligo Mix.
   (d) 1.5 μL T4 DNA Ligase.
3. Add 37 μL master mix to each sample (about 13 μL) and mix well by pipetting.
4. Put the plate to a PCR machine and run 20 °C for 15 min and then hold at 4 °C.

### 3.10 Purify Ligated DNA with AMPure XP Beads (Refer to SureSelect$^{XT}$ Manual Page 33)

1. Warm the AMPure XP beads at room temperature for at least 30 min.
2. Mix the beads to homogeneity.
3. Add 90 μL of beads to each sheared DNA sample well.
4. Seal the plate and mix on a plate mixer.
5. Incubate the samples for 5 min at room temperature.
6. Put the plate in a 96-well magnetic separation device and wait for solutions to be cleared.
7. Centrifuge briefly and put the plate back on the magnetic device and wait until the solution becomes clear (2–4 min).
8. Carefully open the cap and remove and discard the supernatant without disturbing the beads.
9. While the plate is on the magnetic device, add 200 μL of fresh 70% ethanol to each well and wait for 1 min.
10. Remove the ethanol carefully.

11. Repeat **steps 9** and **10**.
12. Briefly Spin the plate and put it back to the magnetic device.
13. Wait for 1 min and remove the residual ethanol without disturbing the beads.
14. Put the plate without caps into a PCR machine and run 37 °C for 2 min to dry the beads.
15. Take the plate out of the PCR machine and add 32 μL nuclease-free water to each well.
16. Seal the wells with caps/film and mix well on a vortex mixer.
17. Briefly spin the plate and let it sit on bench for 2 min.
18. Put the plate in the magnetic device and wait until the solution is cleared.
19. Transfer the solution containing DNA (about 32 μL) to a new plate.

    We recommend stopping here and store samples at −20 °C.

### 3.11 Amplify the Adaptor-Ligated Library (Refer to SureSelect$^{XT}$ Manual Pages 34–36)

1. Thaw reagents on ice and mix well.
2. Prepare PCR master mix for each reaction:
    (a) 21 μL Nuclease-free water.
    (b) 1.25 μL SureSelect Primer.
    (c) 1.25 μL SureSelect ILM Indexing Pre-Capture PCR Reverse Primer.
    (d) 10 μL 5× Herculase II Reaction Buffer.
    (e) 0.5 μL 100 mM dNTP Mix.
    (f) 1 μL Herculase II Fusion DNA Polymerase for one reaction.
3. Transfer the 35 μL PCR master mix to each 15 μL ligated and purified DNA sample (save remaining samples).
4. Mix by pipetting.
5. Run the following PCR profile:
    (a) 1 cycle at 98 °C 2 min.
    (b) 4 cycles at 98 °C 30 s, 65 °C 30 s and 72 °C 1 min.
    (c) 1 cycle at 72 °C 10 min.
    (d) 1 cycle hold at 4 °C.

### 3.12 Purify the Amplified Library by AMPure Beads (Refer to SureSelect$^{XT}$ Manual Page 37)

1. Warm AMPure beads for 30 min at room temperature and mix well.
2. Add 90 μL beads to each well containing 100 μL reaction and mix ten times with pipette.
3. Incubate for 5 min at room temperature.

4. Place the plate to a magnetic device and wait for 3–5 min.
5. Carefully remove the cleared solution with pipette.
6. Add 200 μL fresh 70% ethanol and wait for 1 min.
7. Carefully remove the ethanol with pipette.
8. Repeat the **steps 6** and 7.
9. Cover the plate with strip caps and briefly spin the plate and put back to the magnetic device.
10. Remove residual ethanol without disturbing the beads.
11. Put the plate without caps into a PCR machine and dry the beads at 37 °C for 3–5 min.
12. Add 30 μL nuclease-free water to each well.
13. Seal the plate with strip caps, vortex to mix and briefly spin.
14. Incubate the plate for 2 min on bench.
15. Put the plate back to the magnetic stand and wait for 2–3 min.
16. Remove cleared supernatant to a new plate, each well should have about 30 μL. Discard the beads.

### 3.13 Assess the Quality and Quantity of the Ligated, Amplified and Purified Libraries (Refer to SureSelect$^{XT}$ Manual Page 38)

1. Run 1 μL of each library on 2100 Bioanalyzer using DNA 1000 chip following Bioanalyzer protocols.
2. The peak size 220–278 bp.
3. Determine the concentrations of the DNA libraries by running PicoGreen ds DNA assay (Reference: Life Technologies Quant-iT ds DNA kit).

   We recommend stopping here and store samples at −20 °C.

### 3.14 Hybridize DNA Libraries to Capture Probe (Refer to SureSelect$^{XT}$ Manual Pages 60–65, Tables 30–35)

Test plate sealing methods with strip caps or sealing films by incubating 27 μL of water in the 96-well plate including the wells you will use and the wells in the center and at the four edges at 65 °C for 24 h in the PCR machine to be for hybridization. Evaporation of the water should not exceed 4 μL to pass the test and thus can be used to cover hybridization plate. In addition, change to new caps/film after caps or film is removed from a plate.

1. Thaw reagents under the temperature suggested in the Table 20 on page 60 in the SureSelect$^{XT}$ Manual.
2. Dilute each library to 221 ng/μL dilution of each library with DNase-free water. For libraries ≤221 ng/μL, concentrate DNA in vacuum concentrator below 45 °C and reconstitute it with nuclease-free water to 221 ng/μL. Transfer 3.4 μL of the diluted libraries to a PCR plate.
3. Prepare Hybridization Buffer. For one reaction, you need to mix the following reagents:
   (a) SureSelect Hyb 1 (orange cap or bottle) 6.63 μL.

   (b) SureSelect Hyb 2 (red cap) 0.27 μL
   (c) SureSelect Hyb 3 (yellow cap or bottle) 2.65 μL
   (d) SureSelect Hyb 4 (black cap or bottle) 3.45 μL.
4. Keep the hybridization buffer at room temperature.
5. Prepare SureSelect Block Mix by mixing the components on ice. For each reaction, mix the following components:
   (a) SureSelect Indexing Block 1 (green cap) 2.5 μL.
   (b) SureSelect Block 2 (blue cap) 2.5 μL.
   (c) SureSelect ILM Indexing Block 3 (brown cap) 0.6 μL.
6. Add 5.6 μL SureSelect Block Mix to each library (3.4 μL) and pipette up and down to mix.
7. Seal the plate with the tested caps or film and put the plate in a PCR machine.
8. Run 95 °C for 5 min and hold at 65 °C for at least 5 min in a PCR machine with 105 °C heated lid.
9. Prepare the appropriate dilution of SureSelect RNase Block.
10. For capture library size ≥3.0 Mb, dilute 1 part of the SureSelect RNase Block with 3 parts of nuclease-free water. For capture library size <3.0 Mb, reference Table 33 on page 63 in the manual for the dilution) and keep the mixture at room temperature briefly.
11. Prepare the Capture Library Hybridization Mix.
12. For Capture Library Size ≥3.0Mb, combine:
    (a) Hybridization Buffer mixture from **step 3**: 13 μL.
    (b) 1: 3 diluted RNase Block Mix from **step 5**: 2 μL.
    (c) Capture Library (RNA probe): 5 μL.
13. Maintain the DNA library + Block Mix plate at 65 °C in the PCR machine and add 20 μL of the Capture Library Hybridization Mix from **step 12** to each sample well and pipetting up and down 8–10 times.
14. Seal the plate and incubate the hybridization mixture at 65 °C (lid remain at 105 °C) for 24 h.

### *3.15 Prepare Streptavidin-Coated Beads and Capture Hybridized DNA (Refer to SureSelect$^{XT}$ Manual, Pages 66–68)*

*See* **Note 4**.

1. Prewarm SureSelect Wash Buffer 2 and keep it at 65 °C.
2. Warm the Streptavidin T1 beads at room temperature.
3. Vigorously resuspend the Dynabeads MyOne Streptavidin T1 magnetic beads on a vortex.
4. Add 50 μL of the resuspended beads to each well of a fresh PCR plate.

5. Wash the beads:
   (a) Add 200 μL of SureSelect Binding Buffer.
   (b) Mix by pipetting up and down to resuspend the beads.
   (c) Put the plate or strip tube into a magnetic separator device.
   (d) Wait until the solution is clear, then remove and discard the supernatant.
   (e) Repeat step a through step d two more times for a total of three washes.
6. Resuspend the beads in 200 μL of SureSelect Binding Buffer.
7. While maintaining hybridization at 65 °C, transfer entire hybridization reaction (about 25–29 μL in each well) to the plate containing 200 μL of washed Streptavidin T1 beads (Use multichannel pipette to transfer eight samples at once).
8. Mix well by slowly pipetting 6–10 times until beads are well resuspended.
9. Cap the wells and incubate the plate on a plate mixer and mix vigorously (1400–1800 rpm) for 30 min at room temperature.
10. Briefly spin the plate and put it on a magnetic plate separator and wait for a few min until the solution is cleared.
11. Remove the supernatant with pipette and discard it.
12. Take the plate off the separator and add 200 μL of SureSelect Wash Buffer 1 to each well and mix well with pipette.
13. Incubate the samples for 15 min at room temperature.
14. Briefly spin the plate and put the plate in the separator and wait until solutions are cleared.
15. Remove and discard the supernatant.
16. Take the plate off the separator and resuspend the beads in 200 μL of prewarmed Wash Buffer 2 at 65 °C. Pipette up and down until beads are fully resuspended.
17. Cap the wells, then incubate the sample plate for 10 min at 65 °C in the PCR machine.
18. Put the plate in the magnetic separator. Wait for the solution to clear, then remove and discard the supernatant.
19. Repeat **steps 16–18** two more times with a total of three washes with Wash Buffer 2.
20. Make sure all of the wash buffer has been removed during the final wash.
21. Take the plate off the separator and add 30 μL of nuclease-free water, pipette up and down to resuspend the beads.

22. Keep the samples on ice. The library will be kept on beads at this moment.

### *3.16 Amplify the Captured Libraries with Indexing Primers (Refer to SureSelect$^{XT}$ Manual, Pages 69–72, Tables 36–39)*

1. Thaw PCR reagents and mix well and keep on ice.
2. Prepare one indexing amplification reaction for each DNA library:
   - (a) Nuclease-free water 18.5 μL.
   - (b) 5× Herculase II Reaction Buffer (clear cap) 10 μL.
   - (c) Herculase II Fusion DNA Polymerase (red cap) 1 μL.
   - (d) 100 mM dNTP Mix (green cap) 0.5 μL.
   - (e) SureSelect ILM Indexing Post-Capture Forward PCR Primer (orange cap) 1 μL.
3. Add 31 μL of the PCR reaction mix prepared in **step 2** to each sample well of a fresh PCR plate.
4. Add 5 μL of the appropriate indexing primer to each well, one primer in one well and record which index primer is used for which sample.
5. Pipette up and down to resuspend the beads from **step 21** in Subheading 3.15 and transfer 14 μL beads to the PCR reactions. Now the total volume is 50 μL.
6. Store the remaining library-bound beads at –20 °C for future use, if needed.
7. Transfer the plate to a PCR machine and run the following profile:
   - (a) 1 cycle at 98 °C 2 min.
   - (b) 12 cycles at 98 °C 30 s, 57 °C 30 s, 72 °C 1 min.
   - (c) 1 cycle at 72 °C 10 min.
   - (d) 1 cycle at 4 °C hold.
8. When the PCR reaction is complete, spin the plate briefly and proceed to the next section.

### *3.17 Purify the Amplified Captured Libraries Using AMPure XP Beads*

1. Warm AMPure beads for 30 min at room temperature and mix well.
2. Add 90 μL beads to each well containing 50 μL PCR reaction and mix 10× with pipette.
3. Incubate for 5 min at room temperature.
4. Place the plate to a magnetic device and wait for 3–5 min.
5. Carefully remove the cleared solution with pipette.
6. Add 200 μL fresh 70% ethanol and wait for 1 min.
7. Carefully remove the ethanol with pipette.
8. Repeat the **steps 6** and **7** for a total of two washes.

9. Cover the plate with strip caps and briefly spin the plate and put back to the magnetic device.
10. Remove residual ethanol without disturbing the beads.
11. Put the plate without caps into a PCR machine and dry the beads at 37 °C for 3 min.
12. Add 30 μL nuclease-free water to each well.
13. Seal the plate with strip caps, vortex to mix and briefly spin.
14. Incubate the plate for 2 min on bench.
15. Put the plate back to the magnetic stand and wait for 2–3 min.
16. Remove cleared supernatant (about 30 μL) to a new plate and discard the beads.

We recommend stopping here and store samples at 4 °C overnight or −20 °C for longer time.

### 3.18 Assess Indexed Library DNA Quantity and Quality and Pooling Library for NGS (Refer to SureSelect$^{XT}$ Manual Pages 75–80)

1. Run 1 μL indexed, amplified and purified library on 2100 Bioanalyzer for high sensitivity assay. Our peak range is 215–600 bp.
2. Run quantitative PCR to determine library quantities using KAPA Library Quantification Kits (KAPA Biosystems).
3. Pooling libraries of equal moles to a final 10 nM solution and submit 20 μL for 2× 150 bp paired-end sequencing with Illumina NextSeq 500 platform or higher.

### 3.19 Bioinformatics Analysis of TES Reads

Here, we present a pipeline for evaluating coverage and capture efficiency as well as identifying SNPs, a common type of DNA polymorphisms, using read data generated from TES (Fig. 1b). For downstream analysis, we assume that data are received in Fastq format. Related programs below are operated using Linux

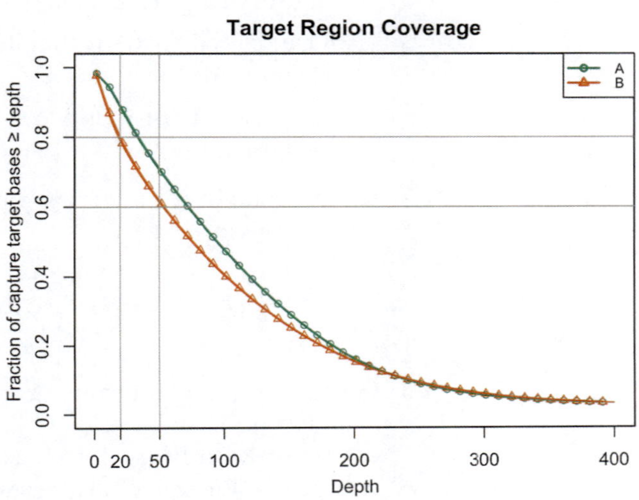


**Fig. 2** Accumulative fraction of target regions covered by sequence reads

command line. For demonstration, we assume 150 bp paired-end reads and Fastq files are received for two samples: "A.R1.fastq", "A.R2.fastq" for sample A; "B.R1.fastq", "B.R2.fastq" for sample B.

Sequence trimming using Trimmomatic v0.36:

1. Concatenate common adapter sequences from Illumina into one file named "adapters.fasta":
   (a) Go to "adapters" directory under trimmomatic installation directory.
   (b) cat NexteraPE-PE.fa TruSeq2-SE.fa TruSeq3-PE.fa TruSeq2-PE.fa TruSeq3-PE-2.fa TruSeq3-SE.fa > adapters.fasta
2. Create a new directory under home directory in which to perform sequence trimming:
   (a) mkdir homedirectory/Trim.
   (b) cd homedirectory/Trim.
3. Copy Fastq files into Trim directory.
4. Copy file "trimmomatic-0.36.jar" from bin directory under trimmomatic installation directory into Trim directory.
5. Run trimmomatic v0.36 to remove adaptor and low-quality sequences, using sample A for demonstration:

```
java -jar trimmomatic-0.36.jar PE A.R1.fastq A.R2.fastq A.R1.paired.fastq A.R1.unpaired.fastq A.R2.paired.fastq A.R2.unpaired.fastq ILLUMINACLIP:adapters.fasta:2:30:20 LEADING:3 TRAILING:3 SLIDINGWINDOW:4:15 MINLEN:50
```

6. Obtain output files:

   "A.R1.paired.fastq" and "A.R2.paired.fastq" are files containing surviving read pairs.

   "A.R1.unpaired.fastq" is the file containing forward only surviving reads.

   "A.R2.unpaired.fastq" is the file containing reverse only surviving reads.
7. Concatenate forward/reverse only surviving read files into one unpaired Fastq file:

```
cat A.R1.unpaired.fastq A.R2.unpaired.fastq > A.unpaired.fastq
```

8. Obtain summary of trimming statistics from history file which includes:

   Input Read Pairs, Both Surviving: (%), Forward Only Surviving: (%), Reverse Only Surviving: (%), Dropped: (%).

9. We suggest checking quality of trimmed reads using Fastqc.
10. For sample B, run same procedure 5–9 and finally obtain following files:

```
"B.R1.paired.fastq", "B.R2.paired.fastq", and "B.unpaired.
fastq"
```

Read mapping using BWA-MEM:

1. Create a new directory under home directory in which to perform read mapping:
   (a) mkdir homedirectory/Mapping.
   (b) cd homedirectory/Mapping.
2. Download genome reference file "Reference.fasta" into directory Mapping (*see* **Note 5**).
3. Index reference file:

```
bwa index reference.fasta
```

4. Align reads to the reference genome:

```
bwa mem reference.fasta A.R1.paired.fastq A.R2.paired.fastq >
A.paired.sam
bwa mem reference.fasta A.unpaired.fastq > A.unpaired.sam
bwa mem reference.fasta B.R1.paired.fastq B.R2.paired.fastq >
B.paired.sam
bwa mem reference.fasta B.unpaired.fastq > B.unpaired.sam
```

5. Convert sam files to bam files using Samtools v1.9:

```
samtools view -S -b -o A.paired.bam A.paired.sam
samtools view -S -b -o A.unpaired.bam A.unpaired.sam
samtools view -S -b -o B.paired.bam B.paired.sam
samtools view -S -b -o B.unpaired.bam B.unpaired.sam
```

6. Merge bam files:

```
samtools merge A.bam A.paired.bam A.unpaired.bam
samtools merge B.bam B.paired.bam B.unpaired.bam
```

7. Sort bam files:

```
samtools sort -o A.sorted.bam A.bam
samtools sort -o B.sorted.bam B.bam
```

8. Obtain mapping statistics, taking sample A as an example:
   (a) Record number of mapped reads *N*1.

```
samtools flagstat A.bam
```

   (b) Calculate number of input reads *N*2 = (total line number of read files)/4. To calculate line number:

```
wc -l A.R1.paired.fastq
wc -l A.R2.paired.fastq
wc -l A.unpaired.sam
```

   (c) Calculate mapping rate = $(N1/N2) \times 100\%$.

Evaluation of coverage and capture efficiency:

For evaluation of probe coverage and capture efficiency, the chromosome location coordinates of probes in the bed file from previous section are extended 150 bp from both directions to make a new bed file named "Probes.bed". For example, a previous location "Chr01 1000 1120" becomes "Chr01 850 1270". The new chromosome locations in this bed file represents probe target regions. The chromosome locations are extended 150 bp, because a probe could hybridize a flanking and partially overlapping DNA fragment which will be subsequently sequenced. The bed file "Probes.bed" and bam files generated at mapping step are used for following steps with Sample A as an example. To visualize overall coverage, a cumulative distribution of the percentage of target bases that are covered by at least 20×, 50×, and so on reads is plotted using Bedtools and R program (http://www.gettinggeneticsdone.com/2014/03/visualize-coverage-exome-targeted-ngs-bedtools.html).

1. Convert bam files to bed format:

```
bedtools bamtobed -i A.sorted.bam > A.sorted.bed
```

2. Obtain read and target region intersections:

```
bedtools intersect -wa -wb -a A.sorted.bed -b Probes.bed >
A-Probes.intersections
```

3. Count the number of reads (O) overlapping with probe target regions:

```
cut -f4 A-Probes.intersections | sort | uniq | wc -l
```

4. Calculate on-target rate ($R$):

$$R = (O/N1) \times 100\%.$$

5. Obtain a summary histogram for all the target regions in the "Probes.bed" file:

```
bedtools coverage -hist -abam A.sorted.bam -b Probes.bed |
grep ^all > A.sorted.bam.hist.all.txt
bedtools coverage -hist -abam B.sorted.bam -b Probes.bed |
grep ^all > B.sorted.bam.hist.all.txt
```

6. Plot the distribution using R script "Coverage.pl" (*see* **Note 6**).

An example figure is provided in Fig. 2. Approximately 80% of the probe target regions are covered by at least 20× reads for Sample B. For Sample A, more probe target regions (about 85%) are covered by at least 20× reads. Similarly, for both Sample A and B, at least 60% of the probe target regions are covered by at least 50× reads. Overall, Sample A has more coverage of reads than Sample B.

SNP calling using TES data:

One common application of TES is to identify SNPs between samples for genotyping. Here we present a pipeline for calling SNPs using TES data. The reference genome and mapping files are used. To improve the accuracy of SNP calling, only uniquely mapped reads are retrieved from the alignment files for SNP calling with "Sample A" as an example.

1. Convert bam files back to sam format:

```
samtools view -h -o A.sam A.bam
```

2. Split sam files for filtering:

```
grep '@' A.sam > A.header
grep -v '@' A.sam > A.body
```

3. Retrieve uniquely mapped reads (mapping quality > 0 and without 'XA:Z' tag):

```
awk '$5>0' A.body > A.body0
grep -v 'XA:Z' A.body0 > A.body0xaz
```

4. Construct new sam file containing uniquely mapped reads:

```
cat A.header A.body0xaz > A.unique.sam
```

5. Convert sam to bam format:

```
samtools view -S -b -o A.unique.bam A.unique.sam
```

6. Sort bam file:

```
Samtools sort -o A.unique.sorted.bam A.unique.bam
```

7. Generate bam index file "A.unique.sorted.bam.bai":

```
samtools index A.unqiue.sorted.bam
```

8. Follow **steps 1**–7 for Sample B to generate below files:

```
B.unique.sorted.bam
B.unique.sorted.bam.bai
```

9. Run Samtools for variant calling:

```
Santools mpileup -ugf reference.fasta -t DPR -q 30 -Q 20 A.
unique.sorted.bam B.unique.sorted.bam -o samtools.raw.bcf
```

10. Convert bcf file to vcf format using Bcftools v1.9:

```
Bcftools call -c -v samtools.raw.bcf > samtools.raw.vcf
```

11. Filter off INDELs:

```
grep -v 'INDEL' samtools.raw.vcf > samtools.raw1.vcf
```

12. Filter SNPs based on read depth:
    (a) A homozygous reference genotype "0/0" with "reference allele depth" $\geq 4$ and "alternate allele depth" $= 0$.
    (b) A homozygous alternate genotype "1/1" with "reference allele depth" $= 0$ and "alternate allele depth" $\geq 4$.
    (c) A heterozygous genotype "0/1" with "reference allele depth" $\geq 2$ and "alternate allele depth" $\geq 2$.

## 4 Notes

1. Perl script "remove_small.pl"
   (Source: http://www.bioinformatics-made-simple.com/2012/07/how-to-filter-sequence-by-their-length.html).

```
#!/usr/bin/perl
use strict;
use warnings;
my $minlen = shift or die "Error: 'minlen' parameter not
provided\n";
{
 local $/=">";
 while(<>) {
 chomp;
 next unless /\w/;
 s/>$//gs;
 my @chunk = split /\n/;
 my $header = shift @chunk;
 my $seqlen = length join "", @chunk;
 print ">$_" if($seqlen >= $minlen);
 }
 local $/="\n";
}
```

2. NanoDrop readings should be $260/280 \geq 1.95$, and $260/230 \geq 2.5$.
3. This setting gives us a target peak size of 150–200 bp. Other specific sample volumes and target peak sizes need to be optimized by adjusting the parameters following the DNA shearing protocol for E220 ultrasonicator (https://covaris.com/wp-content/uploads/pn_010308.pdf).
4. Hybrid capture reagents provided in SureSelect$^{XT}$ Target Enrichment Box 1 are stored at room temperature.

5. For demonstration, we assume that the genome reference of species of interest or a closely related species is available and can be used for analysis. For species without any genome references available. The sequencing reads can be assembled using programs such as Velvet [57] and SPAdes [58], which is not covered in this method.
6. R script "Coverage.pl"

```
print(files <- list.files(pattern=".txt"))
print(labs <- paste("", gsub("prefixToTrash-0|\cr.pe\cr.on
\cr.pos\cr.dedup\cr.realigned\cr.recalibrated\cr.bam\cr.cov
\cr.hist\cr.txt\cr.all\cr.txt", "", files, perl=TRUE),
sep=""))
cov <- list()
cov_cumul <- list()
for (i in 1:length(files)) {
 cov[[i]] <- read.table(files[i])
 cov_cumul[[i]] <- 1-cumsum(cov[[i]][,5])
}
cols2<-c("#1B9E77" ,"#D95F02")
cov <- list()
cov_cumul <- list()
par(new=F)
for (i in 1:length(files)) {
 cov[[i]] <- read.table(files[i])
 cov_cumul[[i]] <- 1-cumsum(cov[[i]][,5])
  plot((cov[[i]][seq(2,401,10),2]), (cov_cumul[[i]][seq
(1,400,10)]), col=cols2[i], lty=1, pch=i,xlab="",ylab="",
xlim=c(0,400),ylim=c(0,1),xaxt='n',yaxt='n',cex=0.8)
  lines((cov[[i]][seq(2,401,10),2]), (cov_cumul[[i]][seq
(1,400,10)]), col=cols2[i], lty=1,lwd=2,xlab="", ylab="")
 par(new=T)
}
plot(cov[[1]][2:401, 2], cov_cumul[[1]][1:400],type='n',
xlab="Depth", ylab="Fraction of capture target bases \u2265
depth", ylim=c(0,1.0), main="Target Region Coverage")
lab<-c("A","B")
for(i in 1:length(cov)) points(cov[[i]][2:401, 2], cov_cumul
[[i]][1:400], type="l",col=cols2[i])
legend("topright",legend=lab, col=c(cols2[1:7]), lty=1,lwd=2,
pch=c(1:7), cex=0.8)
abline(v = 20, col = "gray60")
abline(v = 50, col = "gray60")
abline(h = 0.60, col = "gray60")
abline(h = 0.80, col = "gray60")
axis(1, at=c(20,50), labels=c(20,50))
```

## Acknowledgments

This work was supported by the Florida Peanut Producers Association and National Peanut Board.

### References


1. Schmutz J, Cannon SB, Schlueter J, Ma J, Mitros T, Nelson W, Hyten DL, Song Q, Thelen JJ, Cheng J (2010) Genome sequence of the palaeopolyploid soybean. Nature 463:178
2. Venter JC, Adams MD, Myers EW, Li PW, Mural RJ, Sutton GG, Smith HO, Yandell M, Evans CA, Holt RA (2001) The sequence of the human genome. Science 291:1304–1351
3. Fowler S, Thomashow MF (2002) Arabidopsis transcriptome profiling indicates that multiple regulatory pathways are activated during cold acclimation in addition to the CBF cold response pathway. Plant Cell 14:1675–1690
4. Maher CA, Kumar-Sinha C, Cao X, Kalyana-Sundaram S, Han B, Jing X, Sam L, Barrette T, Palanisamy N, Chinnaiyan AM (2009) Transcriptome sequencing to detect gene fusions in cancer. Nature 458:97
5. Trapnell C, Williams BA, Pertea G, Mortazavi A, Kwan G, Van Baren MJ, Salzberg SL, Wold BJ, Pachter L (2010) Transcript assembly and quantification by RNA-Seq reveals unannotated transcripts and isoform switching during cell differentiation. Nat Biotechnol 28:511
6. Alkan C, Coe BP, Eichler EE (2011) Genome structural variation discovery and genotyping. Nat Rev Genet 12:363
7. Bhatia D, Wing R, Singh K (2013) Genotyping by sequencing, its implications and benefits. Crop Improv 40:101–111
8. Davey JW, Hohenlohe PA, Etter PD, Boone JQ, Catchen JM, Blaxter ML (2011) Genome-wide genetic marker discovery and genotyping using next-generation sequencing. Nat Rev Genet 12:499
9. Van Orsouw NJ, Hogers RC, Janssen A, Yalcin F, Snoeijers S, Verstege E, Schneiders H, van der Poel H, Van Oeveren J, Verstegen H (2007) Complexity reduction of polymorphic sequences (CRoPS™): a novel approach for large-scale polymorphism discovery in complex genomes. PLoS One 2:e1172
10. Van Tassell CP, Smith TP, Matukumalli LK, Taylor JF, Schnabel RD, Lawley CT, Haudenschild CD, Moore SS, Warren WC, Sonstegard TS (2008) SNP discovery and allele frequency estimation by deep sequencing of reduced representation libraries. Nat Methods 5:247
11. Baird NA, Etter PD, Atwood TS, Currey MC, Shiver AL, Lewis ZA, Selker EU, Cresko WA, Johnson EA (2008) Rapid SNP discovery and genetic mapping using sequenced RAD markers. PLoS One 3:e3376
12. Elshire RJ, Glaubitz JC, Sun Q, Poland JA, Kawamoto K, Buckler ES, Mitchell SE (2011) A robust, simple genotyping-by-sequencing (GBS) approach for high diversity species. PLoS One 6:e19379
13. Peterson BK, Weber JN, Kay EH, Fisher HS, Hoekstra HE (2012) Double digest RADseq: an inexpensive method for de novo SNP discovery and genotyping in model and non-model species. PLoS One 7:e37135
14. Sonah H, Bastien M, Iquira E, Tardivel A, Légaré G, Boyle B, Normandeau É, Laroche J, Larose S, Jean M (2013) An improved genotyping by sequencing (GBS) approach offering increased versatility and efficiency of SNP discovery and genotyping. PLoS One 8:e54603
15. Mamanova L, Coffey AJ, Scott CE, Kozarewa I, Turner EH, Kumar A, Howard E, Shendure J, Turner DJ (2010) Target-enrichment strategies for next-generation sequencing. Nat Methods 7:111
16. Mertes F, ElSharawy A, Sauer S, van Helvoort JM, Van Der Zaag P, Franke A, Nilsson M, Lehrach H, Brookes AJ (2011) Targeted enrichment of genomic DNA regions for next-generation sequencing. Brief Funct Genomics 10:374–386
17. Saiki RK, Gelfand DH, Stoffel S, Scharf SJ, Higuchi R, Horn GT, Mullis KB, Erlich HA (1988) Primer-directed enzymatic amplification of DNA with a thermostable DNA polymerase. Science 239:487–491
18. Cho RJ, Mindrinos M, Richards DR, Sapolsky RJ, Anderson M, Drenkard E, Dewdney J, Reuber TL, Stammers M, Federspiel N (1999) Genome-wide mapping with biallelic markers in Arabidopsis thaliana. Nat Genet 23:203
19. Wang DG, Fan J-B, Siao C-J, Berno A, Young P, Sapolsky R, Ghandour G, Perkins N, Winchester E, Spencer J (1998) Large-scale identification, mapping, and

genotyping of single-nucleotide polymorphisms in the human genome. Science 280:1077–1082
20. Tewhey R, Warner JB, Nakano M, Libby B, Medkova M, David PH, Kotsopoulos SK, Samuels ML, Hutchison JB, Larson JW (2009) Microdroplet-based PCR enrichment for large-scale targeted sequencing. Nat Biotechnol 27:1025
21. Nilsson M, Malmgren H, Samiotaki M, Kwiatkowski M, Chowdhary BP, Landegren U (1994) Padlock probes: circularizing oligonucleotides for localized DNA detection. Science 265:2085–2088
22. Porreca GJ, Zhang K, Li JB, Xie B, Austin D, Vassallo SL, LeProust EM, Peck BJ, Emig CJ, Dahl F (2007) Multiplex amplification of large sets of human exons. Nat Methods 4:931
23. Turner EH, Lee C, Ng SB, Nickerson DA, Shendure J (2009) Massively parallel exon capture and library-free resequencing across 16 genomes. Nat Methods 6:315
24. Okou DT, Steinberg KM, Middle C, Cutler DJ, Albert TJ, Zwick ME (2007) Microarray-based genomic selection for high-throughput resequencing. Nat Methods 4:907
25. Gnirke A, Melnikov A, Maguire J, Rogov P, LeProust EM, Brockman W, Fennell T, Giannoukos G, Fisher S, Russ C (2009) Solution hybrid selection with ultra-long oligonucleotides for massively parallel targeted sequencing. Nat Biotechnol 27:182
26. Ballester LY, Luthra R, Kanagal-Shamanna R, Singh RR (2016) Advances in clinical next-generation sequencing: target enrichment and sequencing technologies. Expert Rev Mol Diagn 16:357–372
27. Ng SB, Turner EH, Robertson PD, Flygare SD, Bigham AW, Lee C, Shaffer T, Wong M, Bhattacharjee A, Eichler EE (2009) Targeted capture and massively parallel sequencing of 12 human exomes. Nature 461:272
28. Clark MJ, Chen R, Lam HY, Karczewski KJ, Chen R, Euskirchen G, Butte AJ, Snyder M (2011) Performance comparison of exome DNA sequencing technologies. Nat Biotechnol 29:908
29. Russell J, Mascher M, Dawson IK, Kyriakidis S, Calixto C, Freund F, Bayer M, Milne I, Marshall-Griffiths T, Heinen S (2016) Exome sequencing of geographically diverse barley landraces and wild relatives gives insights into environmental adaptation. Nat Genet 48:1024
30. Mascher M, Schuenemann VJ, Davidovich U, Marom N, Himmelbach A, Hübner S, Korol A, David M, Reiter E, Riehl S (2016) Genomic analysis of 6,000-year-old cultivated grain illuminates the domestication history of barley. Nat Genet 48:1089
31. Grabowski PP, Evans J, Daum C, Deshpande S, Barry KW, Kennedy M, Ramstein G, Kaeppler SM, Buell CR, Jiang Y (2017) Genome—wide associations with flowering time in switchgrass using exome—capture sequencing data. New Phytol 213:154–169
32. Peng Z, Fan W, Wang L, Paudel D, Leventini D, Tillman BL, Wang J (2017) Target enrichment sequencing in cultivated peanut (Arachis hypogaea L.) using probes designed from transcript sequences. Mol Genet Genomics 292:955–965
33. Gasc C, Peyret P (2018) Hybridization capture reveals microbial diversity missed using current profiling methods. Microbiome 6:61
34. McCormack JE, Harvey MG, Faircloth BC, Crawford NG, Glenn TC, Brumfield RT (2013) A phylogeny of birds based on over 1,500 loci collected by target enrichment and high-throughput sequencing. PLoS One 8: e54848
35. Kaur P, Gaikwad K (2017) From genomes to GENE-omes: exome sequencing concept and applications in crop improvement. Front Plant Sci 8:2164
36. Warr A, Robert C, Hume D, Archibald A, Deeb N, Watson M (2015) Exome sequencing: current and future perspectives. G3 Genomes Genet 5:1543–1550
37. Paux E, Roger D, Badaeva E, Gay G, Bernard M, Sourdille P, Feuillet C (2006) Characterizing the composition and evolution of homoeologous genomes in hexaploid wheat through BAC-end sequencing on chromosome 3B. Plant J 48:463–474
38. Weitemier K, Straub SC, Cronn RC, Fishbein M, Schmickl R, McDonnell A, Liston A (2014) Hyb-Seq: combining target enrichment and genome skimming for plant phylogenomics. Appl Plant Sci 2:1400042
39. Comer JR, Zomlefer WB, Barrett CF, Davis JI, Stevenson DW, Heyduk K, JH L-M (2015) Resolving relationships within the palm subfamily Arecoideae (Arecaceae) using plastid sequences derived from next-generation sequencing. Am J Bot 102:888–899
40. Muraya MM, Schmutzer T, Ulpinnis C, Scholz U, Altmann T (2015) Targeted sequencing reveals large-scale sequence polymorphism in maize candidate genes for biomass production and composition. PLoS One 10:e0132120
41. Saintenac C, Jiang D, Akhunov ED (2011) Targeted analysis of nucleotide and copy

number variation by exon capture in allotetraploid wheat genome. Genome Biol 12(9):R88
42. Winfield MO, Allen AM, Burridge AJ, Barker GL, Benbow HR, Wilkinson PA, Coghill J, Waterfall C, Davassi A, Scopes G (2016) High-density SNP genotyping array for hexaploid wheat and its secondary and tertiary gene pool. Plant Biotechnol J 14:1195–1206
43. Wendler N, Mascher M, Nöh C, Himmelbach A, Scholz U, Ruge-Wehling B, Stein N (2014) Unlocking the secondary gene-pool of barley with next-generation sequencing. Plant Biotechnol J 12:1122–1131
44. Neves LG, Davis JM, Barbazuk WB, Kirst M (2013) Whole-exome targeted sequencing of the uncharacterized pine genome. Plant J 75:146–156
45. Galvão VC, Nordström KJ, Lanz C, Sulz P, Mathieu J, Posé D, Schmid M, Weigel D, Schneeberger K (2012) Synteny-based mapping-by-sequencing enabled by targeted enrichment. Plant J 71:517–526
46. Song J, Yang X, Resende MF Jr, Neves LG, Todd J, Zhang J, Comstock JC, Wang J (2016) Natural allelic variations in highly polyploidy Saccharum complex. Front Plant Sci 7:804
47. Yang X, Song J, Todd J, Peng Z, Paudel D, Luo Z, Ma X, You Q, Hanson E, Zhao Z (2019) Target enrichment sequencing of 307 germplasm accessions identified ancestry of ancient and modern hybrids and signatures of adaptation and selection in sugarcane (Saccharum spp.), a 'sweet'crop with 'bitter'genomes. Plant Biotechnol J 17:488–498
48. Haun WJ, Hyten DL, Xu WW, Gerhardt DJ, Albert TJ, Richmond T, Jeddeloh JA, Jia G, Springer NM, Vance CP (2011) The composition and origins of genomic variation among individuals of the soybean reference cultivar Williams 82. Plant Physiol 155:645–655
49. Li W, Godzik A (2006) Cd-hit: a fast program for clustering and comparing large sets of protein or nucleotide sequences. Bioinformatics 22:1658–1659
50. Kent WJ (2002) BLAT—the BLAST-like alignment tool. Genome Res 12:656–664
51. Rice P, Longden I, Bleasby A (2000) EMBOSS: the European molecular biology open software suite. Trends Genet 16:276–277
52. Bolger AM, Lohse M, Usadel B (2014) Trimmomatic: a flexible trimmer for Illumina sequence data. Bioinformatics 30:2114–2120
53. Li H (2013) Aligning sequence reads, clone sequences and assembly contigs with BWA-MEM. arXiv 2013:1303.3997
54. Li H, Handsaker B, Wysoker A, Fennell T, Ruan J, Homer N, Marth G, Abecasis G, Durbin R (2009) The sequence alignment/map format and SAMtools. Bioinformatics 25 (16):2078–2079
55. Quinlan AR, Hall IM (2010) BEDTools: a flexible suite of utilities for comparing genomic features. Bioinformatics 26:841–842
56. Pruitt KD, Tatusova T, Maglott DR (2006) NCBI reference sequences (RefSeq): a curated non-redundant sequence database of genomes, transcripts and proteins. Nucleic Acids Res 35: D61–D65
57. Zerbino DR, Birney E (2008) Velvet: algorithms for de novo short read assembly using de Bruijn graphs. Genome Res 18:821–829
58. Bankevich A, Nurk S, Antipov D, Gurevich AA, Dvorkin M, Kulikov AS, Lesin VM, Nikolenko SI, Pham S, Prjibelski AD (2012) SPAdes: a new genome assembly algorithm and its applications to single-cell sequencing. J Comput Biol 19:455–477

# Chapter 11

# Developing Transposable Element Marker System for Molecular Breeding

R. S. Bhat, K. Shirasawa, Y. Monden, H. Yamashita, and M. Tahara

## Abstract

Transposable element (TE) marker system was developed considering the useful properties of the transposable elements such as their large number in the animal and plant genomes, high rate of insertion polymorphism, and ease of detection. Various methods have been employed for developing a large number of TE markers in several crop plants for genomics studies. Here we describe some of these methods including the recent whole genome search. We also review the application of TE markers in molecular breeding.

**Key words** Transposable element marker system, Transposon display, Suppression PCR, Whole genome in silico search for *AhMITE1*, Application of transposable element marker system in molecular breeding

## 1 Introduction

Transposable elements (TEs) are the stretches of 100 bp to several kilobases DNA which have the ability to transpose within the genome. Genome sequencing studies have identified that TEs make up about 50–70% of eukaryotic genome. For the first time, TEs were discovered by Barbara McClintock more than 60 years ago as a causal factor of mutable loci in maize. Now they are known to induce transposon-mediated variations fueling adaptation and evolution. Largely, two types of TEs have been identified based on the transposition intermediates. Class I TEs transpose through RNA, while class II TEs do so by DNA. This difference results in the increase in the copy number of class I TEs within the genome. This is reflected in several plant species where the retrotransposons predominate over DNA transposons [1–3].

TEs of both the classes have two components; an autonomous element carrying an open reading frame (ORF) coding for transposase required for transposition, and a nonautonomous element without carrying the transposase ORF. However, both

Mukesh Jain and Rohini Garg (eds.), *Legume Genomics: Methods and Protocols*, Methods in Molecular Biology, vol. 2107, https://doi.org/10.1007/978-1-0716-0235-5_11, © Springer Science+Business Media, LLC, part of Springer Nature 2020

autonomous and nonautonomous elements undergo transposition with a duplication of a short genomic sequence [called a target site duplication (TSD)] at the site of insertion [4]. Structurally, class I TEs (specifically called retrotransposons) can either have direct long terminal repeats (LTRs) (as in LTR retrotransposons) or can be free from LTRs (as in non-LTR retrotransposons). Class II transposons generally have short terminal inverted repeat (TIR).

Since the transposons are ubiquitous and widely distributed in the genomes, and the sites of insertion show polymorphism, they have been used to develop TE markers [5–7] for various genetic analysis. Because the TEs influence the organization, stability [8, 9], and evolution of the genome [10], and thereby the transcriptome [11], the TE markers form a special marker system.

### 1.1 Development of Retrotransposon Markers

Several types of retrotransposon markers have been developed. Purugganan, Wessler [12] examined the internal differences in the sequence or structure of *magellan* retrotransposon to examine the relationships within the genus *Zea* and *Tripsacum*. Sequence-specific amplification polymorphisms (S-SAP) was used to study the genetic distribution of *BARE-1*-like retrotransposable elements in the barley genome [13]. Two amplification-based marker detection methods were developed for *BARE-1* in barley [14] based on the position of LTRs which are common to all retrotransposons. Both the techniques detect the polymorphism of LTR retrotransposon insertion sites. The inter-retrotransposon amplified polymorphism (IRAP) amplifies the intervening genomic region between two LTR sequences using one or two PCR primers specific to the LTR sequences, whereas retrotransposon-microsatellite amplified polymorphism (REMAP) amplifies the DNA segment between an LTR sequence and a simple sequence repeat (SSR) using LTR-specific and SSR (such as $(GA)_n$ or $(CT)_n$ sequences)-specific primer combinations. IRAPs were developed in apple [15, 16], wheat [17, 18], flax [19], and *Lallemantia iberica* [20], and REMAPs were generated in apple [15], barley [14, 21], wheat [18], *Lallemantia iberica* [20], almond [22], and so on. Both IRAP and REMAP produce multiple reliable and reproducible bands. But they require LTR sequence information for marker development. Therefore, retrotransposon-based insertion polymorphism (RBIP) markers [23] were developed which target one retrotransposon insertion at a particular site, where the PCR primers specific to the flanking regions of an individual insertion site are used for amplification. Sequence-specific amplified polymorphism (S-SAP) was developed through modification of the AFLP method, which has been widely used in several plant species [13, 16, 24–28]. After genomic DNA is digested with a restriction enzyme, an adaptor is ligated to the digested DNA fragments. PCR amplification for S-SAP is performed with LTR sequence-specific and adaptor sequence-specific primer combinations. Technically,

S-SAP is quite similar to transposon display (TD) which also has been widely used [6, 29–34]. They differ only for the target TE; S-SAP targets LTR retrotransposon while TD targets DNA transposon.

Kalendar et al. [35] developed a novel DNA fingerprinting method, iPBS, which was highly effective for isolating a wide range of LTR retrotransposon families using only sets of primer binding site (PBS) primers, and directly visualizing the polymorphism among cultivars. iPBS was used in *Tetradium ruticarpum* [36] and many Fagaceae species [37]. To screen diverse LTR retrotransposon families at a genome-wide scale, iPBS method was employed using the Illumina HiSeq 2000 sequencing platform to obtain LTR sequences in citrus, apple, and soybean (unpublished data).

Recently, a high-throughput sequencing platform was developed for an efficient screening of LTR retrotransposon families that show high levels of insertion polymorphism among closely related cultivars [38, 39]. This approach was tested in strawberry genome to determine 24 LTR retrotransposon families. Among them, several families were experimentally confirmed for their high levels of insertion polymorphism among closely related cultivars. Additionally, a large number of insertion sites for retrotransposon families that showed diverse insertion patterns were identified using high-throughput sequencing [40–43]. Diversity, distribution, and dynamics of full-length *Copia* and *Gypsy* LTR retroelements were studied in tomato through genome-wide in silico search [44]. A marker system based on TaqMan quantitative PCR (qPCR) combined with retrotransposon-based insertion polymorphism (RBIP) was developed to estimate the dosage of a LTR retrotransposon (*scIvana*) in sugarcane [45].

Apart from LTR retrotransposons, short interspersed nuclear elements (SINEs), a non-LTR retrotransposon, was employed for marker development. SINEs are dispersed repetitive sequences present in most of the plant genomes. Inter-SINE amplified polymorphism (ISAP) was employed for genotype-specific high-resolution fingerprinting in potato [46].

Suppression PCR was used for the whole-genome experimental identification of insertion/deletion polymorphisms of interspersed repeats [47]. The same technique (Fig. 1) was applied for identifying genome-wide polymorphic insertion sites of *PHARE1*, a retrotransposon from the red bean (*Vigna angularis* (Willd.) Ohwi & H. Ohashi) (unpublished). Insertion sites in several cultivars, called tracers, could be efficiently compared to those of a control cultivar, called a driver, and the sites not present in the control were obtained as PCR products.

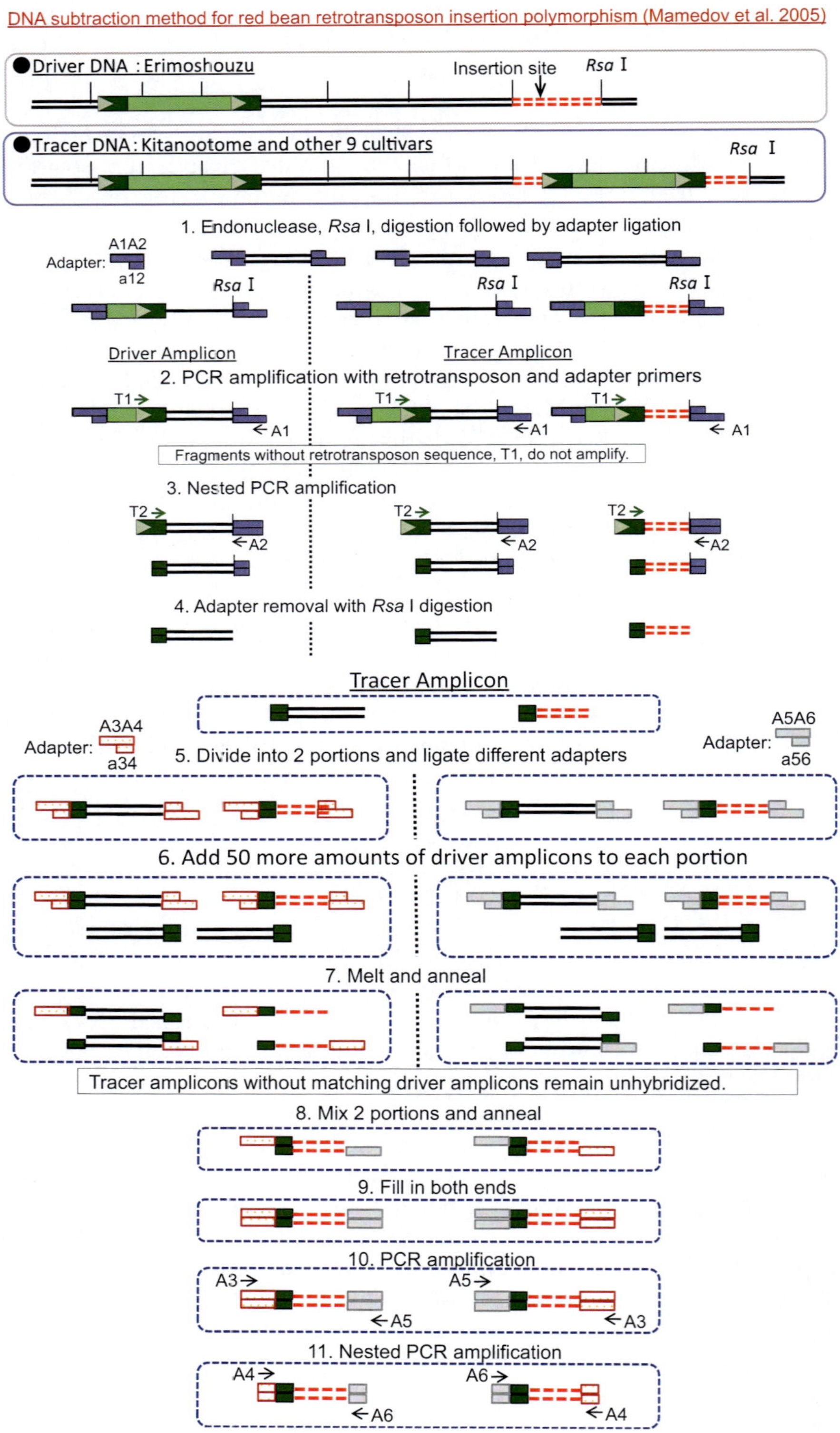


**Fig. 1** Suppression PCR used in red bean

### 1.2 Development of DNA Transposon Markers

Among the DNA transposons, miniature inverted-repeat transposable elements (MITEs) are the most abundant (1000–15,000 per haploid genome) among the genomes. When compared to the nonautonomous elements of class II TEs, MITEs are relatively small (100–500 bp) and have a preference for insertion into 2–3 bp targets that are rich in A and T residues [6]. Various families of MITEs have been described [48]. *Heartbreaker* (*Hbr*) family was the first one to be considered for marker development [6] after it was identified [49] and isolated from maize [48]. Subsequently, the diversity and dynamics of *DcMaster*-like elements of the *PIF/Harbinger* superfamily was studied in carrot [50] and *Medicago truncatula* [51]. *Vulmar/VulMITE* were identified in *Chenopodiaceae* subfamily *Betoideae* [52]. Genomic distribution of MITEs in barley was determined by MITE-AFLP mapping [53], and the novel MITEs were isolated and analyzed for marker utility [54]. TEs were also identified using whole-genome sequencing in banana [55], peanut [1], wheat [56], foxtail millet [57], and so on.

An exhaustive effort has been made in peanut to identify [58] and develop *Arachis hypogaea* miniature inverted-repeat transposable element 1 (*AhMITE1*) markers [59–64]. Before the availability of the genome sequence of peanut (or its progenitors), Shirasawa et al. [62], followed a modified method of transposon display (TD) described by Nunome et al. [65], which is generally used to develop SSR markers, to find out the polymorphic sites of *AhMITE1*. TD was also attempted for TEs other than MITE. TD for *dTph1* (*Ac/Ds* family) in *Petunia hybrida* [33], *DcMaster* (PIF/Harbinger-like) in carrot [66], *Rim2/Hipa* (CACTA family) [67] and *nDart* (*hAT* family) in rice [32], and so on. have been reported in the past.

Alternatively, transposon sites in the genomes could be identified from the whole genome sequence reads obtained by next-generation sequencing technology. Many tools are available for TE discovery and annotation [68]. In peanut, with the availability of the genome sequences of the diploid progenitors of peanut [1], efforts were made to identify the genome-wide distribution of *AhMITE1* [64]. For this, a set of diverse genotypes (33) including the genetically unstable peanut mutants which show hyperactivity of *AhMITE1* [69] were used to discover the *AhMITE1* insertion polymorphic sites (AIPs). Whole genome resequencing (WGRS) reads from these 33 genotypes were analyzed using the computational method polymorphic TEs and their movement detection (PTEMD) [70] for the de novo discovery of AIPs. This study could find thousands of *AhMITE1* copies over the diploid peanut genomes (Fig. 2) which were subsequently used to develop *AhMITE1* markers (Fig. 3).

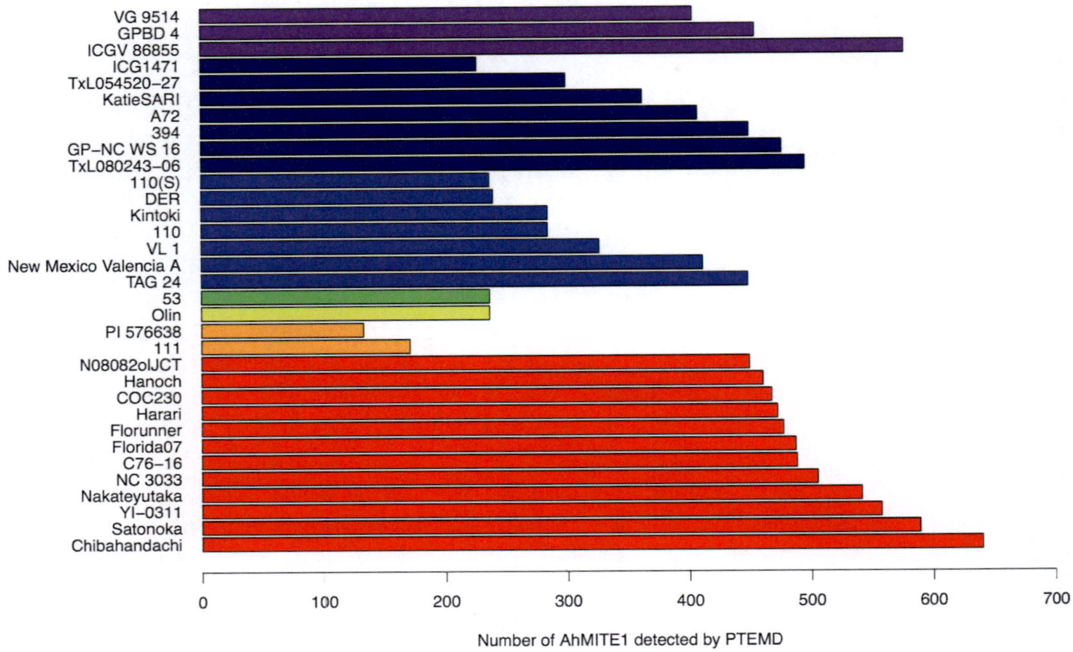


**Fig. 2** Number of *AhMITE1* copies detected in the genomes of 33 peanut lines by the PTEMD method. The bars are colored in accordance with subspecies and varieties: *A. hypogaea* spp. *hypogaea* var. *hypogaea* (red); *A. hypogaea* spp. *hypogaea* var. *hirsuta* (orange); *A. hypogaea* spp. *fastigiata* var. *vulgaris* (yellow); *A. hypogaea* spp. *fastigiata* var. *peruviana* (green); *A. hypogaea* spp. *fastigiata* var. *fastigiata* (blue); *A. hypogaea* (navy blue); and *A. cardenasii* derivatives (purple). Details of the 33 lines are described by Gayathri et al. [64]

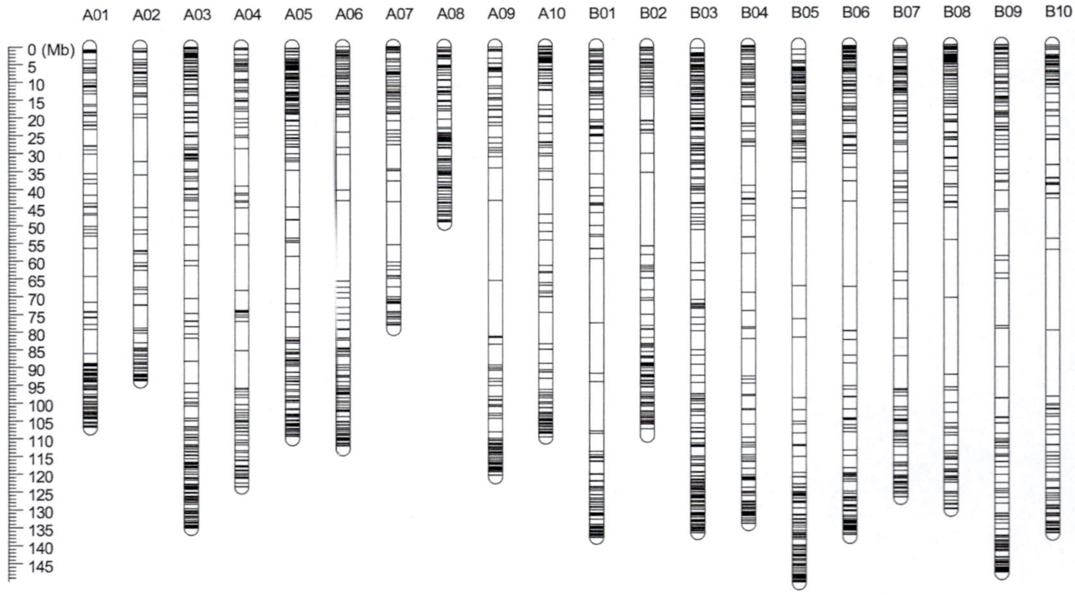


**Fig. 3** Genome positions of the *AhMITE1* markers developed through the PTEMD method. Names of the ten chromosomes of *A. duranensis* (A01–A10) and *A. ipaensis* (B01–B10), both of which are possible diploid ancestors of tetraploid cultivated peanut, *A. hypogaea*, are indicated at the top of the chromosomes

## 2 Material

### 2.1 Suppression PCR

1. Select the appropriate driver cultivar and tracer cultivars.
2. Synthesize the adapter-specific primers [T1 primer (PHARE1-LTR-D0): 5′CCCTTA TCTTACATTATAAATAAAAC TACC3′ and T2 primer (AzCopia-LTR-D1): 5′GCT AAACACATAAGGGAGATTTCTCC3′].
3. *Rsa*I restriction enzyme.

### 2.2 AhMITE1-Display

#### 2.2.1 TE-Enriched Genome Library Construction and Sequencing

1. DNeasy Plant Mini Kit (Qiagen).
2. Restriction enzymes [*Afa*I, *Alu*I, *Hae*III, or *HpyCH4V* (NEB)].
3. Rapid Ligation Buffer (Promega).
4. Forward-strand linker DNA (5′GTTTAGCCTTGTAGCAG AAGC3′) and reverse-strand linker DNA (5′GCTTCTGCTACAAGGCTAAACAAAA3′).
5. T4 DNA Ligase (Promega).

#### 2.2.2 Probe Preparation

1. Oligonucleotide primer (5′AAGGTGGATACTACMATGAAGAT3′) labeled at the 5′ end with biotin.
2. $NH_4$-based Reaction Buffer (Bioline).
3. BIOTAQ DNA Polymerase (Bioline).

#### 2.2.3 Hybridization

1. Dynabeads M-280 Streptavidin (Thermo Fisher Scientific).
2. KOD -Plus- and Buffer for KOD -Plus- Ver. 2 (Toyobo).
3. QIAquick Gel Extraction Kit (Qiagen).
4. PCR Buffer (Takara).
5. TaKaRa Taq DNA Polymerase (Takara).
6. pGEM-T Easy Vector (Promega),
7. Rapid Ligation Buffer (Promega).
8. T4 DNA Ligase (Promega).
9. ElectroTen-Blue Electroporation Competent Cells (Agilent Technologies).
10. DNA sequencing.

### 2.3 Whole Genome In Silico Search for AhMITE1

1. The following *AhMITE1* sequence as reported by Patel et al. [58] is used for whole genome in silico search.

5′GGTGGATACTACAATGAAGATGGCATAATTGTCTTCATA
TGAGTATATTTCTTTTTGACCTTTGGATGATGGATTG
TATGGTTAGATTTTGATATTTATAAAAGTGTTGTTTTT
GTTTAAAGTGTGGCTAAATAAATAAACCACACTTTTAA
CCAAAGCATCTTCATGAGAAGATATTTTTGCCATCTT
CATTGTAGTATCCACC3′

2. The whole-genome sequence reads for as many diverse genotypes as possible is to be collected. They may be collected as reads from target lines with Illumina HiSeq, NextSeq, MiSeq, and so on. Otherwise, the reads can be downloaded from Sequence Read Archives of public DNA databases (e.g., GenBank/NCBI, EMBL-Bank/EBI, and DDBJ).

## 3 Methods

### 3.1 Suppression PCR

The procedure for suppression PCR described below is based on the previous method [47] with a few modifications.

3. Cut the DNA (1000 ng) of a driver cultivar and that of the tester cultivars with *Rsa*I. Ligate the fragments with adapters.
4. Amplify the fragments with the adapter-specific primers.
5. Carryout nested PCR amplification.
6. Remove the adapters with *Rsa*I digestion.
7. Divide the mixture into two portions and ligate with different adapters.
8. Add 12.5 μl of the driver (72 ng/μl) to 2.6 μl of the tracer (6.25 ng/μl).
9. Melt and anneal.
10. Mix two portions and anneal.
11. Fill in both ends.
12. Go for PCR amplification and nested PCR amplification.

### 3.2 AhMITE1-Display

#### 3.2.1 TE-Enriched Genome Library Construction and Sequencing

1. Extract genomic DNA with DNeasy Plant Mini Kit.
2. Digest 500 ng of the genomic DNA with restriction enzymes (*Afa*I, *Alu*I, *Hae*III, or *HpyCH4V*) in 100 μl of 1× reaction buffer for 9 h at 37 °C.
3. Concentrate the DNA with ethanol precipitation method to obtain 15 μl DNA solution in dd$H_2O$ (*see* **Note 1**).
4. Ligate linkers to the end of the DNA in a 50 μl of 1× Rapid Ligation Buffer with 15 μl of the digested DNA, 2.5 μM forward-strand linker DNA, 2.5 μM reverse-strand linker DNA with phosphorylated 5′ ends, and 12 U T4 DNA Ligase with the following thermal cycling condition: 25 cycles of 30 min incubation at 16 °C and 10 min incubation at 68°C (*see* **Note 2**).

#### 3.2.2 Probe Preparation

1. Prepare 50 μl PCR mixture containing 1 ng of plasmid DNA containing *AhMITE1*, 0.5 μM oligonucleotide primers labeled at the 5′ end with biotin, 1× $NH_4$-based Reaction Buffer, 0.8 mM dNTPs, 10 mM $MgCl_2$, and 1.25 U BIOTAQ DNA Polymerase.

2. Amplify DNA with the following thermal cycling condition: 3 min denaturation at 94 °C; 35 cycles of 10 s denaturation at 98 °C and 30 s annealing and extension at 68 °C; and a final 3 min extension at 72 °C (*see* **Note 3**).
3. Dilute the amplified DNA to 1.7 μg/μl as probe stock.

*3.2.3 Hybridization*

1. Hybridize the linker-ligated DNA with the *AhMITE1* probes in a 200 μl reaction buffer containing 6× SSC, 0.05% SDS, and 42.5% formamide for 16 h at 42 °C under gentle continuous agitation.
2. Incubate the hybridized DNA with 60 μg of streptavidin-coated magnetic beads (Dynabeads M-280 Streptavidin) for 2 h at 42 °C under gentle agitation.
3. Wash the DNA-bound beads with 200 μl of 2× SSC in 0.1% SDS for 5 min under room temperature. Repeat this step one more time.
4. Wash the DNA-bound beads with 200 μl of 1× SSC in 0.1% SDS for 5 min at 42 °C. Repeat this step one more time.
5. Suspend the washed beads in 120 μl dd$H_2O$.
6. Prepare 50 μl of 1× Buffer for KOD -Plus- Ver. 2 with 5 μl suspended beads, 0.4 μM the forward-strand linker DNA, 0.16 mM dNTPs, 1 mM $MgSO_4$, and 1 U KOD -Plus- DNA Polymerase.
7. Amplify DNAs with the following thermal cycling condition: 1 min denaturation at 94 °C; 35 cycles of 10 s denaturation at 94 °C, 30 s annealing at 60 °C, and 3 min extension at 68 °C; and a final 7 min extension at 68 °C.
8. Concentrate the DNA with ethanol precipitation method to obtain 15 μl DNA solution.
9. Separate the DNA by electrophoresis on a 1% agarose gel and visualize the DNA with ethidium bromide staining.
10. Extract the DNA ranging from 800 to 1500 bp with QIAquick Gel Extraction Kit to elute in 20 μl of dd$H_2O$.
11. Prepare 20 μl of 1× PCR Buffer with 15 μl of the DNA, 0.1 mM dATP, 2.5 mM $MgCl_2$, and 0.5 U TaKaRa Taq DNA Polymerase to incubate for 30 min at 70 °C.
12. Ligate the DNA with pGEM-T Easy Vector in a 44 μl of 1× Rapid Ligation Buffer with 20 μl of the DNA, 50 ng pGEM-T Easy Vector, and 3 U T4 DNA Ligase for 16 h at 4 °C.
13. Transform the *E. coli* ElectroTen-Blue Electroporation Competent Cells with the plasmid DNA by electroporation.
14. Determine the DNA sequence cloned into the vector by Sanger sequencing.

*3.2.4 Sequence Data Processing*

1. Mask the vector sequences and linkers with CROSS_MATCH program with parameters of -minmatch 10 and -minscore 18.
2. Trim the masked sequences and bases with Phred scores of <20 with TRIM2 program with parameters of -q 20 and -x 10.
3. Mask the *AhMITE1* sequences with CROSS_MATCH program with parameters of -minmatch 12, -penalty -2, and -minscore 20.
4. Design primers on the remaining sequences (flanking sequences of *AhMITE1*) with PRIMER3 program to get the amplicons of 300–600 bp (*see* **Note 4**).

### 3.3 Whole Genome In Silico Search for AhMITE1

1. Collect whole genome sequence reads as paired-end file1.fastq and file2.fastq from target lines with Illumina HiSeq, NextSeq, MiSeq, and so on. Otherwise, obtain the reads from Sequence Read Archives of public DNA databases (e.g., GenBank/NCBI, EMBL-Bank/EBI, and DDBJ).
2. Download reference genome sequences (reference_genome.fasta) from the public DNA databases. In the case of peanut, the references are available in PeanutBase (https://peanutbase.org).
3. Prepare a fasta file for transposon sequences including a single sequence (TEs_sequences.fasta). Multi-FASTA format is not allowed. The *AhMITE1* sequence is available from the NCBI GSS database under the accession numbers DE998420-DE998923 and DH968257-DH968767.
4. Make an index file of the reference sequences with the following command. This step is required only for the first time.

```
<PreTEMD-B.pl -r reference_genome.fasta>
```

5. Run the following command to get the result.
   <PTEMD-B.pl -i reference_genome.fasta -t TEs_sequences.fasta –1 file1.fastq –2 file2.fastq -o TE_insertion_distributions.txt>, where "-2 file2.fastq" should be deleted, if the reads are prepared by single-end mode.
6. The resultant output file includes chromosome names, position of left and right borders, direction of the transposon in the reference, and breakpoint sequence of target site duplications for each insert site.
7. Design primers on the flanking sequences of the insertions with PRIMER3 program to get the amplicons of 300–600 bp.

## 4 Notes

1. DNA digestion can be verified by agarose gel electrophoresis. Smearing DNA patterns ranging from 100 bp to >20 kb can be observed on the gel stained with ethidium bromide.
2. The linker ligation can be confirmed by PCR with Linker1 as primers. DNA ranging from 100 bp to 10 kb can be observed on an agarose gel stained with ethidium bromide.
3. The amplicons (alleles) are expected differ by 205 bp.
4. The flanking sequences can be searched to the genome sequences of peanut ancestors to know the physical positions.

## 5 Applications of TE Markers

Till date, TE markers have been widely developed in various crops for genetic studies (for review [71, 72]). The initial efforts employed the TE markers for phylogenetic analysis and genetic diversity studies in various plants and animals. But with the developments in genome-wide discovery of TE markers, a large number of markers were generated and used for analyzing the population structure and trait mapping because TE markers showed very high level of polymorphism [64, 73].

Retrotransposons were employed for cultivar discrimination and genetic diversity analyses in wheat (*Triticum aestivum*) [74], sweet potato (*Ipomoea batatas*) [75, 76], azuki bean (*Vigna angularis*) [77, 78], cultivated strawberry [79], common bean [41, 80], sweet potato [81, 82], pigeon pea [83], barley [84], flax [19, 85], *Solanum tuberosum* [86], field pea [87], and so on.

Cultivar-specific sequence-characterized amplified region (SCAR) marker was developed for cultivar identification. Such SCAR markers were also developed in apple, strawberry, and citrus. S-SAP method was applied to investigate the *FaRE1* [26, 88] insertion polymorphism among 32 Japanese cultivars. High levels of insertion polymorphisms were observed even among Japanese cultivars which were known to be genetically closely related [79]. A rapid, efficient and highly reliable cultivar discrimination method was reported by Monden et al. [89] using single tag hybridization (STH) chromatographic printed array strip (PAS) for DNA signal detection of the PCR amplicons in strawberry cultivars. Apart from LTRs, a non-LTR-RT, SINE was employed for rapid and robust plant genotyping [46, 90]. Genetic diversity and variability analyses were carried out using DNA transposons in maize [91, 92], rice [93], wheat [94], Chenopodiaceae subfamily Betoideae [52], and so on.

Currently, both class I and class II TE markers are being used for linkage and association mapping owing to their genome-wide distribution and high level of polymorphism. Since the retrotransposon based molecular markers showed an extremely high proportion of simplex markers (~90%) in comparison with other molecular markers in polyploid species, they were used for linkage map construction in sweet potato [95, 96], lentil [97], sunflower [98], Japanese gentian [99], rye [100], and so on.

In peanut, *AhMITE1* markers (AhTE) have been extensively used for molecular mapping. AhTE and SSR markers were used on the $F_2$ mapping populations (Satonoka × Kintoki and YI-0311 × Nakateyutaka) to map high oleate [63]. Further, a reference genetic map was constructed using AhTE and SSR markers from the three RIL populations derived from diploid (AA and BB) and tetraploid (AABB) peanut [101]. The genetic map exclusively using AhTE markers was reported from the RILs derived from TMV 2 and its EMS-derived mutant, TMV 2-NLM. The parents were genotyped with 369 AhTE markers, of which 105 showed polymorphism. The first genetic map constructed using 110 AhTE markers on a recombinant inbred line (RIL) mapping population of TMV 2 × TMV 2-NLM was used to map the QTL for branching pattern, number of primary and secondary branches, protein content, days to 50% flowering, content of oleic acid, test weight, and pod width [102]. The same set of AhTE markers were also used on an association panel of independent mutants derived from a few related parents. Marker–trait associations were detected for taxonomic, agronomic, nutritional, and productivity traits including foliar disease resistance. AhTE0050 showed strong association with the number of primary branches in both the mapping schemes (RIL and mutant population), indicating the scope of its application in selections. These studies also helped to identify the candidate gene for a few traits. Two genic markers (AhTE0357 and AhTE0391) with *AhMITE1* activity in the genes *Araip.TG1BL* (B02 chromosome) and *Aradu.7N61X* (A09 chromosome), respectively, showed strong linkage with taxonomic, agronomic and nutritional traits. AhTE0391 showed strong association with the contents of protein, oil, oleic acid, linoleic acid, and palmitic acid. The marker locus corresponding to AhTE0391 coincided with the gene *Aradu.7N61X*. The *AhMITE1* transposition site was present at 2129 bp of the second intronic region of *Aradu.7N61X*.

AhTE markers along with SSR and SNP markers helped in fine-mapping and QTL dissection to identify the candidate genes for late leaf spot and rust resistance in peanut [73], and background selection in marker-assisted backcrossing [103]. Similarly, DNA transposons were used for genetic mapping in rice [104–106], *Antirrhinum majus* [107], barley [53, 108], and so on. In maize, an association mapping was attempted using MITE markers

[92]. TE markers have been successfully used for studying the population structure. SCAR markers developed from CACTA-TD showed high level of polymorphism as well as improved discriminating ability in maize [109]. RBIP was used to study the population structure in pea [110]. These studies indicated that the TE markers can be useful in breeding programs to develop high yielding germplasm.

Other applications of TE markers include mutation breeding in sorghum [111], understanding species boundaries and geographic structure in Madagascan coffee species [112], analyzing somatic polymorphism in *Vitis vinifera* [113], and estimating the outcrossing rate of barley landraces and wild barley populations [114].

In conclusion, since a large portion of the genome is made up of transposable elements, numerous TE markers can be developed in future. Considering the polymorphism exhibited by these TE markers, it looks promising to develop TE markers in large numbers using genome-wide analysis. They can be used for genetic as well as genomics studies to find their association with the important traits. Insertion of TEs both in genic and intergenic regions is known to regulate gene expression [9]. Therefore, TE markers can be useful in identifying marker–trait associations as well as identifying the gene(s) for crop improvement.

## References

1. Bertioli DJ, Cannon SB, Froenicke L, Huang G, Farmer AD, Cannon EK, Liu X, Gao D, Clevenger J, Dash S, Ren L, Moretzsohn MC, Shirasawa K, Huang W, Vidigal B, Abernathy B, Chu Y, Niederhuth CE, Umale P, Araujo AC, Kozik A, Do Kim K, Burow MD, Varshney RK, Wang X, Zhang X, Barkley N, Guimaraes PM, Isobe S, Guo B, Liao B, Stalker HT, Schmitz RJ, Scheffler BE, Leal-Bertioli SC, Xun X, Jackson SA, Michelmore R, Ozias-Akins P (2016) The genome sequences of *Arachis duranensis* and *Arachis ipaensis*, the diploid ancestors of cultivated peanut. Nat Genet 48 (4):438–446. https://doi.org/10.1038/ng.3517
2. Consortium IBGS (2012) A physical, genetic and functional sequence assembly of the barley genome. Nature 491(7426):711
3. Schnable PS, Ware D, Fulton RS, Stein JC, Wei F, Pasternak S, Liang C, Zhang J, Fulton L, Graves TA (2009) The B73 maize genome: complexity, diversity, and dynamics. Science 326(5956):1112–1115
4. Wessler SR (2006) Transposable elements and the evolution of eukaryotic genomes. Proc Natl Acad Sci 103(47):17600–17601
5. Kumar A, Hirochika H (2001) Applications of retrotransposons as genetic tools in plant biology. Trends Plant Sci 6(3):127–134
6. Casa AM, Brouwer C, Nagel A, Wang L, Zhang Q, Kresovich S, Wessler SR (2000) The MITE family *Heartbreaker* (*Hbr*): molecular markers in maize. Proc Natl Acad Sci 97 (18):10083–10089. https://doi.org/10.1073/pnas.97.18.10083
7. Izsvak Z, Ivics Z, Shimoda N, Mohn D, Okamoto H, Hackett PB (1999) Short inverted-repeat transposable elements in teleost fish and implications for a mechanism of their amplification. J Mol Evol 48(1):13–21
8. Chen ZJ, Ni Z (2006) Mechanisms of genomic rearrangements and gene expression changes in plant polyploids. BioEssays 28 (3):240–252
9. Lu C, Chen J, Zhang Y, Hu Q, Su W, Kuang H (2012) Miniature inverted–repeat transposable elements (MITEs) have been accumulated through amplification bursts and play important roles in gene expression and species diversity in *Oryza sativa*. Mol Biol Evol 29 (3):1005–1017
10. Casacuberta JM, Santiago N (2003) Plant LTR-retrotransposons and MITEs: control

of transposition and impact on the evolution of plant genes and genomes. Gene 311:1–11

11. Marcon HS, Domingues DS, Silva JC, Borges RJ, Matioli FF, de Mattos Fontes MR, Marino CL (2015) Transcriptionally active LTR retrotransposons in *Eucalyptus* genus are differentially expressed and insertionally polymorphic. BMC Plant Biol 15(1):198
12. Purugganan M, Wessler S (1995) Transposon signatures: species-specific molecular markers that utilize a class of multiple-copy nuclear DNA. Mol Ecol 4(2):265
13. Waugh R, McLean K, Flavell A, Pearce S, Kumar A, Thomas B, Powell W (1997) Genetic distribution of *Bare-1*-like retrotransposable elements in the barley genome revealed by sequence-specific amplification polymorphisms (S-SAP). Mol Gen Genet 253(6):687–694
14. Kalendar R, Grob T, Regina M, Suoniemi A, Schulman A (1999) IRAP and REMAP: two new retrotransposon-based DNA fingerprinting techniques. Theor Appl Genet 98 (5):704–711
15. Antonius-Klemola K, Kalendar R, Schulman AH (2006) TRIM retrotransposons occur in apple and are polymorphic between varieties but not sports. Theor Appl Genet 112 (6):999–1008
16. Du D, Du X, Mattia MR, Wang Y, Yu Q, Huang M, Yu Y, Grosser JW, Gmitter FG (2018) LTR retrotransposons from the *Citrus* x *clementina* genome: characterization and application. Tree Genet Genomes 14(4):43
17. Belyayev A, Kalendar R, Brodsky L, Nevo E, Schulman AH, Raskina O (2010) Transposable elements in a marginal plant population: temporal fluctuations provide new insights into genome evolution of wild diploid wheat. Mob DNA 1(1):6
18. Nasri S, Mandoulakani BA, Darvishzadeh R, Bernousi I (2013) Retrotransposon insertional polymorphism in Iranian bread wheat cultivars and breeding lines revealed by IRAP and REMAP markers. Biochem Genet 51 (11–12):927–943
19. Smýkal P, Bačová-Kerteszová N, Kalendar R, Corander J, Schulman AH, Pavelek M (2011) Genetic diversity of cultivated flax (*Linum usitatissimum* L.) germplasm assessed by retrotransposon-based markers. Theor Appl Genet 122(7):1385–1397
20. Cheraghi A, Rahmani F, Hassanzadeh-Ghorttapeh A (2018) IRAP and REMAP based genetic diversity among varieties of *Lallemantia iberica*. Mol Biol Res Commun 7 (3):125–132. https://doi.org/10.22099/mbrc.2018.29924.1327
21. Kalendar R, Tanskanen J, Immonen S, Nevo E, Schulman AH (2000) Genome evolution of wild barley (*Hordeum spontaneum*) by *BARE-1* retrotransposon dynamics in response to sharp microclimatic divergence. Proc Natl Acad Sci 97(12):6603–6607
22. Sorkheh K, Dehkordi MK, Ercisli S, Hegedus A, Halász J (2017) Comparison of traditional and new generation DNA markers declares high genetic diversity and differentiated population structure of wild almond species. Sci Rep 7(1):5966
23. Flavell AJ, Knox MR, Pearce SR, Ellis TN (1998) Retrotransposon-based insertion polymorphisms (RBIP) for high throughput marker analysis. Plant J 16(5):643–650
24. Konovalov FA, Goncharov NP, Goryunova S, Shaturova A, Proshlyakova T, Kudryavtsev A (2010) Molecular markers based on LTR retrotransposons *BARE-1* and Jeli uncover different strata of evolutionary relationships in diploid wheats. Mol Gen Genomics 283 (6):551–563
25. Lou Q, Chen J (2007) *Ty1-copia* retrotransposon-based SSAP marker development and its potential in the genetic study of cucurbits. Genome 50(9):802–810
26. Melnikova NV, Kudryavtseva AV, Speranskaya AS, Krinitsina AA, Dmitriev AA, Belenikin MS, Upelniek VP, Batrak ER, Kovaleva IS, Kudryavtsev AM (2012) The *FaRE1* LTR-retrotransposon based SSAP markers reveal genetic polymorphism of strawberry (*Fragaria* × *ananassa*) cultivars. J Agric Sci 4(11):111
27. Petit M, Guidat C, Daniel J, Denis E, Montoriol E, Bui Q, Lim K, Kovarik A, Leitch A, Grandbastien MA (2010) Mobilization of retrotransposons in synthetic allotetraploid tobacco. New Phytol 186(1):135–147
28. Syed N, Sureshsundar S, Wilkinson M, Bhau B, Cavalcanti J, Flavell A (2005) *Ty1-copia* retrotransposon-based SSAP marker development in cashew (*Anacardium occidentale* L.). Theor Appl Genet 110 (7):1195–1202
29. Hirano R, Naito K, Fukunaga K, Watanabe KN, Ohsawa R, Kawase M (2011) Genetic structure of landraces in foxtail millet (*Setaria italica* (L.) P. Beauv.) revealed with transposon display and interpretation to crop evolution of foxtail millet. Genome 54(6):498–506
30. Karki S, Tsukiyama T, Okumoto Y, Rizal G, Naito K, Teraishi M, Nakazaki T, Tanisaka T (2009) Analysis of distribution and

proliferation of *mPing* family transposons in a wild rice (*Oryza rufipogon* Griff.). Breed Sci 59(3):297–307

31. Shan X, Ou X, Liu Z, Dong Y, Lin X, Li X, Liu B (2009) Transpositional activation of *mPing* in an asymmetric nuclear somatic cell hybrid of rice and *Zizania latifolia* was accompanied by massive element loss. Theor Appl Genet 119(7):1325
32. Takagi K, Ishikawa N, Maekawa M, Tsugane K, Iida S (2007) Transposon display for active DNA transposons in rice. Genes Genet Syst 82(2):109–122
33. Van den Broeck D, Maes T, Sauer M, Zethof J, De Deukeleire P, D'hauw M, Van Montagu M, Gerats T (1998) Transposon display identifies individual transposable elements in high copy number lines. Plant J 13 (1):121–129
34. Zhang X, Feschotte C, Zhang Q, Jiang N, Eggleston WB, Wessler SR (2001) *P* instability factor: an active maize transposon system associated with the amplification of *Tourist-like MITEs* and a new superfamily of transposases. Proc Natl Acad Sci 98 (22):12572–12577
35. Kalendar R, Antonius K, Smýkal P, Schulman AH (2010) iPBS: a universal method for DNA fingerprinting and retrotransposon isolation. Theor Appl Genet 121(8):1419–1430
36. Xu JY, Zhu Y, Yi Z, Wu G, Xie GY, Qin MJ (2018) Molecular diversity analysis of *Tetradium ruticarpum* (WuZhuYu) in China based on inter-primer binding site (iPBS) markers and inter-simple sequence repeat (ISSR) markers. Chin J Nat Med 16(1):1–9. https://doi.org/10.1016/S1875-5364(18)30024-4
37. Coutinho JP, Carvalho A, Martin A, Lima-Brito J (2018) Molecular characterization of Fagaceae species using inter-primer binding site (iPBS) markers. Mol Biol Rep 45 (2):133–142. https://doi.org/10.1007/s11033-018-4146-3
38. Monden Y, Fujii N, Yamaguchi K, Ikeo K, Nakazawa Y, Waki T, Hirashima K, Uchimura Y, Tahara M (2014) Efficient screening of long terminal repeat retrotransposons that show high insertion polymorphism via high-throughput sequencing of the primer binding site. Genome 57 (5):245–252
39. Monden Y, Yamaguchi K, Tahara M (2014) Application of iPBS in high-throughput sequencing for the development of retrotransposon-based molecular markers. Curr Plant Biol 1:40–44
40. Monden Y, Takai T, Tahara M (2014) Characterization of a novel retrotransposon *TriRe-1* using nullisomic-tetrasomic lines of hexaploid wheat. Sci Rep Fac Agric 103:21–30
41. Yamane F, Hirashima Y, Shindo A, Tahara M, Fujii N, Ikeo K, Yamashita Y (2012) Cultivar identification markers of common beans based on the retrotransposon insertion site sequences obtained by next-generation DNA sequencing. DNA Test 4:67–74
42. Monden Y, Takai T, Tahara M, Umeno Y, Nakamura R (2014) High-throughput development of DNA markers for wheat cultivar discrimination based on an active retrotransposon TriRe-1 insertion polymorphism. DNA Polymorph 22:60–65
43. Monden Y, Yamamoto A, Shindo A, Tahara M (2014) Efficient DNA fingerprinting based on the targeted sequencing of active retrotransposon insertion sites using a bench-top high-throughput sequencing platform. DNA Res 21(5):491–498
44. Paz RC, Kozaczek ME, Rosli HG, Andino NP, Sanchez-Puerta MV (2017) Diversity, distribution and dynamics of full-length *Copia* and *Gypsy* LTR retroelements in *Solanum lycopersicum*. Genetica 145 (4–5):417–430. https://doi.org/10.1007/s10709-017-9977-7
45. Metcalfe CJ, Oliveira SG, Gaiarsa JW, Aitken KS, Carneiro MS, Zatti F, Van Sluys M-A (2015) Using quantitative PCR with retrotransposon-based insertion polymorphisms as markers in sugarcane. J Exp Bot 66 (14):4239–4250
46. Wenke T, Seibt KM, Döbel T, Muders K, Schmidt T (2015) Inter-SINE amplified polymorphism (ISAP) for rapid and robust plant genotyping. In: Plant genotyping. Springer, New York, NY, pp 183–192
47. Mamedov IZ, Arzumanyan ES, Amosova AL, Lebedev YB, Sverdlov ED (2005) Whole-genome experimental identification of insertion/deletion polymorphisms of interspersed repeats by a new general approach. Nucleic Acids Res 33(2):e16–e16
48. Zhang Q, Arbuckle J, Wessler SR (2000) Recent, extensive and preferential insertion of members of the miniature inverted-repeat transposable element family *Heartbreaker* into genic regions of maize. Proc Natl Acad Sci 97(3):1160–1165
49. Johal GS, Briggs SP (1992) Reductase activity encoded by the *HM1* disease resistance gene in maize. Science 258(5084):985–987
50. Grzebelus D, Simon PW (2009) Diversity of *DcMaster*-like elements of the *PIF/Harbinger*

superfamily in the carrot genome. Genetica 135(3):347–353

51. Grzebelus D, Lasota S, Gambin T, Kucherov G, Gambin A (2007) Diversity and structure of *PIF/Harbinger*-like elements in the genome of *Medicago truncatula*. BMC Genomics 8(1):409
52. Grzebelus D, Stawujak K, Mitoraj J, Szklarczyk M (2011) Dynamics of *Vulmar/VulMITE* group of transposable elements in *Chenopodiaceae* subfamily *Betoideae*. Genetica 139(9):1209–1216
53. Takahashi H, Akagi H, Mori K, Sato K, Takeda K (2006) Genomic distribution of MITEs in barley determined by MITE-AFLP mapping. Genome 49(12):1616–1620
54. Lyons M, Cardle L, Rostoks N, Waugh R, Flavell AJ (2008) Isolation, analysis and marker utility of novel miniature inverted repeat transposable elements from the barley genome. Mol Gen Genomics 280 (4):275–285
55. Hřibová E, Neumann P, Matsumoto T, Roux N, Macas J, Doležel J (2010) Repetitive part of the banana (*Musa acuminata*) genome investigated by low-depth 454 sequencing. BMC Plant Biol 10(1):204
56. Wanjugi H, Coleman-Derr D, Huo N, Kianian SF, Luo M-C, Wu J, Anderson O, Gu YQ (2009) Rapid development of PCR-based genome-specific repetitive DNA junction markers in wheat. Genome 52(6):576–587
57. Yadav CB, Bonthala VS, Muthamilarasan M, Pandey G, Khan Y, Prasad M (2015) Genome-wide development of transposable elements-based markers in foxtail millet and construction of an integrated database. DNA Res 22(1):79–90
58. Patel M, Jung S, Moore K, Powell G, Ainsworth C, Abbott A (2004) High-oleate peanut mutants result from a MITE insertion into the *FAD2* gene. Theor Appl Genet 108 (8):1492–1502
59. Bhat RS, Patil VU, Chandrashekar TM, Sujay V, Gowda MVC, Kuruvinashetti MS (2008) Recovering flanking sequence tags of miniature inverted-repeat transposable element by thermal asymmetric interlaced-PCR in peanut. Curr Sci 95(4):452–453
60. Gowda MVC, Bhat RS, Motagi BN, Sujay V, Varshakumari BS (2010) Association of high-frequency origin of late leaf spot resistant mutants with *AhMITE1* transposition in peanut. Plant Breed 129(5):567–569
61. Gowda MVC, Bhat RS, Sujay V, Kusuma P, Varshakumari BS, Varshney RK (2011) Characterization of *AhMITE1* transposition and its association with the mutational and evolutionary origin of botanical types in peanut (*Arachis* spp.). Plant Syst Evol 291 (3–4):153–158
62. Shirasawa K, Hirakawa H, Tabata S, Hasegawa M, Kiyoshima H, Suzuki S, Sasamoto S, Watanabe A, Fujishiro T, Isobe S (2012) Characterization of active miniature inverted-repeat transposable elements in the peanut genome. Theor Appl Genet 124 (8):1429–1438. https://doi.org/10.1007/s00122-012-1798-6
63. Shirasawa K, Koilkonda P, Aoki K, Hirakawa H, Tabata S, Watanabe M, Hasegawa M, Kiyoshima H, Suzuki S, Kuwata C, Naito Y, Kuboyama T, Nakaya A, Sasamoto S, Watanabe A, Kato M, Kawashima K, Kishida Y, Kohara M, Kurabayashi A, Takahashi C, Tsuruoka H, Wada T, Isobe S (2012) *In silico* polymorphism analysis for the development of simple sequence repeat and transposon markers and construction of linkage map in cultivated peanut. BMC Plant Biol 12(1):80. https://doi.org/10.1186/1471-2229-12-80
64. Gayathri M, Shirasawa K, Varshney RK, Pandey MK, Bhat RS (2018) Development of new *AhMITE1* markers through genome-wide analysis in peanut (*Arachis hypogaea* L.). BMC Res Notes 11(1):10. https://doi.org/10.1186/s13104-017-3121-8
65. Nunome T, Negoro S, Miyatake K, Yamaguchi H, Fukuoka H (2006) A protocol for the construction of microsatellite enriched genomic library. Plant Mol Biol Rep 24 (3–4):305
66. Grzebelus D, Jagosz B, Simon PW (2007) The DcMaster transposon display maps polymorphic insertion sites in the carrot (*Daucus carota* L.) genome. Gene 390(1):67–74
67. Kwon S-J, Park K-C, Kim J-H, Lee JK, Kim N-S (2005) *Rim 2/Hipa CACTA* transposon display: a new genetic marker technique in *Oryza* species. BMC Genet 6(1):15
68. Goerner-Potvin P, Bourque G (2018) Computational tools to unmask transposable elements. Nat Rev Genet 19:688–704
69. Hake AA, Shirasawa K, Yadawad A, Nadaf HL, Gowda MVC, Bhat RS (2018) Genome-wide structural mutations among the lines resulting from genetic instability in peanut (*Arachis hypogaea* L.). Plant Gene 13 (March):1–7. https://doi.org/10.1016/j.plgene.2017.11.001
70. Kang H, Zhu D, Lin R, Opiyo SO, Jiang N, Shiu S-H, Wang G-L (2016) A novel method for identifying polymorphic transposable elements via scanning of high-throughput short

reads. DNA Res 23(3):241–251. https://doi.org/10.1093/dnares/dsw011

71. Nadeem MA, Nawaz MA, Shahid MQ, Doğan Y, Comertpay G, Yıldız M, Hatipoğlu R, Ahmad F, Alsaleh A, Labhane N (2018) DNA molecular markers in plant breeding: current status and recent advancements in genomic selection and genome editing. Biotechnol Biotechnol Equip 32 (2):261–285
72. Monden Y, Tahara M (2015) Plant transposable elements and their application to genetic analysis via high-throughput sequencing platform. Hort J 84(4):283–294
73. Shirasawa K, Bhat RS, Khedikar YP, Sujay V, Kolekar RM, Yeri SB, Sukruth M, Cholin S, Asha B, Pandey MK, Varshney RK, Gowda MVC (2018) Sequencing analysis of genetic loci for resistance for late leaf spot and rust in peanut (*Arachis hypogaea* L.). Front Plant Sci:9. https://doi.org/10.3389/fpls.2018.01727. https://www.frontiersin.org/articles/10.3389/fpls.2018.01727/abstract
74. Takai T, Tahara M (2011) Discovery of the retrotransposon showing genome insertion polymorphisms among wheat cultivars. DNA Polymorph 20:80–90
75. Monden Y, Yamamoto A, Tahara M (2013) Development of DNA markers for anthocyanin content purple sweet potato using active retrotransposon insertion polymorphisms. DNA Polymorph 21:47–54
76. Tahara M, Yamashita H, Ooe N (2007) Cultivar identification based on retrotransposon insertion polymorphisms applied for sweet potato products. DNA Polymorph 15:122–125
77. Nakagawa A, Yamashita H, Tahara M, Ooyama Y (2009) Retrotransposon DNA marker for Azuki cultivar Shumari identification. DNA Polymorph 17:85–91
78. Yamashita H (2008) Retrotransposon DNA marker for Azuki cultivar identification. DNA Polymorph 16:82–87
79. Akitake H, Tahara M, Monden Y, Takasaki K, Futo S (2013) Strawberry cultivar identification by retrotransposon insertion polymorphisms. DNA Polymorph 21:64–72
80. Tanaka Y, Shindo A, Tahara M, Yamashita Y (2011) Species identification marker for common bean (*Phaseolus vulgaris*) based on a retrotransposon insertion site. DNA Polymorph 19:82–87
81. Yamashita H, Tahara M (2006) A LINE-type retrotransposon active in meristem stem cells causes heritable transpositions in the sweet potato genome. Plant Mol Biol 61 (1–2):79–84
82. Tahara M, Aoki T, Suzuka S, Yamashita H, Tanaka M, Matsunaga S, Kokumai S (2004) Isolation of an active element from a high-copy-number family of retrotransposons in the sweetpotato genome. Mol Gen Genomics 272(1):116–127
83. Maneesha UKC (2017) Analysis of genetic diversity in pigeon pea germplasm using retrotransposon-based molecular markers. J Genet 96(4):551–561
84. Bonchev G, Dusinsky R, Hauptvogel P, Svec M (2017) Patterns of evolutionary trajectories and domestication history within the genus *Hordeum* assessed by REMAP markers. J Mol Evol 84(2–3):116–128. https://doi.org/10.1007/s00239-016-9779-z
85. Galindo-González L, Mhiri C, Grandbastien M-A, Deyholos MK (2016) *Ty1*-copia elements reveal diverse insertion sites linked to polymorphisms among flax (*Linum usitatissimum* L.) accessions. BMC Genomics 17 (1):1002
86. Sharma V, Nandineni MR (2014) Assessment of genetic diversity among Indian potato (*Solanum tuberosum* L.) collection using microsatellite and retrotransposon based marker systems. Mol Phylogenet Evol 73:10–17
87. Jing R, Vershinin A, Grzebyta J, Shaw P, Smýkal P, Marshall D, Ambrose MJ, Ellis TN, Flavell AJ (2010) The genetic diversity and evolution of field pea (*Pisum*) studied by high throughput retrotransposon based insertion polymorphism (RBIP) marker analysis. BMC Evol Biol 10(1):44
88. He P, Ma Y, Zhao G, Dai H, Li H, Chang L, Zhang Z (2010) *FaRE1*: a transcriptionally active *Ty1-copia* retrotransposon in strawberry. J Plant Res 123(5):707–714
89. Monden Y, Takasaki K, Futo S, Niwa K, Kawase M, Akitake H, Tahara M (2014) A rapid and enhanced DNA detection method for crop cultivar discrimination. J Biotechnol 185:57–62
90. Ray DA (2007) SINEs of progress: mobile element applications to molecular ecology. Mol Ecol 16(1):19–33
91. Kavar T, Meglič V, Rozman L (2007) Diversity of Slovenian maize (*Zea mays*) populations by *Hbr* (*MITE*) markers and morphological traits. Russ J Genet 43 (9):989–995
92. Zerjal T, Rousselet A, Mhiri C, Combes V, Madur D, Grandbastien M-A, Charcosset A, Tenaillon MI (2012) Maize genetic diversity

and association mapping using transposable element insertion polymorphisms. Theor Appl Genet 124(8):1521–1537

93. Park K-C, Lee JK, Kim N-H, Shin Y-B, Lee J-H, Kim N-S (2003) Genetic variation in *Oryza* species detected by MITE-AFLP. Genes Genet Syst 78(3):235–243
94. Yaakov B, Ceylan E, Domb K, Kashkush K (2012) Marker utility of miniature inverted-repeat transposable elements for wheat biodiversity and evolution. Theor Appl Genet 124 (7):1365–1373
95. Monden Y, Tahara M (2017) Genetic linkage analysis using DNA markers in sweetpotato. Breed Sci 67(1):41–51. https://doi.org/10.1270/jsbbs.16142
96. Monden Y, Hara T, Okada Y, Jahana O, Kobayashi A, Tabuchi H, Onaga S, Tahara M (2015) Construction of a linkage map based on retrotransposon insertion polymorphisms in sweet potato via high-throughput sequencing. Breed Sci 65 (2):145–153
97. Rey-Banos R, Saenz de Miera LE, Garcia P, Perez de la Vega M (2017) Obtaining retrotransposon sequences, analysis of their genomic distribution and use of retrotransposon-derived genetic markers in lentil (*Lens culinaris* Medik.). PLoS One 12(4):e0176728. https://doi.org/10.1371/journal.pone.0176728
98. Ali SG, Darvishzadeh R, Ebrahimi A, Bihamta MR (2018) Identification of SSR and retrotransposon-based molecular markers linked to morphological characters in oily sunflower (*Helianthus annuus* L.) under natural and water-limited states. J Genet 97 (1):189–203
99. Nakatsuka T, Yamada E, Saito M, Hikage T, Ushiku Y, Nishihara M (2012) Construction of the first genetic linkage map of Japanese gentian (Gentianaceae). BMC Genomics 13 (1):672
100. Tenhola-Roininen T, Kalendar R, Schulman AH, Tanhuanpää P (2011) A doubled haploid rye linkage map with a QTL affecting α-amylase activity. J Appl Genet 52 (3):299–304
101. Shirasawa K, Bertioli DJ, Varshney RK, Moretzsohn MC, Leal-Bertioli SC, Thudi M, Pandey MK, Rami JF, Fonceka D, Gowda MV, Qin H, Guo B, Hong Y, Liang X, Hirakawa H, Tabata S, Isobe S (2013) Integrated consensus map of cultivated peanut and wild relatives reveals structures of the A and B genomes of *Arachis* and divergence of the legume genomes. DNA Res 20 (2):173–184. https://doi.org/10.1093/dnares/dss042
102. Hake AA, Shirasawa K, Yadawad A, Sukruth M, Patil M, Nayak SN, Lingaraju S, Patil PV, Nadaf HL, Gowda MVC, Bhat RS (2017) Mapping of important taxonomic and productivity traits using genic and non-genic transposable element markers in peanut (*Arachis hypogaea* L.). PLoS One 12(10): e0186113. https://doi.org/10.1371/journal.pone.0186113. eCollection 2017
103. Kolekar RM, Sukruth M, Shirasawa K, Nadaf HL, Motagi BN, Lingaraju S, Patil PV, Bhat RS (2017) Marker-assisted backcrossing to develop foliar disease resistant genotypes in TMV 2 variety of peanut (*Arachis hypogaea* L.). Plant Breed 136(6):948–953. https://doi.org/10.1111/pbr.12549
104. Kwon S, Hong S, Son J, Lee J, Cha Y, Eun M, Kim N (2006) *CACTA* and *MITE* transposon distributions on a genetic map of rice using $F_{15}$ RILs derived from Milyang 23 and Gihobyeo hybrids. Mol Cells 21(3):360
105. Kwon S, Yu J, Park Y, Son J, Kim N, Lee J (2015) Genetic analysis of seed-shattering genes in rice using an F. Genet Mol Res 14 (1):1347–1361
106. Monden Y, Naito K, Okumoto Y, Saito H, Oki N, Tsukiyama T, Ideta O, Nakazaki T, Wessler SR, Tanisaka T (2009) High potential of a transposon *mPing* as a marker system in *japonica* × *japonica* cross in rice. DNA Res 16 (2):131–140
107. Schwarz-Sommer Z, Gübitz T, Weiss J, Gómez-di-Marco P, Delgado-Benarroch L, Hudson A, Egea-Cortines M (2010) A molecular recombination map of *Antirrhinum majus*. BMC Plant Biol 10(1):275
108. Chang RY, O'Donoughue LS, Bureau TE (2001) Inter-MITE polymorphisms (IMP): a high throughput transposon-based genome mapping and fingerprinting approach. Theor Appl Genet 102(5):773–781
109. Roy NS, Park K-C, Lee S-I, Im M-J, Ramekar RV, Kim N-S (2018) Development of CACTA transposon derived SCAR markers and their use in population structure analysis in *Zea mays*. Genetica 146(1):1–12
110. Smýkal P, Hýbl M, Corander J, Jarkovský J, Flavell AJ, Griga M (2008) Genetic diversity and population structure of pea (*Pisum sativum* L.) varieties derived from combined retrotransposon, microsatellite and morphological marker analysis. Theor Appl Genet 117(3):413–424
111. Im S, Kwon S, Ryu J, Jeong S, Kim J, Ahn J, Kim S, Jo Y, Choi H, Kang S (2016)

Development of a transposon-based marker system for mutation breeding in sorghum (*Sorghum bicolor* L.). Genet Mol Res 15(3): PMID:27706735

112. Roncal J, Guyot R, Hamon P, Crouzillat D, Rigoreau M, Konan ONG, Rakotomalala J-J, Nowak MD, Davis AP, de Kochko A (2016) Active transposable elements recover species boundaries and geographic structure in Madagascan coffee species. Mol Gen Genomics 291(1):155–168

113. Carrier G, Le Cunff L, Dereeper A, Legrand D, Sabot F, Bouchez O, Audeguin L, Boursiquot J-M, This P (2012) Transposable elements are a major cause of somatic polymorphism in *Vitis vinifera* L. PLoS One 7(3):e32973

114. Abdel-Ghani AH, Parzies HK, Omary A, Geiger HH (2004) Estimating the outcrossing rate of barley landraces and wild barley populations collected from ecologically different regions of Jordan. Theor Appl Genet 109 (3):588–595

# Chapter 12

# Machine Learning-Based Annotation of Long Noncoding RNAs Using PLncPRO

Niraj K. Khemka, Urminder Singh, Anuj K. Dwivedi, and Mukesh Jain

## Abstract

Long noncoding RNAs (lncRNAs) are noncoding RNAs with transcript length more than 200 nucleotides. Although poorly conserved, lncRNAs are expressed across diverse species, including plants and animals, and are known to be involved in regulation of various biological processes. To understand their biological significance, we first need to identify the lncRNAs accurately. However, distinguishing lncRNAs from coding transcripts is still a challenging task. Here, we describe a machine learning-based approach to accurately identify the plant lncRNAs. We describe the usage of plant long noncoding RNA prediction by random forests (PLncPRO), which employs machine learning-based random forest algorithm to recognize the lncRNAs from the set of given transcript sequences. Stepwise instructions have been provided to use PLncPRO to annotate the lncRNA sequences.

**Key words** lncRNA, Machine learning, Noncoding RNA, RNA-seq, PLncPRO

## 1 Introduction

Long noncoding RNAs (lncRNAs) are noncoding RNAs with transcript length more than 200 nucleotides [1]. Although lncRNAs do not code for a protein, they are involved in regulation of wide variety of biological processes, such as differentiation, developmental timing and stress response in plant [1, 2]. LncRNAs, such as *COLDAIR* and *COOLAIR* regulate flowering time [3, 4], while *IPS1* and *DRIR* were shown to be involved in stress tolerance in plants [5, 6]. The experimental evidences provide clues that many biological processes are regulated by lncRNAs. The biological significance of lncRNAs has led to recognition of this as an active area of research.

Accurate identification of lncRNAs is the first step toward further understanding the roles of these lncRNAs in various biological contexts. Next-generation sequencing along with computational tools has made it feasible to sequence the transcriptome under different conditions and identify expressed transcripts

Mukesh Jain and Rohini Garg (eds.), *Legume Genomics: Methods and Protocols*, Methods in Molecular Biology, vol. 2107, https://doi.org/10.1007/978-1-0716-0235-5_12, 

in various organisms including plants [7–9]. The role of lncRNAs in regulation of inheritance and alternative splicing has been documented [10, 11]. LncRNAs are involved in different developmental processes and abiotic stress responses in different plants [12–15]. RNA-seq data could be leveraged to identify lncRNA population under different biological contexts [16].

PLncPRO is a free open-source machine learning-based computational tool written in python [17]. PLncPRO extracts 71 features based on homology search and sequence makeup from each input transcript sequence and uses random forest models to either build a classification model or to classify input transcript sequences as lncRNAs or mRNAs as described previously [17]. In order to build a new classification model, PLncPRO requires known lncRNA and mRNA transcripts as input for training the classification model. This new model can be used to predict unknown transcripts as lncRNAs or mRNAs. PLncPRO was specifically developed to fulfil the need for a lncRNA prediction tool, which could be trained on data from different organisms especially plants. PLncPRO can be trained on species-specific data to achieve higher accuracy and data from a closely related species could be used in case of newly sequenced species. PLncPRO was shown to attain high accuracy to predict lncRNAs from human, mouse and different plant species. PLncPRO comes with pretrained classification models for different plant species along with consensus dicot and monocot models. Protein domain family database Pfam can also be used along with annotated protein database (SwissProt) to identify lncRNAs more confidently.

Previous studies have reported lncRNAs in legume species, like soybean, Medicago and chickpea which are involved in biological processes related to development and stress responses [17–20]. However, lncRNA annotation is still limited in other legumes. This chapter describes stepwise method to accurately predict lncRNAs using PLncPRO from a given set of transcript sequences from any legume species.

## 2 Materials

Prepare all prerequisites required to run PLncPRO.

### 2.1 Hardware

Workstation (preferably Intel Core i7 or better processor and at least 8 GB memory).

### 2.2 Software (See Notes 1–3)

1. Linux-based operating system.
2. Python 2.7.11 or higher with scikit-learn, bio-python, and sci-py.
3. PLncPRO (http://ccbb.jnu.ac.in/plncpro/).

4. GNU C Library (glibc ≧ 2.14) to compile Framefinder program.
5. NCBI BLAST package (ftp://ftp.ncbi.nlm.nih.gov/blast/executables/blast+/LATEST/).
6. Hmmer package (http://eddylab.org/software/hmmer/hmmer.tar.gz).
7. EMBOSS package. (ftp://emboss.open-bio.org/pub/EMBOSS/EMBOSS-6.5.7.tar.gz).
8. Perl script (https://github.com/nirajkhe/script/blob/master/fetch_fasta_exclude_ids.pl).

### 2.3 Databases

1. Annotated protein database, SwissProt (ftp://ftp.uniprot.org/pub/databases/uniprot/current_release/knowledgebase/complete/uniprot_sprot.fasta.gz).
2. Protein family database, Pfam-A (ftp://ftp.ebi.ac.uk/pub/databases/Pfam/current_release/Pfam-A.fasta.gz).

### 2.4 Data

Transcriptome sequence obtained after assembly (de novo/reference) of RNA-seq data.

## 3 Methods

Follow the stepwise protocol for the prediction of lncRNAs (Fig. 1).

### 3.1 Software Installation and Database Setup

1. Download and setup PLncPRO with all its dependencies as described in its instruction manual (http://ccbb.jnu.ac.in/plncpro/docs/manual.pdf) available online with the program.
2. Setup SwissProt database to be used with BLASTX with makeblastdb command.
3. Setup Pfam-A database to be used with hmmscan (*see* **Notes 2** and **3**).

### 3.2 Building New Classification Model with PLncPRO

Plant-specific models of Arabidopsis, rice, soybean, and few other plants along with consensus monocot and dicot models available in PLncPRO can be used. If user wants to build a new model using new training datasets, it can be done using “build.py” script (*see* **Note 4**).

python build.py -p mRNA_sample.fa -n lncRNA_sample.fa -o sample_cp -m sample_model_cp -d sprotdb/sprotdb -t 4

Here, mRNA_sample.fa contains mRNA sequences and lncRNA_sample.fa contains known lncRNA sequences to train the new model. This step outputs relative feature rankings also, which indicate the importance of various features in building the

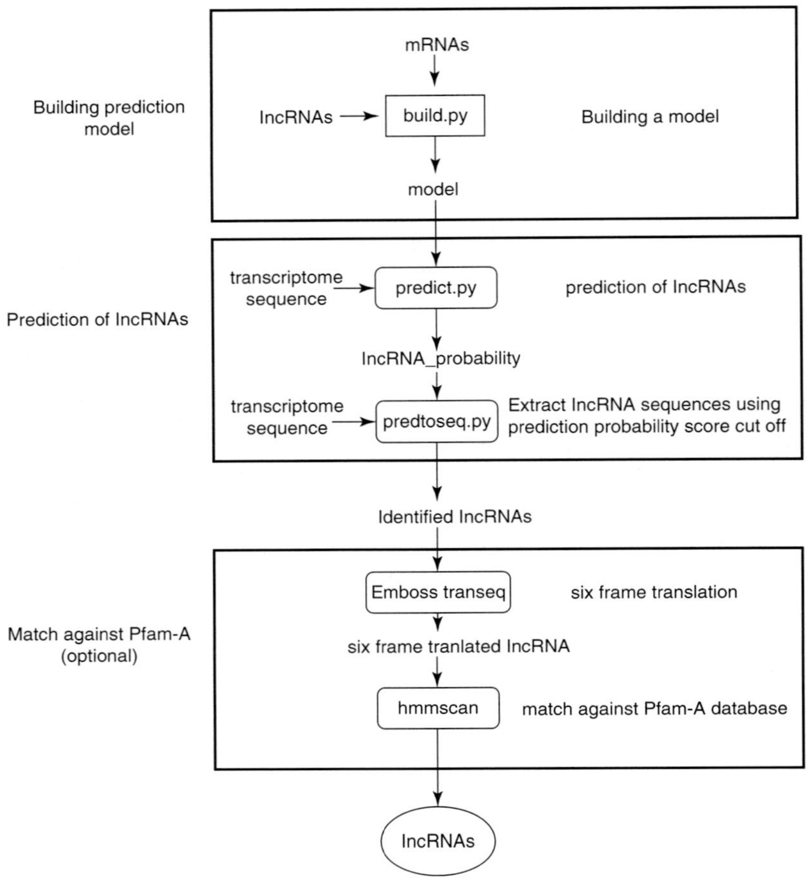


**Fig. 1** Flowchart for identification of lncRNAs from transcriptome data

model. A representative graph showing relative ranking of features (top 10) in prediction model is shown in Fig. 2.

### 3.3 Prediction of lncRNAs Using Prediction Model

python prediction.py -i fasta_sequence.fa -o lncrna_folder -p plncpro_prediction -m sample_cp/sample_model_cp -d sprotdb/sprotdb -t 4

The "prediction.py" script takes a fasta file of transcript sequences as an input and predicts the transcripts representing lncRNAs using a classification model. In the above command line, the model used is "sample_model_cp" which was built in **step 1**. User can use species-specific model or consensus of monocot/dicot model provided in PLncPRO, depending on the species being analyzed. For legume species, the consensus dicot model can be used. The output of this step is a text file ("plncpro_prediction") in the output folder ("lncrna_folder") which contains the predicted result. The output file is a tab-delimited file with four columns (Fig. 3). The first column is the transcript ID. The second column contains the predicted labels (i.e., 1 for mRNA and 0 for lncRNA).

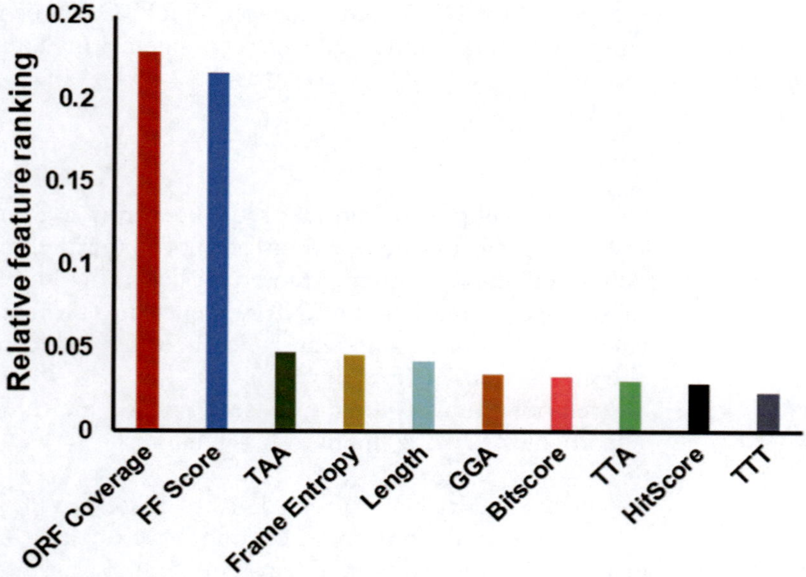


**Fig. 2** Ranking of top 10 features in a prediction model

| Transcript Id | Label | Coding probability | Non-coding probability |
|---|---|---|---|
| XLOC_000001 | 0 | 0.168 | 0.832 |
| XLOC_000002 | 0 | 0.006 | 0.994 |
| XLOC_000003 | 1 | 0.731 | 0.269 |
| XLOC_000004 | 0 | 0.014 | 0.986 |
| XLOC_000005 | 1 | 0.88 | 0.22 |

**Fig. 3** Sample output of lncRNA prediction probability scores. Labels showing 0 represent predicted lncRNAs at given probability score

The third and fourth columns contain original probabilistic score for each transcript to be a mRNA or lncRNA, respectively.

### *3.4 Extraction of lncRNA Sequences*

The "predstoseq.py" script is used to extract the sequences of lncRNAs predicted in **step 2** from the prediction file "plncpro_-prediction" and the input fasta sequence file "fasta_sequence.fa". The user can specify a threshold value between 0 and 1, with -*s* option. Default value is 0.5 and higher value can be used to reduce false positives (*see* **Note 5**). The output of this step is the file "plncpro_lncRNA.fa" containing the predicted lncRNA sequences in fasta format.

python predstoseq.py -f fasta_sequence.fa -o plncpro_lncRNA.fa -p lncrna_folder/plncpro_prediction -s 0.5

### 3.5 Removal of lncRNAs Showing Similarity with Known Protein Domains (Optional)

Selection of lncRNAs predicted from PlncPRO can be made more stringent by filtering the sequences containing any known protein domain available in Pfam-A database.

#### 3.5.1 Translation of lncRNAs into Six Frames

To consider all possible protein sequences from a given transcript sequence, perform six frame translation of all the identified putative lncRNAs in the above step. Transeq tool from the Emboss package can be used to translate the RNA sequence into protein sequences. The output is six frame translated lncRNAs (plncpro_lncRNA.fa_6_frams_translate.fa).

/emboss/transeq -frame 6 -seq plncpro_lncRNA.fa -outseq plncpro_lncRNA.fa_6_frams_translate.fa

#### 3.5.2 Identify the lncRNAs Having Conserved Pfam-A Domain

To check the presence of any known conserved protein family domain, the translated protein sequences from the lncRNAs are analyzed in the conserved protein database (Pfam-A) using hmmscan utility of hmmer package.

Download the hidden Markov models (HMMs) of all the available domains from Pfam database (https://pfam.xfam.org/) and build a local pfam database using hmmpress.

hmmscan --tblout plncpro_lncRNA.fa_6_frams_translate.fa_pfam_output /database/Pfam-A.hmm plncpro_lncRNA.fa_6_frams_translate.fa

Here, input is six frame translated sequences (plncpro_lncRNA.fa_6_frams_translate.fa) and output is plncpro_lncRNA.fa_6_frams_translate.fa_pfam_output, which contains sequence ids with significantly matched protein domain.

#### 3.5.3 Remove lncRNAs Showing Match with Pfam Database

Process the hmmscan output file (plncpro_lncRNA.fa_6_frams_translate.fa_pfam_output) to remove the sequences showing match against Pfam domains using an e-value cutoff. Steps required to process the files are given below:

1. Remove header from hmmscan output using grep.

   grep -v '#' plncpro_lncRNA.fa_6_frams_translate.fa_pfam_output >plncpro_lncRNA.fa_6_frams_translate.fa_pfam_output_header_rm

   (a) Extract ids having match with Pfam with e-value ≤0.01 using awk (*see* **Note 6**).

   awk '{if($5<= 0.01)print $3}' plncpro_lncRNA.fa_6_frams_translate.fa_pfam_output_header_rm >plncpro_lncRNA.fa_6_frams_translate.fa_pfam_output_match_id_0.01

   (b) Remove the frame number from each matched id using sed.

sed 's/_ [1–6]//g' plncpro_lncRNA.fa_6_frams_translate.fa_pfam_output_match_id_0.001 >plncpro_lncRNA.fa_6_frams_translate.fa_pfam_output_match_id_0.001_2

(c) Sort and extract the uniquely match ids with Pfam domain.

sort plncpro_lncRNA.fa_6_frams_translate.fa_pfam_output_match_id_0.001_2 | uniq >plncpro_lncRNA.fa_6_frams_translate.fa_pfam_output_match_id_0.001_uniq

(d) Extract the final lncRNA sequences.

Remove the ids having significant match with Pfam database and extract the sequences of final lncRNAs using perl script. Output file "Final_lncRNA_sequences.fa" contains high-confidence lncRNA sequences.

perl fetch_fasta_exclude_ids.pl plncpro_lncRNA.fa_6_frams_translate.fa_pfam_output_match_id_0.001_uniq fasta_sequence.fa > Final_lncRNA_sequences.fa

LncRNAs obtained can be used for further analysis and functional characterization.

## 4 Notes

1. Install the additional python modules required by PLncPRO using pip.
2. Setup all the installations and databases in a single directory to avoid any path related problems to run the scripts.
3. Run the program/scripts from the same folder where program is installed via giving proper path of inputs like sequences/model/database to perform the analysis.
4. Already available model for the specific plant or monocot/dicot model can be used for lncRNA prediction. New model can also be constructed for any species using available scripts.
5. Noncoding probability cutoff can be chosen to extract high-confidence lncRNAs.
6. The e-value cutoff of Pfam database match can be set depending on the desired stringency.

## Acknowledgments

Funding from the Department of Biotechnology, Government of India, New Delhi under the Challenge Programme on Chickpea Functional Genomics (BT/AGR/CG-PhaseII/01/2014) and Tata Innovation Fellowship Scheme (BT/HRD/35/01/04/2019) is gratefully acknowledged.

## References


1. Chekanova JA (2015) Long non-coding RNAs and their functions in plants. Curr Opin Plant Biol 27:207–216
2. Liu X, Hao L, Li D et al (2015) Long non-coding RNAs and their biological roles in plants. Genomics Proteomics Bioinformatics 13(3):137–147
3. Heo JB, Sung S (2011) Vernalization-mediated epigenetic silencing by a long intronic non-coding RNA. Science 331:76–79
4. Csorba T, Questa JI, Sun Q et al (2014) Antisense *COOLAIR* mediates the coordinated switching of chromatin states at FLC during vernalization. Proc Natl Acad Sci U S A 111:16160–16165
5. Franco-Zorrilla JM, Valli A, Todesco M et al (2007) Target mimicry provides a new mechanism for regulation of microRNA activity. Nat Genet 39:1033–1037
6. Qin T, Zhao H, Cui P et al (2017) A nucleus-localized long non-coding RNA enhances drought and salt stress tolerance. Plant Physiol 175:1321–1336
7. Jain M (2012) Next-generation sequencing technologies for gene expression profiling in plants. Brief Funct Genomics 11(1):63–70
8. Jain M, Misra G, Patel RK et al (2013) A draft genome sequence of the pulse crop chickpea (*Cicer arietinum* L.). Plant J 74:715–729
9. Garg R, Singh VK, Rajkumar MS et al (2017) Global transcriptome and co-expression network analyses reveal cultivar-specific molecular signatures associated with seed development and seed size/weight determination in chickpea. Plant J 91:1088–1107
10. Xu W, Yang T, Wang B et al (2018) Differential expression networks and inheritance patterns of long non-coding RNAs in castor bean seeds. Plant J 95:324–340
11. Kiegle EA, Garden A, Lacchini E et al (2018) A genomic view of alternative splicing of long non-coding RNAs during rice seed development reveals extensive splicing and lncRNA gene families. Front Plant Sci 9:115
12. Kindgren P, Ard R, Ivanov M et al (2018) Transcriptional read-through of the long non-coding RNA SVALKA governs plant cold acclimation. Nat Commun 9(1):4561
13. Wang A, Hu J, Gao C et al (2019) Genome-wide analysis of long non-coding RNAs unveils the regulatory roles in the heat tolerance of Chinese cabbage (*Brassica rapa* ssp. *chinensis*). Sci Rep 9:5002
14. Di C, Yuan J, Wu Y et al (2014) Characterization of stress-responsive lncRNAs in *Arabidopsis thaliana* by integrating expression, epigenetic and structural features. Plant J 80:848–861
15. Yuan J, Li J, Yang Y et al (2018) Stress-responsive regulation of long non-coding RNA polyadenylation in *Oryza sativa*. Plant J 93:814–827
16. Çiftçi YO (2018) Role of next-generation RNA-Seq data in discovery and characterization of long non-coding RNA in plants. In: Next generation plant breeding. IntechOpen, London
17. Singh U, Khemka N, Rajkumar MS et al (2017) PLncPRO for prediction of long non-coding RNAs (lncRNAs) in plants and its application for discovery of abiotic stress-responsive lncRNAs in rice and chickpea. Nucleic Acids Res 45:e183
18. Golicz AA, Singh MB, Bhalla PL (2018) The long intergenic noncoding RNA (lincRNA) landscape of the soybean genome. Plant Physiol 176(3):2133–2147
19. Wang T-Z, Liu M, Zhao M-G et al (2015) Identification and characterization of long non-coding RNAs involved in osmotic and salt stress in *Medicago truncatula* using genome-wide high-throughput sequencing. BMC Plant Biol 15:131
20. Khemka N, Singh VK, Garg R et al (2016) Genome-wide analysis of long intergenic non-coding RNAs in chickpea and their potential role in flower development. Sci Rep 6:33297

Chapter 13

# Methods for Identification and Validation of G-Quadruplex Sequences in Legumes

Surabhi Singh, Tanya Mathur, Khushboo Gupta, and Rohini Garg


## Abstract

Among the different secondary structures that DNA can adopt, G-quadruplex is a noncanonical form that has recently started to garner attention about the possible layers of regulation they could introduce in cellular processes. Here, we outline how the presence of G-quadruplexes can be probed in legumes and other plant genomes. This chapter describes various in silico approaches that can be utilized to identify putative G-quadruplex forming sequences (GQSes) and validate their formation through in vitro experimental approaches.


**Key words** G-quadruplex, Cellular processes, Legumes, Annotation

## 1 Introduction

G-quadruplex (GQS) is a noncanonical three-dimensional structure that is formed by stacking of guanine quartets on top of each other. Guanines in these quartets are bonded by Hoogsteen type hydrogen bonding. GQSes usually contain bulges and loops and are classified based on number of guanine repeats (2-G2, 3-G3, and 4-G4 etc.) and number of nucleotides in the loop [1]. These structures can adopt parallel, antiparallel or hybrid topologies depending upon the orientation of the loops. GQSes could form under physiological conditions like presence of $K^+$ and $Na^+$ ions and are thermodynamically stable. Their stability is further dependent on loop length [2].

GQSes have been shown to be involved in different cellular functions associated to both DNA processes (telomere elongation, recombination, and transcription) and RNA posttranscriptional mechanisms (including pre-mRNA processing, mRNA turnover, targeting, and translation) [3–5]. GQSes have also been found in the promoter regions of c-myc, regulating the transcription of this proto oncogene [6]. Instead of being randomly located in the genome, GQSes are found in the specific regions of the genome,

Mukesh Jain and Rohini Garg (eds.), *Legume Genomics: Methods and Protocols*, Methods in Molecular Biology, vol. 2107, https://doi.org/10.1007/978-1-0716-0235-5_13, 

such as near transcription start site (TSS), intergenic regions (including 5′UTR and 3′UTR regions) and genic regions [7, 8]. Presence of GQSes has been implicated in the replication fork stalling leading to chromosome breaks compromising the genomic stability [9].

Functional characterization of several GQSes has been probed in human cells and yeast for gene regulation and genome maintenance, but their roles in plants remain obscure. Putative G-quadruplex motifs were also identified in many plant species and their role in gene transcription or translation has been contemplated, however detailed experimental verification is still needed [10–13]. Few reporter-based studies suggest role of GQSes present in 5′UTR to acting as repressor of translation [14]. Due to these roles played by GQSes in physiological and molecular processes, it is a productive exercise to probe the abundance and functionality of these secondary structures in plants. In order to identify putative G-quadruplex forming sequences from a global perspective, one needs to search putative G-quadruplex forming sequences in plant genomes through bioinformatics screening using prediction-based algorithms, such as QuadBase TetraPlexFinder, QuadParser, QGRS Mapper, G4P Calculator, and QGRS-Conserve [15–19]. These putative sequences can be validated for GQS formation in vitro using various biophysical methods like UV-based melting curve analysis, CD spectroscopy and other complex methods like X-ray diffraction and NMR spectroscopy [20–23]. Identification of GQSes in legumes will be helpful in understanding the role of these DNA structural alterations in regulating various cellular processes, gene regulation, and plant development.

## 2 Materials

### 2.1 Computational Requirements and Software

1. Windows/Linux operating system.
2. QuadBase2 [15] http://quadbase.igib.res.in/TetraPlexFinder

### 2.2 CD Spectroscopy

1. 10× TE buffer—Add 100 ml of 1 M Tris Buffer (pH 8.0), 20 ml of 0.5 M EDTA and make up the volume to 1 l with distilled water, autoclave and store the solution at room temperature.
2. Commercially synthesized oligos of specific sequences (gel purified grade oligos should be used from IDT or any other vendor).
3. CD spectroscopy instrument—Chirascan Spectro polarimeter (Applied Biosciences).
4. Quartz cuvette.

**2.3 Gel Electrophoresis**

1. 30% acrylamide-bisacrylamide solution [19:1, acrylamide: bisacrylamide] (Himedia) (*see* **Note 1**).
2. 0.5 M EDTA solution.
3. 5× TBE—Use tris base, boric acid and 0.5 M EDTA to make stock solution.
4. 2.5 M KCl.
5. Ammonium persulfate—10% solution in water.
6. Gel electrophoresis running buffer—Add 200 ml of 5× TBE and 20 ml of 2.5 M KCl and make up the volume to 1 liter with autoclaved water.
7. GelRed Stain (Biotium). For gel staining use at 3× concentration in water.
8. Tetramethyl ethylenediamine (TEMED) (Sigma).
9. 6× Loading dye (without SDS)—0.25% bromophenol blue, 0.25% xylene cyanol, 30% glycerol in water.
10. Vertical gel electrophoresis apparatus (Biorad) and Gel documentation system (Protein Simple) for visualization of gels.

## 3 Methods

**3.1 Identification of Genome-Wide Putative G-Quadruplex Forming Sequences**

For identification of GQ forming sequences, various bioinformatics tools can be used, like Quadparser [16], G4Hunter [24] and QuadBase2 [15]. These are algorithm-based tools and are widely used in predicting the GQS forming nucleotide sequences in the genome of an organism. The G4Hunter tool gives a scoring function that reflects GQS propensity (based on G richness and G skewness). It assigns values between −4 and 4 (0 for A and T, 1 for G, 2 for GG, 3 for GGG, 4 for more than 4G's, and a negative value for C) [24]. On the other hand, Quadparser tool follows "Folding rule" which is based on strand stoichiometry, the number of stacked tetrads in the quadruplex structure, presence of mutations or deletions and the length and composition of loops [16]. QuadBase2 is a web server-based tool to mine GQ motifs using three modules: EuQuad, TetraPlexFinder and ProQuad [15]. EuQuad allows mining the quadruplexes from eukaryotes, whereas ProQuad allows search from prokaryotes. TetraPlexFinder allows mining of user provided sequences for identification of GQSes. The configuration tab in TetraplexFinder, allows selection of per-configured stem and loop length combination, such as $G_2L_{1-10}$ (low stringency), $G_3L_{1-10}$ (medium stringency) and $G_3L_{1-10}$ (high stringency), where $G_2$ or $G_3$ represents number of G repeats and $L_{1-10}$ represents loop length of a minimum 1 base and a maximum of 10 bases. Identification of nonoverlapping motifs can be done by "non-greedy" approach and is more preferred because one genomic location will be counted once, leading to a more useful statistical analysis [15]. For genome-wide identification of GQSes, command-based programs, such as Quadparser [25] and G4Hunter [24] can be used, and for the identification of GQS motifs in selected sequences, web server tools such as QGRS mapper [17] or TetraPlexFinder (available in QuadBase2 [15]) can be used. For identification of GQSes using TetraPlexFinder following steps could be followed:

1. Click on the QuadBase2 TetraplexFinder link given in Subheading 2.
2. A graphical user interface will open consisting of a tab to insert user specified sequences in fasta format (*see* **Note 2**).
3. Load or paste the sequences and define search parameters like, stem length, minimum loop length, maximum loop length, algorithm type (greedy or non-greedy), strand information and bulge size (0–3).
4. Click the submit tab after the selection of all the parameters.

The output page consists of statistics for the summary of GQS search; a table containing sequencewise counts of the number of GQS motifs along with strand information is generated. User can click on each sequence to view the detailed results as well as can download the same in BED file format for further analysis. Bed format files can be viewed with file editing tools such as notepad$^{++}$ or gedit.

### 3.2 Validation of G-Quadruplex Structure Formation

Different biophysical and biochemical methods have been developed to confirm the formation of G-quadruplex in the putative GQS motifs identified by computational analysis. For instance, GQS topology can be studied experimentally by CD spectroscopy, wherein different conformations of GQSes can be identified based on the peaks obtained in the absorption spectrum. A positive band at 260 nm in CD spectroscopy shows parallel GQS, whereas a negative band near 260 nm and positive one at 295 and 240 nm implies antiparallel GQSes [1, 21, 26]. Similarly, thermostability of GQSes can be contemplated by UV spectroscopy. The melting of GQS in the UV range at 295 nm shows lesser absorbance implicating hyperchromatic shift which is a characteristic feature for GQS formation [20]. Even though, these methods are appropriate for finding GQS formation in short oligonucleotide sequences, for finding GQS in context of longer sequences, biochemical methods like DNA polymerase stop assay can be applied. This involves blockage of polymerase on encountering GQS formation leading to halts in the primer extension [27, 28]. Genome wide identification of GQSes could also be done by using Chip-seq or next generation sequencing technology [28].

For validating GQS formation, CD spectroscopy experiment could be followed by native PAGE [29]. As native PAGE is free from denaturants, it allows sequences to retain their conformation. For analyzing GQS formation, folding is induced in putative GQSes in test ($K^+/Na^+$) and control ($Li^+$) ionic conditions which are then subjected to SDS-free native polyacrylamide gel electrophoresis. Under conditions where G-quadruplex formation is induced, there is presence of an additional band which show varied mobility compared to the band of unfolded oligo under control conditions. By analyzing shift in the band as compared to the band corresponding to the control conditions, it is possible to determine the molecularity of the structure formed. If the additional band is retarded and lies above the linear/control band, it implies the presence of an intermolecular GQ structure causing it to become bulkier than the free linear sequence [29].

#### *3.2.1 CD Spectroscopy for G-Quadruplex Structure Determination*

1. Once putative GQSes have been identified, these sequences can be obtained in the form of commercially synthesized oligos for determining G-quadruplex formation in vitro.
2. After obtaining synthesized oligos, reconstitute oligos to 100 μM stock solution in 1× TE buffer and use it to induce G-quadruplex formation by preparing sample mix of 30 μM oligos and different concentrations of salts (such as 150 μM of LiCl/KCl/NaCl) in 20 μl (*see* **Note 3**).
3. Heat the prepared oligos at 95 °C for 10 min to denature the DNA and let the samples cool down to room temperature (in the heating block itself to slowly bring samples to room temperature) to fold into secondary structures. Store the control sample (i.e., with LiCl salt solution) at −20 °C and test samples (with KCl/NaCl) at 4 °C overnight.
4. Prepare three salt solutions of 150 μM salt (LiCl/KCl/NaCl) and 50 mM Tris–HCl (pH 7.5).
5. Prepare final sample by diluting 20 μl of G-quadruplex induced oligo sample prepared in **step 2** with 380 μl of 50 mM Tris–HCl for total sample volume of 400 μl.
6. Prepare blank for each salt solution by using 20 μl of salt solution prepared in **step 4** with 380 μl of 50 mM Tris–HCl for total sample volume of 400 μl (you will need to prepare three blanks one each for LiCl, KCl and NaCl).
7. Assign the CD parameters in the spectrophotometer as scanning wavelength range of 200–320 nm, bandwidth of 1 nm, path length of 1 mm and number of scans per sample to 5 (*see* **Note 4**).
8. Load prepared CD blank sample into a cuvette of width 1 mm and place the cuvette into the scanning chamber of the spectrophotometer and scan the sample. Repeat the process for

other two blanks and samples in the corresponding salt solution. Always ensure to subtract the correct blank readings from the sample readings. Repeat the above step with other samples.

9. Once done with sample read outs, smoothen the sample readings and transfer the data to Microsoft Excel or a similar program and plot the blank subtracted data as line graphs (*see* **Note 5**).

*3.2.2 Gel Electrophoresis for Structure Determination*

1. Prepare 15% native PAGE by adding 5 ml of 30% acrylamide solution, 1 ml 5× TBE buffer, 40 μl 2.5 M KCl, 100 μl 10% APS, 10 μl TEMED and 3.880 ml autoclave MQ water to make final volume to 10 ml (*see* **Notes 6** and 7).
2. Assemble the gel casting apparatus and pour the above prepared acrylamide solution in between the gel casting plates. Gently place the comb without making any air bubbles and let it solidify for some time. Once done, remove the plates from the casting apparatus and transfer the gel to the running chamber.
3. Fill the running chamber with prepared running buffer till the mark and between the plates than remove the comb. For native PAGE analysis, prepare samples as described in **steps 2** and **3** of (Subheading 3.2.1) and add 6× DNA gel loading dye (to a final of 1× concentration) and load the samples into the wells (*see* **Note 8**).
4. Run the gel at 50 V, 400 mA at 4 °C till the dye front reaches 3/4th of the gel length.
5. Remove the gel from the plates and post-stain the gel for different time points (15 min, 30 min, 1 h to overnight) using GelRed 3× stain. Finally, visualize the stained gel in Gel imaging system at 302 nm (*see* **Note 9**).

## 4 Notes

1. Ensure protection by wearing gloves, lab coats, and safety goggles while handling chemicals.
2. Maximum of 20 Mb of input data can be uploaded.
3. Various salt concentrations needs to be tested for induction for GQS formation, such as 10 nm, 100 nm, 1 μM, 10 μM, 100 μM, and 1 mM. The concentration of salt that will induce GQS formation can vary for different sequences.
4. Ensure cuvette is rinsed thoroughly and during data acquisition, make sure the blank solution has the same salt as the sample solution.

5. It is recommended that for a particular sample, all control and test readings in different ionic conditions (LiCl/KCl/NaCl) are plotted on the same graph for better comparison. Smoothing of data is done in the CD data analysis software and same smoothening should be applied to all samples for comparison.
6. Percentage of gel will vary according to the oligo length. Higher percentage will be preferred for smaller oligos and vice versa.
7. While preparing the mixture for casting acrylamide gel, APS and TEMED should be added at the end, and the prepared mixture should immediately be mixed and poured.
8. Running buffer should be chilled beforehand to ensure temperature is maintained.
9. Remove the gel from the cassette carefully by removing gel from one side at a time and ensuring it does not break, hold the gel gently, and transfer to a box containing GelRed stain. If using UV transilluminator for visualization, ensure protection by wearing UV glasses or masks, and cover exposed skin.

## Acknowledgments

SS and KG acknowledge Shiv Nadar University for Research Fellowship and TM acknowledges SNU for funding OUR project. RG acknowledges Shiv Nadar University and Early Career Research Award from the Science and Engineering Research Board, Government of India, for funding support.

### References

1. Burge S, Parkinson GN, Hazel P, Todd AK, Neidle S (2006) Quadruplex DNA: sequence, topology and structure. Nucleic Acids Res 34:5402–5415
2. Bugaut A, Balasubramanian S (2008) A sequence-independent study of the influence of short loop lengths on the stability and topology of intramolecular DNA G-quadruplexes. Biochemistry 47:689–697
3. Biffi G, Tannahill D, Balasubramanian S (2012) An intramolecular G-quadruplex structure is required for binding of telomeric repeat-containing RNA to the telomeric protein TRF2. J Am Chem Soc 134:11974–11976
4. Balasubramanian S, Hurley LH, Neidle S (2011) Targeting G-quadruplexes in gene promoters: a novel anticancer strategy? Nat Rev Drug Discov 10:261
5. Shahid R, Bugaut A, Balasubramanian S (2010) The BCL-2 5′ untranslated region contains an RNA G-quadruplex-forming motif that modulates protein expression. Biochemistry 49:8300–8306
6. Siddiqui-Jain A, Grand CL, Bearss DJ, Hurley LH (2002) Direct evidence for a G-quadruplex in a promoter region and its targeting with a small molecule to repress c-MYC transcription. Proc Natl Acad Sci U S A 99:11593–11598
7. Bugaut A, Rodriguez R, Kumari S, Hsu STD, Balasubramanian S (2010) Small molecule-mediated inhibition of translation by targeting a native RNA G-quadruplex. Org Biomol Chem 8:2771–2776
8. Kumari S, Bugaut A, Balasubramanian S (2008) Position and stability are determining factors for translation repression by an RNA G-quadruplex-forming sequence within the 5′ UTR of the NRAS proto-oncogene. Biochemistry 47:12664–12669

9. Paeschke K, Capra JA, Zakian VA (2011) DNA replication through G-quadruplex motifs is promoted by the *Saccharomyces cerevisiae* Pifl DNA helicase. Cell 145:678–691
10. Mullen MA, Olson KJ, Dallaire P, Major F, Assmann SM, Bevilacqua PC (2010) RNA G-Quadruplexes in the model plant species *Arabidopsis thaliana*: prevalence and possible functional roles. Nucleic Acids Res 38:8149–8163
11. Garg R, Aggarwal J, Thakkar B (2016) Genome-wide discovery of G-quadruplex forming sequences and their functional relevance in plants. Sci Rep 6:28211
12. Griffin BD, Bass HW (2018) Plant G-quadruplex (G4) motifs in DNA and RNA; abundant, intriguing sequences of unknown function. Plant Sci 269:143–147
13. Andorf CM, Kopylov M, Dobbs D, Koch KE, Stroupe ME, Lawrence CJ, Bass HW (2014) G-quadruplex (G4) motifs in the maize (*Zea mays* L.) genome are enriched at specific locations in thousands of genes coupled to energy status, hypoxia, low sugar, and nutrient deprivation. J Genet Genomics 41:627–647
14. Kwok CK, Ding Y, Shahid S, Assmann SM, Bevilacqua PC (2015) A stable RNA G-quadruplex within the 5′-UTR of *Arabidopsis thaliana* ATR mRNA inhibits translation. Biochem J 467:91–102
15. Dhapola P, Chowdhury S (2016) QuadBase2: web server for multiplexed guanine quadruplex mining and visualization. Nucleic Acids Res 44: W277–W283
16. Huppert JL, Balasubramanian S (2005) Prevalence of quadruplexes in the human genome. Nucleic Acids Res 33:2908–2916
17. Kikin O, D'Antonio L, Bagga PS (2006) QGRS mapper: a web-based server for predicting G-quadruplexes in nucleotide sequences. Nucleic Acids Res 34:W676–W682
18. Eddy J, Maizels N (2006) Gene function correlates with potential for G4 DNA formation in the human genome. Nucleic Acids Res 34:3887–3896
19. Frees S, Menendez C, Crum M, Bagga PS (2014) QGRS-conserve: a computational method for discovering evolutionarily conserved G-quadruplex motifs. Hum Genomics 8:8
20. Mergny JL, Phan AT, Lacroix L (1998) Following G-quartet formation by UV-spectroscopy. FEBS Lett 435:74–78
21. Paramasivan S, Rujan I, Bolton PH (2007) Circular dichroism of quadruplex DNAs: applications to structure, cation effects and ligand binding. Methods 43:324–331
22. Parkinson GN, Lee MP, Neidle S (2002) Crystal structure of parallel quadruplexes from human telomeric DNA. Nature 417:876
23. Webba da Silva M (2007) NMR methods for studying quadruplex nucleic acids. Methods 43:264–277
24. Brázda V, Kolomazník J, Lýsek J, Bartas M, Fojta M, Šťastný J, Mergny JL (2019) G4Hunter web application: a web server for G-quadruplex prediction. Bioinformatics 24:1711
25. Wong HM, Stegle O, Rodgers S, Huppert JL (2010) A toolbox for predicting g-quadruplex formation and stability. J Nucleic Acids 2010:564946. https://doi.org/10.4061/564946
26. Randazzo A, Spada GP, da Silva MW (2013) Circular dichroism of quadruplex structures. Top Curr Chem 330:67–86
27. Hurley LH, Salazar M (1999) A DNA polymerase stop assay for G-quadruplex-interactive compounds. Nucleic Acids Res 27:537–542
28. Marsico G, Chambers VS, Sahakyan AB, McCauley P, Boutell JM, Antonio MD, Balasubramanian S (2019) Whole genome experimental maps of DNA G-quadruplexes in multiple species. Nucleic Acids Res 47:3862–3874
29. Sun D, Hurley LH (2010) Biochemical techniques for the characterization of G-quadruplex structures: EMSA, DMS footprinting, and DNA polymerase stop assay. Methods Mol Biol 608:65–79

# Chapter 14

# Method for Characterization of Root Traits in Chickpea Germplasm for Legume Genomics and Breeding

Yinglong Chen, Tao Zhou, and Kadambot H. M. Siddique

## Abstract

A semi-hydroponic phenotyping platform was constructed using inexpensive and easily obtained materials for characterizing root trait variability in a large set of chickpea (*Cicer arietinum*) germplasm. The system was designed to accommodate a large number of plants in a small area allowing relatively deeper root development, and thus serves as a high-throughput phenotyping tool for studying root dynamic growth. The root trait quantitative platform could provide accurate phenotyping data for parameterizing root models and for genome-wide association analyses or mapping studies of quantitative trait loci.

**Key words** Semi-hydroponic system, Phenotyping, Root traits, Chickpea

## 1 Introduction

Chickpea (*Cicer arietinum*) has been a predominant grain legume crop and is grown on about 12 million ha of land from temperate to subtropical regions of the world [1, 2]. The cultivation of chickpea covers some major agroecological environments, including stored soil moisture systems in South Asia, in-season rainfall in Mediterranean regions, alkaline sands in North India, alluvial soils in northwest India and Nepal, and lower water-holding-capacity soils in southern Australia [3–5]. However, the sustainable production of chickpea is challenged by some abiotic stress, such as drought and low-fertility soils, which are also the main factors restricting the production of other major crops in many countries.

Alteration to root architecture may improve desirable agronomic traits such as yield, drought tolerance, and resistance to nutrient deficit and thus a genetic improvement in root system architecture is important for breeding program [6–8]. For example, through genetic improvement efforts, the estimating 19% of the reduced production of chickpea by drought stress could be recovered [9]. However, wide-scale use of root-related genetic information in breeding is hampered by relatively small mapping

Mukesh Jain and Rohini Garg (eds.), *Legume Genomics: Methods and Protocols*, Methods in Molecular Biology, vol. 2107, https://doi.org/10.1007/978-1-0716-0235-5_14, © Springer Science+Business Media, LLC, part of Springer Nature 2020

populations and inaccurate phenotyping [10]. Obtaining accurate phenotyping and characterization of root-related traits from huge numbers of plants are very difficult, especially in soil condition. Thus, we have designed and constructed an inexpensive, space-saving, and high-throughput phenotyping platform for characterization of root trait variability among a large set of chickpea collection. The methods of the semi-hydroponic system were previously published in Functional Plant Biology (2011) [11] and the phenotyping data of various root traits in 270 chickpea genotypes were published in Journal of Experimental Botany (2017) [12].

## 2 Materials

The phenotyping system consists of support framework, growth unit, and irrigation system.

### 2.1 The Phenotyping System (Fig. 1)

1. Bin: a commonly-used mobile plastic bin (240 L, top 751 mm × 580 mm, height 1080 mm; Silverlock Packaging, Perth, WA, Australia) (*see* **Note 1**).
2. Support framework: consisted of a top frame (400 mm × 600 mm) connected perpendicularly with two U-shaped support bars (420 mm deep). Outer frame made of 4 pieces of stainless-steel wire (diameter 6 mm); horizontal rods made of 9 pieces of stainless-steel wire (diameter 5 mm); 1 piece of central dividing rod made of stainless-steel wire (diameter 5 mm).
3. Growth unit: 20 panels, acrylic sheet-one standard sheet (1200 mm × 2400 mm) makes 20 panels (260 mm × 480 mm); 10 m, black calico cloth (1.35 m wide, Textile Traders, Australia).
4. Irrigation system: 1 water pump (Pond Max PM1000F, pump flow rate 550 L $h^{-1}$); 1 timer controller; 20 drippers (Philmac Pty Ltd., Adelaide, SA, Australia); 5 m polyvinyl chloride (PVC) Drip Irrigation Tube (diameter 4 mm).
5. Other: 22 Cable ties; 160 Flodback clips (25 mm) (Corporate Express Australia Ltd., Sydney, NSW, Australia); 20 Flodback clips (15 mm) (Corporate Express Australia Ltd).

### 2.2 Solution

1. Prepare the solution using DI water and analytical grade regents. Prepare and store all reagents at room temperature.
2. Solution containing (in μM): $K_2SO_4$ (600), $MgSO_4{\cdot}7H_2O$ (200), $CaCl_2{\cdot}2H_2O$ (600), $KH_2PO_4$ (20), $H_3BO_3$ (5), $ZnSO_4{\cdot}7H_2O$ (0.75), $MnSO_4{\cdot}H_2O$ (0.75), $CoSO_4{\cdot}7H_2O$ (0.2), $CuSO_4{\cdot}5H_2O$ (0.2), $Na_2MoO_4{\cdot}H_2O$ (0.03), FeNaEDTA (20), $NH_4NO_3$ (1000) and pH 5.5 (*see* **Note 2**).

**Fig. 1** The semi-hydroponic phenotyping platform with chickpea plants grown in for 35 days in a temperature-controlled glasshouse at The University of Western Australia, Perth, WA

### *2.3 Data Collection*

1. Potable photographing system including a digital camera (no specific requirements on the brand and type), lighting system a camera holder, and an enclosed frame.
2. Root scanner (Epson Perfection V700, Long Beach, CA, USA).
3. Images analyzers (WinRHIZO, V.2009 Pro, Regent Instruments, Montreal, QC, Canada).

## 3 Method

### *3.1 Support Framework (Stainless Steel)*

1. Connect the top frame (400 × 600 mm) perpendicularly with two U-shaped support bars (420 mm deep) to make a supporting framework, (Fig. 1).
2. Using nine parallel rods and one central dividing rod to divide the top rectangular frame into 20 equal areas for inserting the growth units (*see* Subheading 3.2).
3. Bend the bottom wires of the support bars in a zig–zag and separate them evenly to hold the growth units.

### *3.2 Growth Unit*

1. Use black calico cloth (570 × 900 mm) to wrap around the 5 mm thick acrylic panel (260 × 480 mm) to form a plant growth unit.
2. Attach the cloth to the panel using four 15 mm fold-back butterfly clips on both vertical sides (Fig. 1).

3. Cut the lower part of the cloth over each panel (about 10 cm) in half to make two columns for training roots if grow plants for a longer time.

### 3.3 Irrigation System

1. Connect the fountain pump and drippers with rigid and flexible PVC tubing.
2. Secure the rigid PVC tubing to the middle bar of the support frame with cable ties.
3. Branch the flexible tubes to form the rigid tubing onto each dividing bar (Fig. 1).
4. A timer can be used to control the pump system for periodic water supplying.

### 3.4 Growth Unit Assembly

1. Place the growth units in the chambers against the horizontal rods and site on the zig–zagged wire frame at the base (20 growth units per bin).
2. Adjust the dippers to locate in the middle of the growth units and to face the growth unit.
3. Keep the plant growth units wet during the experiment using the set up pumping system in the bin.
4. The growth unit allows easy imaging down to the bottom of the plate, and the extensions of cloth allow room for a total length of root growth of ~1.2 m.
5. Add 30 L solution to the bin (*see* **Note 1**).

### 3.5 Plant Growth Condition and Managements

1. Place the phenotyping systems in a temperature-controlled glasshouse.
2. Glasshouse conditions: the average daily temperature 22/16 °C (day/night), and adequate light intensity and quality for the daytime (Fig. 1).
3. Set the periodic pumping as 10 min on and 5 min off during the daytime and 5 min on and 10 min off during the night time.
4. Surface-sterilize seeds in 30% v/v $H_2O_2$ for 20 min and wash with deionized water.
5. Germinate seeds in washed river sand.
6. Carefully transplant emerged plants at about 7 days after sowing into the growth pouches.
7. Refresh the nutrient solution weekly.
8. Randomize the location and position of each bin systems weekly to minimize environmental effects.
9. Record shoot growth weekly.
10. Access the plants 5 weeks after sowing (*see* **Note 3**).

**Fig. 2** A portable photographic system (photo box) was used for imaging root system

***3.6 Root Measurements***

1. Photograph the whole plants using a portable photographic system (Fig. 2).
2. Divide the plants into shoot and root parts.
3. Photograph root system without the shoot (*see* **Note 4**) (Fig. 3).
4. Measure root depth manually.
5. Cut the root system into 20-cm sections starting from the base for measuring root morphological traits in a laboratory.
6. Scan the root subsamples in grey scale at 300 dpi using a desktop scanner.
7. Analyze root images from the scanner using WinRHIZO Pro software (*see* **Note 5**).
8. Measured traits in WinRHIZO include: total root length, root diameter, root surface area, root volume, and root diameter class length.

## 4 Notes

1. The size of this bin allows for studying deep root systems (up to 1.2 m) and many genotypes within a limited space (40 plants in

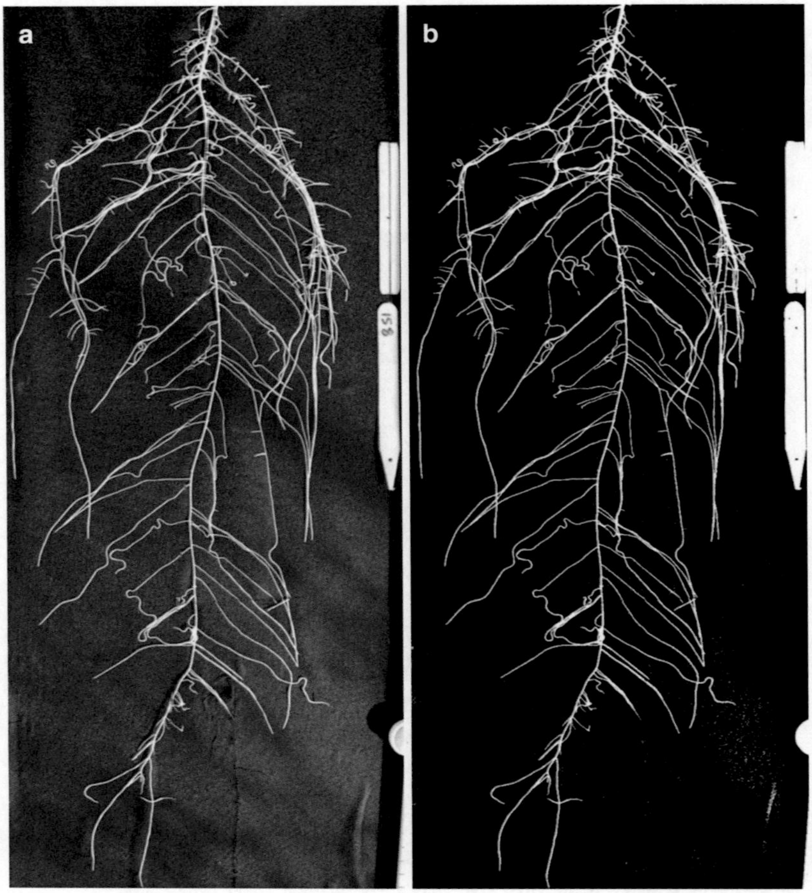


**Fig. 3** An example image of one chickpea genotype grown in the semi-hydroponic phenotyping platform for 35 days. (**a**) Root photo and (**b**) root image after calibration with DIRT software

approx. 0.5 $m^2$). As an added bonus, the bin is mobile (wheels), which helps in moving and repositioning during the course of the experiment.

2. The pH of the nutrient solution can be monitored weekly, and the solution can be renewed once every 2 weeks.
3. Termination of the root growth assessment is determined when the taproots of the plants reached the bottom of the bin system. We found that plants (e.g. narrow-leafed lupin, wheat, barley, and chickpea) using the same phenotyping platform grown for 4–6 weeks showed maximum root trait diversity among the genotypes tested.
4. The photographs are used for measuring root angle, and lateral root number etc., using the Digital imaging of root traits (DIRT) [13, 14], or manually if plants number is not high.

5. During the scanning, the debris removal filter is set to discount objects less than 1 $cm^2$ with a length/width ratio less than 10. The roots can be partitioned into 15 diameter classes at 0.15 mm intervals and root length for each root diameter class is computed.

## Acknowledgement

This research was supported by the Australian Research Council, the CGIAR Research Program on Grain Legumes in collaboration with The International Centre for Agricultural Research in the Dry Areas (ICARDA) (CRP-Grain Legumes competitive grants), Chinese Academy of Sciences "100 Talent" Program, "111 Program" (B12007). Mr. Ray Scott of The University of Western Australia provided assistance in constructing the bin system.

### References

1. Siddique KHM, Johansen C, Turner NC et al (2012) Innovations in agronomy for food legumes. A review. Agron Sustain Dev 32:45–64
2. Foyer CH, Lam HM, Nguyen HT et al (2016) Neglecting legumes has compromised human health and sustainable food production. Nat Plants 2:16112
3. Siddique KHM, Sedgley RH (1986) Chickpea (*Cicer arietinum* L.), a potential grain legume for South-Western Australia: seasonal growth and yield. Crop Pasture Sci 37:245–261
4. Berger JD, Turner NC (2007) The ecology of chickpea. In: Yadav SS, Chen W (eds) Chickpea breeding and management. Wallingford, UK, CABI International
5. Kashiwagi J, Krishnamurthy L, Purushothaman R et al (2015) Scope for improvement of yield under drought through the root traits in chickpea (*Cicer arietinum* L.). Field Crop Res 170:47–54
6. Tuberosa R, Sanguineti MC, Landi P et al (2002) Identification of QTLs for root haracteristics in maize grown in hydroponics and analysis of their overlap with QTLs for grain yield in the field at two water regimes. Plant Mol Biol 48:697–712
7. Beebe SE, Rojas-Pierce M, Yan X et al (2006) Quantitative trait loci for root architecture traits correlated with phosphorus acquisition in common bean. Crop Sci 46:413–423
8. Ghanem ME, Hichri I, Smigocki AC et al (2011) Root-targeted biotechnology to mediate hormonal signalling and improve crop stress tolerance. Plant Cell Rep 30:807–823
9. Varshney RK, Hiremath PJ, Lekha P et al (2009) A comprehensive resource of drought- and salinity-responsive ESTs for gene discovery and marker development in chickpea (*Cicer arietinum* L.). BMC Genomics 10:1
10. de Dorlodot S, Forster B, Pages L, Price A et al (2007) Root system architecture: opportunities and constraints for genetic improvement of crops. Trends Plant Sci 12:474–481
11. Chen Y, Ghanem ME, Siddique KHM (2017) Characterising root trait variability in chickpea (*Cicer arietinum* L.) germplasm. J Exp Bot 68:1987–1999
12. Chen Y, Dunbabin VM, Diggle AJ, Siddique KHM, Rengel Z (2011) Development of a novel semi-hydroponic phenotyping system for studying root architecture. Funct Plant Biol 38:355–363
13. Bucksch A, Burridge J, York LM, Das A, Nord E, Weitz JS, Lynch JP (2014) Image-based high-throughput field phenotyping of crop roots. Plant Physiol 166(2):470–486
14. Das A, Schneider H, Burridge J, Ascanio AK, Wojciechowski T, Topp CN, Lynch JP, Weitz JS, Bucksch A (2015) Digital imaging of root traits (DIRT): a high-throughput computing and collaboration platform for field-based root phenomics. Plant Methods 11(1):51

# Chapter 15

# Methods for Screening Legume Crops for Abiotic Stress Tolerance through Physiological and Biochemical Approaches

**Ragini Sinha, Meenu Bala, Madan Kumar, Tilak Raj Sharma, and Anil Kumar Singh**

## Abstract

Legume crops are subjected to a wide range of abiotic stresses, which stimulate an array of physiological, biochemical, and molecular responses. However, different genotypes may exhibit significant variations between individual responses, which can determine their tolerance or susceptibility to these stresses. The present chapter suggests a broad range of assays that can help in understanding stress perception by plants at cellular and molecular levels. The genotypes may be sorted depending on their tolerance potential, by broadly analysing morphological, physiological, biochemical, and enzyme kinetics parameters. These assays are very beneficial in revealing the mechanism of stress perception and response in varied plant types, and have helped in discriminating contrasting genotypes. Here, we have described detailed protocols of assays which may be carried out to assess tolerance or susceptibility to abiotic stresses. The analysis, as a whole, can help researchers understand the effect of abiotic stresses on plant biochemical pathways, be it photosynthesis, redox homeostasis, metabolite perturbation, signaling, transcription, and translation. These protocols may be beneficial in identification of suitable donors for breeding programs, as well as for identifying promising candidate genes or pathways for developing stress tolerant legume crops through genetic engineering.

**Key words** Abiotic stress, Biochemical, Enzyme kinetics, Legumes, Morphological, Physiological, Stress tolerance

## 1 Introduction

In the present scenario of global climate change and ever-growing population, legume crops offer an immediate solution to ensure global food and nutritional security. Owing to their natural capacity to fix atmospheric nitrogen, legume crops are advantageous for agriculture [1]. However, the legume crop production is under threat due to erratic climate conditions and environmental stresses, like drought [2], salinity [3], cold [4], and heat [5]. Abiotic stresses affect broad cellular processes, like carbon partitioning,

Mukesh Jain and Rohini Garg (eds.), *Legume Genomics: Methods and Protocols*, Methods in Molecular Biology, vol. 2107, https://doi.org/10.1007/978-1-0716-0235-5_15, © Springer Science+Business Media, LLC, part of Springer Nature 2020

photosynthesis, carbohydrate and lipid metabolism, osmotic balance and gene expression [6–8]. Abiotic stresses can affect plant, in a both general and specific manner; for example, salinity stress affects physiological processes and causes ion toxicity [9]; drought stress damages photosynthesis and limits metabolic process and nutrient availability [10, 11]; low temperature stress affects osmotic homeostasis [12]; UV radiation damages DNA and affects the synthesis of secondary metabolites; and heavy metals specifically affect mitochondrial respiration [13]. Reactive oxygen species (ROS) production and subsequent oxidative stress is the most prevalent consequence of most abiotic stresses [14].

As a part of adaptation, plants activate multiple responses at morphophysiological, biochemical, enzymatic, and molecular levels. These responses against abiotic stresses have been extensively reviewed in plants [15, 16], as well as in legumes, like chickpea, Medicago, soybean, and others [4, 17–25]. In the present chapter, we have described some standard laboratory experiments and their protocols, which are helpful in deciphering the extent of stress perception and response in plants, especially legumes (Fig. 1).

### 1.1 Morpho-physiological Analysis

Parameters like relative water content (RWC), stomatal density, leaf area index (LAI), and photosynthetic efficiency indicate the effect on overall photosynthetic apparatus of the plants [26]. Relative water content (RWC) is the appropriate measure of plant water status in terms of the physiological consequence of cellular water deficit [27]. It is the oldest method of measuring water deficit in the leaf which is expressed as a percentage either on a fresh weight or a dry weight basis. It estimates the current water content of the sampled tissue relative to the maximal water content it can hold at full turgidity. When light energy is absorbed by the chlorophyll molecules it can have one of three fates: (a) it can be used to drive photosynthesis (photochemistry), (b) excess energy can be dissipated as heat, and/or (c) it can be reemitted as long-wavelength light via a process called chlorophyll fluorescence. Increase in the efficiency of any one of these can decrease the efficiency of other two. Thus, by measuring chlorophyll fluorescence it is possible to measure the efficiency of photochemistry [28]. It indicates the extent to which PSII is using the energy absorbed by chlorophyll and the extent to which it is being damaged by excess of light. In healthy unstressed leaves, $F_v/F_m$ is highly consistent with the value of ≈0.8. A lower value indicates that a proportion of PSII reaction center is damaged, a phenomenon called photoinhibition, often observed in plants under stress conditions. Cellular membrane modification and dysfunction due to stress is well expressed in increased permeability and leakage of ions and other molecules, which can be readily measured by the efflux of electrolytes. Hence, the change in permeability of membranes or injury can be readily quantified by measuring cellular electrolyte

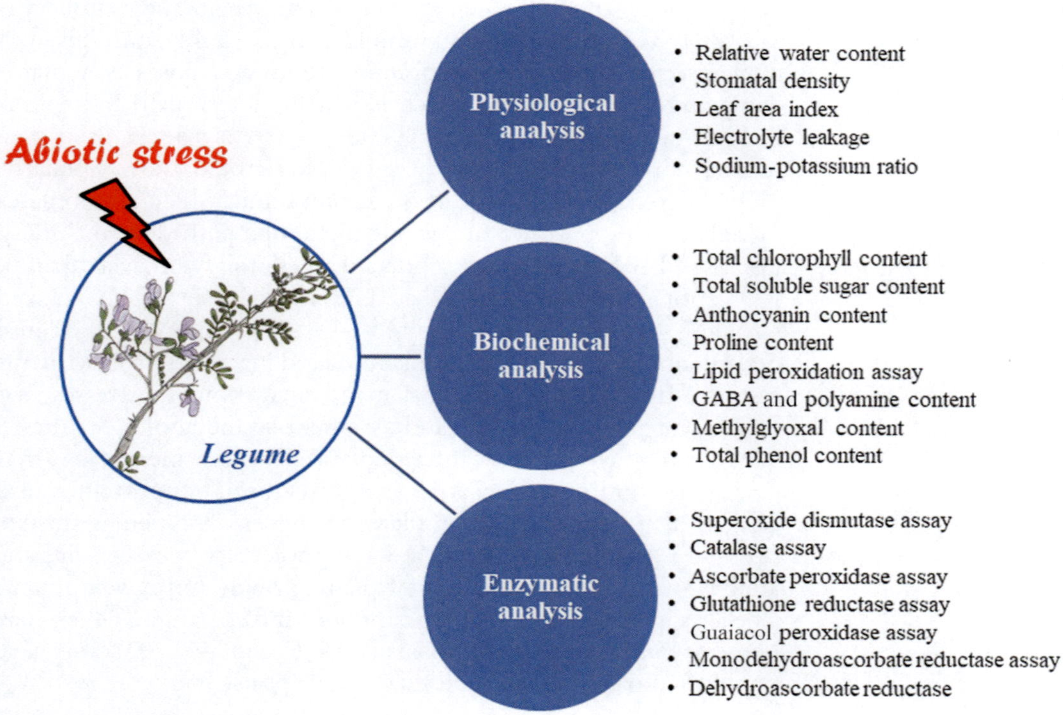


**Fig. 1** Generalized assays for estimating plant tolerance ability toward abiotic stresses

leakage from affected leaf tissue into an aqueous medium. The general protocol involves the application of stress to the leaf followed by the measurement of electrolyte leakage in terms of cell membrane stability (CMS) using electrical conductivity meter. $Na^+/K^+$ ratio and electrolyte leakage values are particularly important in deciding ion toxicity caused by stress, particularly salinity. The electrolyte leakage is also related to $K^+$ efflux from plant cells, which is maintained by plasma membrane cation conductance. It is also associated with ROS and programmed cell death [29].

### 1.2 Biochemical Analysis

Factors like total chlorophyll content and total soluble sugar content can also be correlated to photosynthesis and provide indication regarding source–sink ratio. Adverse effects of abiotic stresses on chlorophyll pigments have been reported [30]. Metabolites like anthocyanin and proline are osmolytes, which help in maintaining vital properties of the cell [31]. Proline may also serve as major source of nitrogen during post stress metabolism and may supply energy during growth and survival [32]. In addition, proline also functions as an ROS scavenger, a redox buffer, and a molecular chaperone, which stabilizes proteins and membrane during cell damage by stress [33, 34]. Likewise, anthocyanin estimation may help in understanding photoprotection activities [35], neutralization of ROS [36], stress signaling [37], and xenohormesis [38]. ROS also damages biomolecules like lipids, proteins, and nucleic acids [39]. Peroxidation of membrane lipids is a major damaging effect of ROS, which is usually determined by malondialdehyde (MDA) estimation. Thus, MDA is widely used as a marker of oxidative lipid injury (i.e., lipid peroxidation). Similarly, methylglyoxal (MG) is a dicarbonyl compound, which accumulates in cells as a by-product of various metabolic pathways, including glycolysis [40]. It also plays important roles in signaling via calcium, ROS, potassium and abscisic acid. Recently, role of MG in cell signaling has also been suggested [41, 42]. The MG can cause detrimental effects by generating ROS and inactivating antioxidant systems. It has been shown that many environmental stresses that cause oxidative stress, often induce phenolic metabolite synthesis, in order to scavenge ROS intermediates. Another metabolite, that is, γ amino butyric acid (GABA), a four-carbon, nonprotein amino acid, is also shown to accumulate during environmental stresses [43, 44]. Polyamines (PAs) are low-molecular weight aliphatic amines involved in regulation of plant growth and development. The PAs include putrescine, spermidine, and spermine; which may be present as free, soluble, or insoluble forms [45]. PAs also have established roles in plant abiotic stress response [46].

### *1.3 Enzyme Kinetics Assay*

ROS like hydrogen peroxide ($H_2O_2$), superoxide ($O_2^{\cdot -}$), and hydroxyl radicals (·OH) are generally present in the cellular compartments. However, under stress conditions, production of ROS increases rapidly, causing their accumulation at higher concentrations, which results in to significant damage to DNA, proteins, lipids, and photosynthetic pigments. The major ROS scavenging pathways include superoxide dismutase (SOD, found in nearly all cellular compartments), the ascorbate–glutathione cycle (found in chloroplast, cytosol, mitochondria, apoplast, and peroxisomes), glutathione peroxidase (GPX), and catalase (CAT, found in peroxisomes) [47]. Thus, enzyme activity assays of SOD, CAT, GR, APX, guaiacol peroxidase, monodehydroascorbate reductase (MDHAR), and dehydroascorbate reductase (DHAR) can provide insights in to ROS homeostasis status of plants.

## 2 Materials

Prepare all solutions using ultrapure water (e.g., milli Q Type I water to attain a resistivity of 18 MΩ cm at 25 °C) and analytical grade reagents. For preparation of all standard buffers, please refer Green and Sambrook [48]. Prepare and store all reagents at room temperature (unless indicated otherwise). Carefully follow all waste disposal regulations, when disposing of waste materials.

### 2.1 Morpho-physiological Analysis

RWC can be measured by either whole leaf method or leaf disc method.

#### 2.1.1 Relative Water Content (RWC)

1. Fresh leaf material.
2. Petri plates.
3. Milli Q Type I water.
4. Leaf punch cutter (for making leaf discs).
5. Polythene bags.
6. Filter paper.
7. Hot air oven.

#### 2.1.2 Stomatal Density

1. Replica fluid: cellulose diacetate or polymethyl methacrylate.
2. Leaves.
3. Brush.
4. Forceps.
5. Needles.
6. Glass slides.
7. Coverslips.
8. Microscope.
9. Ocular meter.
10. Stage micrometer.

#### 2.1.3 Sodium, Potassium Content

1. Nitric acid (1%, v/v) (*see* **Note 1**).
2. Flame photometer.
3. Hot air oven.

### 2.2 Biochemical Analysis

#### 2.2.1 Total Chlorophyll Content

1. Pestle and mortar.
2. 100% DMSO
3. Spectrophotometer.

#### 2.2.2 Total Soluble Sugar Content

1. Pestle and mortar.
2. 80% methanol (v/v) in milli Q Type I water, keep at 4 °C.
3. 5% phenol (w/v) in milli Q Type I water.
4. 95% sulfuric acid solution (v/v) in milli Q Type I water (*see* **Note 1**).
5. Spectrophotometer.

#### 2.2.3 Anthocyanin Content

1. Pestle and mortar.
2. 1% HCl in 100% methanol, prepare fresh and store at 4 °C (*see* **Note 1**).

3. 100% chloroform.
4. Spectrophotometer.

#### 2.2.4 Proline Content

1. Pestle and mortar.
2. 3% sulfosalicylic acid solution (w/v) in milli Q Type I water.
3. Acid ninhydrin reagent: 1.25 g ninhydrin in 30 ml glacial acetic acid and 20 ml 6 M phosphoric acid. Dissolve by warming and agitation.
4. 6 M phosphoric acid: Orthophosphoric acid stock (14.615 M), dilute in milli Q Type I water.
5. Glacial acetic acid.
6. 100% toluene.
7. Spectrophotometer.

#### 2.2.5 Lipid Peroxidation Assay

1. Pestle and mortar.
2. 0.1% (w/v) solution of trichloroacetic acid (TCA), in milli Q Type I water.
3. 0.5% (w/v) solution of thiobarbituric acid (TBA), in milli Q Type I water.
4. Spectrophotometer.

#### 2.2.6 Endogenous Gamma Aminobutyric Acid (GABA) and Polyamine Contents

1. Pestle and mortar.
2. 2 M $AlCl_3$(w/v) in milli Q type I water.
3. 2 M NaOH (w/v) in milli Q type I water.
4. 1 M KOH (w/v) in milli Q type I water.
5. 0.1 M sodium tetraborate (w/v), (pH = 10.0); adjust the pH with 1 N NaOH.
6. 6% phenol (w/v) in milli Q type I water.
7. 5% sodium hypochlorite.
8. 60% ethanol in milli Q type I water.
9. Benzoyl chloride.
10. 5% (v/v) perchloric acid.
11. Saturated NaCl solution.
12. Ice-cold diethyl ether.
13. Methanol.
14. UV-Vis spectrophotometer.

#### 2.2.7 Methylglyoxal (MG) Content

1. Pestle and mortar.
2. 0.5 M perchloric acid in milli Q Type I water.
3. Saturated solutions of potassium carbonate.

4. Charcoal (10 mg/ml).
5. 7.2 mM 1,2-diaminobenzene.
6. Spectrophotometer.

*2.2.8 Total Phenol Content*

1. Pestle and mortar.
2. 95% (v/v) solution of methanol in milli Q Type I water.
3. 10% (v/v) Folin-Ciocalteau (F-C) reagent.
4. 700 mM $Na_2CO_3$ in milli Q Type I water.
5. Prepare solutions of 50 μM to 2.5 mM gallic acid in 95% (v/v) methanol.
6. Spectrophotometer.

### 2.3 Enzyme Kinetics and Protein Estimation

Enzyme Extraction Buffer

1. 100 mM sodium phosphate buffer pH 7.8: mix 89.6 ml of 1 M $Na_2HPO_4$ with 10.4 ml of 1 M $NaH_2PO_4$ to make 1 M of phosphate buffer pH-7.8. Dilute to 100 mM as required.
2. 0.1 mM EDTA (pH 8.0): To make 0.5 M EDTA, dissolve 18.6 gm in 100 ml milli Q type I water. Adjust pH to 8.0 with NaOH. Dilute to 100 mM as required.
3. 1% (w/v) polyvinylpyrrolidone.
4. 0.5% (v/v) Triton X-100.

Protein Quantification

1. Bradford reagent.
2. 20% trichloroacetic acid (TCA).
3. 1.5 M sodium chloride (NaCl).
4. UV-Vis spectrophotometer.

*2.3.1 Superoxide Dismutase Activity*

1. 100 mM phosphate buffer, pH 7.8.
2. 100 mM L-methionine in milli Q Type I water, keep at 4 °C.
3. 57 μM nitroblue tetrazolium (NBT), in milli Q Type I water, keep at 4 °C.
4. 4.4% riboflavin in milli Q Type I water.

*2.3.2 Catalase Activity*

1. 50 mM phosphate buffer pH 7.0: Mix 57.7 ml of 1 M $Na_2HPO_4$ with 42.3 ml of 1 M $NaH_2PO_4$ to make 1 M of phosphate buffer pH-7.0. Dilute to 50 mM as required.
2. 20 mM $H_2O_2$, store at 4 °C.

*2.3.3 Glutathione Reductase Activity*

1. 50 mM phosphate buffer pH 7.6; mix 84.5 ml of 1 M $Na_2HPO_4$ with 15.5 ml of 1 M $NaH_2PO_4$ to make 1 M of phosphate buffer pH-7.6. Dilute to 50 mM as required.

2. 3 mM EDTA.
3. 0.1 mM NADPH, keep at 4 °C.
4. 1 mM GSSG, keep at 4 °C.

*2.3.4 Ascorbate Peroxidase Activity*

1. 100 mM phosphate buffer pH 7.0: Mix 57.7 ml of 1 M $Na_2HPO_4$ with 42.3 ml of 1 M $NaH_2PO_4$ to make 1 M of phosphate buffer pH 7.0. Dilute to 100 mM as required.
2. 5 mM ascorbate, keep at 4 °C.
3. 0.5 mM $H_2O_2$, keep at 4 °C.

*2.3.5 Guaiacol Peroxidase Activity*

1. 100 mM phosphate buffer, pH 7.0: Mix 57.7 ml of 1 M $Na_2HPO_4$ with 42.3 ml of 1 M $NaH_2PO_4$ to make 1 M of phosphate buffer, pH 7.0. Dilute to 100 mM as required.
2. 20 mM guaiacol.
3. 12.3 mM $H_2O_2$, keep at 4 °C.

*2.3.6 Monodehydroascorbate Reductase Activity*

1. 150 mM phosphate buffer pH 7.6.
2. 0.1 mM EDTA.
3. 0.25% Triton X-100.
4. 30 mM ascorbate.
5. 3 mM NADH.
6. 0.25 Units ascorbate oxidase.

*2.3.7 Dehydroascorbate Reductase Activity*

1. 100 mM phosphate buffer pH 7.0.
2. 1 mM EDTA.
3. 15 mM reduced GSH.
4. 2 mM Dehydroascorbate.

## 3 Methods

### 3.1 Morpho-physiological Analysis

*3.1.1 Relative Water Content*

Whole Leaf Method

1. Select uppermost fully expanded leaves of different plant (*see* **Note 2**).
2. Put the leaf samples carefully in polythene bags and bring it to lab, immediately (*see* **Note 3**).
3. In lab, remove the petioles/sheaths and record the fresh weight, quickly.
4. Float the leaf samples in petri plates containing doubled distilled water for 4 h (*see* **Note 4**).
5. After 4 h, remove the leaf surface moisture by blotting with filter paper, and record the turgid weight, immediately.

6. Dry the leaf sample in oven at 80-85 °C and record dry weight (*see* **Note 5**).
7. Calculate the RWC using the formula described below:

$$\text{RWC}(\%) = \left[\frac{\text{Fresh weight} - \text{Dry weight}}{\text{Turgid weight} - \text{Dry weight}}\right] \times 100$$

Leaf Disc Method

1. Collect the leaf samples from the field in polythene bags, and immediately remove the petioles/sheaths.
2. Take out 25 discs from each leaf sample (1 cm diameter) with the help of a leaf punch cutter and record their fresh weight using weighing balance.
3. Repeat **steps 4**–7 of the whole leaf method.

#### *3.1.2 Stomatal Density Estimation*

Stomatal density or stomatal frequency is measured following replica method [49, 50].

##### Preparation of Replica Slide

1. Take suitable plant material, wash it gently, and dry properly.
2. Put few drop of replica fluid material on the leaf surface uniformly, allow it to dry and carefully peel off using forceps.
3. Keep the replica in a manner that imprinted side should be on upper side and add few drops of water for proper spreading, cover it with cover slip. Now, slides are ready for calculating the stomatal density (i.e., number of stomata/mm$^2$).

##### Calibration of Ocular Micrometer

1. Place the stage micrometer (SM) on the stage of the microscope and focus it at a given magnification (e.g., 40×) (*see* **Note 6**).
2. Place the ocular micrometer (OM) in one of the eye pieces of microscope. Align the scale of OM with that of SM and calibrate the ocular scale for different magnifications (20×, 60×, 100×, etc.) (*see* **Note 7**).

##### Observation and Calculation

1. Remove the SM after calibration at desired magnification and place the replica slide having sample.
2. Count the total number of stomata as visible in circular view field of microscope.
3. Measure the diameter of view field by ocular scale.
4. Calculate the area of circle under microscopic view field using the formula: $\pi r^2$.where $r$ is the radius of the circular view field.

   Calculate the stomatal density per mm$^2$ area of leaf surface.

### 3.1.3 Leaf Area Index (LAI) Determination

Leaf Area Index (LAI) is the ratio of leaf area (one sided only) to the proportion of ground area covered by the plant. Leaf Area Index is estimated by using the formula given below.

$$\text{Leaf Area Index (LAI)} = \frac{\text{Leaf area (cm}^2)}{\text{proportion of land area covered by plant (cm}^2)}$$

1. Collect the leaf sample from uniform plant population.
2. Keep it in polythene bag and sprinkle water drops if necessary to maintain leaf turgidity and to avoid leaf shriveling.
3. Remove the leaf sheath immediately and take leaf area ($cm^2$) with the help of leaf area meter. Calculate the ground area (*see* **Note 8**).

   Calculate the LAI using the above formula.

### 3.1.4 Photosynthetic Efficiency ($F_0/F_v/F_m$ Values)

1. Dark adapt the leaf (*see* **Note 9**) using leaf clips on the plant in the field condition for 20 min.
2. Insert the fluorescence probe into the leaf clip and then open the shutter of the leaf clip.
3. Apply measuring light (*see* **Note 10**) and measure the ground state fluorescence value $F_0$.
4. Then apply a saturating flash (*see* **Note 11**) of light and measure maximum fluorescence value, $F_m$ (*Kautsky Effect*).
5. Calculate the quantum efficiency of Photosystem II = $Fv/F_m = (F_m - F_0)/F_m$.

   where $F_v$ (variable fluorescence) represents the difference between the $F_0$ and $F_m$.

### 3.1.5 Estimation of Electrolyte Leakage as an Index of Membrane Injury

1. To measure membrane injury owing to some experimental condition, harvest leaf samples from Control (*C*) and treated (*T*) leaves.
2. Cut plant tissue (100 mg), into small pieces and place in glass vials that can accommodate a conductivity electrode.
3. Wash the sample for 2–3 times with deionized water to remove electrolytes adhering to the leaf surface.
4. Keep at least 5 vials as replicates), labeled as $C_n$ and $T_n$ respectively ($n = 1, 2, 3, 4, 5$).
5. Add 20 ml of doubled distilled water to each vial and ensure that all plant materials are fully submerged.
6. Place all the vials at about 10 °C (generally, on the lowest shelf of refrigerator) for 24 h.
7. After incubation take out the vials and allow equilibrating to room temperature.

8. Measure the conductivity of all samples, recorded as $C_{n1}$ and $T_{n1}$, respectively. Cap all the vials and autoclave for 15 min to kill all tissues.
9. Allow the vials to equilibrate to room temperature and take care that volume of water is same in all the vials.
10. Measure the conductivity of all samples, recorded as $C_{n2}$ and $T_{n2}$, respectively.

*Calculation*

$$\text{Cell Membrane Stability(CMS)\%} = \left\{ \frac{\left[1 - \left(\frac{T}{T_{n2}}\right)\right]}{\left[1 - \left(\frac{C_{n1}}{C_{n2}}\right)\right]} \right\} 100$$

$$\text{Membrane injury(\%)} = 100 - \text{CMS\%}$$

#### 3.1.6 Sodium, Potassium Content

1. Dry the plant samples in hot air oven at 60 °C for 48 h.
2. Extract the dried samples in 2 ml of 1% nitric acid solution for 24 h at 70 °C.
3. Sodium and potassium content may be estimated using flame photometry [51].

### 3.2 Biochemical Analysis

#### 3.2.1 Total Chlorophyll Content

1. Total chlorophyll is estimated following Hiscox and Israelstam [52].
2. Grind 100 mg of leaf tissue in liquid nitrogen and suspend in 7 ml DMSO to make slurry (the volume may be scaled up or down depending on the sample availability).
3. Incubate the slurry at 65 °C for 30 min.
4. Adjust the final volume to 10 ml with DMSO.
5. Measure absorbance at 645 nm and 663 nm.
6. Total chlorophyll is estimated using following formula

$$\text{Total Chlorophyll}\left(\frac{g}{l}\right) = 0.0202(A_{645\text{nm}}) + 0.00802\ (A_{663\text{nm}})$$

#### 3.2.2 Total Soluble Sugar Content

1. Total soluble sugar content is estimated following Dubois et al. [53].
2. Grind 200 mg of plant tissue in liquid nitrogen and homogenize in 5 ml 80% methanol (the volume may be scaled up or down depending on the sample availability).
3. Incubate the slurry at 70 °C for 30 min.
4. Mix 1 ml of extract with 1 ml of 5% phenol and 95% sulfuric acid.
5. Incubate in dark for 15 min and measure absorbance at 490 nm.

6. Estimate the sugar content (μg/g), using glucose standard curve.
7. For preparing standard curve, take OD at 490 nm with different dilutions (0–50 μg) of glucose.

#### 3.2.3 Anthocyanin Content

1. Total anthocyanin content is estimated following Kim et al. [54].
2. Grind 100 mg of plant tissue in liquid nitrogen and add 5 ml of extraction buffer (1% HCl in methanol) (the volume may be scaled up or down depending on the sample availability).
3. Incubate the mixture overnight at 4 °C.
4. Next day, add 200 μl of water and chloroform, each; mix and centrifuge.
5. Measure the absorbance of supernatant at 530 nm and 657 nm.
6. Anthocyanin content (per g FW) can be determined using following formula:

$$\text{Anthocyanin content (per g FW)} = [A_{530\text{nm}} - (0.25 * A_{657\text{nm}})] * 5$$

#### 3.2.4 Proline Content

1. Total proline content is estimated following Bates et al. [55].
2. Grind 500 mg of plant tissue in liquid nitrogen and homogenize in 10 ml of 3% aqueous sulfosalicylic acid and centrifuge (the volume may be scaled up or down depending on the sample availability).
3. Mix 0.5 ml of supernatant with equal volumes of prepared acid-ninhydrin reagents and glacial acetic acid.
4. Incubate the mixture in boiling water bath for 1 h.
5. Measure the absorbance of toluene phase at 520 nm and 657 nm.
6. Proline content can be quantified using proline standard curve and proline content [(μmole)/g FW]can be determined by equation:

$$\text{Proline content} = \left( \frac{\left\{ \frac{\mu\text{g proline}}{\text{ml}} * \text{ml Toluene} \right\}}{115.5 \frac{\mu\text{g}}{\mu\text{M}}} \right) * \left( \text{g} \frac{\text{sample}}{5} \right)$$

#### 3.2.5 Lipid Peroxidation Assay

1. Lipid peroxidation was performed following Heath and Packer [56].
2. Grind 100 mg of plant tissue in liquid nitrogen and homogenize in 500 μl, 0.1% (w/v) TCA and centrifuge (the volume may be scaled up or down depending on the sample availability).

3. Mix supernatant with 1.5 ml, 0.5% TBA and incubate at 95 °C for 25 min, stop the reaction by incubating on ice for 5 min.
4. Measure the absorbance at 532 nm and 600 nm.
5. Lipid peroxidation can be estimated by MDA content (μM) by calculating difference in absorbance using extinction coefficient of 156 $mM^{-1}$ $cm^{-1}$.

#### 3.2.6 Endogenous Gamma Aminobutyric Acid (GABA) and Polyamine Contents Estimation

The endogenous GABA content is determined by Berthelot reaction with some modifications [57].

1. Grind 100 mg of plant tissue in liquid nitrogen, homogenize with 1 ml methanol.
2. Centrifuge at 5000 × *g* for 15 min and discard the supernatant, repeat above **steps 2** and **3** times.
3. Redissolve the sediment in 1.5 ml milli Q Type I water and incubate at 50 °C for 2 h, and then centrifuge at 7000 × *g* for 15 min.
4. Mix 1 ml supernatant with 0.1 ml $AlCl_3$ (2 M). Cool the mixture to room temperature and then centrifuge at 12,000 × *g* for 10 min.
5. Mix 0.3 ml 1 M KOH to the supernatant (0.5 ml) and shake for 5 min followed by centrifugation at 12,000 × *g* for 5 min.
6. The resulting supernatant can be used to measure the content of GABA.
7. Prepare the reaction solutions including 0.5 ml sodium tetraborate (0.1 M) (pH = 10.0), 0.4 ml 6% phenol and 0.6 ml 5% sodium hypochlorite. To this, add0.3 ml supernatant, mix and keep the mixture into boiling water bath for 10 min. Immediately place it in ice bath for 5 min.
8. Finally, shake the solution with 2 ml 60% ethanol and measure the absorbance in 645 nm.
9. GABA content (mg/g DW) can be calculated using standard curve.

Quantification of free, insoluble, and soluble conjugated polyamines, like putrescine, spermine, and spermidine is done according to Duan et al. [58].

1. Homogenize 100 mg of leaves with 1 ml precooled perchloric acid (5%, v/v), incubate at 4 °C for 1 h.
2. Centrifuge at 12,000 × *g* for 30 min at 4 °C.
3. Use the supernatant to determine free and soluble conjugated polyamines and keep the pellet to determine insoluble bound polyamines.

4. Benzoylate the supernatant by adding 500 μl supernatant with 2 ml NaOH (2 M) and 10 μl benzoyl chlorides and incubate at 37 °C for 30 min.
5. Terminate the reaction by adding 2 ml of saturated NaCl solution.
6. Add 2 ml of cold diethylether and remove nearly 1 ml of ether phase by evaporation.
7. Redissolve the dried product in 100 μl of methanol and determine the polyamines content by high-performance liquid chromatography.

#### *3.2.7 Methylglyoxal (MG) Content Estimation*

1. MG content is estimated following Yadav et al. [59].
2. Homogenize 300 mg of plant tissue in 3 ml of 0.5 M perchloric acid (the volume may be scaled up or down depending on the sample availability) and incubate on ice for 15 min.
3. Centrifuge the mixture at 11,000 × *g* for 10 min at 4 °C.
4. If color appears in extract, decolorize by adding charcoal (10 mg/ml), keep for 15 min at room temperature.
5. Centrifuge the mixture at 11,000 × *g* for 10 min, take the supernatant and neutralize by adding saturated solution of potassium carbonate. Let it rest for 15 min at room temperature and centrifuge the mixture again at 11000 × *g* for 10 min.
6. Use this supernatant for MG assay.
7. In 1 ml reaction mixture add, 250 μl of 7.2 mM 1,2-diaminobenzene, 100 μl of 5 M perchloric acid and 650 μl of sample extract. Immediately, measure the absorbance at 336 nm.
8. The final concentration was calculated from standard curve and expressed as μM/g FW.
9. For preparing standard curve add different concentrations of pure MG (10, 25, 50 and 100 μM), derivatize with 1,2 diaminobenzene. For 1 ml reaction mixture, add 250 μl of 7.2 mM 1,2-diaminobenzene, 100 μl of 5 M perchloric acid and milli Q Type I water. Add standards and immediately measure the absorbance at 336 nm.

#### *3.2.8 Total Phenol Content Estimation*

1. Total phenol content is estimated following Ainsworth and Gillespie [60].
2. Homogenize 500 mg of plant tissue in 2 ml of ice-cold 90% methanol (the volume may be scaled up or down depending on the sample availability).
3. Incubate the mixture for 48 h at room temperature.

4. Add 200 μl of 10% (v/v) Folin-Ciocalteau (F-C) reagent, 100 μl of sample, and vortex thoroughly. Be careful in adding F-C reagent to prevent air oxidation of phenols.
5. Add 800 μl 700 mM $Na_2CO_3$ into each tube and incubate the assay tubes at room temperature for 2 h.
6. Measure the absorbance at 765 nm.
7. Total phenol can be quantified as mg/g DW, using gallic acid standard curve.

### 3.3 Enzyme Kinetics

For the protein estimation and activities of antioxidant enzymes; the enzyme extract is prepared by homogenizing 0.5 g of plant tissue with 1.0 ml of extraction buffer under ice cold conditions. Centrifuge the homogenate at 15,000 × *g* for 20 min and store the supernatant (enzyme extract) at 4 °C.

Although there are many ways for protein quantification, but Bradford assay [61] is most commonly followed method. Precipitate 20 μl of enzyme extract with equal volume of 20% TCA. Dissolve the precipitate in 1.5 M NaCl, mix 1 ml Bradford reagent and take OD at 595 nm. Protein concentration can be plotted using BSA V (Bovine serum albumin fraction V) standard curve.

#### 3.3.1 Superoxide Dismutase Activity

1. The SOD activity is determined following Beauchamp and Fridovich [62].
2. In the test cuvette, prepare the reaction mixture (2.5 ml), by adding 100 mM phosphate buffer, 100 mM L-methionine, 57 μM NBT and 100 μl enzyme extract.
3. Add 400 μl of 4.4% riboflavin and take OD immediately at 560 nm.
4. Take final OD after 7 min.
5. One unit of SOD activity can be measured as 50% reduction of NBT.SOD activity can be expressed as mol UA/mg protein.

$$\%\ \text{inhibition}\ (y) = \frac{(\Delta A_{min}\text{blank} - \Delta A_{min}\text{test}) * 100}{\Delta A_{min}\text{blank}}$$

$\Delta A$ min = change in absorbance per minute.

*y* (%) of inhibition is produced by 100 μl of sample.

Hence, 50% inhibition is produced by $(50 \times 100)/y = z$ μl of sample

#### 3.3.2 Catalases Activity

1. The CAT activity is determined following Luck [63].
2. Prepare the reaction mixture (2.5 ml), by adding 50 mM phosphate buffer, 20 mM $H_2O_2$ and 100 μl enzyme extract.
3. Decrease in absorbance was measured for 3 min at 240 nm.
4. One unit of catalase activity is the amount of enzyme required to oxidize 1 mM of $H_2O_2$ per min.

5. CAT activity can be expressed as Units/min/gFW.

$$\text{Unit Activity} = \frac{\Delta A_{min} \times \text{total volume (ml)}}{\varepsilon \times \text{volume of sample (ml)}}$$

where, extinction coefficient ($\varepsilon$) = $6.93 \times 10^{-3}$ mM$^{-1}$ cm$^{-1}$

$$\text{Specific activity} = \frac{\text{Unit activity}}{\text{Protein content (mg/g FW)}}$$

*3.3.3 Glutathione Reductase Activity*

1. The GR activity is determined following Schaedle and Bassham [64].
2. Prepare the reaction mixture (3 ml), containing 1.8 ml of 50 mM phosphate buffer (pH 7.6), 300 μl each of EDTA (3 mM), NADPH (0.1 mM), oxidized glutathione (GSSG) (1 mM), and enzyme extract.
3. Decrease in absorbance was measured per min at 340 nm.
4. One unit of the enzyme activity is defined as the amount of enzyme required to oxidize 1.0 μM of NADPH/min/g FW.

$$\text{Unit Activity} = \frac{\Delta A_{min} \times \text{total volume (ml)}}{\varepsilon \times \text{volume of sample (ml)}}$$

where, extinction coefficient ($\varepsilon$) = $6.22 \times 10^{-3}$ mM$^{-1}$ cm$^{-1}$

$$\text{Specific activity} = \frac{\text{Unit activity}}{\text{Protein content (mg/g FW)}}$$

*3.3.4 Ascorbate Peroxidase Activity*

1. The APX activity is determined following Nakano and Asada [65].
2. Prepare the reaction mixture (2.5 ml), containing 1.5 ml of 100 mM phosphate buffer (pH 7.0), 300 μl ascorbate (5 mM), 600 μl $H_2O_2$ (0.5 mM), and 100 μl enzyme extract.
3. Decrease in absorbance was measured per min at 290 nm.
4. One unit of the enzyme activity was calculated as the amount of enzyme required to oxidize 1.0 μM of ascorbate/min/g FW

$$\text{Unit Activity} = \frac{\Delta A_{min} \times \text{total volume (ml)}}{\varepsilon \times \text{volume of sample (ml)}}$$

where, extinction coefficient ($\varepsilon$) = $2.8 \times 10^{-3}$ mM$^{-1}$ cm$^{-1}$

$$\text{Specific activity} = \frac{\text{Unit activity}}{\text{Protein content (mg/g FW)}}$$

*3.3.5 Guaiacol Peroxidase Activity*

1. The guaiacol peroxidase is estimated according to the method given by Putter [66].

2. Prepare the reaction mixture, containing 3 ml phosphate buffer, 50 μl guaiacol solution, 30 μl $H_2O_2$, and 100 μl enzyme extract.
3. The rate of formation of guaiacol dehydrogenation product (GDHP) was measured 436 nm.
4. One unit of enzyme activity is defined as the amount of enzyme catalyzing the formation of 1.0 μM of GDHP/min/g FW

$$\text{Unit Activity} = \frac{\Delta A_{min} \times \text{total volume (ml)}}{\varepsilon \times \text{volume of sample (ml)}}$$

where, extinction coefficient $(\varepsilon) = 25 \times 10^{-3}\ \text{mM}^{-1}\ \text{cm}^{-1}$

$$\text{Specific activity} = \frac{\text{Unit activity}}{\text{Protein content (mg/g FW)}}$$

*3.3.6 Monodehydroascorbate Reductase (MDHAR) Activity*

1. The MDHAR activity is determined following Hossain et al. [67].
2. In the test cuvette, prepare the reaction mixture, containing 1.8 ml phosphate buffer, 300 μl EDTA, 200 μl NADH, 250 μl ascorbate, 0.25 Units ascorbate oxidase, and 300 μl enzyme extract.
3. Decrease in absorbance was measured per min at 340 nm.
4. One unit of the enzyme activity is defined as the amount of enzyme required to oxidize 1.0 μM of NADH/min/g FW

$$\text{Unit Activity} = \frac{\Delta A_{min} \times \text{total volume (ml)}}{\varepsilon \times \text{volume of sample (ml)}}$$

where, extinction coefficient $(\varepsilon) = 6.22 \times 10^{-3}\ \text{mM}^{-1}\ \text{cm}^{-1}$

$$\text{Specific activity} = \frac{\text{Unit activity}}{\text{Protein content (mg/g FW)}}$$

*3.3.7 Dehydroascorbate Reductase Activity*

1. The DHAR activity is determined following Dalton et al. [68].
2. Prepare the reaction mixture (3 ml), containing 1.8 ml phosphate buffer, 300 μl EDTA, 500 μl reduced GSH, 300 μl dehydroascorbate and 100 μl enzyme extract.
3. Increase in absorbance was measured per min at 265 nm.
4. One unit of the enzyme activity is defined as the amount of enzyme catalyzing the formation of 1.0 μM of ascorbate/min/g FW.

$$\text{Unit Activity} = \frac{\Delta A_{min} \times \text{total volume (ml)}}{\varepsilon \times \text{volume of sample (ml)}}$$

where, extinction coefficient $(\varepsilon) = 14 \times 10^{-3}\ \text{mM}^{-1}\ \text{cm}^{-1}$

$$\text{Specific activity} = \frac{\text{Unit activity}}{\text{Protein content (mg/g FW)}}$$

### 3.4 Conclusions

The adverse effects of environmental stresses on growth and productivity of a wide variety of model and crop plants have been explored. To attain global food and nutritional security, development of improved legume crop varieties is becoming increasingly important. Thus, high-throughput phenotyping technologies may be employed for phenotyping of donor genotypes under control and stressed conditions. During any abiotic or environmental stress, tolerant/resistant plants exhibit a basic defense that is generally orchestrated by a very well organized and conserved cellular defense response that involves a networking and interactions of ROS, compatible solutes, polyamines, and downstream genes and transcription factors. The present chapter describes about the methods for qualitative and quantitative estimation of such factors. However, in order to categorize a cultivar as stress tolerant/resistant or susceptible, an in-depth analysis of these factors and their interactions is equally important. Thus, this section highlights the role of these factors in plant defense, which may be corroborated with previous findings and may also help in commenting the stress tolerance ability of a crop cultivar.

#### 3.4.1 Morpho-physiological Data Analysis

When the plants are subjected to environmental stress, immediately, it is perceived at cellular membrane that gets damaged at first. The primary response of all the plants toward moisture stress is closure of stomata, to avoid water loss via transpiration. Thus, a genotype with lower stomatal density has higher osmotic stress tolerance due to its ability to reduce loss of water through transpiration. The electrolyte leakage (EL) is used as a measure of cell membrane stability (CMS) and as a parameter to screen genotypes for abiotic stress tolerance. The EL takes place due to changes in lipid-protein configuration [69] and leads to leakage of ions and impairs cellular functions [70]. On the basis of determined CMS, membrane injury (MI, %) is calculated, the genotype having less MI is considered as tolerant as compared to others. Following membrane damage, the stress affects photosynthetic ability of plants, due to the disruption of chloroplast membranes by direct or indirect effects, such as photooxidation, which denatures the chlorophyll molecules [71–73]. Photosynthesis is also restricted by reduced stomatal conductance and reduction of chlorophyll molecule as determined by $F_v/F_m$ value. The photosystem II (PSII) is very sensitive to temperature and is greatly affected by high temperature stress, which may damage its components [74]. Thus, a tolerant genotype may have higher PS II activity and efficient light reactions as reported. Likewise, RWC is an indicator of water absorbed by the plants and consumed through transpiration. In general, genotypes with higher RWC are more tolerant owing to stable osmoregulation [75]. Higher RWC is correlated with increased concentration of photosynthetic pigments, CMS, osmolyte concentration, and antioxidant activities [76]. Higher RWC and closure of stomata (thus inhibiting gaseous exchange) provide stress tolerance [77]. RWC can also be correlated with TSS, as enhanced RWC can increase the TSS content [78], and TSS plays an important role in osmotic adjustments of plants [79]. Thus, any genotype that is able to sustain its basal photosynthetic activities and carbon assimilation during stress conditions can be considered as tolerant.

### *3.4.2 Biochemical Data Analysis*

*Reactive oxygen Species*: All abiotic stresses trigger production of ROS and reactive carbonyl species (RCS) (i.e., MDA and MG). ROS and MG also function as signaling molecules for regulating various physiological processes. MDA, on the other hand, can initiate lipid peroxidation, causing protein dysfunction and resulting in ROS proliferation [80, 81]. Lipid peroxidation acts as a marker of oxidative stress induced injury during abiotic stress [82]. It also indicates a relationship between drought and oxidative stress [83]. Significantly higher MDA content under drought stress indicates vulnerability to oxidative damage caused by drought stress [84, 85]. Thus, tolerant genotypes should have lesser lipid peroxidation and lower MDA content as reported earlier in legumes like pea [86] and lentil [25]. Also the tolerant genotypes should neutralize ROS by employing detoxifying enzymes, like SODs, CATs, and peroxidases (PODs). Enzymatic defense is usually considered as the most effective [87]. SOD converts $O^{2\cdot-}$ to $H_2O_2$, which is further reduced to water by CAT and PODs. Enzymes such as SOD, POD, and CAT either directly scavenge the ROS or protect plants indirectly by managing nonenzymatic defense [88]. Similarly, another pathway of Ascorbate-GSH is required for APX that involves DHAR, MDHAR, and GR. APX is the key enzyme for $H_2O_2$ removal. However, an enhanced SOD activity was observed in drought tolerant genotype, as against drought susceptible in lentil [24, 25]. In response to ROS, an increased content of MDA has been reported which is an indicator of drought induced oxidative damage [82]. Therefore, maintenance of the higher levels of the antioxidants is one of the strategies employed by the plants to counter the negative effects of ROS [89]. Various studies demonstrated stress tolerance against drought, salt, and high light in plants by overexpressing genes encoding antioxidant enzymes [90–98].

MG is a cytotoxic molecule, which is produced as a by-product of glycolysis and can form adducts with proteins and nucleic acids [40]. Thus, tolerant cultivars should ideally have lower MG content; however, MG can be detoxified by glyoxalase pathway consisting of glyoxalase I and II enzymes. Genetic manipulations of glyoxalase systems can provide tolerance to multiple abiotic stresses [99–101]

*Compatible Solutes/osmolytes*: Molecules like sugars, amino acids, and their derivatives, such as raffinose, trehalose, inositol, mannitol, proline, and glycine betaine (GB), help in maintaining vital properties of the cell under optimal conditions [102–104]. However, during stress, these molecules act as osmoprotectant against dehydration, ROS and RCS [105]. In addition to role as osmoprotectant, proline also stabilizes cellular redox potential [106, 107]. Thus, it can be inferred that accumulation of such solutes is always beneficial for plants during stress conditions. Polyamines (PAs) form another major group of compatible solutes. The role of PAs in imparting protection against a wide range of stresses has been well elucidated [39, 46, 108, 109]. As the stress intensity increases, PAs promote osmoregulation and preserve RWC, by causing accumulation of proline and GABA as reported in various legume crops [110–112]. Likewise, GABA also plays role in stress signaling [113], it can also trap MDA via direct reaction with the aldehyde group [81]. GABA accumulates through GABA shunt pathway and also via PA degradation pathway [114]. Higher accumulation of GABA is correlated with higher tolerance toward abiotic stresses.

A wide range of plant responses against stress could be generalized into morphological, physiological and biochemical responses. These could be helpful in sorting genotypes based on their tolerance and susceptibility to abiotic stresses. In the climate change scenario, improving crop tolerance to abiotic stresses is a major goal of present day research. However, it is also important to consider and analyze other interrelated parameters of growth and metabolism of plants.

## 4 Notes

1. Take all preventive measures while working with concentrated acid; add acid to water, keep chill in ice.
2. Fully expanded leaves are physiologically functional. Since the plant water status is influenced by the solar radiation and temperature, to minimize the error, fix a time of the day preferably between 11 AM and noon, for sampling.
3. Samples should be collected in three replications and sealed properly to minimize water loss.
4. Samples were hydrated to full turgidity by floating on water. Close the petri dish and put it at the lower shelf of a lab refrigerator at about 10 °C.

5. The dry weight should be recorded after proper drying till the weight becomes constant.
6. Stage micrometer (SM) is a calibrated scale having 100 divisions of 1 mm; therefore, 1 division is equal to 0.01 mm or 10 μm.
7. For calibration of ocular micrometer, observe through the eye pieces and carefully count the divisions of OM coinciding with divisions of SM.

   For example, 4.0 divisions of OM = 50 divisions of SM, that is, 500 μm.

   1 division of OM = 12.5 μm.
8. The ground area can be calculated based on the row spacing and length of crop row harvested. For example, if 100 cm wheat row harvested where the row spacing was 22 cm, then the ground area of the sample will be $22 \times 100 = 2200\ \text{cm}^2$.
9. Dark adapted condition means all PSII centers are open and no nonphotochemical quenching is present.
10. The measuring light excites chlorophyll but is not enough to induce electron transport through PSII. It gives $F_0$ value, the minimal level of fluorescence, and reaction centers are said to be open.
11. Saturating flash of light of about 8000 μmol $m^{-2}\ s^{-1}$.

## Acknowledgments

AKS acknowledges funding support by Indian Council of Agricultural Research, New Delhi in the form of projects IXX12585 and IXX12644;Science and Engineering Research Board (SERB), Department of Science and Technology (DST), Government of India for the project under "Scheme for Young Scientists" (SB/YS/LS-255/2013). RS acknowledges DST-SERB for the National-Postdoctoral Fellowship (PDF/2016/000924), and DST, Govt. of India for funding under the Women Scientists Scheme-A (SR/WOS-A/LS-160/2018). MB acknowledges University Grant Commission (UGC), New Delhi for JRF.

### References

1. Foyer CH, Lam H, Nguyen HT, Siddique KHM, Varshney RK, Colmer TD et al (2016) Neglecting legumes have compromised health and sustainable food production. Nat Plants 2:1–10. https://doi.org/10.1038/nplants.2016.112
2. Kunert KJ, Vorster BJ, Fenta BA, Kibido T, Davis TS (2016) Drought stress responses in soybean roots and nodules. Front Plant Sci 7:1015. https://doi.org/10.3389/fpls.2016.01015
3. Vadez V, Krishnamurthy L, Serraj R, Gaur PM, Upadhyaya HD (2007) Large variation in salinity tolerance in chickpea is explained by differences in sensitivity at the reproductive stage. Field Crop Res 104:123–129. https://doi.org/10.1016/j.fcr.2007.05.014

4. Hekneby M, Antol MC, Manuel S (2006) Frost resistance and biochemical changes during cold acclimation in different annual legumes. J Environ Exp Bot 55:305–314. https://doi.org/10.1016/j.envexpbot.2004.11.010
5. Valdés-lópez O, Batek J, Gomez-hernandez-N, Stacey G, Sayre R (2016) Soybean roots grown under heat stress show global changes in their transcriptional and proteomic profiles. Front Plant Sci 7:517. https://doi.org/10.3389/fpls.2016.00517
6. Hasegawa PM, Bressan RA, Zhu JK, Bohnert HJ (2000) Plant cellular and molecular responses to high salinity. Annu Rev Plant Physiol Plant Mol Biol 51:463–499. https://doi.org/10.1146/annurev.arplant.51.1.463
7. Munns R (2002) Comparative physiology of salt and water stress. Plant Cell Environ 25:239–250. https://doi.org/10.1046/j.0016-8025.2001.00808.x
8. Rosa M, Hilal M, González JA, Prado FE (2004) Changes in soluble carbohydrates and related enzymes induced by low temperature during early developmental stages of quinoa (Chenopodium quinoa) seedlings. J Plant Physiol 161:683–689. https://doi.org/10.1078/0176-1617-01257
9. Tester M, Davenport R (2003) $Na^+$ tolerance and $Na^+$ transport in higher plants. Ann Bot 91:503–527. https://doi.org/10.1093/aob/mcg058
10. Bohnert HJ, Gong Q, Li P, Ma S (2006) Unraveling abiotic stress tolerance mechanisms—getting genomics going. Curr Opin Plant Biol 9:180–188. https://doi.org/10.1016/j.pbi.2006.01.003
11. Hu YC, Schmidhalter U (2005) Drought and salinity: a comparison of their effects on mineral nutrition of plants. J Plant Nutr Soil Sci 168:541–549. https://doi.org/10.1002/jpln.200420516.541
12. Browse J, Xin Z (2001) Temperature sensing and cold acclimation. Curr Opin Plant Biol 4:241–246. https://doi.org/10.1016/S1369-5266(00)00167-9
13. Hall JL (2002) Cellular mechanisms for heavy metal detoxification and tolerance. J Exp Bot 53:1–11. https://doi.org/10.1093/jexbot/53.366.1
14. Foyer CH, Noctor G (2005) Oxidant and antioxidant signalling in plants : a re-evaluation of the concept of oxidative stress in a physiological context. Plant Cell Environ 28:1056–1071. https://doi.org/10.1111/j.1365-3040.2005.01327.x
15. Zhu JK (2016) Abiotic stress signaling and sesponses in plants. Cell 167:313–324. https://doi.org/10.1016/j.cell.2016.08.029
16. He M, He C-Q, Ding N-Z (2018) Abiotic stresses: general defenses of land plants and chances for engineering multistress tolerance. Front Plant Sci 9:1–18. https://doi.org/10.3389/fpls.2018.01771
17. Merchan F, de Lorenzo L, Rizzo SG, Manyani H, Frugier F, Sousa C et al (2007) Identification of regulatory pathways involved in the reacquisition of root growth after salt stress in Medicago truncatula. Plant J 51:1–17. https://doi.org/10.1111/j.1365-313X.2007.03117.x
18. Jain D, Chattopadhyay D (2010) Analysis of gene expression in response to water deficit of chickpea (Cicer arietinum L.) varieties differing in drought tolerance. BMC Plant Biol 10:24. https://doi.org/10.1186/1471-2229-10-24
19. Mhadhbi H, Fotopoulos V, Mylona PV, Jebara M, Elarbi M (2011) Antioxidant gene—enzyme responses in Medicago truncatula genotypes with different degree of sensitivity to salinity. Physiol Plant 141:201–214. https://doi.org/10.1111/j.1399-3054.2010.01433.x
20. Filippou P, Antoniou C, Fotopoulos V (2011) Effect of drought and rewatering on the cellular status and antioxidant response of Medicago truncatula plants. Plant Signal Behav 6:270–277. https://doi.org/10.4161/psb.6.2.14633
21. Chen LM, Zhou XA, Li WB, Chang W, Zhou R, Wang C et al (2013) Genome-wide transcriptional analysis of two soybean genotypes under dehydration and rehydration conditions. BMC Genomics 14:687. https://doi.org/10.1186/1471-2164-14-687
22. Araújo SS, Beebe S, Crespi M, Delbreil B, Gonzalez EM, Gruber V et al (2015) Abiotic stress responses in legumes: strategies used to cope with environmental challenges. Crit Rev Plant Sci 34:237–280. https://doi.org/10.1080/07352689.2014.898450
23. Muscolo A, Junker A, Klukas C, Weigelt-Fischer K, Riewe D, Altmann T (2015) Phenotypic and metabolic responses to drought and salinity of four contrasting lentil accessions. J Exp Bot 66:5467–5480. https://doi.org/10.1093/jxb/erv208
24. Singh D, Singh CK, Taunk J, Sewak R, Tomar S, Chaturvedi AK (2017) Transcriptome analysis of lentil (Lens culinaris Medikus) in response to seedling drought stress.

BMC Genomics 18:206. https://doi.org/10.1186/s12864-017-3596-7

25. Sinha R, Sharma TR, Singh AK (2019) Validation of reference genes for qRT-PCR data normalisation in lentil (Lens culinaris) under leaf developmental stages and abiotic stresses. Physiol Mol Biol Plants 25:123–134. https://doi.org/10.1007/s12298-018-0609-1
26. Guidi L, Lo Piccolo E, Landi M (2019) Chlorophyll fluorescence, photoinhibition and abiotic stress: does it make any difference the fact to be a C3 or C4 species? Front Plant Sci 10:1–11. https://doi.org/10.3389/fpls.2019.00174
27. Barrs HD, Weatherley PE (1962) A re-examination of the relative turgidity techniques for estimating water deficits in leaves. Aust J Biol Sci 15:413–428. https://doi.org/10.1071/BI9620413
28. Kautsky H, Apple W, Amann H (1960) Chlorophyll fluorescence and carbon assimilation. Part XIII. The fluorescence and the photochemistry of plants. Biochem Z 332:277–292
29. Demidchik V, Straltsova D, Medvedev SS, Pozhvanov GA, Sokolik A, Yurin V (2014) Stress-induced electrolyte leakage: the role of $K^+$-permeable channels and involvement in programmed cell death and metabolic adjustment. J Exp Bot 65:1259–1270. https://doi.org/10.1093/jxb/eru004
30. Singh J, Thakur JK (2018) Photosynthesis and abiotic stress in plants. In: Vats S (ed) Biotic and abiotic stress tolerance in plants. Springer Nature, Singapore. https://doi.org/10.1007/978-981-10-9029-5
31. Yoshiba Y, Kiyosue T, Katagiri T, Ueda H, Mizoguchi T, Yamaguchi-Shinozaki K et al (1995) Correlation between the induction of a gene for d 1-pyrroline-5-carboxylate synthetase and the accumulation of proline in Arabidopsis thaliana under osmotic stress. Plant J 7:751–760. https://doi.org/10.1046/j.1365-313X.1995.07050751.x
32. Gupta B, Huang B (2014) Mechanism of salinity tolerance in plants: physiological, biochemical, and molecular characterization. Int J Genomics 2014:701596. https://doi.org/10.1155/2014/701596
33. Krasensky J, Jonak C (2015) Drought, salt, and temperature stress-induced metabolic rearrangements and regulatory networks. J Exp Bot 63:1593–1608. https://doi.org/10.1093/jxb/err460
34. Hossain MS, Dietz K (2016) Tuning of redox regulatory mechanisms, reactive oxygen species and redox homeostasis under salinity stress. Front Plant Sci 7:548. https://doi.org/10.3389/fpls.2016.00548
35. Tattini M, Landi M, Brunetti C, Giordano C, Remorini D, Gould KS, Guidi L (2014) Epidermal coumaroyl anthocyanins protect sweet basil against excess light stress: multiple consequences of light attenuation. Physiol Plant 152:585–598. https://doi.org/10.1111/ppl.12201
36. Nakabayashi R, Yonekura-Sakakibara K, Urano K, Suzuki M, Yamada Y, Nishizawa T et al (2014) Enhancement of oxidative and drought tolerance in *Arabidopsis* by over accumulation of antioxidant flavonoids. Plant J 77:367–379. https://doi.org/10.1111/tpj.12388
37. Pourcel L, Irani NG, Koo AJ, Bohorquez-Restrepo A, Howe GA, Grotewold E (2013) A chemical complementation approach reveals genes and interactions of flavonoids with other pathways. Plant J 74:383–397. https://doi.org/10.1111/tpj.12129
38. Howitz KT, Sinclair DA (2008) Xenohormesis: sensing the chemical cues of other species. Cell 133:387–391. https://doi.org/10.1016/j.cell.2008.04.019
39. Gill SS, Tuteja N (2010) Plant physiology and biochemistry reactive oxygen species and antioxidant machinery in abiotic stress tolerance in crop plants. Plant Physiol Biochem 48:909–930. https://doi.org/10.1016/j.plaphy.2010.08.016
40. Mustafiz A, Singh AK, Pareek A, Sopory SK, Singla-Pareek SL (2011) Genome-wide analysis of rice and Arabidopsis identifies two glyoxalase genes that are highly expressed in abiotic stresses. Funct Integr Genomics 11:293–305. https://doi.org/10.1007/s10142-010-0203-2
41. Hoque TS, Hossain MA, Mostofa MG, Burritt DJ (2016) Methylglyoxal: an emerging signaling molecule in plant abiotic stress responses and tolerance. Front Plant Sci 7:1–11. https://doi.org/10.3389/fpls.2016.01341
42. Kaur C, Singla-pareek SL, Sopory SK (2014) Glyoxalase and methylglyoxal as biomarkers for plant stress tolerance. Crit Rev Plant Sci 33:429–456. https://doi.org/10.1080/07352689.2014.904147
43. Bown AW, Shelp BJ (1997) The metabolism and functions of y-Aminobutyric acid. Plant Physiol 1:1–5. https://doi.org/10.1104/pp.115.1.1
44. Narsai R, Rocha M, Geigenberger P, Whelan J, Van Dongen JT (2011) Comparative analysis between plant species of

transcriptional and metabolic responses to hypoxia. New Phytol 190:472–487. https://doi.org/10.1111/j.1469-8137.2010.03589.x
45. Bouchereau A, Aziz A, Larher F (1999) Polyamines and environmental challenges: recent development. Plant Sci 140:103–125. https://doi.org/10.1016/S0168-9452(98)00218-0
46. Minocha R, Majumdar R, Minocha SC (2014) Polyamines and abiotic stress in plants: a complex relationship. Front Plant Sci 5:1–17. https://doi.org/10.3389/fpls.2014.00175
47. Mittler R (2002) Oxidative stress, antioxidants and stress tolerance. Trends Plant Sci 7:405–410. https://doi.org/10.1016/S1360-1385(02)02312-9
48. Green MR, Sambrook J (2014) Molecular cloning, a laboratory manual. Cold Spring Harbor Laboratory Press, New York
49. Meidner H, Mansfieid TA (1968) Physiology of stomata. McGraw Hill, New York
50. Wolf DD, Carson EW, Parrish DJ (1979) A replica method of determining stomatal and epidermal cell intensity. J Agro Edu 8:52–54
51. Williams CH, Twine V (1960) In: Peach K, Tracey MV (eds) Modern methods of plant analysis. Springer-Verlag Berlin, Berlin/Heidelberg, pp 3–5
52. Hiscox JD, Israelstam GF (1979) A method for the extraction of chlorophyll from leaf tissue without maceration. Can J Bot 57:1332–1334. https://doi.org/10.1139/b79-163
53. DuBois M, Gilles KA, Hamilton JK, Rebers PA, Smith F (1956) Colorimetric method for determination of sugars and related substances. Anal Chem 28:350–356. https://doi.org/10.1021/ac60111a01
54. Kim J, Yi H, Choi G, Shin B, Song P, Choi G (2003) Functional characterization of phytochrome interacting factor 3 in phytochrome mediated light signal transduction. Plant Cell 15:23992407. https://doi.org/10.1105/tpc.014498
55. Bates LS, Waldren RP, Teare ID (1973) Rapid determination of free proline for water - stress studies. Plant Soil 39:205–207. https://doi.org/10.1007/BF00018060
56. Heath RL, Packer L (1968) Photoperoxidation in isolated chloroplasts: I. kinetics and stoichiometry of fatty acid peroxidation. Arch Biochem Biophys 125:189–198. https://doi.org/10.1016/0003-9861(68)90654-1
57. Guijin Z, Bown AW (1997) The rapid determination of γ amino-butyric acid. Phytochemistry 44:1007–1009. https://doi.org/10.1016/S0031-9422(96)00626-7
58. Duan JJ, Li J, Guo S, Kang Y (2008) Exogenous spermidine affects polyamine metabolism in salinity-stressed Cucumis sativus roots and enhances short-term salinity tolerance. J Plant Physiol 165:1620–1635. https://doi.org/10.1016/j.jplph.2007.11.006
59. Yadav SK, Singla-Pareek SL, Reddy MK, Sopory SK (2005) Transgenic tobacco plants overexpressing glyoxalase enzymes resist an increase in methylglyoxal and maintain higher reduced glutathione levels under salinity stress. FEBS Lett 579:6265–6271. https://doi.org/10.1016/j.febslet.2005.10.006
60. Ainsworth EA, Gillespie KM (2007) Estimation of total phenolic content and other oxidation substrates in plant tissues using Folin-Ciocalteu reagent. Nat Protoc 2:875–877. https://doi.org/10.1038/nprot.2007.102
61. Bradford MM (1976) A rapid and sensitive method for the quantitation of microgram quantities of protein utilizing the principle of protein-dye binding. Anal Biochem 72:248–254. https://doi.org/10.1006/abio.1976.9999
62. Beauchamp C, Fridovich I (1971) Superoxide dismutase: improved assays and an assay applicable to acrylamide gels. Anal Biochem 44:276–287. https://doi.org/10.1016/0003-2697(71)90370-8
63. Luck H (1974) In: Bergmeyer HU, Bergmeyer J, Grassl M (eds) Methods in enzymatic analysis. Academic, New York, p 885
64. Schaedle M, Bassham JA (1977) Chloroplast glutathione reductase. Plant Physiol 59:1011–1012. https://doi.org/10.1104/pp.59.5.1011
65. Nakano Y, Asada K (1981) Hydrogen peroxide is scavenged by ascorbate-specific peroxidase in spinach chloroplasts. Plant Cell Physiol 22:867–880. https://doi.org/10.1093/oxfordjournals.pcp.a076232
66. Putter J (1974) Peroxidase. In: Bergmeyer HU (ed) Methods of enzymatic analysis. Verlag Chemie, Weinhan, pp 685–690
67. Hossain M, Nakano Y, Asada K (1984) Monodehydroascorbate reductase in spinach chloroplasts and its participation in the regeneration of ascorbate for scavenging hydrogen peroxide. Plant Cell Physiol 25:385–395. https://doi.org/10.1093/oxfordjournals.pcp.a076726
68. Dalton DA, Russell SA, Hanus FJ, Pascoet GA, Evans HJ (1986) Enzymatic reactions of

ascorbate and glutathione that prevent peroxide damage in soybean root nodules. Proc Natl Acad Sci U S A 83:3811–3815. https://doi.org/10.1073/pnas.83.11.3811

69. Earnshaw MJ (1993) Stress indicators: electrolyte leakage. In: Hendry GAF, Grime JP (eds) Methods in comparative ecology. Chapman and Hall, London, pp 152–154
70. Conde A, Chaves MM, Geros H (2011) Membrane transport, sensing and signaling in plant adaptation to environmental stress. Plant Cell Physiol 52:1583–1602. https://doi.org/10.1093/pcp/pcr107
71. Kotak S, Larkindale J, Lee U, von Koskull-Doring P, Vierling E, Scharf KD (2007) Complexity of the heat stress response in plants. Curr Opin Plant Biol 10:310–316. https://doi.org/10.1016/j.pbi.2007.04.011
72. Ristic Z, Bukovnik U, Prasad PVV (2007) Correlation between heat stability of thylakoid membranes and loss of chlorophyll in winter wheat under heat stress. Crop Sci 47:2067–2073. https://doi.org/10.2135/cropsci2006.10.0674
73. Djanaguiraman M, Prasad PVV, Seppanen M (2010) Selenium protects sorghum leaves from oxidative damage under high temperature stress by enhancing antioxidant defense system. Plant Physiol Biochem 48:999–1007. https://doi.org/10.1016/j.plaphy.2010.09.009
74. Monakhova OF, Chernyadev II (2002) Protective role of kartolin-4 in wheat plants exposed to soil drought. Appl Biochem Microbiol 38:373–380. https://doi.org/10.1023/A:1016243424428
75. Gunasekera D, Berkowitz GA (1992) Evaluation of contrasting cellular-level acclimation responses to leaf water deficits in three wheat genotypes. Plant Sci 86:1–12. https://doi.org/10.1016/0168-9452(92)90173-J
76. Moussa HR, Abdel-Aziz SM (2008) Comparative response of drought tolerant and drought sensitive maize genotypes to water stress. Aust J Crop Sci 1:31–36
77. Thameur A, Lachiheb B, Ferchichi A (2012) Drought effect on growth, gas exchange and yield in two strains of local barley Ardhaoui, under water deficit conditions in southern Tunisia. J Environ Manag 113:495–500. https://doi.org/10.1016/j.jenvman.2012.05.02
78. Koingshofer HL, Oppert HG (2015) Regulation of invertase activity in different root zones of wheat (Triticum aestivum L.) seedlings in the course of osmotic adjustment under water deficit conditions. J Plant Physiol 183:130–137. https://doi.org/10.1016/j.jplph.2015.06.005
79. Zhou Q, Yu BJ (2009) Accumulation of inorganic and organic osmolytes and their role in osmotic adjustment in NaCl-stresses vetiver grass seedlings. Russ J Plant Physiol 56:678–685. https://doi.org/10.1134/S1021443709050148
80. Farmer EE, Davoine C (2007) Reactive electrophile species. Curr Opin Plant Biol 10:380–386. https://doi.org/10.1016/j.pbi.2007.04.019
81. Deng Y, Xu L, Zeng X, Li Z, Qin B, He N (2010) New perspective of GABA as an inhibitor of formation of advanced lipoxidation end-products: it's interaction with malondialdehyde. J Biomed Nanotechnol 6:318–324. https://doi.org/10.1166/jbn.2010.1130
82. Moller IM, Jensen PE, Hansson A (2007) Oxidative modifications to cellular components in plants. Annu Rev Plant Biol 58:459–481. https://doi.org/10.1146/annurev.arplant.58.032806.103946
83. Munne-Bosch S, Jubany-Mari T, Alegre L (2001) Drought induced senescence is characterised by a loss of antioxidant defences in chloroplasts. Plant Cell Environ 24:1319–1327. https://doi.org/10.1046/j.1365-3040.2001.00794.x
84. Sultan MARF, Hui L, Yang LJ, Xian ZH (2012) Assessment of drought tolerance of some *Triticum* L. species through physiological indices. Czech J gen. Czech J Genet Plant Breed 48:178–184. https://doi.org/10.17221/21/2012-CJGPB
85. Siddique MH, Al-Khaishany MY, Al-Qutami MA, Al-Whaibi MH, Grover A, Ali HM et al (2015) Responses of different genotypes of Faba bean plant to drought stress. Int J Mol Sci 16:10214–10227. https://doi.org/10.3390/ijms160510214
86. Moran JF, Becana M, Iturbe-Ormaetxe I, Frechilla S, Klucas RV, Aparicio-Trejo P (1994) Drought induces oxidative stress in pea plants. Planta 194:346–352. https://doi.org/10.1007/BF00197534
87. Farooq M, Aziz T, Basra SMA, Cheema MA, Rehman H (2008) Chilling tolerance in hybrid maize induced by seed priming with salicylic acid. J Agron Crop Sci 194:161–168. https://doi.org/10.1111/j.1439-037X.2008.00300.x
88. Anjum SA, Wang LC, Farooq M, Hussain M, Xue LL, Zou CM (2011) Brassinolide application improves the drought tolerance in maize through modulation of enzymatic

antioxidants and leaf gas exchange. J Agron Crop Sci 197:177–185. https://doi.org/10.1111/j.1439-037X.2010.00459.x

89. Sharma P, Dubey RS (2005) Drought induces oxidative stress and enhances the activities of antioxidant enzyme in growing rice seedling. Plant Growth Regul 46:209–221. https://doi.org/10.1007/s10725-005-0002-2

90. Pang CH, Li K, Wang BS (2011) Overexpression of SsCHLAPXs confers protection against oxidative stress induced by high light in transgenic Arabidopsis thaliana. Physiol Plant 143:355–366. https://doi.org/10.1111/j.1399-3054.2011.01515.x

91. Cao S, Du XH, Li LH, Liu YD, Zhang L, Pan X et al (2017) Overexpression of Populus tomentosa cytosolic ascorbate peroxidase enhances abiotic stress tolerance in tobacco plants. Russ J Plant Physiol 64:224–234. https://doi.org/10.1134/s1021443717020029

92. Roxas VP, Lodhi SA, Garrett DK, Mahan JR, Allen RD (2000) Stress tolerance in transgenic tobacco seedlings that overexpress glutathione S-transferase/glutathione peroxidase. Plant Cell Physiol 41:1229–1234. https://doi.org/10.1093/pcp/pcd051

93. Qi YC, Liu WQ, Qiu LY, Zhang SM, Ma L, Zhang H (2010) Overexpression of glutathione S-transferase gene increases salt tolerance of *Arabidopsis*. Russ J Plant Physiol 57:233–240. https://doi.org/10.1134/s102144371002010x

94. Kumar S, Trivedi PK (2018) Glutathione *S*-transferases: role in combating abiotic stresses including arsenic detoxification in plants. Front Plant Sci 9:751. https://doi.org/10.3389/fpls.2018.00751

95. Shafi A, Chauhan R, Gill T, Swarnkar MK, Sreenivasulu Y, Kumar S et al (2015) Expression of SOD and APX genes positively regulates secondary cell wall biosynthesis and promotes plant growth and yield in *Arabidopsis* under salt stress. Plant Mol Biol 87:615–631. https://doi.org/10.1007/s11103-015-0301-6

96. Shafi A, Gill T, Sreenivasulu Y, Kumar S, Ahuja PS, Singh AK (2015) Improved callus induction, shoot regeneration, and salt stress tolerance in *Arabidopsis* overexpressing superoxide dismutase from Potentilla atrosanguinea. Protoplasma 252:41–45. https://doi.org/10.1007/s00709-014-0653-9

97. Shafi A, Gill T, Zahoor I, Ahuja PS, Sreenivasulu Y, Kumar S et al (2019) Ectopic expression of *SOD* and *APX* genes in *Arabidopsis* alters metabolic pools and genes related to secondary cell wall cellulose biosynthesis and improve salt tolerance. Mol Biol Rep 46:1985–2002. https://doi.org/10.1007/s11033-019-04648-3

98. Shafi A, Pal AK, Sharma V, Kalia S, Kumar S, Ahuja PS, Singh AK (2017) Transgenic potato plants overexpressing SOD and APX exhibit enhanced lignification and starch biosynthesis with improved salt stress tolerance. Plant Mol Biol Rep 35:504–518. https://doi.org/10.1007/s11105-017-1041-3

99. Singla-Pareek SL, Reddy MK, Sopory SK (2003) Genetic engineering of the glyoxalase pathway in tobacco leads to enhanced salinity tolerance. Proc Natl Acad Sci U S A 100:14672–14677. https://doi.org/10.1073/pnas.2034667100

100. Singla-Pareek SL, Yadav SK, Pareek A, Reddy MK, Sopory SK (2006) Transgenic tobacco overexpressing glyoxalase pathway enzymes grow and set viable seeds in zinc-spiked soils. Plant Physiol 140:613–623. https://doi.org/10.1104/pp.105.073734

101. Gupta BK, Sahoo KK, Ghosh A, Tripathi AK, Anwar K, Das P et al (2018) Manipulation of glyoxalase pathway confers tolerance to multiple stresses in rice. Plant Cell Environ 41:1186–1200. https://doi.org/10.1111/pce.12968

102. Yoshiba Y, Nanjo T, Miura S, Yamaguchi-Shinozaki K, Shinozaki K (1999) Stress-responsive and developmental regulation of D1-pyrroline-5-carboxylate synthetase 1 (*P5CS1*) gene expression in Arabidopsis thaliana. Biochem Biophys Res Commun 261:766–772. https://doi.org/10.1006/bbrc.1999.1112

103. Renault H, El-Amrani A, Berger A, Mouille G, Soubigou Taconnat L, Bouchereau A et al (2013) γ-Aminobutyric acid transaminase deficiency impairs central carbon metabolism and leads to cell wall defects during salt stress in *Arabidopsis* roots. Plant Cell Environ 36:1009–1018. https://doi.org/10.1111/pce.12033

104. Muscolo A, Sidari M, Anastasi U, Santonoceto C, Maggio A (2014) Effect of drought stress on germination of four lentil genotypes. J Plant Interact 9:354–363. https://doi.org/10.1080/17429145.2013.835880

105. Szabadous L, Savoure A (2010) Proline: a multifunctional amino acid. Trends Plant Sci 15:89–97. https://doi.org/10.1016/j.tplants.2009.11.009

106. Yancey PH (2005) Organic osmolytes as compatible, metabolic and counteracting cytoprotectants in high osmolarity and other

stresses. J Exp Biol 208:2819–2830. https://doi.org/10.1242/jeb.01730

107. Ashraf M, Foolad MR (2007) Roles of glycine betaine and proline in improving plant abiotic stress resistance. Environ Exp Bot 59:206–216. https://doi.org/10.1016/j.envexpbot.2005.12.006

108. Liu JH, Kitashiba H, Wang J, Ban Y, Moriguchi T (2007) Polyamines and their ability to provide environmental stress tolerance to plants. Plant Biotechnol 24:117–126. https://doi.org/10.5511/plantbiotechnology.24.117

109. Alcázar R, Altabella T, Marco F, Bortolotti C, Reymond M, Koncz C et al (2010) Polyamines: molecules with regulatory functions in plant abiotic stress tolerance. Planta 231:1237–1249. https://doi.org/10.1007/s00425-010-1130-0

110. Espasandin FD, Maiale SJ, Calzadilla P, Ruiz OA, Sansberro PA (2014) Transcriptional regulation of 9-cis-epoxycarotenoid dioxygenase (NCED) gene by putrescine accumulation positively modulates ABA synthesis and drought tolerance in Lotus tenuis plants. Plant Physiol Biochem 76:29–35. https://doi.org/10.1016/j.plaphy.2013.12.018

111. Nayyar H, Satwinder K, Kumar S, Singh KJ, Dhir KK (2005) Involvement of polyamines in the contrasting sensitivity of chickpea (Cicer arietinum L.) and soybean (Glycine max (L.) Merrill.) to water deficit stress. Bot Bull Acad Sin Taipei 46:333–338

112. Yong B, Xie H, Li Z, Li Y-P, Zhang Y, Nie G et al (2017) Exogenous application of GABA improves PEG-induced drought tolerance positively associated with GABA-shunt, polyamines, and proline metabolism in white clover. Front Physiol 8:1107. https://doi.org/10.3389/fphys.2017.01107

113. Bouche N, Fromm H (2004) GABA in plants: just a metabolite? Trends Plant Sci 9:110–115. https://doi.org/10.1016/j.tplants.2004.01.006

114. Xing SG, Jun YB, Hau ZW, Liang LY (2007) Higher accumulation of γ-aminobutyric acid induced by salt stress through stimulating the activity of diamine oxidases in Glycine max (L.) Merr. Roots. Plant Physiol Biochem 45:560–566. https://doi.org/10.1016/j.plaphy.2007.05.007

# Chapter 16

# An Efficient Method of Mitochondrial DNA Isolation from *Vigna radiata* for Genomic Studies

**Pooja Singh, Ranjan Kumar Sahoo, Mallesham Bulle, and Kapuganti Jagadis Gupta**


## Abstract

Isolation of mitochondrial DNA from root tissues of mung bean (*Vigna radiata*) is quite tedious, complex, and often results in low yield. Hence here we show a simple, rapid, and improved protocol for isolation of mitochondrial DNA from root tissues of hydroponically grown mung bean plants. This method involves purification of mitochondria and subsequent isolation of DNA from obtained purified mitochondria. For this purpose, mitochondria were isolated using a discontinuous Percoll gradient centrifugation followed by RNase I treatment to the isolated DNA to remove any traces of RNA contamination. The mitochondrial DNA was isolated from mitochondrial samples by commonly used CTAB method. The specificity of isolated mitochondrial DNA was confirmed using mtDNA-specific genes (*NAD1* and *COX3*). β-Actin primer was used to check the nuclear DNA contamination. PCR amplification was observed in mtDNA specific genes *NAD1* and *COX3* except nuclear encoded β-actin gene suggesting that mitochondrial DNA was not contaminated by nuclear DNA.


**Key words** β-Actin, DNA, Mitochondria, Mung bean, Percoll gradient

## 1 Introduction

Mitochondria are considered as energy-power houses of cell as these organelles play key roles in regulation of various cellular activities and energy metabolism. In this organelle, transfer of electrons via various oxidoreductase respiratory complexes (complexes I, II, III, IV or AOX pathway) to oxygen takes place in a systemic manner accompanied by movement of protons from matrix to inner membrane space leading to generation of proton gradient [1, 2] This proton gradient is being used for production of energy in the form of ATP which is important for fulfilling various cellular energy requirements [1, 2]. Accumulating evidence suggest that plant mitochondria not only play role in energy production but

Pooja Singh and Ranjan Kumar Sahoo contributed equally to this work.

Mukesh Jain and Rohini Garg (eds.), *Legume Genomics: Methods and Protocols*, Methods in Molecular Biology, vol. 2107, https://doi.org/10.1007/978-1-0716-0235-5_16, © Springer Science+Business Media, LLC, part of Springer Nature 2020

also involved in various alternative functions such as programmed cell death (PCD) (similar to animal mitochondria); synthesis of vitamins, cofactors; and generation of reactive oxygen species (ROS) [3]. Mitochondria also play important role in conferring male sterility in various species and also play role in retrograde signaling [4–8]. This semi independent organelle contains its own genome, called as mitochondrial DNA (mtDNA), which is compatimentized in the matrix of this organelle [9]. The mitochondrial genome contains a slightly modified genetic code which makes it contrast to the nuclear genome of cell [10]. Intriguingly mitochondrial DNA has quite extraordinary special features such as variation in its size, physical form, capacity of coding, complex patterns of organization, and quite different mechanism of gene expression in comparison to nuclear DNA [11–14]. In the course of evolution mitochondria likely have originated from alpha purple bacteria and carries hallmark such as *N*-formylmethionyl-tRNA (fMet-tRNA) of its bacterial origin [15]. The (fMet-tRNA), is an initiator of protein synthesis in bacteria and interestingly quite similar initiation system is present in mitochondria [16]. mtDNA is abundant in content (based on mitochondrial number and stage of development). Till date extensive nucleotide sequence has been determined for mtDNA of numerous plant and animal systems [17]. Comparative analysis of both plant and animal mtDNA and its structure revealed highly conserved nature among animals. In contrast to animals, in plants mitochondrial DNA is highly dynamic [18, 19].

One of the interesting features of mitochondrial genome is its size difference in animal and plants [20]. Animal mitochondrial genome size is much smaller and comparatively less complex than plants (i.e., 15–20 kbp) in length and plant mitochondrial genomes are in a range between 200 and 2000 kbp, whereas in algae the size is 13 and 96 kb [20]. Despite of its large size (In comparison to animals) it is very surprising that plant mitochondrial genomes do not contain significantly more genes than animal mitochondrial genomes. This is due presence of noncoding regions, repeats and large introns, adding to this complexity the origin of non coding regions are not completely known [20].

Due to large sizes of plant mitochondrial genome, the isolation of mitochondrial DNA becomes increasingly complex. The best approach is to isolate pure, intact plant mitochondria and subsequent isolation of mitochondrial DNA. Several different protocols have been reported for isolation of mitochondrial DNA from various crops such as sunflower [21], wheat [22], and rice [23]. Here in this study, we report a simple, rapid, reliable, and reproducible protocol for isolation of mitochondrial DNA from mung bean roots.

## 2 Materials

### 2.1 Equipment and Miscellaneous

1. 96-Well Thermal Cycler (Applied Biosystems™ Veriti™).
2. Blotting paper.
3. Centrifuge.
4. Disposable syringe (10 ml).
5. Filter paper (Whatman grade).
6. Laboratory bottle.
7. Magnetic stirrer block.
8. Marker pen.
9. Microcentrifuge tube (both 1.5 and 2 ml).
10. Microcentrifuge tube stand.
11. Mixer grinder.
12. Nitrile gloves.
13. Oak Ridge tubes (50 ml; Tarsons, India).
14. Paintbrush.
15. Pipette.
16. Plastic beakers.
17. Plastic trays.
18. Sandwich of Miracloth and cotton.
19. Scissor.
20. Spatula.
21. Spectrophotometer.
22. Surgical blade.
23. Thermo block.
24. Timer.
25. Tips and cut tips.
26. Tissue paper.
27. Water bath.

### 2.2 Chemicals/ Solutions/ Miscellaneous

1. Absolute ethanol.
2. Cetyl trimethylammonium bromide (CTAB, 2%).
3. dNTP Mix.
4. EDTA (Ethylenediaminetetraacetic acid EDTA, 20 mM).
5. HEPES.
6. Hoagland media.
7. $KH_2PO_4$.
8. L-Cysteine.

9. $MgCl_2$.
10. Monopotassium phosphate ($KH_2PO_4$).
11. NaCl.
12. Percoll.
13. Phenol–Chloroform–Isoamyl alcohol (25:24:1).
14. Prechilled Isopropanol.
15. Prechilled Sodium acetate.
16. PVP.
17. Seed sterilization solution (0.4% Mercuric chloride).
18. Skim milk powder.
19. Standard Taq Reaction Buffer.
20. Sucrose.
21. Taq DNA polymerase.
22. Tris HCl.
23. β-mercaptoethanol.

## 3 Methods

### *3.1 Mung Bean Seed Sterilization, Germination, and Plant Growth*

1. Weigh 30 g of healthy mung bean seeds and place in a beaker and wash them thoroughly with milliQ water to remove any dust particles and impurities from seeds.
2. Once the seeds are well washed then dip the seeds by placing them in 0.1% $HgCl_2$ and incubate them for 15 min with slow continuous shaking (50 rpm).
3. After incubation discard the $HgCl_2$ solution from beaker containing seeds and rinse the seeds thoroughly for 4–5 times with autoclaved distilled water.
4. Imbibe the sterile mung bean seeds in autoclaved distilled water for a period of 4 h.
5. Once seeds are imbibed then keep the seeds in dark for 48 h at room temperature for facilitating germination.
6. Take hydroponic tray and fill with freshly prepared 1/2 strength Hoagland solution. Initially for a week use 1/2 strength Hoagland later use full strength solution.
7. Place the seeds on tray holes (Fig. 1) and keep them for 2 weeks.
8. Once plants are grown then cut the roots of all mung bean plants using sharp blade and use immediately for isolation of mitochondria (Fig. 1).

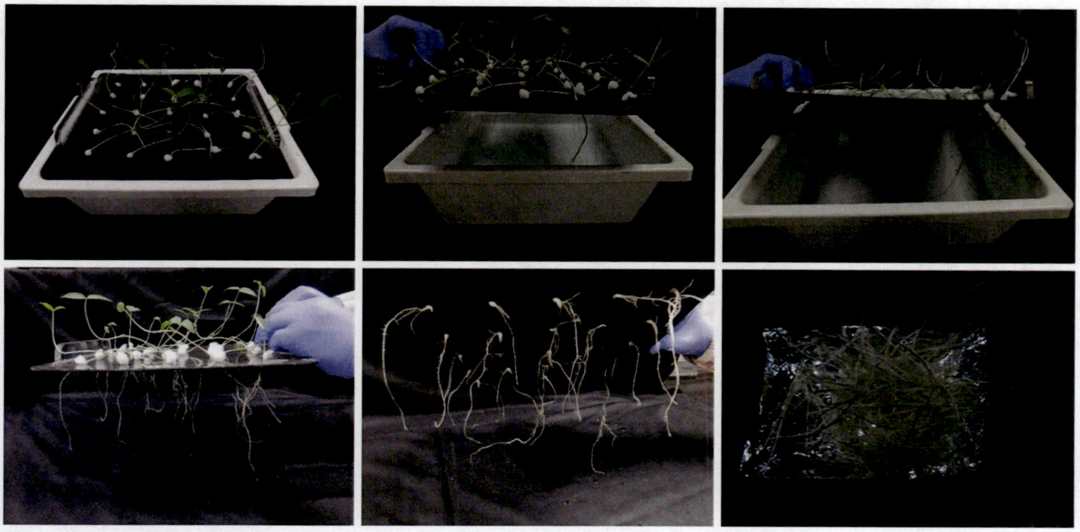

**Fig. 1** Phenotype of mung bean plants grown on hydroponics

### 3.2 Preparation of Solutions and Percoll Gradient for Mitochondrial Isolation

1. Buffer preparation: Prepare 500 ml of extraction buffer, 100 ml of suspension buffer, and 100 ml gradient buffer as described in Table 1 (*see* **Note 1**).
2. Discontinuous Percoll gradient preparation.
   (a) Using a measuring cylinder and 5 ml pipette, prepare different concentrations of sucrose and Percoll (According to table) and mix thoroughly to achieve desired concentrations of 60%, 45%, 28% and 5% of Percoll (Fig. 2).
   (b) Take clean Oakridge tubes and add 3 ml of 60%, 4 ml of 45%, 4 ml of 28%, and 4 ml of 5% Percoll very slowly with the help of 10 ml syringe (needle less). Keep these tubes in stand and place in ice carefully to be used during the separation of mitochondria.

### 3.3 Mitochondria Extraction Procedure from Mung Bean Roots

It is recommended to conduct all steps of isolation of mitochondria at 4 °C (*see* **Note 2**). For better clarity we divided the whole procedure into 4 major steps such as extraction step, centrifugation, Percoll gradient centrifugation and Percoll removal steps (Fig. 3).

#### 3.3.1 Extraction

1. Cut the roots from the mung bean seedlings. Finely chop them into 1–2 cm slices and weigh 50 g of roots (Fig. 1) (*see* **Note 3**).
2. In a grinder mixer container add 150 ml prechilled extraction buffer and root slices and grind the roots (Several pulses can be given for better grinding). Repeat this step.
3. Filter the obtained homogenate through 8 layers of Miracloth–cotton sandwich layer and collect the filtrate in 1 L centrifuge bottle and if require squeezing is needed to obtain complete extract (Fig. 3).

**Table 1**
**Composition of extraction, suspension, and gradient buffer for mung bean mitochondria isolation**

| Extraction buffer composition | Suspension buffer composition | Gradient buffer composition |
|---|---|---|
| 100 mM HEPES (pH 7.2–7.4) | 100 mM HEPES (pH 7.2–7.4) | 20 mM HEPES (pH 7.2–7.4) |
| 0.3 M sucrose | 0.3 M sucrose | 250 mM sucrose |
| 1 mM EDTA | 1 mM EDTA | 0.1% Skim milk powder or BSA |
| 2 mM $MgCl_2$ | 0.5 mM $KH_2PO_4$ | |
| 5 mM $KH_2PO_4$ | 2 mM $MgCl_2$ | |
| 1% Skim milk powder (alternatively BSA can be used) | 1% Skim milk powder (alternatively BSA can be used) | |
| 0.6% PVP (freshly prepared) | | |
| 4 mM L-cysteine | | |

| Suspension buffer (ml) | Percoll (ml) | Final volume (ml) | Percoll gradient (%) |
|---|---|---|---|
| 19  | 1.0  | 20  | 5 |
| 14.4  | 5.6  |  | 28 |
| 11  | 9  | 20  | 45 |
| 8  | 12  | 20  | 60 |

**Fig. 2** Percoll gradient preparation

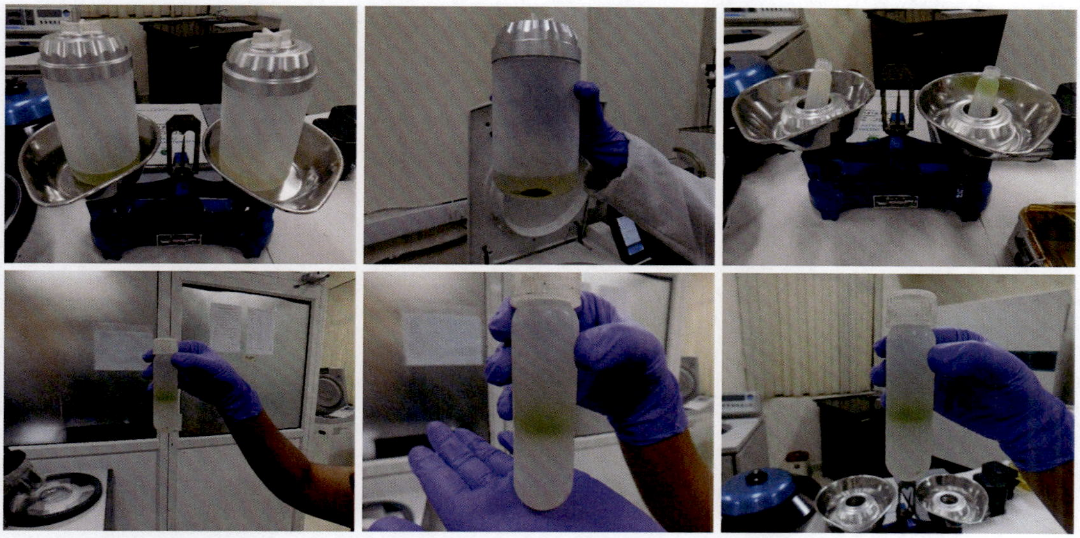

**Fig. 3** Important steps in mitochondrial isolation from mung bean

#### 3.3.2 Centrifugation

1. Centrifugation has three stages. First and second step should be performed at speed of 500 × *g* and 2000 × *g* and third stage should be done at 12,000 × *g* speed.
2. First and second step should be performed for 10 min and 15 min respectively.
3. Third step should be longer and should be done at 12,000 × *g* for 45 min.
4. In first two steps, collect the supernatant while in third step discard the supernatant and add 4 ml suspension buffer to the pellet.

#### 3.3.3 Percoll Gradient Centrifugation

1. Mix the pellet obtained from third centrifugation with 2–3 ml of suspension buffer. Dissolve it slowly with paint brush. Using 1 ml cut tip take the dissolved pellet and slowly load into discontinuous Percoll gradient containing Oakridge tube (Figs. 3 and 4) (*see* **Note 4**).
2. Centrifuge the gradient tubes for 30,000 × *g* for 1 h.
3. After centrifugation, yellowish or creamy thick white layer of mitochondrial fraction settle at interface between 45% and 28% discontinuous Percoll gradient (Figs. 3 and 4).
4. Carefully collect the mitochondrial fraction using a cut tip (exclude green part) (*see* **Note 5**) and place into 50 ml Oakridge tube and add 5 ml of suspension buffer to it.

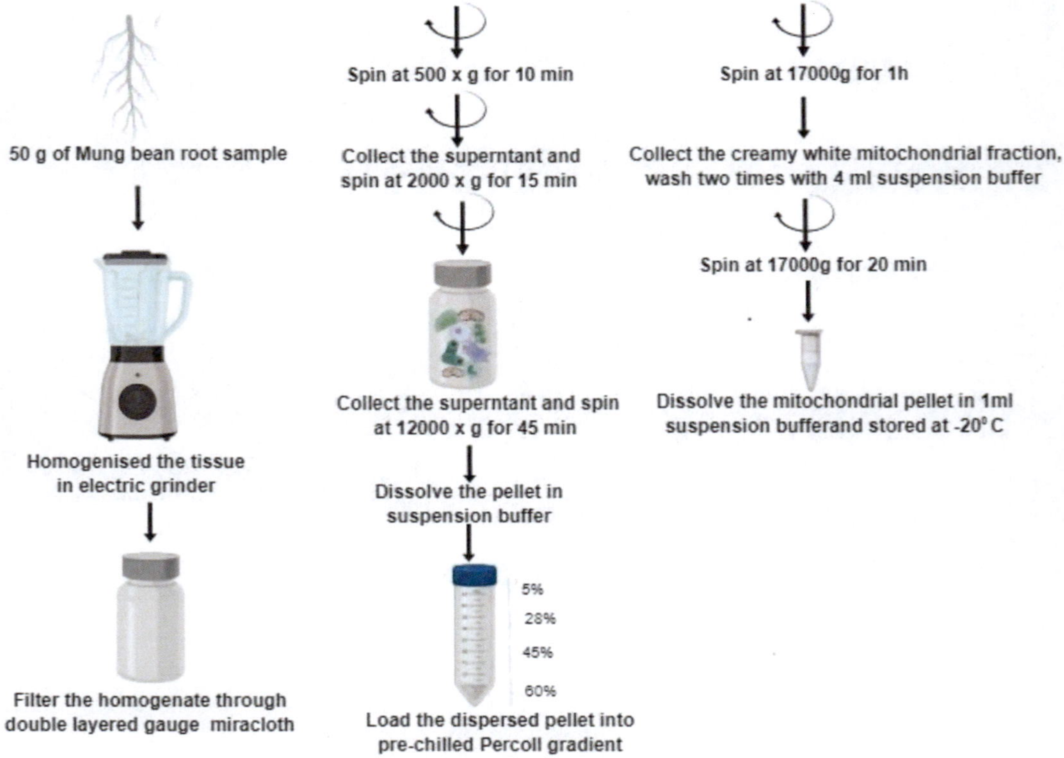


**Fig. 4** Illustration of steps in mitochondrial isolation from mung bean

*3.3.4 Percoll Removal and Obtaining Pure Mitochondria*

1. Centrifuge the obtained mung bean mitochondrial suspension at 17,000 × *g* for 25 min at 4 °C to remove excess of Percoll.
2. Remove the supernatant carefully and add 5 ml of suspension buffer, gently tap it with hand and again centrifuge at 17,000 × *g* for 25 min at 4 °C.
3. At this point a strong concentrated mitochondrial pellet can be seen at the bottom of Oakridge tube.
4. Dissolve this final mitochondrial pellet using paint brush into 1 ml suspension buffer for use immediately or it can be preserved for several months at −20 °C.
5. Determine the protein concentration by using commonly used Bradford method.

### 3.4 Isolation of Mitochondrial DNA (mtDNA)

Isolate mitochondrial DNA using cetyl trimethyl ammonium bromide (CTAB) method with minor modifications [24].

1. In this procedure take 500 μl mung bean mitochondrial samples which contains approx. 40 μg of mitochondria in 2 ml of micro centrifuge tube.

**Table 2**
**List of primers used**

| Gene name | Primer sequence |
|---|---|
| *NAD1* | F 5′-GAAACCGCTAACAGCAGAGG-3′<br>R 5′-GAACGCTTTGGCTCTTTTTG-3′ |
| *COX3* | F 5′-CGTTCATCGTATCGGAGGTT-3′<br>R5′-ATGATGAGCCCAAGTTACGG-3′ |
| *Actin 3* | F 5′-ATTCCTTCTTTTCGCAGCAA-3′<br>R 5′-CCCAACATACGCATCCTTCT-3′ |

**Table 3**
**Purity of mtDNA isolated from mung bean mitochondria**

| Sample ID | Concentration (ng/μl) | $A_{260/280}$ |
|---|---|---|
| Sample 1 | 128.3 | 1.85 |
| Sample 2 | 133.6 | 1.84 |
| Sample 3 | 162.6 | 1.96 |

2. Add 600 μl of CTAB buffer to the mung bean mitochondria containing tubes and mix vigorously by vortexing.

   Follow the well-known CTAB procedure for DNA isolation [24].

#### 3.5 Purity of Isolated Mitochondrial DNA

1. Check the purity of the isolated mitochondrial DNA using the mitochondrial specific primers.

   The details sequences of primers used in this study for the amplification of mitochondrial genes are presented in Table 2.

#### 3.6 Quality Check of Mitochondrial DNA Via PCR

The mitochondrial DNA isolated from root samples of mung bean using the protocol has high purity without nuclear DNA contamination. The Nano drop profile supports the high quality of the DNA (Table 3). The presence of mitochondrial DNA in the isolated sample is confirmed by PCR amplification with mitochondrial gene specific primers (*NAD1* and *COX3*). The genes amplified for mitochondrial DNA producing intense amplification profiles at expected product sizes (Fig. 5). Primers for *NAD1* and *COX3* produced a band of ~200 bp. In case of β-actin there was no amplification found (Fig. 5). The amplification of all two mitochondrial DNA specific primers confirmed the presence of mitochondrial DNA but no amplification of β-actin primers confirmed the absence of nuclear DNA contamination in mitochondrial

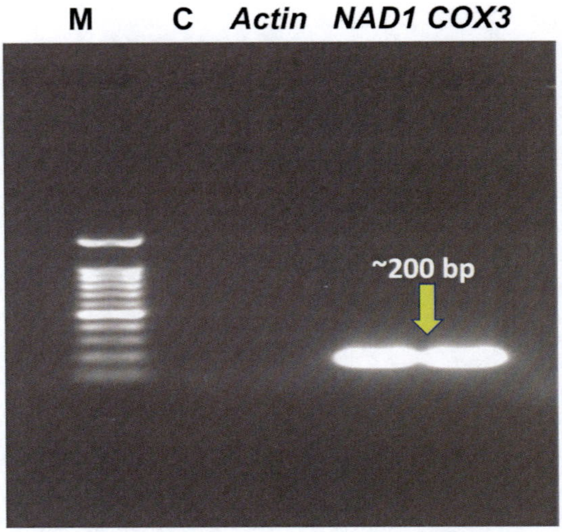


**Fig. 5** PCR amplification of mtDNA with *β Actin*, *NAD1*, *and COX3*

isolate proving the efficiency of method with reduced nuclear DNA contamination. This protocol can be used for legume genomic studies.

## 4 Notes

1. PVP takes time to dissolve; hence, stir until it dissolves completely.
2. Always perform whole isolation procedure at 4 °C.
3. Use sharp blade to chop the roots.
4. Using cut tip is necessary for prevention of mitochondrial damage.
5. Extensive care is needed to exclude green parts.

## Acknowledgments

PS is supported by a senior research fellowship from UGC, India.

### References

1. Vanlerberghe GC (2013) Alternative oxidase: A mitochondrial respiratory pathway to maintain metabolic and signaling homeostasis during abiotic and biotic stress in plants. Int J Mol Sci 14:6805–6847
2. Zorova LD, Popkov VA, Plotnikov EY, Silachev DN, Pevzner IB, Jankauskas SS, Babenko VA, Zorov SD, Balakireva AV, Juhaszova M, Sollott SJ, Zorov DB (2018) Mitochondrial membrane potential. Anal Biochem 552:50–59

3. Kowaltowski AJ (2000) Alternative mitochondrial functions in cell physiopathology: beyond ATP production. Braz J Med Biol Res 33:241–250
4. Horn R, Gupta KJ, Colombo N (2014) Mitochondrion role in molecular basis of cytoplasmic male sterility. Mitochondrion 19:198–205
5. Van Aken O, Van Breusegem F (2015) Licensed to kill: mitochondria, chloroplasts, and cell death. Trends Plant Sci 20:754–766
6. Liberatore KL, Dukowic-Schulze S, Miller ME, Chen C, Kianian SF (2016) The role of mitochondria in plant development and stress tolerance. Free Radic Biol Med 100:238–256
7. Kim YJ, Zhang D (2018) Molecular control of male fertility for crop hybrid breeding. Trends Plant Sci 23:53–65
8. Siqueira JA, Hardoim P, Ferreira PCG, Nunes-Nesi A, Hemerly AS (2018) Unraveling interfaces between energy metabolism and cell cycle in plants. Trends Plant Sci 23:731–747
9. Taanman JW (1999) The mitochondrial genome structure, transcription, translation and replication. Biochim Biophys Acta 1410:103–123
10. Jacob JE, Vanholme B, Van Leeuwen T, Gheysen G (2009) A unique genetic code change in the mitochondrial genome of the parasitic nematode *Radopholus similis*. BMC Res Notes 2:192–202
11. Michaels GS, Hauswirth WW, Laipis PJ (1982) Mitochondrial DNA copy number in bovine oocytes and somatic cells. Dev Biol 94:246–251
12. Shuster RC, Rubenstein AJ, Wallace DC (1988) Mitochondrial DNA in anucleate human blood cells. Biochem Biophys Res Commun 155:1360–1365
13. Robin ED, Wong R (1988) Mitochondrial DNA molecules and virtual number of mitochondria per cell in mammalian cells. J Cell Physiol 136:507–513
14. Wiesner RJ, Rüegg JC, Morano I (1992) Counting target molecules by exponential polymerase chain reaction: copy number of mitochondrial DNA in rat tissues. Biochem Biophys Res Commun 183:553–559
15. Gray MW (1992) The endosymbiont hypothesis revisited. Int Rev Cytol 141:233–357
16. Galper JB, Darnell JE (1969) The presence of N-formyl-methionyl-tRNA in HeLa cell mitochondria. Biochem Biophys Res Commun 34:205–214
17. Ladoukakis ED, Zouros E (2017) Evolution and inheritance of animal mitochondrial DNA: rules and exceptions. J Biol Res 24:2–9
18. Wolstenholme DR (1992) Animal mitochondrial DNA: structure and evolution. Int Rev Cytol 141:173–216
19. Galtier N (2011) The intriguing evolutionary dynamics of plant mitochondrial DNA. BMC Biol 9:61–63
20. Gualberto JM, Newton KJ (2017) Plant mitochondrial genomes: dynamics and mechanisms of mutation. Annu Rev Plant Biol 68:225–252
21. Triboush SO, Danilenko NG, Davydenko OG (1998) A method for isolation of chloroplast DNA and mitochondrial DNA from sunflower. Plant Mol Biol Rep 16:183–189
22. Li WQ, Zhang GS, Wang K, Niu N, Pan DL (2007) An efficient method for isolation of mitochondrial DNA in wheat. Yi Chuan 29:771–775
23. Chaudhary A, Chaudhary S, Ghosh A, Vuduthala S, Singh KM, Chikara SK (2015) A Rapid, low cost, and efficient method for isolation of high quality mitochondrial DNA from *Oryza sativa*. J Crop Sci Biotechnol 18:155–160
24. Murray MG, Thompson WF (1980) Rapid isolation of high molecular weight plant DNA. Nucleic Acids Res 8:4321–4325

# Chapter 17

# Embryonic Explant and Plumular Meristem Transformation Methods for Development of Transgenic Pigeon Pea

**Shreeparna Ganguly, Arnab Purohit, Rituparna Kundu Chaudhuri, Sampa Das, and Dipankar Chakraborti**


## Abstract

A reliable pigeon pea transformation system can assist the rapid improvement of this important grain legume through transgenic development. Here we describe two methods of *Agrobacterium tumefaciens*-mediated pigeon pea transformation. In the tissue culture based embryonic explant transformation method, microshoot grafting was included to obtain rapid root induction, while the other method was culture independent and designated as plumular meristem transformation. Both methods drastically enhanced the transformation frequency and have the potential to provide reasonable solutions for maximum transgenic recovery in biotechnological breeding programs.


**Key words** Antibiotic screening, Decapitated plumule, GUS analysis, In vitro transformation, Pigeon pea regeneration, Plumular meristem transformation, Shoot grafting, Transformation frequency

## 1 Introduction

Among the pulses, pigeon pea is popular as sixth most important protein rich grain legume in the semiarid tropics. This legume crop caters to protein requirement of the vegetarian population of Indian subcontinent [1]. Modern tools of recombinant DNA technology and genetic transformation play crucial roles in sustainable development of agricultural systems and improvement of food and nutritional security in the developing countries [2]. The introduction of any defined genetic material to a new host often leads to a protective activity against the different constraints that limit the crop yield [3]. Despite the immense importance in sustainable agriculture, pigeon pea production has remained static over the last few decades due to prevalence of several biotic and abiotic stresses along with the nonavailability of efficient management practices [4]. Additionally, narrow genetic diversity in primary gene pool coupled with low utilizations of genetic resources have considered as the major impediments for the genetic improvement

Mukesh Jain and Rohini Garg (eds.), *Legume Genomics: Methods and Protocols*, Methods in Molecular Biology, vol. 2107, https://doi.org/10.1007/978-1-0716-0235-5_17, © Springer Science+Business Media, LLC, part of Springer Nature 2020

of pigeon pea through conventional breeding [5]. A number of in vitro tissue culture-based transformation methods were introduced for the development of transgenic pigeon pea but the potential outcome of these techniques had not been satisfactory due to lack of efficiency and reproducibility [5]. Poor tissue culture response coupled with recalcitrant nature of this grain legume have been considered as the major hurdle for in vitro transformation method. The use of suitable media combinations along with specific growth regulators effectively influenced the explant growth and maturity. Modified Murashige and Skoog (MS) medium was preferably used for culture-based pigeon pea transgenic development strategies [6]. However, the reported media combinations were found to be highly variable and cultivar specific, which represents a significant bottleneck in the overall recovery of plants through in vitro techniques. Additionally, successful root induction of in vitro regenerated shoots has always been a major concern for recovery and establishment of pigeon pea plantlets. As a result, rooting of putative transformants in the presence of selective agents was unpredictable [7]. Past reports revealed that, this failure of in vitro root induction resulted in the loss of 30–80% regeneration efficiency [7, 8].

Here we report two methods of *Agrobacterium tumefaciens*-mediated pigeon pea transformation, which can be routinely used for transgenic research. The first method is tissue culture based, where we introduced a novel shoot grafting strategy after embryonic explant transformation to provide a solution to overcome in vitro rooting problem [8] (Fig. 1a–d). This grafting technique was adapted from the previously reported transgenic development method applied in cotton and chickpea [9, 10, 11]. Rapid establishment of plantlets in soil with well-developed root system enabled the quick recovery of transgenic shoots, which ensured improved success rate of transgenic development.

In culture-based development of transgenic pigeon pea an antibiotic-mediated stringent selection is required to eliminate the untransformed cells, which may also be involved in shoot multiplication process and thus regenerate nontransgenic shoots. Due to the abundance of chimeric tissues, maximum number of regenerated shoots coming out from transformed explant could not withstand the selection pressure and were unable to survive during selection based transgenic shoot recovery. Additionally due to recalcitrant nature of this grain legume and nonavailability of adequate amount of regenerable tissues, the highest possible transformation frequency was not achieved by grafting-mediated in vitro transgenic development strategy.

Although the problem of in vitro root induction was partially solved following the grafting strategy, there was a need for further improvement of transformation frequency. This criterion directed us to develop the second method, which is a non–tissue culture-

**Fig. 1** Different stages of *Agrobacterium tumefaciens*-mediated transformation in pigeon pea. (**a**) Decapitated embryonic axis explants used for in vitro tissue culture method. (**b**) Multiple shoots regeneration from an explant after 21 days of

based plumular meristem transformation method in pigeon pea. This alternative method was the combination of in vitro infection and cocultivation of explants with *A. tumefaciens*, followed by tissue culture independent plant establishment [12] (Fig. 1g–j). The method was proved to be effective to overcome the recalcitrance nature of pigeon pea, with 72% transformation frequency, which was followed by a stringent antibiotic selection of transgenic plants in $T_1$ generation to remove the chimeras.

## 2 Materials

Prepare all the required solutions in double distilled water using specified analytical grade reagents and autoclave them before use. Keep the autoclaved media in sterile place under room temperature and store the stock solutions of salt, vitamins, hormones and amino acids required for the preparation of plant growth medium at 4 °C (*see* **Note 1**).

### 2.1 Pigeon Pea Cultivar and Binary Vector

1. Use seeds of pigeon pea cultivar ICPL 87119 (ASHA) or UPAS 120, although other cultivars may also be used.
2. EHA 105 strain of *A. tumefaciens* containing binary vector pBI121, harboring *nptII* selectable marker and *gus* reporter gene.
3. *E. coli* DH5α strain.

### 2.2 Media

1. MS major solution (20×): 33 g/l $NH_4NO_3$, 38 g/l $KNO_3$, 8.8 g/l $CaCl_2·2H_2O$, 7.4 g/l $MgSO_4·7H_2O$, and 3.4 g/l $KH_2PO_4$.
2. MS minor solution (200×): 0.166 g/l KI, 1.24 g/l $H_3BO_3$, 4.46 g/l $MnSO_4·4H_2O$, 1.72 g/l $ZnSO_4·7H_2O$, 0.05 g/l $Na_2MoO_4·2H_2O$, 0.005 g/l $CuSO_4·5H_2O$, and 0.005 g/l $CoCl_2·6H_2O$.

**Fig. 1** (continued) incubation on 1 mg $l^{-1}$ BAP and 0.2 mg $l^{-1}$ NAA medium, containing 100 mg $l^{-1}$ kanamycin. (**c**) Grafted plant in Hoagland agar medium. (**d**) Established $T_0$ transformants after 12 weeks of soil transfer. (**e**) Regenerated untransformed leaf as negative control for GUS experiment. (**f**) GUS expression in regenerated transformed leaflets. (**g**) Pricking of explants for plumular meristem transformation. (**h**) Growth of *A. tumefaciens* along with explants after 3 days of cocultivation. (**i**) Multiple shooting of putative $T_0$ transformants after 3 weeks of transformation. (**j**) Established pigeon pea transgenic events after 2 months of transformation. (**k**) Decapitated untransformed explants used as negative control in GUS analysis. (**l**) GUS expression in decapitated transformed explant. Bar represents 1 cm

3. Fe-EDTA solution (100×): 2.78 g/l $FeSO_4 \cdot 7H_2O$ and 3.72 g/l $Na_2EDTA \cdot 2H_2O$ (*see* **Note 2**).
4. Hoagland micronutrient solution: 2.86 g/l $H_3BO_3$, 1.82 g/l $MnCl_2 \cdot 4H_2O$, 0.22 g/l $ZnSO_4 \cdot 7H_2O$, 0.09 g/l $Na_2MoO_4 \cdot 2H_2O$, and 0.09 g/l $CuSO_4 \cdot 5H_2O$ [13].
5. Kanamycin: Dissolve 100 mg in 1 ml sterile water. Store at −20 °C.
6. Rifampicin: Dissolve 20 mg in 1 ml ethanol. Store at −20 °C.
7. Acetosyringone: 20 mg/ml of 70% ethanol. Dissolve 20 mg in 700 μl ethanol and finally make up the volume with 300 μl sterile water. Store at 4 °C.
8. Cefotaxime: Dissolve 250 mg in 1 ml sterile water. Store at 4 °C.
9. Regeneration medium (RM): Take 500 ml water in the beaker and add 50 ml MS Major solution, 20 ml MS minor solution, 20 ml Fe-EDTA solution, B5 vitamins [14] (100 mg/l Myo-inositol, 10 mg/l Thiamine–HCl, 4 mg/l Nicotinic acid, 1 mg/l Pyridoxine, 570 mg/l L-Proline), 10 mM MES buffer along with 3% (w/v) sucrose. Add 1 mg/l 6-benzylaminopurine (BAP) and 0.2 mg/l α-naphthaleneacetic acid (NAA) and make up the volume to 1 l with water. Adjust the pH 5.8 with 1 N NaOH, distribute in conical flasks and add 0.8% plant agar to each flask before autoclaving.
10. Elongation medium (EM): Take 500 ml water in the beaker and add 50 ml MS Major solution, 20 ml MS minor solution, 20 ml Fe-EDTA solution, B5 vitamins [14] (100 mg/l Myo-inositol, 10 mg/l Thiamine–HCl, 4 mg/l Nicotinic acid, 1 mg/l Pyridoxine, 570 mg/l L-Proline), 10 mM MES buffer along with 3% (w/v) sucrose. Add 0.5 mg/l BAP and 0.5 mg/l gibberellic acid to make a 1 l elongation medium as in previous step. Distribute the media and add 0.8% plant agar before autoclaving.
11. Hoagland medium: Take 500 ml sterile water and add 0.25 g/l $KNO_3$, 0.59 g/l $Ca(NO_3)_2 \cdot 4H_2O$, 0.61 g/l $MgSO_4 \cdot 7H_2O$, 0.43 g/l $KH_2PO_4$, 1 ml Hoagland micronutrient solution, and 20 ml Fe-EDTA solution from their respective stocks [13]. Make up the media volume to 1 l with water. Adjust the pH to 5.8 with 1 N NaOH and add 0.8% plant agar before autoclaving.
12. Yeast extract broth (YEB) medium: Take 500 ml water in a sterile beaker and add 5 g of each nutrients (viz., Beef extract, peptone and sugar, 1 g yeast extract, and 0.49 g/l $MgSO_4$). Make up the volume to 1 l with water and adjust the pH to 7.0 before autoclaving.
13. Soilrite (Keltech Energies Ltd., Bangalore, India).

### 2.3 Polymerase Chain Reaction

1. Autoclaved PCR reaction tubes.
2. Details of oligonucleotide primers: Use *nptII* forward primer, 5′ GAGGCTATTCGGCTATGACTG 3′ and reverse primer, 5′ ATCGGGAGCGGCGATACCGTA 3′; to get the amplicon size 700 bp. Use *gus* forward primer, 5′ CAGTGAAGGGC-GAACAGTTC 3′ and reverse primer, 5′ GTTCAGGCACAG-CACATCAA 3′ to get 572 bp amplified product.

### 2.4 GUS Analysis

1. 1 mg/ml potassium ferrocyanide.
2. 1 mg/ml potassium ferricyanide.
3. 0.5 M EDTA.
4. 0.1 M Phosphate buffer: 2.32 ml 1 M $Na_2HPO_4$, 2.69 ml 1 M $NaH_2PO_4$, 44.99 water, pH 6.8.
5. GUS buffer: 10 mg 5-bromo-4-chloro-3-indolyl glucuronide (X-Gluc) (dissolved in 500 μl DMSO), 100 μl 10% Triton X-100 (v/v), 200 μl 0.5 M EDTA, 125 μl each potassium ferrocyanide and potassium ferricyanide solution, and 8.95 ml 0.1 M phosphate buffer.
6. Prechilled 80% acetone.
7. Chloral hydrate solution: 4 g chloral hydrate, 1 ml glycerol, and 2 ml water.

### 2.5 DNA Extraction

1. DNA isolation buffer: Add 10 ml 2 M Tris–HCl pH 8.0, 10 ml 0.5 M EDTA, 30 ml 5 M NaCl along with 2.5% w/v CTAB powder and mix it well to make slurry. Make up the volume to 100 ml with water. After autoclaving freshly add 0.2% β-mercaptoethanol (v/v) and 1% polyvinylpyrrolidone, MW 40,000.
2. 10:1 Tris EDTA buffer: 10 mM Tris–Cl, 1 mM EDTA, pH 8.0.
3. Phenol–chloroform: Autoclave all the glass materials before use. Preheat the measuring cylinder, amber bottle and molecular biology grade phenol at 60 °C hot water bath for 1 h. Take 50 ml phenol and 50 ml chloroform in the bottle and add 100 ml 10:1 TE buffer (pH 8.0). Finally add a pinch of hydroxyquinoline and allow it for overnight saturation in the magnetic stirrer. Next day change the buffer solution very carefully and keep it for overnight saturation. Finally store it at 4 °C (*see* **Note 3**).
4. 24:1 chloroform and isoamyl alcohol.
5. 3 M sodium acetate (pH 5.2).
6. 70% ethanol.

### 2.6 Southern Blotting

1. DNA gel electrophoresis.
2. Hybridization barrels.

3. Nitrocellulose membrane (Nylon membrane, Hybond N$^+$, (Amersham Biosciences).
4. Qiagen gel extraction and purification kit (Qiagen, Germany).
5. Whatman filter paper (GE healthcare, UK).
6. 50× TAE: 242 g/l Tris base, 57.1 ml/l glacial acetic acid, and 100 ml 0.5 M EDTA, pH 8.0.
7. 10:1 TE buffer (pH 8.0).
8. Denaturation solution: 0.5 M NaOH, 1.5 M NaCl, pH 12.5.
9. Neutralization solution: 0.5 M Tris–HCl, 1.5 M NaCl, pH 7.5.
10. 20× Saline Sodium Citrate (SSC): 0.3 M Sodium citrate, 3.0 M NaCl, pH 7.0.
11. All the above-mentioned solutions are autoclaved and stored at room temperature.
12. 0.8% agarose gel.
13. 6× DNA loading dye: 0.25% bromophenol blue, 0.25% xylene cyanol FF, and 30% (v/v) glycerol.
14. 10% SDS.
15. Prehybridization solution: 5× SSC, 0.5% (w/v) skimmed milk, 0.1% (w/v) *N*-Lauryl-sarcosine, and 0.02% SDS.
16. Washing solution I: 2× SSC, 0.1% SDS.
17. Washing solution II: 0.1× SSC, 0.1% SDS.
18. Rediprime II DNA labeling system (Amersham Biosciences).
19. [$\alpha$-$^{32}$P]-dCTP, (PerkinElmer Cetus).
20. X ray developer and fixer (*see* **Note 4**).
21. Kodak film (*see* **Note 5**).
22. DIG DNA labeling kit (Roche) (*see* **Note 6**).
23. Church buffer: 5× SSC, 0.1% (w/v) *N*-Lauryl-sarcosine, and 0.02% SDS.
24. Washing solution III: 0.5× SSC, 0.1× SDS.
25. Washing buffer: 0.1 M Maleic acid, 0.15 M NaCl (v/v), Tween 20, pH 7.5.
26. Maleic acid buffer: 0.1 M Maleic acid, 0.15 M NaCl, pH 7.5.
27. Detection buffer: 0.1 M Tris–HCl, 0.1 M NaCl, pH 9.5.
28. Blocking solution (10×): Dissolve 10% (w/v) blocking reagent (provided with DIG DNA labeling kit) in maleic acid buffer under constant stirring on a heating block and autoclave it before use. Store the solution at 4 °C. Prepare the 1× working solution by diluting the stock solution 1:10 with maleic acid buffer.

29. Antibody solution: Centrifuge anti-digoxigenin alkaline phosphatase (provided with DIG DNA labeling kit) for 5 min at 8445 × *g*, 4 °C and add in the blocking solution at 1:5000 dilution (*see* **Note 7**).
30. Color substrate solution: Add 200 μl NBT/BCIP to 10 ml detection buffer.

## 3 Methods

### 3.1 Tissue Culture Based Embryonic Explant Transformation

Carry out all regeneration and transformation steps in the laminar air flow cabinet and maintain the explants in plant growth chamber at 22 ± 2 °C under a 16 h photoperiod, with a light intensity of 45 μ mol $m^{-2}$ $s^{-1}$ (*see* **Note 8**).

#### 3.1.1 Regeneration and Grafting Method

1. Surface sterilize pigeon pea cultivar ICPL 87119 seeds by keeping in 0.05% $HgCl_2$ along with Tween 20 for 10 min and wash with autoclaved water for four times. Then transfer the seeds to 0.1% $HgCl_2$ solution and keep it for another 15 min. After washing, imbibe the seeds in 100 ml sterile water for overnight and drain out the water on next day.
2. After removing the seed coat, keep them in basal medium for 4–5 days. To prepare the embryonic explant for *A. tumefaciens*-mediated transformation, excise the plumule tip and eliminate the radicle of the germinated seed (Fig. 1a).
3. Prepare the *A. tumefaciens* culture by growing a single colony in YEB medium for 16–18 h. After adjusting the $OD_{600}$ at 0.8–1.0, pellet down the culture by centrifugation at 2111 × *g* for 5 min. Resuspend the pellet in 20 ml RM, supplemented with 150 μM acetosyringone (Himedia, India) and incubate the explants in it for 45 min. Remove the excess bacterial suspension and allow the in vitro cocultivation on 150 μM acetosyringone supplemented solid RM for 3 days.
4. Wash the explants with autoclaved water and maintain them on 100 mg $l^{-1}$ kanamycin supplemented RM for 6 weeks along with reculturing at regular interval (*see* **Note 9**). Transfer the well regenerated explants with multiple shoots to the kanamycin supplemented EM and follow the similar reculturing for 4 weeks. Finally select the elongated established transgenic shoots for grafting (Fig. 1b).
5. For grafting prepare the root stock by culturing the germinated pigeon pea seeds in Hoagland agar medium for 3 weeks. Select the established transgenic shoots and cut the base in "v" shape to properly fit into the base of vertically split stem. Give support by a sterile Teflon ring and keep it for 7 days to heal the tissue of the grafted region (Fig. 1c).

6. After 1 week of grafting, transfer the plants to liquid Hoagland solution on filter paper bridge and keep it for 5–7 days. Then transfer the plants to plastic pots containing autoclaved Soilrite with covering. Finally place the plants in pots containing soil, sand and organic manure (1:1:1) (Fig. 1d).

#### *3.1.2 Analysis of Putative Transformants*

1. Histochemical GUS analysis: Perform the histochemical studies after 3, 6 and 9 weeks of antibiotic selection using X-Gluc as the substrate to confirm the GUS activity in regenerated putative transformants containing transgenes. Transfer the leaf samples in prechilled (−20 °C) 80% acetone and place it in −20 °C for 20 min. After washing the samples for three times add 300–500 μl GUS buffer to the samples.
2. Allow the infiltration of the GUS buffer along with the tissue inside a vacuum desiccator for 3 h and finally keep them for overnight incubation at 37 °C.
3. Next day remove the GUS buffer and add chloral hydrate solution with 1 h incubation.
4. Observe the GUS reaction in the form of blue color and preserve the samples at 4 °C in chloral hydrate solution (Fig. 1e, f) (*see* **Note 10**).

#### *3.1.3 Genomic DNA Isolation*

1. Grind the 100 mg fresh leaf tissues of the transgenic and nontransgenic pigeon pea plants in liquid nitrogen (*see* **Note 11**) and add 1 ml freshly prepared DNA extraction buffer to it. Add 1–2 μl RNase A before transfer the solution to a 2 ml polypropylene tube and mix it well until to make it slurry.
2. Incubate the mixture at 60 °C in water bath for 30 min. Add equal volume of phenol–chloroform (1:1) and mix by gentle inversion for about 10 min. Allow centrifugation for 10 min at 4750 × *g* at 25–30 °C. Transfer the upper clear aqueous layer to another tube.
3. After that add equal volume of chloroform–isoamyl alcohol (24:1) and gently mix it for 5 min.
4. Spin at 4750 × *g* for 5 min at 25–30 °C. Take the upper clear aqueous layer and add 1/10th volume of 3 M sodium acetate, pH 5.2 and double volumes chilled ethanol and incubate it in room temperature for 20–25 min. Mix it well before centrifugation at 14272 × *g* for 10 min. Discard the supernatant and wash with 70% ethanol.
5. Dry it in a vacuum for 15 min and dissolve in 20 μl TE (10:1) buffer.
6. Run the gel with 0.8% agarose to determine the quality of the DNA.

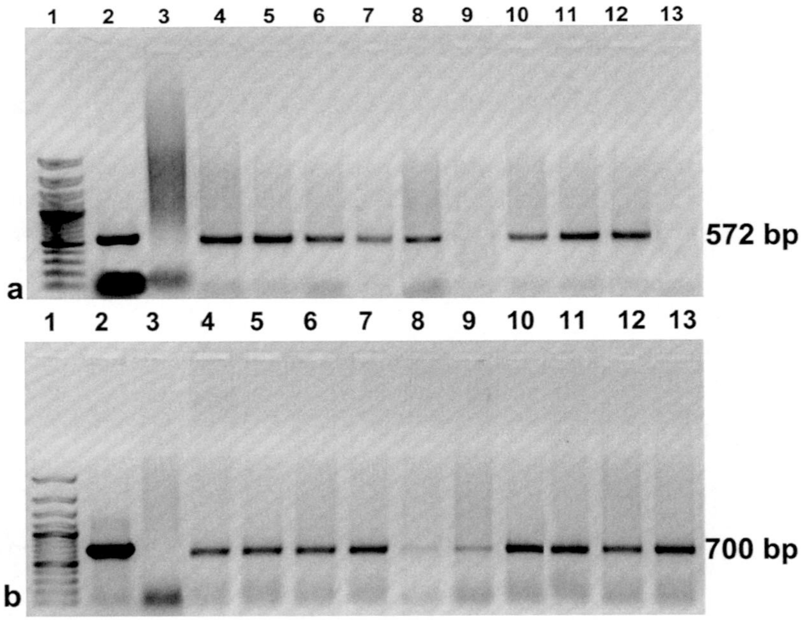


**Fig. 2** Representative PCR profile of established transgenic pigeon pea plants. *Lanes* 1, 2 and 3 DNA ladder, pBI121 plasmid as positive control and untransformed pigeon pea DNA as negative control, respectively. (**a**) *Lane* 4–13 amplification of *gus* specific amplicon in tissue culture established putative $T_0$ transgenic events. (**b**) *Lane* 4–13 amplification of *nptII* specific fragment of selected $T_1$ transgenic events generated by plumular meristem transformation

#### 3.1.4 PCR Analysis

1. Allow the PCR screening of the isolated DNA samples along with binary plasmid DNA as positive control.
2. Run the PCR for detection of *gus* gene using 24 μl reaction mixture containing 1× *Taq* Buffer (Thermo Scientific MA, USA), 100 μM each dNTPs, 1.5 mM $MgCl_2$, 200 nM forward and reverse primers, and 1 unit *Taq* polymerase enzyme. Add 100 ng DNA of each $T_0$ plant samples along with negative control (Fig. 2a).
3. PCR conditions are: 1 cycle of 5 min at 95 °C, followed by 35 cycles at 95 °C for 45 s, 57 °C for 45 s, and 72 °C for 1 min and a final extension of 72 °C for 10 min.

#### 3.1.5 Selection of $T_1$ Transgenics

1. Obtain the $T_1$ seeds from individual $T_0$ plants.
2. Imbibe the $T_1$ seeds overnight in water and after removal of seed coat allow 5 h incubation in 100 mg/l kanamycin. Treat the nontransformed seeds under same conditions. After 5 h treatment, sow the seeds on sterile Soilrite.
3. Kanamycin resistance healthy seedlings in comparison to nontransformed ones are selected for further molecular analysis. They are compared on the basis of their morphological differences (viz., plant height, number of leaves and leaflets, formation of roots).

*3.1.6 Analysis of $T_1$ Transgenics*

Established pigeon pea $T_1$ plants are used for transgenic analysis. PCR positive events are chosen for Southern blot analysis to confirm the transgene copy number.

Radioactive Southern blot analysis.

1. Take 40 μg isolated genomic DNA to prepare the 60 μl reaction mixture with appropriate restriction enzyme and keep it in 37 °C water bath for overnight digestion.
2. Then on next day precipitate the DNA by 1/10th volume of 3 M Na-acetate and equal volume chilled ethanol and finally dissolve in 20 μl DNase free water. Incubate the digested DNA samples at 65 °C for 5 min and immediately place in ice for 5 min. Add 2 μl of 6× DNA loading dye to the DNA samples and allow to separate in a 0.8% agarose gel for 6 h at 40 V.
3. After the electrophoretic separation of the DNA samples wash the gel properly with sterile water and incubate it on a shaker in depurination solution for 20 min at room temperature. Then wash it for two times, each for 10 min.
4. Again transfer it in denaturation solution for 30 min and wash it for two times. Give final incubation in neutralization solution for 40 min in same condition followed by two washes each of 20 min (*see* **Note 12**).
5. Carry out the transfer of DNA samples to the nylon membrane in transfer apparatus. Fill the tray with 10× SSC and place a 3 mm Whatman filter paper wick soaked with 10× SSC on the support (*see* **Note 13**). Place the gel on top of the wick avoiding any air bubble (*see* **Note 14**). Cut the membrane, three filter papers and stack of blotting paper according to the size of the gel. Equilibrate membrane and three filter paper in 2× SSC solution. Place the membrane on gel and place the filter papers on it. Place a stack of blotting papers on it. Keep a glass plate on top of stack and place ~1 kg weight on it to equalize the pressure distribution (*see* **Note 15**). Continue the transfer for 12 h then remove the membrane and mark the lanes (*see* **Note 16**). Finally cross-link the membrane using UV radiation at 0.16 Joules (*see* **Note 17**).
6. Prepare the probe with the eluted sequence obtained from PCR amplification. Elute the sequence using gel extraction kit. 100 ng DNA is preferably used for radiolabeling. Add 5 μl radioactive solution for probe preparation (*see* **Note 18**).
7. Prepare the prehybridization solution and keep it at 68 °C for 1 h.
8. Allow the membrane to soak in 2× SSC and keep inside a hybridization tube avoiding trapping of air bubbles.
9. Add prehybridization solution in the tube and keep in hybridization oven at 68 °C for 3 h. Beside this, boil the probe in water

bath for 5 min and immediately cool on ice. Finally add the probe in prehybridization solution, and allow it for 16 h incubation in same condition.

10. Next day wash it with solution I for 1 h at room temperature followed by solution II for 1 h at 68 °C. Expose the blot to the Kodak film at −80 °C for 10 days and then develop (*see* **Note 19**).

### 3.2 Plumular Meristem Transformation

The process of tissue culture independent plumular meristem transformation for transgenic pigeon pea development is described stepwise. Carry out the *A. tumefaciens* infection in laminar air flow cabinet and maintain the explants in plant growth chamber at 22 ± 2 °C under a 16 h photoperiod, with a light intensity of 45 μ mol $m^{-2}$ $s^{-1}$.

#### 3.2.1 Transformation Method

1. Take healthy and light brown seeds for surface sterilization following the previously mentioned steps. Germinate seeds in sterile cotton bed soaked with water. Three days old seedlings are usually more preferable for use as explants for *A. tumefaciens*-mediated plumular meristem transformation.
2. Prepare the *A. tumefaciens* culture carrying the binary plasmid pBI121 in 50 ml YEB medium containing 100 mg/l kanamycin and 50 mg/l rifampicin in similar growth condition mentioned earlier.
3. After adjusting the OD at 0.8–1.0, centrifuge it at 2111 × *g* for 5 min. After that resuspend the pellet in 20 ml of 150 μM acetosyringone supplemented liquid RM (mentioned earlier). Prepare explants by removing 3–4 mm of shoot apical meristem (decapitation) with a sterile scalpel along with the pricking at apical region and cotyledonary nodes with the help of a sterile needle (24 gauge), but keep the root intact (*see* **Note 20**). Allow incubation of pricked explants in *A. tumefaciens* containing RM for 30 min. Transfer the plants on 150 μM acetosyringone supplemented RM containing 0.8% agar for 3 days cocultivation (Fig. 1g–h).
4. Then repeated washing is needed to be done with sterile water containing 250 mg/l cefotaxime to remove the excess *Agrobacterium* present on seed surface, followed by placing them in plastic pots containing sterile Soilrite.
5. After 4–5 weeks transfer the established putative $T_0$ transformants to 25–30 diameter pots containing soil, sand and organic manure for obtaining $T_1$ seeds (Fig. 1i, j).

#### 3.2.2 GUS Analysis

GUS expression study should be done during cocultivation period to determine the transformation frequency. Take untransformed explants as negative control (Fig. 1k, l).

*3.2.3 DNA Isolation and PCR Screening*

Isolate the genomic DNA of putative $T_0$ transgenic plants according to previously mentioned method (*see* Subheading 3.1.3). Perform the PCR analysis of genomic DNA for the presence of transgenes using primers mentioned earlier. Use DNA of binary plasmid and untransformed plant as positive and negative controls, respectively.

*3.2.4 Selection and Establishment of $T_1$ Transgenics*

Obtain the $T_1$ seeds from established $T_0$ plants. Allow them to incubate for 5 h in 100 mg/l kanamycin. Following the previously mentioned selection method select those Kanamycin resistant $T_1$ events and perform PCR analysis (Fig. 2b).

*3.2.5 Nonradioactive Method for Southern Blot Analysis*

Perform the DNA sample preparation, gel electrophoresis, and blotting according to the methods mentioned earlier.

1. Prepare the church buffer freshly for prehybridization and incubate it for 1 h with gentle agitation at specific temperature calculated from the provided manual. Add 10 ng (6 μl) of probe in a 1.5 ml microfuge tube along with 9 μl sterile water to make up the volume to 15 μl. Boil it for 10 min in water bath and immediately place in ice for 5 min. For labeling the probe add 2 μl hexanucleotide mix, 2 μl DIG dNTP and 1 μl Klenow fragment in the reaction mixture (*see* **Note 21**). Allow it for 2 h incubation at room temperature.
2. After 1 h incubation of prehybridization solution place the membrane in it. To stop the probe labeling reaction add 2 μl 0.2 M EDTA solution or keep it in 65 °C water bath for 10 min followed by quick chill in ice. Add the probe in prehybridized solution and keep it for overnight.
3. Next day wash the membrane with wash I solution for 40 min in two steps and then with wash III solution for 30 min at 65 °C in a shaker. Finally give a stringent wash for 10 min with maleic acid containing washing buffer solution.
4. Keep the membrane for 1 h blocking with 100 ml freshly prepared 1× blocking solution at 25 °C. After that add 8 μl antibody from the provided vial in 50 ml blocking reagent and allow the incubation for another 1 h.
5. Wash the membrane again with stringent washing buffer (according to the manufacturer's instructions) two times, each for 15 min. Then add 20 ml detection buffer to equilibrate the membrane for 5 min. Then add 650 μl NBT/BCIP substrate in 20 ml detection buffer and keep it in dark at room temperature for signal development (*see* **Note 22**). Finally add 10:1 TE to stop the reaction (*see* **Note 23**) (Fig. 3).

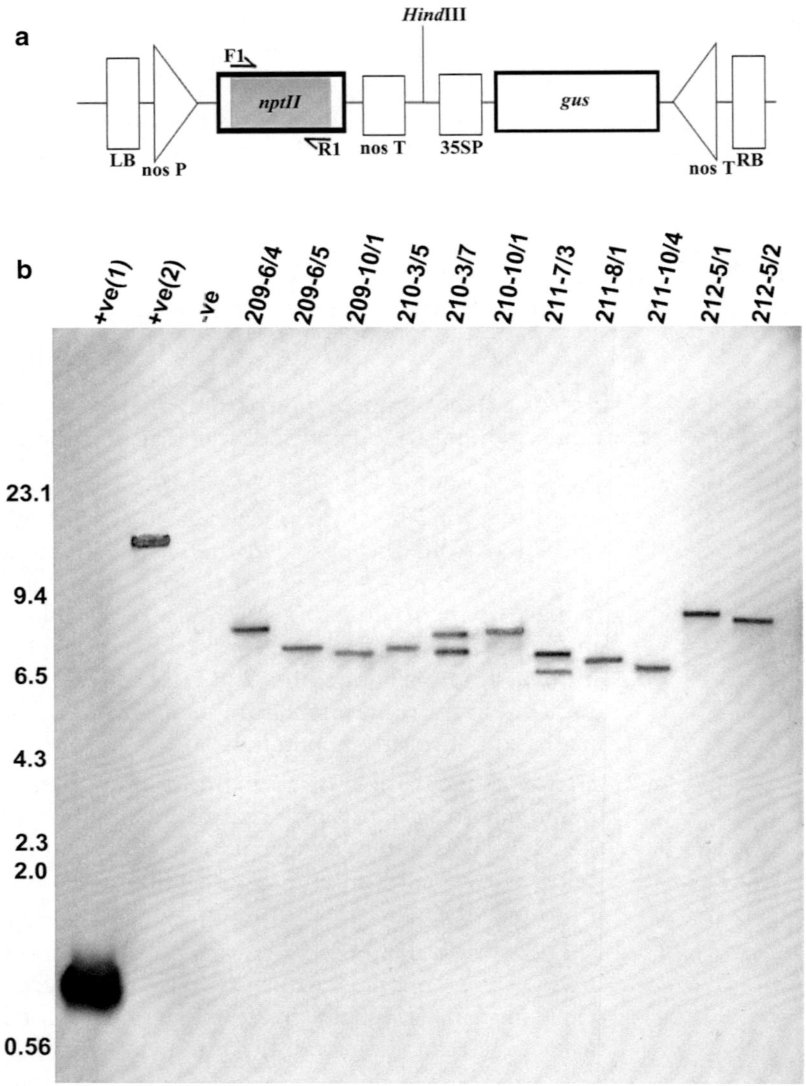


**Fig. 3** Representative Southern blot analysis of established $T_1$ transgenic pigeon pea events. (**a**) Schematic representation of *Hin*dIII recognition site and 700 bp probe sequence of *nptII* gene (marked in gray) obtained from pBI121 using F1 and R1 primers. (**b**) *Lanes* +ve (**1**) and +ve (**2**), PCR purified 700 bp *nptII* specific amplicon and *Hin*dIII-digested 14.7 kb pBI121 plasmid, respectively, as positive controls. *Lane* −ve, *Hin*dIII-digested genomic DNA of untransformed pigeon pea plant as negative control. *Lane* 4–15 *Hin*dIII-digested genomic DNA of $T_1$ transgenic events (reproduced from ref. 12 with permission from Springer Nature)

## 4 Notes

1. Stock of PGRs and other nutrients of media preparation should be stored at 4 °C for long time use.
2. Prepare 200 ml of $FeSO_4{\cdot}7H_2O$ solution and $Na_2$ $EDTA{\cdot}2H_2O$ solution separately and then mix them well in boiling condition.
3. Always wear gloves during the preparation of phenol–chloroform. Handle phenol very carefully because it is corrosive and penetrate rapidly through the skin and also cause the major internal damage. If it is exposed to the skin immediately wash with water and take first aid.
4. Always weigh the X ray developer and fixer in dark and prepare them in amber bottle. Use the warm water to dissolve them quickly.
5. Always keep the film in dark place.
6. DIG DNA labeling kit should be preserved in −20 °C.
7. A centrifugation step of 5 min at 8445 × *g* is always required to add the antibody solution in blocking reagent.
8. Pigeon pea explants should be kept in plant growth chamber at 25 °C under a 16 h photoperiod, with a light intensity of 45 μmol $m^{-2}$ $s^{-1}$.
9. All the regeneration and transformation steps should be done in the Laminar Air Flow with all sterile glass materials. Any type of contamination can damage the explants.
10. Preserve all the GUS stained tissue samples in chloral hydrate solution at 4 °C for long time.
11. Handle the liquid nitrogen very carefully during DNA isolation and always wear a glove to crush the samples and keep a safe distance from your eye.
12. Use gloves to treat the gel with acid and alkali solution and always take a scoop to place the gel for overnight transfer.
13. Give enough buffer in the tank and wick should not be dried during the transfer.
14. Try to squeeze out all the air bubbles by rolling a glass pipette while setting the blot. Any air bubbles between the gel and membrane will damage the transfer.
15. Always give sufficient weight on the top of the filter papers to give the pressure for capillary transfer and set the balance. Excessive weight may crush the gel that will hamper the blotting.
16. Try to mark the position of the wells on membrane with a pencil.

17. The membrane should not be dried completely before cross-linking. Do not expose the membrane for long time to the UV-radiation and always keep a safe distance from body.
18. Handle the radioactive solutions very carefully. Always wear gloves while handling the radioactive solutions and keep a safe distance from body. Use separate set of glass and plastic goods for the radioactive room and discard the wastes in labeled containers only.
19. Always cover the hybridized membrane with a transparent wrap to prevent the direct contact with the film and store in −80 °C before developing.
20. Use sterile forceps to handle the pigeon pea explants and use the needle very carefully for pricking. The young meristematic region of pigeon pea is very sensitive thereby multiple pricking of meristem can damage the explants.
21. Keep the constituents of DIG DNA labeling kit always in −20 °C.
22. The color starts to develop within few minutes and usually take 16 h to complete the reaction. Do not expose membrane to light until the complete development of DNA bands. Only a short exposure can be allowed to monitor the color development.
23. Dry the membrane at room temperature before storage. During long time storage band color may fade out. Keep the membrane in 10:1 TE buffer for 5 min to retrieve the color.

## Acknowledgements

The authors thank Indian Council of Agricultural Research (Project code NFBSFARA/PB2010/2010-11) for financial assistance, WBDBT (Project No. 335/WBBDC/IP-2/2013) for infrastructural grant, and St. Xavier's College (Autonomous), Kolkata for providing infrastructure. Swami Vivekananda Merit-cum-Means Scholarship Scheme of the Higher Education, Science & Technology and Biotechnology Department (No. 52-Edn (B) l 5B-l s / 2017) for fellowship grant is acknowledged.

### References

1. FAO (2016) Food and Agricultural Organization of the United Nation, FAO Statistical Database. http://faostat.fao.org. Accessed 3 January 2019
2. Popelka JC, Terryn N, Higgins TJV (2004) Gene technology for grain legumes: can it contribute to the food challenge in developing countries? Plant Sci 167(2):195–206
3. Dennis ES, Ellis J, Green A, Llewellyn D, Morell M, Tabe L, Peacock WJ (2007) Genetic contributions to agricultural sustainability. Philos Trans R Soc B 363(1491):591–609

4. Varshney RK, Chen W, Li Y, Bharti AK, Saxena RK, Schlueter JA et al (2012) Draft genome sequence of pigeonpea (*Cajanus cajan*), an orphan legume crop of resource-poor farmers. Nat Biotechnol 30:83–89
5. Ghosh G, Purohit A, Chaudhuri RK, Chakraborti D (2014) Advances in genetic transformation of important pulse crop pigeonpea. Open Access Biotechnol 12:5
6. Murashige T, Skoog F (1962) A revised medium for rapid growth and bioassay with tobacco tissue cultures. Physiol Plant 15:473–497
7. Ghosh G, Purohit A, Ganguly S, Chaudhuri RK, Chakraborti D (2014) In vitro shoot grafting on rootstock: an effective tool for *Agrobacterium*-mediated transformation of pigeonpea (*Cajanus cajan* (L.) Millsp). Plant Biotechnol 31:301–308
8. Krishna G, Reddy PS, Ramteke PW, Bhattacharya PS (2010) Progress of tissue culture and genetic transformation research in pigeonpea [*Cajanus cajan*, L. Millsp.]. Plant Cell Rep 29:1079–1095
9. Murfet IC (1971) Flowering in *Pisum*. Reciprocal grafts between known genotypes. Aust J Biol Sci 24:1089–1101
10. Luo J, Gould JH (1999) *In vitro* shoot-tip grafting improves recovery of cotton plants from culture. Plant Cell Tissue Organ Cult 57:211–213
11. Chakraborti D, Sarkar A, Das S (2006) Efficient and rapid in vitro plant regeneration system for Indian cultivars of chickpea (*Cicer arietinum* L.). Plant Cell Tissue Organ Cult 86:117–123
12. Ganguly S, Ghosh G, Purohit A, Chaudhuri RK, Chakraborti D (2018) Development of transgenic pigeonpea using high throughput plumular meristem transformation method. Plant Cell Tissue Organ Cult 135:73–83
13. Hoagland DR, Arnon DI (1950) The water-culture method for growing plants without soil. Circ. 347. California Agricultural Experimental Station, College of Agriculture, University of California, Berkley
14. Gamborg OL, Miller RA, Ojima K (1968) Nutrient requirements of suspension cultures of soybean root cells. Exp Cell Res 50:151–158

# Chapter 18

# Optimization of Hairy Root Transformation for the Functional Genomics in Chickpea: A Platform for Nodule Developmental Studies

**Drishti Mandal, Deevita Srivastava, and Senjuti Sinharoy**

## Abstract

Chickpea is a major protein source in low socio-economic classes and cultivated in marginal soil without fertilizer or irrigation. As a result of its root nodule formation capacity chickpea can directly use atmospheric nitrogen. Chickpea is recalcitrant to stable transformation, particularly root regeneration efficiency of chickpea is low. The composite plant-based system with a non-transformed shoot and transformed root is particularly important for root biologist and this approach has already been used successfully for root nodule symbiosis, arbuscular mycorrhizal symbiosis, and other root-related studies. Use of fluorescent marker-based approach can accurately identify the transformed root from its non-transgenic counterpart. RNAi-based gene knockout, overexpression of genes, promoter *GUS* analysis to understand tissue specific expression and localization of protein can be achieved using the hairy root-based system. We have already published a hairy root-based transformation and composite plant regeneration protocol of chickpea. Here we are describing the recent modification that we have made to increase the transformation frequency and nodule morphology. Further, we have developed a pouch based artificial system, large number of plants can be scored for its nodule developmental phenotype, by using this system.

**Key words** *A. rhizogenes*, Chickpea, Composite plants, Co-transformation, Hairy root transformation, Root nodule symbiosis

## 1 Introduction

Chickpea (*Cicer arietinum*) is the second largest cultivated legume. India is the topmost producer and exporter of chickpea. Chickpea seeds contain a moderate amount of protein estimated around 20–22% and serve as an important and inexpensive source of protein majorly in the southern part of Asia [1]. Chickpea belongs to the inverted repeat lacking clade (IRLC) and forms indeterminant nodule upon its interaction with *Mesorhizobium sp.* [2, 3]. Limited information is available on the root nodule development in chickpea. However, genomic, transcriptomic, and proteomic data, and

Mukesh Jain and Rohini Garg (eds.), *Legume Genomics: Methods and Protocols*, Methods in Molecular Biology, vol. 2107, https://doi.org/10.1007/978-1-0716-0235-5_18, 

resequencing of 429 chickpea accessions laid down enormous resources and potential for future research [1, 4–7].

Root nodule symbiosis (RNS) is a pillar behind sustainable agriculture. RNS research has seen a huge advancement in the last two decades due to the use of less time-consuming hairy root transformation and composite plant-based approach [8, 9]. *Agrobacterium rhizogenes* contains *root locus* containing *Ri-plasmid*, which promotes the formation of hairy roots. If the *A. rhizogenes* strain contains an additional binary vector, then a fraction of hairy roots will obtain the binary vector. These binary vector containing roots are called co-transformed roots. By using the co-transformation approach, overexpression or downregulation of a specific gene, promoter *GUS* assay to determine tissue-specific expression, protein localization in roots or nodules, and *CRISPR/Cas9*-mediated gene knockout can be achieved in a time-effective manner [10, 11]. To identify the co-transformation event, usually a binary vector is used which contains a non-destructive co-transformation marker, such as red fluorescent protein (RFP) or green fluorescent protein (GFP). Use of these fluorescence markers ensures the co-transformation event and by this approach non-transgenic (lateral) or hairy root (without co-transformtion with binary vector) can easily be eliminated (by chopping) from the composite plant for downstream functional analysis. Additionally, while performing hairpin based downregulation of genes the biggest advantage of using hairy root based composite plant system is that the hairpin RNA does not spread efficiently systemically to non-cotransformed roots and non-transgenic shoot, thus the root-specific downregulation of a particular gene can be achieved [10, 12]. Until now hairy root based transformation has been successfully used for *Lotus corniculatus*, *Trifolium repens*, *Vigna aconitifolia*, *Casuarina glauca*, *Glycine max*, *Vicia hirsuta*, *Lotus japonicus*, *Trifolium pretense*, *Medicago truncatula*, *Sesbania rostrata*, *Phaseolus spp.*, *Arachis hypogaea*, *Aeschynomene indica*, *Pisum sativum*, *Discaria trinervis*, *Vigna unguiculata*, *Solanum lycopersicum* [9, 13–29].

An efficient hairy root transformation protocol and a cognate rhizobium strain is the first and foremost requirement for root nodule developmental studies. Recently, we have published a hairy root transformation protocol for crop legume chickpea [2]. Here we are summarizing the protocol with the modification that further improved the transformation efficiency and morphology of the generated nodules (summarized in Table 1). The workflow of the transformation and the applications are summarized in Fig. 1. We also have generated a pouch-based system to grow a large number of chickpea ecotypes in a smaller space (summarized in Fig. 2). Nodule development has been standardized in this pouch-based system. The pouch-based system will also be useful for creation of nutrient-depleted condition especially accurate phosphate depleted condition, and the application of any expensive chemical(s) for modulation of root and nodule architecture.

**Table 1**
***A. rhizogenes* ARqua1-mediated transformation of different chickpea ecotypes with binary vector *pKGW-Red-Root***

| Genotype | Region | Method of transformation (fraction of transformed plants) | DsRed-positive hairy roots per co-transformed plants (%) | Reference |
|---|---|---|---|---|
| BDG 256[a] | India | 19/20 | $49 \pm 20$ | [2] |
| BDG 256[a] | India | 21/25 | $53 \pm 23$[b] | [2] |
| BDG 256[a] | India | 36/40 | $59 \pm 19$ | In this study |
| GNG 1488[a] | India | 36/40 | $44 \pm 29$ | In this study |
| RSG 931[a] | India | 36/40 | $50 \pm 26$ | In this study |
| PUSA 372[a] | India | 39/40 | $46 \pm 23$ | In this study |
| GNG 1581[a] | India | 35/40 | $18 \pm 11$ | In this study |
| GNG 1958[a] | India | 34/40 | $31 \pm 20$ | In this study |
| BGD 112[a] | India | 35/40 | $36 \pm 16$ | In this study |

The fraction of co-transformed composite plants and numbers of co-transformed hairy roots are counted based on DsRed fluorescence positive hairy roots per plant. 3 weeks after infection fraction of co-transformed roots are determined. All the transformation is conducted by cutting and in vitro transformation protocol
[a]Desi variety
[b]% after antibiotic selection

## 2 Materials

### 2.1 General

1. Chickpea seeds: BDG 256 (also known as BG 256), BGD 112, PUSA 372, GNG 1958, GNG 1488, GNG 1581, RSG 931.
2. *A. rhizogenes* strain ARqua1 cells to prepare competent cells.
3. *Mesorhizobium ciceri* CC1192 strain.
4. Antibiotics: spectinomycin (100 mg/mL stock solution), streptomycin (100 mg/mL stock solution) and kanamycin (50 mg/mL stock solution).
5. Commercial bleach solution (RIN, Hindustan Unilever Limited), antifungal powder (Vitavax Power, Dhanuka), Filter paper (Whatman), petri plates for germinating seeds.
6. Square petri plate for tissue culture.
7. Soil testing kit (Himedia, Cat. No. K054L), vermiculite and fire clay balls (leca), trays and pots for nodulation.
8. Filter paper (Whatman, 460 × 570 mm, Cat. No. 1001917) for preparation of pouch.

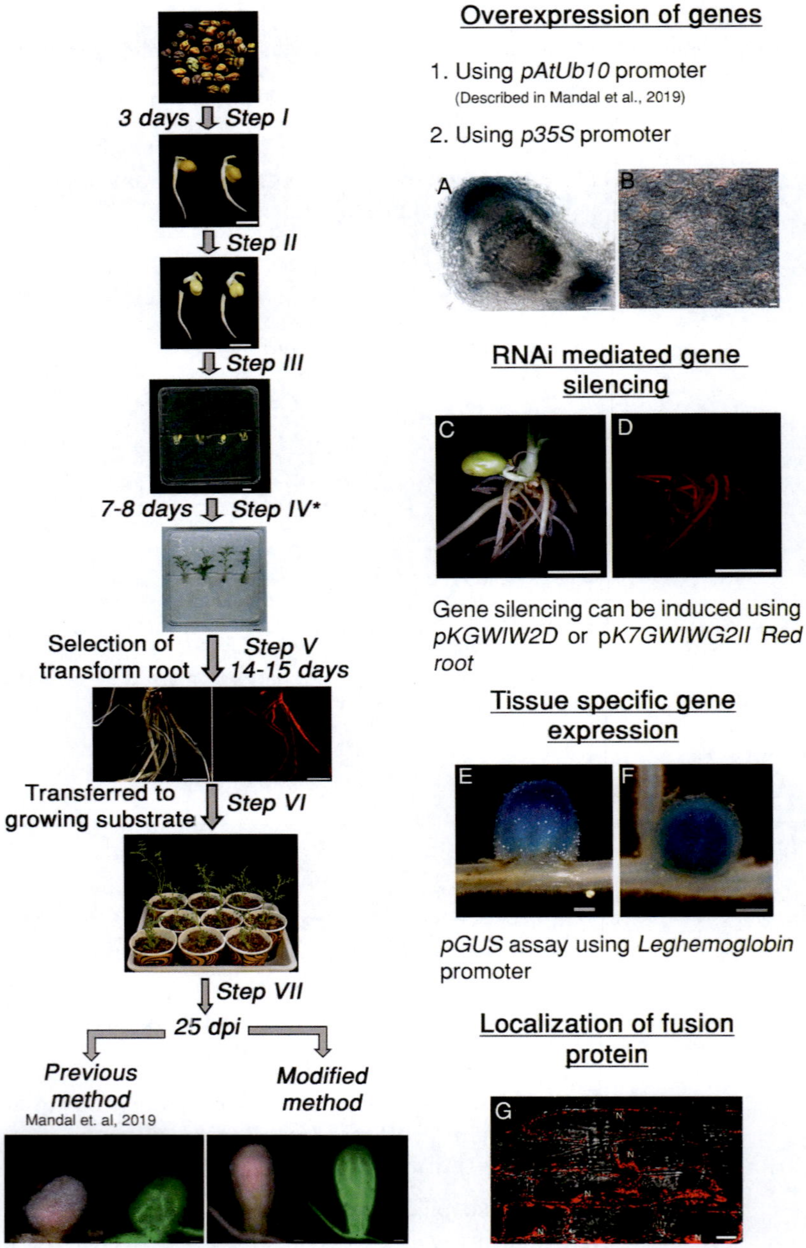


**Fig. 1** Flowchart of *A. rhizogenes*-mediated hairy roots transformation and composite plant generation and its applications. Left-hand panel depicts the modified protocol for hairy root transformation in chickpea. Chickpea seeds are germinated (depicted in Step I). For transformation the seed coats are removed and a slant cut is made in the radicles at 10–15 mm from the root tip (depicted in Step II). Germinated seedlings are infected with *A. rhizogenes* and plated on Fahraeus medium (depicted in Step III). After 7–8 days roots are cut and placed in new square petri plate containing solidified Fahraeus medium (depicted in Step IV). After 14–15 days, transformed roots are screened under the red channel of the stereomicroscope (depicted in Step V). The plants are transferred to growing substrate (depicted in Step VI). Composite plants are inoculated

9. Aminoethoxyvinylglycine (AVG) (Sigma, Cat No. 06665) for nodulation.
10. 6-Benzylaminopurine (6-BAP) (Duchefa, Cat. No. B0904) for spontaneous nodulation.

### 2.2 Media

1. Luria Bertani Media: 10 g tryptone, 10 g sodium chloride, and 5 g yeast extract dissolved in 1 L deionized water. The final volume is adjusted to 1 L with deionized water. For plate preparation, 15 g of agar is added per litre of LB.
2. Tryptone Yeast Extract Media: 3 g yeast extract, 6 g tryptone, and 0.38 g calcium chloride ($CaCl_2{\cdot}2H_2O$) dissolved in 1 L deionized water.
3. Yeast Mannitol medium: 1 g yeast extract, 10 g mannitol, 0.5 g dipotassium hydrogen phosphate, 0.2 g magnesium sulfate, 0.1 g sodium chloride, 0.025 g congo red, 20 g agar dissolved in 1 L deionized water. (Himedia, Cat No. M721).
4. Fähraeus Medium [14]: 0.9 mM/L calcium chloride ($CaCl_2{\cdot}2H_2O$), 3 mM/L magnesium sulfate ($MgSO_4$), 0.7 mM/L potassium dihydrogen phosphate ($KH_2PO_4$), 0.25 mM/L disodium hydrogen phosphate ($Na_2HPO_4$), 20 μM/L ferric citrate ($C_6H_5FeO_7$), 1 mg/mL manganese chloride ($MnCl_2$), 1 mg/mL copper sulfate ($CuSO_4$), 1 mg/mL zinc chloride ($ZnCl_2$), 1 mg/mL sodium molybdate ($Na_2MoO_4$), 1 mg/mL boric acid ($H_3BO_3$). The pH is adjusted to 7.4 with 1 M potassium hydroxide (KOH). Final volume is adjusted to 1 L with deionized water. 5 g/L CleriGar (Himedia Cat No. PCT0905) is added to it.

**Fig. 1** (continued) with *M. ciceri* CC1192 after 22–23 days (Step VII). After 25 dpi the transformed nodules are harvested. Transformation is conducted either by the method described in [2] or with modification. Transformed nodules are shown below. The right-hand panel describes the applications of hairy root transformation. Overexpression of gene can be achieved by composite plant approach. (**a**, **b**) The hairy root is generated with a binary vector that contains a *35S* promoter-*GUS* construct. Transformed nodules are hand sectioned. The blue color is due to GUS staining using X-Gluc (5-Bromo-4-chloro-3-indolyl β-D-glucuronide) and represents *35S* promoter expression inside the infected cells. (**a**) Section of whole nodule. (**b**) Nitrogen fixation zone. RNAi-mediated gene silencing can be achieved by composite plant approach. (**c**, **d**) Roots after transforming with *pK7GWIWG2-II Red root* vector [32]. (**c**) Bright field image. (**d**) Under red filter. Tissue-specific gene expression can be achieved by composite plant approach. (**e**, **f**) The hairy root is generated with a destination vector that contains chickpea *pLeghemoglobin-GUS* construct. The blue color is due to *GUS* staining and represents promoter expression at 25 dpi nodules. (**e**) 550 bp *Ca-leghemoglobin* promoter and (**f**) 1.5 kb *Ca-leghemoglobin* promoter. (**g**) Localization of protein in chickpea hairy root using *pUB-COX4-mCherry* construct [33]. Days post inoculation is abbreviated as dpi, Scale bars step I–V = 1 cm, Step VII = 1 mm, (**a**, **e** and **f**) = 1 mm, (**b** and **g**) = 10 μm, (**c** and **d**) = 10 mm. The transformed root and nodules micrograph are captured under either bright field or red or green channel of stereomicroscope. (* represents the modification from Mandal et al., 2019)

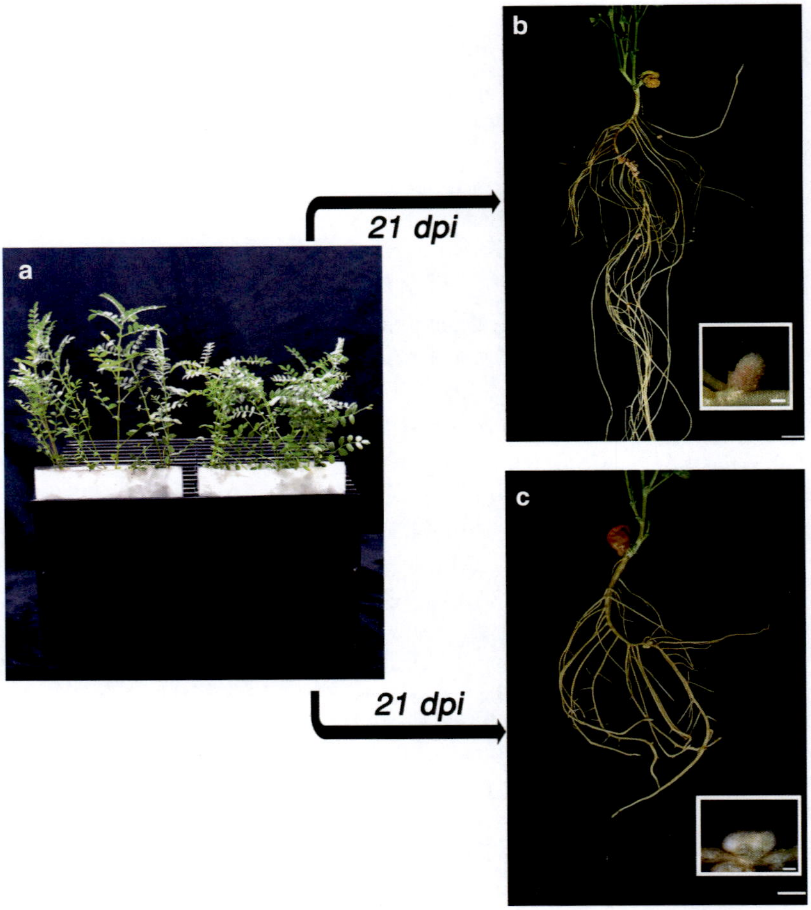


**Fig. 2** Standardization of pouch-based system in chickpea. (**a**) Setup for pouch-based system. (**b**) *M. ciceri* CC1192 induced root nodule after 21 dpi. (**c**) 6-Benzylaminopurine induced spontaneous nodulation at 21 dpi. Inset representing enlarged view of individual nodule. Scale bars (**b** and **c**) = 1 cm, inset (**b** and **c**) = 1 mm

5. Broughton and Dilworth (B & D) medium: 250 μM potassium dihydrogen phosphate ($KH_2PO_4$), 500 μM potassium sulfate ($K_2SO_4$), 125 μM magnesium sulfate ($MgSO_4$), 1 μM manganese (II) sulfate ($MnSO_4 \cdot H_2O$), 2 μM boric acid ($H_3BO_3$), 0.5 μM zinc sulfate ($ZnSO_4 \cdot 7H_2O$), 0.2 μM copper sulfate ($CuSO_4 \cdot 5H_2O$), 0.1 μM cobaltous sulfate heptahydrate ($CoSO_4 \cdot 7H_2O$), 0.1 μM sodium molybdate ($Na_2MoO_4$), 10 μM ferric citrate, 1 mM calcium chloride ($CaCl_2 \cdot 2H_2O$). For full nitrogen, 2 mM potassium nitrate ($KNO_3$) and 2 mM ammonium nitrate ($NH_4NO_3$) should be added. All solutions should be prepared separately. Individual items should be filter-sterilized and stored at 4 °C when needed for longer storage. 50 × stock solution is prepared by mixing $KH_2PO_4$, $K_2SO_4$, $MgSO_4$, ferric citrate, and micronutrients. The working stock

of 0.5 × B & D is prepared from the stock solution and to it, $CaCl_2 \cdot 2H_2O$ is added. The pH is adjusted to 6.8 with 1 M potassium hydroxide (KOH).

## 3 Methods

### *3.1 A. rhizogenes ARqua1 Competent Cell Preparation*

1. Grow a 5 mL seed culture from a single colony of *A. rhizogenes* ARqua1 by shaking (200 rpm) overnight at 28 °C in Luria–Bertani Broth (LB) with streptomycin (100 μg/mL).
2. Grow a large culture by inoculating 1% of the ARqua1 starter culture in 50 mL of LB medium up to an OD 600 of ~ 0.8–1 at 28 °C.
3. Harvest the cells by centrifugation at 3000 × *g* for 10 min at 4 °C (*see* **Note 1**).
4. Discard the supernatant and resuspend the cells in 25 mL of 20 mM $CaCl_2$.
5. Incubate in ice for 30 min.
6. Harvest the cells by centrifugation at 3000 × *g* for 5 min at 4 °C.
7. Discard the supernatant and resuspend the cells in 25 mL of 20 mM $CaCl_2$.
8. Incubate in ice for 1 h.
9. Harvest the cells by centrifugation at 3000 × *g* for 5 min at 4 °C, and discard the supernatant.
10. Resuspend the cells in 10 mL solution of 20 mM $CaCl_2$ containing 15% glycerol.
11. Aliquot the cells in small tubes of 500 μL each, freeze in liquid nitrogen and keep in −80 °C for future use.

### *3.2 A. rhizogenes Transformation Using the Modified Freeze–Thaw Method [30]*

1. Thaw one aliquot of the −80 °C stored cells on ice before the transformation.
2. Add 500 ng of binary vector DNA (volume 2–5 μL) (*see* **Note 2**) to 500 μL of competent cells.
3. Refreeze the content in liquid nitrogen for 5 min and then thaw at 37 °C water bath for 5 min.
4. Quickly cool it in ice for 5 min.
5. Dilute the cells with 1 mL of LB medium, and keep it in incubator shaker for 4 h at 28 °C with vigorous shaking (200 rpm).
6. Next, centrifuge the cells at 2500 × *g* for 5 min, and discard around 0.8–1 mL of supernatant from the top (*see* **Note 1**).

Resuspend the bacterial pellet, and plate on LB medium plate with appropriate antibiotics selection.

7. Incubate the plates at 28 °C for 2 days.
8. Identify the positive transformants by colony PCR screening.

### 3.3 Composite Plant Generation Using Hairy Root Transformation

*A. rhizogenes*-based hairy root transformation is a less time-consuming alternative of stable transformation where we obtained composite plants with wild type shoot and transgenic roots. Here we describe *A. rhizogenes*-mediated hairy root transformation method for chickpea in detail.

1. Two days prior to transformation, grow *Agrobacterium* from its glycerol stock in LB agar plate with suitable antibiotics (spectinomycin and streptomycin, both at 100 μg/L) at 28 °C to produce a lawn in the plate.
2. Surface-sterilize the chickpea seeds by 30% commercial bleach solution containing active sodium hypochlorite 1.2% for 3 min. Wash the seeds thoroughly to remove the bleach with sterile water for 8–10 times. Soak them in antifungal solution (0.1 g/ 100 mL) (*see* **Note 3**), and keep in room temperature for 4 h. Next, wash them with sterile water for 5–6 times to completely remove the antifungal solution. Place the seeds in a petri dish with sterile wet Whatman filter paper (soaked in sterile deionized water) for germination. Keep the plates in 20–22 °C temperature in dark for germination.
3. For transformation, use only the seedlings with a radicle length of approximately 40 mm after 3 days (Fig. 1, step I).
4. Remove the seed coats with sterile forceps for transformation (Fig. 1, step II).
5. Make a slant cut in the radicles at 10–15 mm from the root tip with a sterile scalpel, holding the seedlings by the cotyledon with sterile forceps (Fig. 1, step II).
6. Rub the cut surface on the *A. rhizogenes* strain ARqua1 bacterial lawn that contains the appropriate construct.
7. After infection, transfer the seedlings to a square petri plate containing solid Fahraeus medium [14] without antibiotics. Cut the upper half portion of the media and discarded from the plate prior to seedlings transfer (*see* **Note 4**). Position the infected seedlings inside the media with sterile forceps (Fig. 1, step III).
8. Seal the plates with surgical tape (*see* **Note 5**) and cover the lower part (2/3 part) with aluminum foil. Place the plates longitudinally in a growth chamber at 22 °C under 16 h/8 h light and dark cycle, allowing the roots to grow gravitropically.

9. Production of roots is first noticed after 4–5 days. After 7–8 days, cut these roots off and place the remaining seedling portion in a new square petri plate containing solidified Fahraeus medium without antibiotics (*see* **Note 6**), after removal of the agar from the half portion of the plate as mentioned in Subheading 3.3, **step** 7. The media free space is created in the plate for the shoot part.
10. Seal the plates with surgical tape and cover the lower part (2/3 part) with aluminum foil. Place the plates longitudinally in a growth chamber at 22 °C under 16 h/8 h light and dark cycle, allowing the roots to grow gravitropically.
11. Screen the transformed roots under stereo fluorescence microscope after 14–15 days. Transfer the plants with transformed roots in pots with the growing substrate.
12. Mix 3 parts of vermiculite, and 1 part of fire clay ball (Leca) to prepare the substrate for growing plants in pots. Wash this mix with distilled water for removal of any trace of nitrogen. Test the growing substrate for ammoniacal nitrogen and nitrate nitrogen using soil testing kit (*see* **Note** 7). Redo these steps if necessary. Finally, autoclave this mix twice and use after complete cooling.
13. Carefully remove the plants from the plates with the intact root system, and then wash thoroughly to remove the residual amount of agar (*see* **Note 8**). Transfer the seedlings to pots containing growing substrate soaked with 0.5× B & D medium with full nitrogen. Place the pots on trays and cover by clear plastic wraps to maintain high humidity inside. Place the trays in the growth chamber under 22 °C temperature and 16 h/8 h light and dark cycle. Water the plants from bottom four times a week with sterile water. After 5 days, remove the plastic wraps to lower the moisture content and prevent fungal growth.

### 3.4 Nodulation Assay

1. After 20 days, water the composite plants (in the pot) with excess distilled water. Transfer the plants into new pots and tray for removal of nitrogen. Maintain the plants in this nitrogen free condition for the next 2–3 days prior to inoculation with *Mesorhizobium* strain for nodulation.
2. For inoculation, grow the *M. ciceri* CC1192 strain from its glycerol stock in yeast mannitol agar plate for 2 days at 28 °C. Next, prepare a *M. ciceri* CC1192 starter culture in TY medium at 28 °C. Inoculate 1% of this starter culture to a larger volume of secondary culture and grow overnight at 28 °C, with shaking at 200 rpm, up to OD 600 ~ 0.7–0.9 (*see* **Note 9**).

3. When the culture reaches the desired OD, harvest the cells at 2500 × *g* for 10 min and resuspend the pellet in 0.5× nitrogen-free B & D medium to OD 600 ~ 0.03–0.05.
4. Inoculate each pot with 50 mL of this bacterial culture in B & D medium. Harvest the plants for evaluation of nodulation 25 days post inoculation (dpi). In between, bottom-water the plants four times per week with sterile water.
5. For the overexpression of the gene inside nodule both *p35S* or *pUB* promoter can be used (*see* **Note 10**).

### 3.5 Artificial Growth Pouch Preparation for Nodulation and Hormonal Treatment

#### 3.5.1 Artificial Medium: Pouch Preparation

1. Cut Whatman filter paper of 460 × 570 mm using a surgical scalpel and fold in a "M" shape. The dimension of this setup is 13 × 18 cm.
2. Make approximately five incisions in the folded region of the filter paper with a surgical scalpel (*see* **Note 11**). Transfer this setup inside an autoclavable plastic bag (15 × 20 cm) and seal three sides of this autoclavable plastic using heat sealing machine. This creates an autoclavable unit containing the "M" shaped filter paper.
3. Autoclaved this assembly twice (*see* **Note 12**).
4. Prepare a black plexiglass box to maintain the vertical position of the pouch and soil like darkness in the root growing area (Fig. 2a).
5. Transfer the pouches to the box and then insert the chickpea radicles inside the incisions made in the "M" shaped filter paper.

#### 3.5.2 Protocol for Nodulation

1. Surface-sterilize chickpea seeds (ecotype BDG 256) as mentioned in Subheading 3.3, **step 2**.
2. After 3 days, seedlings develop radicles. Transfer these in the folded region of the pouch by fixing radicle in an incision.
3. Soak the artificial medium with 30 mL 0.5× B & D medium without any nitrogen source and along with AVG (50 μM).
4. Keep the whole set up in a growth chamber under 22 °C temperature and 16 h/8 h light and dark cycle.
5. Inoculate the plants with *M. ciceri* CC1192 strain after 7–8 days. *M. ciceri* CC1192 inoculum can be prepared as described in Subheading 3.4.
6. Count the numbers of the nodules 21 dpi. Water the plants with sterile water and AVG (50 μM) thrice a week till 21 days (*see* **Note 13**).

*3.5.3 Protocol for Hormonal Treatment*

1. Sterilize and germinate chickpea seeds (ecotype BDG 256) as mentioned in Subheading 3.3, **step 2**.
2. Soak the pouches with 30 mL of 0.5× B & D medium without any nitrogen and transfer radicles to the pouches as describe in Subheading 3.5.2, **step 2**.
3. After 7 days, add 30 mL 0.5× B & D medium and 6-Benzylaminopurine (10 μM) to the artificial media.
4. Count the number of nodules at 21 dpi. Water the plants with sterile water thrice a week till 21 days (*see* **Note 13**).

## 4 Notes

1. A low centrifugation speed should be maintained, if the speed is too high then it will rupture the cells.
2. High concentration of DNA in small volume is needed to be used. The diluted DNA sample will reduce the frequency of transformation.
3. Concentration of fungicides should be maintained meticulously. Fungicides have an adverse impact on the development of nodule and sometime nodule looks unhealthy when higher amount of fungicide is used.
4. It is suggested to remove half of the media toward the shooting part to prevent fungal infection in shoot due to high moister content in the medium. Removal of the upper part of the medium helps us to maintain the plants for a longer period of time inside the square plate without shoot rotting.
5. The use of surgical tape increases aeration. If parafilm is used, several cuts have to be made in the parafilm to allow gas exchange.
6. Though after antibiotic selection a slightly greater frequency of transformed roots can be obtained, but the use of antibiotics may stress the plant and it will result in early degradation of the bacteria inside the root nodule. So we routinely screen the transformed roots under the stereo fluorescence microscope without using antibiotic.
7. If the soil has any nitrogen it will decrease the nodulation efficiency.
8. It is important to remove agar properly from roots before transferring to pots, to prevent fungal contamination.
9. When the OD of secondary culture reaches ~1, incubation of culture should be stopped and culture should be inoculated to the plant. Prolong growth of the culture leads to less cell viability and decrease nodule number.

10. We have seen that *p35S* can express inside the chickpea nodule, unlike *Medicago* [31]. Hence *p35S*-mediated gene overexpression can be achieved inside the transformed nodule but transformation with *pUB* promoter gives a better result.
11. The incision size should be such that the radicle from germinating seedling can penetrate down in the pouch.
12. It will be helpful to autoclave the artificial medium (pouch) twice to ensure any further contamination.
13. The plant should be watered at a definite time interval. As in this set up instead of soil, plants are only grown based on the nutrient and water availability in the liquid inside the pouches hence, drying of pouch due to water deficiency adversely affect nodule formation and plant health.

## Acknowledgments

We thank J. Terpolilli, Murdoch University, Australia for providing *M. ciceri* CC1192; Senthil-Kumar Muthappa, NIPGR, New Delhi for providing chickpea seeds (BGD 112, PUSA 372, GNG 1958, GNG 1488, GNG 1581, RSG 931); Michael Udvardi, Nobel Research Institute, Oklahoma, USA for *A. rhizogenes* ARqua1 strain; Maria J. Harrison, Boyce Thompson Institute for Plant Research Ithaca, USA for providing pAtUb driven subcellular marker construct; M. DasGupta, Department of Biochemistry, University of Calcutta, for providing *pCAMBIA2301* vector; NIPGR for their confocal facilities; and CIF-NIPGR; NIPGR-DELCON for their support. This work is supported by core research grant from National Institute of Plant Genome Research, Ramalingwaswami Re-entry grant, DBT (BT/RLF/Re-entry/41/2013).

### References

1. Jain M, Misra G, Patel RK, Priya P, Jhanwar S, Khan AW et al (2013) A draft genome sequence of the pulse crop chickpea (*Cicer arietinum* L.). Plant J 74:715–729
2. Mandal D, Sinharoy S (2019) A toolbox for nodule development studies in chickpea: a hairy-root transformation protocol and an efficient laboratory strain of Mesorhizobium sp. Mol Plant-Microbe Interact 32:367–378
3. Montiel J, Downie JA, Farkas A, Bihari P, Herczeg R, Balint B et al (2017) Morphotype of bacteroids in different legumes correlates with the number and type of symbiotic NCR

peptides. Proc Natl Acad Sci U S A 114:5041–5046

4. Varshney RK, Thudi M, Roorkiwal M, He W, Upadhyaya HD, Yang W et al (2019) Resequencing of 429 chickpea accessions from 45 countries provides insights into genome diversity, domestication and agronomic traits. Nat Genet 51:857–864
5. Pandey A, Chakraborty S, Chakraborty N (2018) Nuclear proteome: isolation of intact nuclei, extraction of nuclear proteins, and 2-DE analysis. Methods Mol Biol 1696:41–55
6. Ramalingam A, Kudapa H, Pazhamala LT, Weckwerth W, Varshney RK (2015) Proteomics and metabolomics: two emerging areas for legume improvement. Front Plant Sci 6:1116
7. Varshney RK, Song C, Saxena RK, Azam S, Yu S, Sharpe AG et al (2013) Draft genome sequence of chickpea (*Cicer arietinum*) provides a resource for trait improvement. Nat Biotechnol 31:240–246
8. Georgiev MI, Agostini E, Ludwig-Muller J, Xu J (2012) Genetically transformed roots: from plant disease to biotechnological resource. Trends Biotechnol 30:528–537
9. Stougaard J, Petersen TE, Marcker KA (1987) Expression of a complete soybean leghemoglobin gene in root nodules of transgenic *Lotus corniculatus*. Proc Natl Acad Sci U S A 84:5754–5757
10. Limpens E, Ramos J, Franken C, Raz V, Compaan B, Franssen H et al (2004) RNA interference in agrobacterium rhizogenes-transformed roots of Arabidopsis and *Medicago truncatula*. J Exp Bot 55:983–992
11. Sinharoy S, Pislariu CI, Udvardi MK (2015) A high-throughput RNA interference (RNAi)-based approach using hairy roots for the study of plant-rhizobia interactions. Methods Mol Biol 1287:159–178
12. Kumagai H, Kouchi H (2003) Gene silencing by expression of hairpin RNA in Lotus japonicus roots and root nodules. Mol Plant-Microbe Interact 16:663–668
13. Kereszt A, Li D, Indrasumunar A, Nguyen CD, Nontachaiyapoom S, Kinkema M et al (2007) Agrobacterium rhizogenes-mediated transformation of soybean to study root biology. Nat Protoc 2:948–952
14. Boisson-Dernier A, Chabaud M, Garcia F, Becard G, Rosenberg C, Barker DG (2001) Agrobacterium rhizogenes-transformed roots of *Medicago truncatula* for the study of nitrogen-fixing and endomycorrhizal symbiotic associations. Mol Plant-Microbe Interact 14:695–700
15. Diaz CL, Spaink HP, Kijne JW (2000) Heterologous rhizobial lipochitin oligosaccharides and chitin oligomers induce cortical cell divisions in red clover roots, transformed with the pea lectin gene. Mol Plant-Microbe Interact 13:268–276
16. Stiller J, Martirani L, Tuppale S, Chian R-J, Chiurazzi M, Gresshoff PM (1997) High frequency transformation and regeneration of transgenic plants in the model legume Lotus japonicus. J Exp Bot 48:1357–1365
17. Diouf D, Gherbi H, Prin Y, Franche C, Duhoux E, Bogusz D (1995) Hairy root nodulation of *Casuarina glauca*: a system for the study of symbiotic gene expression in an actinorhizal tree. Mol Plant-Microbe Interact 8:532–537
18. Quandt HJ, Puhler A, Broer I (1993) Transgenic root nodules of *Vicia hirsuta*. A fast and efficient system for the study of gene expression in indeterminate-type nodules. Mol Plant-Microbe Interact 6:699–703
19. Lee NG, Stein B, Suzuki H, Verma DP (1993) Expression of antisense nodulin-35 RNA in *Vigna aconitifolia* transgenic root nodules retards peroxisome development and affects nitrogen availability to the plant. Plant J 3:599–606
20. Cheon CI, Lee NG, Siddique ABM, Bal AK, Verma DPS (1993) Roles of plant homologs of Rabi p and Rab7p in the biogenesis of the peribacteroid membrane, a subcellular compartment formed de novo during root nodule symbiosis. EMBO J 12:4125–4135
21. Díaz C, Melchers LS, Hooykaas PJJ, Lugtenberg BJJ, Kijne JW (1989) Root lectin as a determinant of host-plant specificity in the rhizobium-legume symbiosis. Nature 338:579–558
22. Mellor KE, Hoffman AM, Timko MP (2012) Use of ex vitro composite plants to study the interaction of cowpea (*Vigna unguiculata* L.) with the root parasitic angiosperm *Striga gesnerioides*. Plant Methods 8:22
23. Imanishi L, Vayssieres A, Franche C, Bogusz D, Wall L, Svistoonoff S (2011) Transformed hairy roots of Discaria trinervis: a valuable tool for studying actinorhizal symbiosis in the context of intercellular infection. Mol Plant-Microbe Interact 24:1317–1324
24. Clemow SR, Clairmont L, Madsen LH, Guinel FC (2011) Reproducible hairy root transformation and spot-inoculation methods to study root symbioses of pea. Plant Methods 7:46
25. Bonaldi K, Gherbi H, Franche C, Bastien G, Fardoux J, Barker D et al (2010) The nod

factor-independent symbiotic signaling pathway: development of agrobacterium rhizogenes-mediated transformation for the legume *Aeschynomene indica*. Mol Plant-Microbe Interact 23:1537–1544

26. Sinharoy S, Saha S, Chaudhury SR, Dasgupta M (2009) Transformed hairy roots of Arachis hypogea: a tool for studying root nodule symbiosis in a non-infection thread legume of the Aeschynomeneae tribe. Mol Plant-Microbe Interact 22:132–142
27. Estrada-Navarrete G, Alvarado-Affantranger X, Olivares JE, Diaz-Camino C, Santana O, Murillo E et al (2006) Agrobacterium rhizogenes transformation of the Phaseolus spp.: a tool for functional genomics. Mol Plant-Microbe Interact 19:1385–1393
28. Van-de-Velde W, Mergeay J, Holsters M, Goormachtig S (2003) Agrobacterium rhizogenes-mediated transformation of Sesbania rostrata. Plant Sci 165:1281–1288
29. Ron M, Kajala K, Pauluzzi G, Wang D, Reynoso MA, Zumstein K et al (2014) Hairy root transformation using agrobacterium rhizogenes as a tool for exploring cell type-specific gene expression and function using tomato as a model. Plant Physiol 166:455–469
30. Hofgen R, Willmitzer L (1988) Storage of competent cells for agrobacterium transformation. Nucleic Acids Res 16:9877
31. Auriac MC, Timmers AC (2007) Nodulation studies in the model legume *Medicago truncatula*: advantages of using the constitutive EF1alpha promoter and limitations in detecting fluorescent reporter proteins in nodule tissues. Mol Plant-Microbe Interact 20:1040–1047
32. Karimi M, Inze D, Depicker A (2002) GATEWAY vectors for agrobacterium-mediated plant transformation. Trends Plant Sci 7:193–195
33. Ivanov S, Harrison MJ (2014) A set of fluorescent protein-based markers expressed from constitutive and arbuscular mycorrhiza-inducible promoters to label organelles, membranes and cytoskeletal elements in *Medicago truncatula*. Plant J 80:1151–1163

# Chapter 19

# CRISPR/Cas9-Based Gene Editing in Soybean

**Aili Bao, Lam-Son Phan Tran, and Dong Cao**

## Abstract

CRISPR/Cas9 (clustered regularly interspaced short palindromic repeat/CRISPR associated Cas9)-based gene editing is a robust tool for functional genomics research and breeding programs in various crops. In soybean, a number of laboratories have obtained mutants by CRISPR/Cas9 system; however, there has been not yet a detailed method for the CRISPR/Cas9-based gene editing in soybean. Here, we describe the procedures for constructing the CRISPR/Cas9 plasmid suitable for soybean gene editing and the modified protocols for *Agrobacterium*-mediated soybean transformation and regeneration from cotyledonary node explants containing the Cas9/sgRNA (single guide RNA) transgenes.

**Key words** *Agrobacterium*-mediated soybean transformation, Cotyledonary node, CRISPR/Cas9, Gene editing, Soybean, Tissue culture

## 1 Introduction

Soybean (*Glycine max*) is an important legume crop with abundant protein and oil in its seeds, and thus is grown widely for food, feed materials, and industrial purposes [1]. As the great economic value, soybean has attracted global research interest in dissecting its gene function and improving its agronomic traits by soybean transformation technologies. There are two most frequently used methods of soybean transformation for the delivery of transgenes into cells, namely, *Agrobacterium*-mediated transformation and particle bombardment [2]. Although these two methods have successfully been applied to produce genetically modified soybeans over the past few decades, achieving a stable transformation platform for soybean is still regarded as a challenge due to soybean recalcitrance to transformation and regeneration [2]. In general, *Agrobacterium*-mediated transformation is a more preferred choice because of its simple protocol, low copy insertion, low operation cost, and high reproducibility [3].

Traditional genetic engineering breeding technologies can modulate gene expression by the integration of exogenous DNA

Mukesh Jain and Rohini Garg (eds.), *Legume Genomics: Methods and Protocols*, Methods in Molecular Biology, vol. 2107, https://doi.org/10.1007/978-1-0716-0235-5_19, 

into the plant genome, resulting in either gene silencing or overexpression [4–6]. However, they cannot overcome some technical shortcomings, such as the inaccuracy and incompleteness of gene knockout, uncontrollable expression levels of foreign genes and randomness of insertion sites in the genome [7]. Given that soybean is a crop with low transformation efficiency, a robust gene targeting method is necessary to be used for functional studies. Recently, the CRISPR/Cas9 (clustered regularly interspaced short palindromic repeat/CRISPR associated Cas9)-based gene editing has been widely used in various crops with its advantages of high efficiency, preciseness, and simplicity [8, 9]. Using CRISPR/Cas9 technology, researchers can knock out unfavorable genes or negative regulatory genes to achieve precise improvement of target traits, thereby obtaining stress-resistant or high-quality breeding materials [8, 9]. Besides, the mechanism of action of the CRISPR/Cas9 system makes the edit loci of target genes and insertion loci of exogenous transgenes separating. Therefore, breeders can remove all the exogenous transgenes, including sgRNA (single guide RNA)/Cas9 elements by segregation, and acquire genome-edited, transgene-free progenies [10].

In addition to the conventional CRISPR/Cas9 system integrated into plant genome assisted by T-DNAs, there is another CRISPR/Cas9 technology, called CRISPR/Cas9 ribonucleoproteins (RNPs), which can be used to perform DNA-free gene editing [11–13]. However, RNPs may not yet be suitable for soybean, at least for now, as there is neither a reliable routine method for regenerating soybean plants from protoplasts or embryo cells nor reports about RNP applications in soybean. On the contrary, the plasmid-dependent CRISPR/Cas9 system has been successfully applied in soybean research according to several reports [14–22]. In soybean, the first mutagenesis generated by the CRISPR/Cas9 system was obtained in hairy roots by *A. rhizogenes*-mediated transformation in 2015 [14–17]. Such hairy roots are effective to assess the mutagenesis efficiency but they do not produce plantlets. In the same year, Li et al. [18] obtained the whole soybean plants mutated by targeted mutagenesis using the CRISPR/Cas9 system coupled with particle bombardment. Later on, the studies of soybean genome editing have been carried out mainly through mutants regenerated from explants containing Cas9/sgRNA transgenes [19–22]. For instance, using the CRISPR/Cas9 system and *Agrobacterium*-mediated transformation, Cai et al. [21] found that loss-of-function of *GmFT2a*, encoding florigen, could delay the flowering time. They also designed a dual-sgRNA/Cas9 system to delete large fragments in soybean and confirmed that these mutations could be transmitted to the progenies [22]. Taken together, these successful applications in soybean indicate the great value of the CRISPR/Cas9 system for soybean breeding.

In this chapter, we describe a detailed procedure for constructing the CRISPR/Cas9 plasmid suitable for soybean gene editing, followed by the delivery of the CRISPR/Cas9 system into soybean by *Agrobacterium*-mediated transformation and regeneration technologies, which have worked well in our laboratory.

## 2 Materials

### 2.1 Construction of the CRISPR/Cas9 Plasmid

1. Vectors: pYLsgRNA-AtU3d/LacZ, pYLsgRNA-AtU3b, pYLsgRNA-AtU6-1, pYLsgRNA-AtU6-29, pYLCRISPR/Cas9P35S-B (*see* **Note 1**).
2. Enzymes and kits: *Bsa*I-HF (NEB, R3535s), 10× CutSmart buffer (NEB, R3535s), 10× T4 DNA ligase buffer containing10 mM ATP (NEB, M0202s), T4 DNA ligase (NEB, M0202s), KOD-Plus (TOYOBO, KOD-201), PCR product purification kit (TRANSGEN), FastPure plasmid mini kit (Vazyme).
3. Bacterial strains: Chemically competent DH5α *Escherichia coli* cells, *Agrobacterium tumefaciens* strains EHA105 competent cells.
4. LB liquid medium: 5 g/L yeast extract, 10 g/L tryptone, 5 g/L NaCl, (pH 7.0, adjusted with NaOH). Autoclave 20 min. Before use, add appropriate antibiotics.
5. LB agar plates: LB liquid medium containing 15 g/L agar. Autoclave 20 min and cool to about 55 °C. When necessary, supplement appropriate antibiotics to the medium before pouring into petri plates.
6. 50 mg/mL kanamycin and 30 mg/mL rifampicin (stock solutions).
7. Primes: Target adapters (Table 1), Universal site-specific primers (Table 2).
8. 1× TE buffer (pH 8.0): Add 5 mL 1 M Tris–HCl buffer (pH 8.0) and 1 mL 0.5 M EDTA (pH 8.0) to a beaker, then adjust the final volume with $ddH_2O$ to 500 mL. Autoclave 20 min.
9. 50% glycerol (autoclaved for 20 min).
10. Other required lab tools and equipment: PCR thermocycler, water bath, shaking and static incubator, Nanodrop spectrophotometer, centrifuge, agarose gel equipment, DNA ladders, 50-mL centrifuge tubes, 0.2-mL PCR tubes, 1.5-mL and 2-mL microcentrifuge tubes.

**Table 1**
**Target sequences and the target adaptors designed for construction of different sgRNA expression cassettes (*see* also Note 4)**

| pYLsgRNA-AtU3/U6 intermediate vectors | Target sequences | Target adaptors |
|---|---|---|
| pYLsgRNA-AtU3d/LacZ<br>pYLsgRNA-AtU3b | 5′-$AN_{19}$-3′<br><br>5′-(C/T/G)$N_{19}$-3′ | 5′-gtca$N_{19}$-3′<br>3′-$N_{19}$caaa-5′<br>5′-gtca$N_{20}$-3′<br>3′-$N_{20}$caaa-5′ |
| pYLsgRNA-AtU6-1<br>pYLsgRNA-AtU6-29 | 5′-$GN_{19}$-3′<br><br>5′-(C/T/A)$N_{19}$-3′ | 5′-attg$N_{19}$-3′<br>3′-$N_{19}$caaa-5′<br>5′-attg$N_{20}$-3′<br>3′-$N_{20}$caaa-5′ |

**Table 2**
**Site-specific primers used for cloning of sgRNA expression cassettes**

| Position | Primer | Sequence (5′–3′) |
|---|---|---|
| Site B-L | Uctcg-B1′ | TTCAGAGGTCTCTCTCGACTAGTATGGAATCGGCAGCAAAGG |
| Site 2 | gRctga-B2 | AGCGTGGGTCTCGTCAGGGTCCATCCACTCCAAGCTC |
| Site 2 | Uctga-B2′ | TTCAGAGGTCTCTCTGACACTGGAATCGGCAGCAAAGG |
| Site 3 | gRaaga-B3 | AGCGTGGGTCTCGTCTTCACTCCATCCACTCCAAGCTC |
| Site 3 | Uaaga-B3′ | TTCAGAGGTCTCTAAGACTTTGGAATCGGCAGCAAAGG |
| Site 4 | gRgact-B4 | AGCGTGGGTCTCGAGTCCTTTCCATCCACTCCAAGCTC |
| Site 4 | Ugact-B4′ | TTCAGAGGTCTCTGACTACATGGAATCGGCAGCAAAGG |
| Site B-R | gRcggt-BL | AGCGTGGGTCTCGACCGACGCGTATCCATCCACTCCAAGCTC |
| Sequencing Primers | SP-L<br>SP-R | GTCGTGCTCCACATGTTGACCGG<br>CCCGACATAGATGCAATAACTTC |

### *2.2 Transformation and Regeneration of Soybean Plantlets*

1. LB liquid medium:

   Before use, add 30 mg/L rifampicin and 50 mg/L kanamycin.

2. Resuspension medium (RSM):

   0.31 g/L Gamborg's B-5 Basal Salt Mixture, 3.9 g/L 2-(*N*-morpholino)ethanesulfonic acid (MES), 30 g/L sucrose (pH 5.4, adjusted with NaOH). Autoclave 20 min, cool to about 55 °C, and then supplement several filter-sterilized solutions containing 1000× Gamborg's B5 vitamins, 40 mg/L acetosyringone (AS), 1.67 mg/L 6-benzylaminopurine (6-BA) and 0.25 mg/L gibberellic acid ($GA_3$) to the medium.

3. Germination medium (GM):

   0.31 g/L Gamborg's B-5 Basal Salt Mixture, 20 g/L sucrose, 8 g/L agar (pH 5.8, adjusted with NaOH). Autoclave 20 min, cool to about 55 °C and then add filter-sterilized 1000× Gamborg's B5 vitamins. Pour the mixture into petri plates.

4. Cocultivation medium (CCM):

   0.31 g/L Gamborg's B-5 Basal Salt Mixture, 3.9 g/L MES, 30 g/L sucrose, 8 g/L agar (pH 5.4, adjusted with NaOH). Autoclave 20 min, cool to about 55 °C and then add several filter-sterilized solutions to the medium containing 1000× Gamborg's B5 vitamins, 400 mg/L cysteine, 40 mg/L AS, 1.67 mg/L 6-BA, 154 mg/L dithiothreitol, 158 mg/L $Na_2S_2O_3$, and 0.25 mg/L $GA_3$. Pour the mixture into petri plates. After the medium solidifies, place a piece of autoclaved filter paper on the medium of each plate.

5. Washing medium (WM):

   3.1 g/L Gamborg's B-5 Basal Salt Mixture, 0.59 g/L MES, 30 g/L sucrose (pH 5.7, adjusted with NaOH). Autoclave 20 min, cool to about 55 °C, and then add several filter-sterilized solutions to the medium containing 1000× Gamborg's B5 vitamins, 1.11 mg/L 6-BA, 400 mg/L timentin (Tim), and 100 mg/L cefotaxime (Cef).

6. Shoot induction medium (SIM):

   3.1 g/L Gamborg's B-5 Basal Salt Mixture, 0.59 g/L MES, 30 g/L sucrose, 8 g/L agar (pH 5.7, adjusted with NaOH). Autoclave 20 min, cool to about 55 °C, and then add several filter-sterilized solutions to the medium containing 1000× Gamborg's B5 vitamins, 1.11 mg/L 6-BA, 400 mg/L Tim, 100 mg/L Cef, and 5 mg/L glufosinate. Pour the mixture into petri plates.

7. Shoot elongation medium (SEM):

   4.43 g/L Murashige and Skoog (MS) basal medium with vitamins, 0.59 g/L MES, 30 g/L sucrose, 8 g/L agar (pH 5.7, adjusted with NaOH). Autoclave 20 min, cool to about 55 °C, and then add several filter-sterilized solutions to the medium containing 1 mg/L zeatin riboside (ZR), 50 mg/L asparagine (Asp), 100 mg/L L-pyroglutamic acid (Glu), 400 mg/L Tim, 0.5 mg/L $GA_3$, 2 mg/L glufosinate, 100 mg/L Cef, and 0.1 mg/L indole-3-acetic acid. Pour the mixture into 100-mL autoclaved vials (~30 mL/vial).

8. Rooting medium (RM):

   2.22 g/L Murashige and Skoog (MS) basal medium with vitamins, 0.59 g/L MES, 20 g/L sucrose, 8 g/L agar (pH 5.6, adjusted with NaOH). Autoclave 20 min, cool to about 55 °C, and then add several filter-sterilized solutions to the medium containing 1 mg/L indole-3-butyric acid, 50 mg/L Asp, 100 mg/L Glu, 400 mg/L Tim, and 100 mg/L Cef. Pour the mixture into 250-mL autoclaved vials (~70 mL/vial).

9. Other required lab tools and equipment:

   Mature soybean seeds, NaOCl, 12 N HCl, 250-mL beakers, desiccator, fume hood, super-clean workbench, artificial climate chamber, and greenhouse. 15-mm × 90-mm plastic petri plates, 100-mL and 250-mL autoclaved vials. Reagents for genomic DNA extraction, *Taq* DNA polymerase kit (TaKaRa, DRR001A) for PCR, specific PCR primers for the target sites and QuickStix Strips (EnviroLogix, AS013).

## 3 Methods

### 3.1 Construction of the CRISPR/Cas9 Plasmid

The method described here has been modified for multiplex gene editing in soybean using a robust CRISPR/Cas9 vector system [23, 24]. Taking four targets as an example, we describe in this chapter the protocol of constructing a CRISPR/Cas9 plasmid suitable for soybean genome editing (Fig. 1):

#### 3.1.1 Ligate the Target Adaptors into the pYLsgRNA-AtU3/U6 Plasmids (Fig. 1a)

1. Select sgRNA targeting sequences by using the web tool CRISPR-GE (http://skl.scau.edu.cn/) [25] (*see* **Note 2**), which can show all optional target sequences (20 bp) followed by a PAM (protospacer adjacent motif) and automatically generate the forward and reverse target adaptors according to the selected U3/U6 promoters used to express the sgRNA (*see* **Note 3**). The target adaptors contain the target sequence and overhangs specific for the corresponding nonpalindromic ends of the pYLsgRNA-AtU3/U6 intermediate vectors digested by *Bsa*I (*see* **Note 4**, Table 1 and Fig. 1a).
2. Anneal the target adaptors by mixing 1 μL forward target adaptor (100 μM), 1 μL reverse target adaptor (100 μM), and 98 μL 0.5× TE buffer, followed by incubating the mixture at 90 °C for 30 s and annealing at room temperature.
3. Construct each sgRNA expression cassette separately by preparing the following 10 μL restriction-ligation solution: 1 μL pYLsgRNA-U3/U6 vector (20 ng/μL), 1 μL 10× CutSmart buffer (NEB), 0.5 μL 10× T4 DNA ligase buffer (NEB, containing10 mM ATP), 5 U *Bsa*I-HF (NEB), 20 U T4 DNA ligase (NEB), 0.5 μL annealed target adaptors (**step 2**), and water up to 10 μL.
4. Perform the restriction–ligation reaction according to the thermal cycling procedure without using a heated lid: 37 °C for 30 s, (37 °C for 5 min, 20 °C for 5 min) × 5 cycles (*see* **Note 5**).

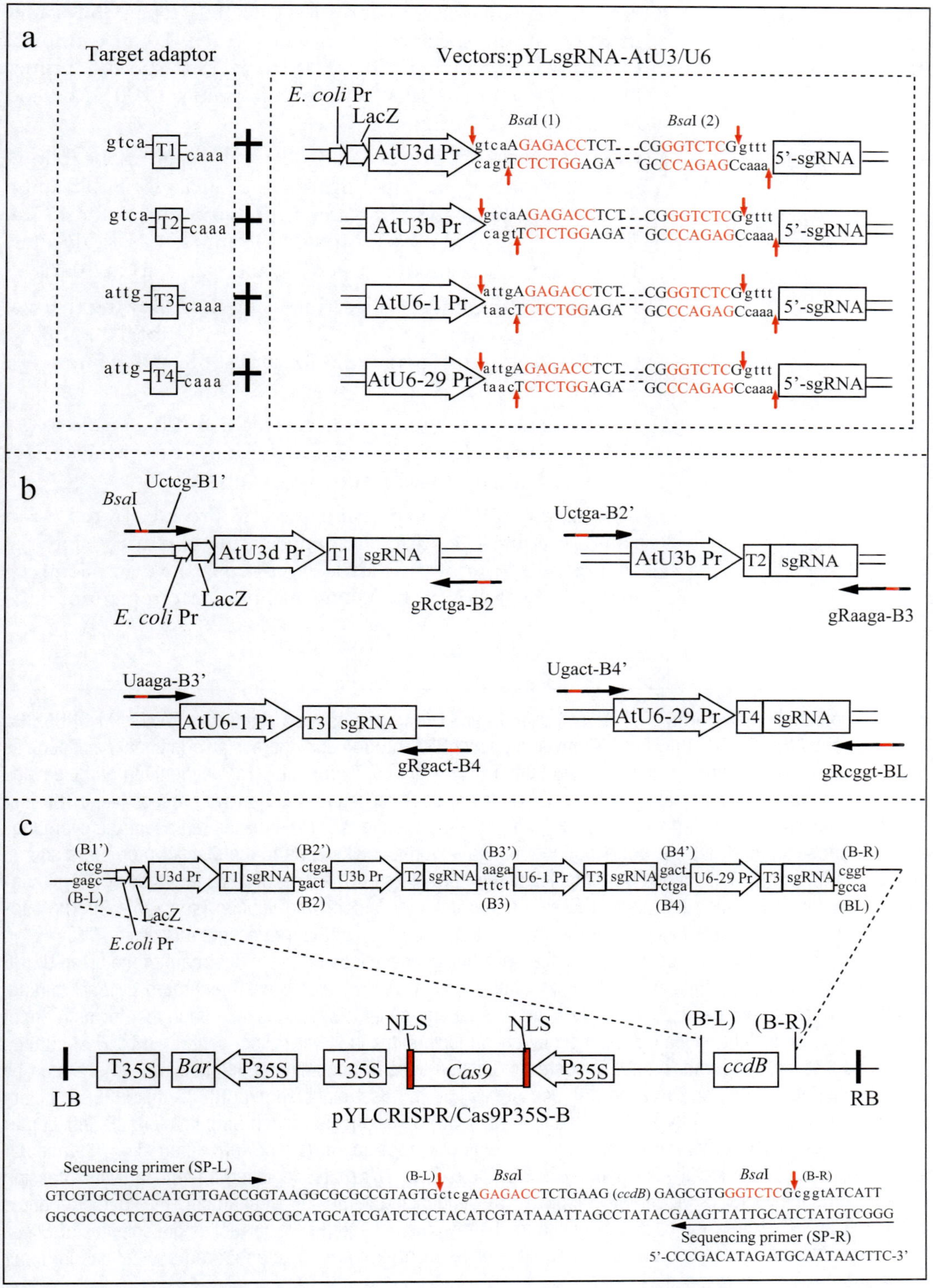


**Fig. 1** Schematic explanation of the construction of CRISPR/Cas9 plasmid. (**a**) Insertion of the annealed target adaptors into the *Bsa*I-digested pYLsgRNA-AtU3/U6 plasmids (major parts of the plasmids are illustrated in

*3.1.2 Clone the sgRNA Expression Cassettes by PCR (Fig. 1b)*

1. Before clone the sgRNA expression cassette, prepare the primer mixture of site-specific primer pairs (each 10 μM/primer) (Table 2) by mixing 10 μL forward site-specific primer (100 μM), 10 μL reverse site-specific primer (100 μM), and 80 μL 0.5× TE buffer.
2. Clone each sgRNA expression cassette by preparing the following 50 μL reaction: 1 μL ligation product (*see* Subheading 3.1.1, **step 4**) as the template, 5 μL 10× buffer for KOD, 5 μL dNTP (2 mM), 2 μL $MgSO_4$ (25 mM), 1 U KOD-Plus, 1.5 μL the primer mixture (**step 1**), and water up to 50 μL.
3. Perform the amplification reaction using the following thermal cycling procedure: 95 °C for 2 min, (95 °C for 10 s, 58 °C for 15 s, 68 °C for 35 s) × 25–35 cycles, 68 °C for 5 min (*see* **Note 6**).
4. Load 3 μL amplification products (**step 3**) onto an agarose gel (1.0%) and run electrophoresis for checking the specificity, size and approximate concentration of the product.
5. Mix all the sgRNA expression cassette products (**step 3**) in roughly equal quantities, and purify the mixture using the PCR product purification kit as described in the manufacturer's instructions, followed by a concentration determination.

**Fig. 1** (continued) linear form). The *Arabidopsis* U3 and U6 promoters are depicted as "AtU3d Pr, AtU3b Pr, AtU6-1 Pr and AtU6-29 Pr". Only in the first vector, pYLsgRNA-AtU3d/LacZ, there is an additional *LacZ* gene as a cloning selection marker. "*E. coli* Pr" represents *Escherichia coli* promoter. The recognition and cleavage sites of *Bsa*I enzyme are depicted as red nucleotides in upper cases and red arrows, respectively. The four target sequences are depicted as T1, T2, T3, and T4. The nucleotides in lower cases represent the overhangs to the target sequences and the correspondent nonpalindromic sticky ends to the distinct promoters and an end to the same sgRNA sequence. (**b**) Amplification of the four sgRNA expression cassettes by PCR with site-specific primers containing the distinct *Bsa*I-cut sites for the Golden Gate ligation (*see* Table 2). The site-specific primers are depicted as black arrows with a small red section that represents a *Bsa*I site. "U####" in the primer names indicates a used forward site-specific sequence located in the upstream of the given U3/U6 promoter, with distinct nonpalindromic sticky ends (nucleotides in lower cases) generated by *Bsa*I cutting, "gR####" in the primer names indicates a used reverse site-specific sequence located in the downstream of the sgRNA, with distinct nonpalindromic sticky ends (nucleotides in lower cases) generated by *Bsa*I cutting, "Bn'" and "Bn" in the primer names indicate the ligation sites of nonpalindromic sticky ends generated by *Bsa*I. (**c**) Insertion of four sgRNA expression cassettes into the *Bsa*I-digested pYLCRISPR/Cas9P35S-B binary vector (major part of the vector is illustrated in linear form) by the single Golden Gate ligation. LB, left-border sequence; RB, right-border sequence; $P_{35S}$, cauliflower mosaic virus *35S* promoter; $T_{35S}$, terminator sequence; NLS, nuclear localization sequence; B-L and B-R, two *Bsa*I sites for inserting four sgRNA expression cassettes into the vector. SP-L and SP-R indicate the sequencing primers used to amplify the ligated products of multiple sgRNA expression cassettes within the binary plasmid and detect correct single colonies of *E. coli* transformants. The recognition and cleavage sites of *Bsa*I enzyme are depicted as red nucleotides in upper cases and red arrows, respectively

*3.1.3 Assemble the sgRNA Expression Cassettes into the pYLCRISPR/Cas9P35S-B Binary Vector by Golden Gate Ligation [26] (Fig. 1c)*

1. Assemble the sgRNA expression cassettes into the CRISPR/Cas9 expression vector by mixing the following 15 μL Golden Gate reaction: 1.5 μL 10 × CutSmart buffer (NEB), 0.75 μL 10× T4 DNA ligase buffer (NEB, containing10 mM ATP), 10 U *Bsa*I-HF (NEB), 35 U T4 DNA ligase (NEB), 60–100 ng pYLCRISPR/Cas9P35S-B, 60–70 ng mixture of sgRNA expression cassettes (*see* Subheading 3.1.2, **step 5**) (*see* **Note** 7), and water up to 15 μL.
2. Perform the Golden Gate reaction using the following thermal cycling procedure: 37 °C for 10 min, (37 °C for 2 min, 10 °C for 3 min, 20 °C for 5 min) × 10–15 cycles, 37 °C for 5 min.

*3.1.4 Introducing the Binary Vector into the* E. coli *and Agrobacterium*

1. Transform *E. coli* competent cells: Thaw a tube of 50 μL *E. coli* competent cells on ice. Then add 15 μL ligation product (*see* Subheading 3.1.3, **step 2**) and mix them by gently pipetting up and down. Incubate the mixture on ice for 30 min, in a 42 °C water bath for 45 s and then immediately on ice for 2 min. Subsequently, add 500 μL of LB liquid medium and incubate the tubes in a shaker at 37 °C for 1 h. Plate 200 μL of the bacterial cells on LB agar plates containing 50 μg/mL kanamycin (for pYLCRISPR/Cas9P35S-B). Incubate the plates in an incubator overnight at 37 °C.
2. Confirm the correct single colonies of *E. coli* transformants by agarose gel analysis and sequencing with the sequencing primers SP-L and SP-R (*see* Table 2 and Fig. 1c).
3. Isolate the CRISPR/Cas9 plasmid from the culture of correct colonies using the plasmid mini kit following the manufacturer's instructions.
4. Transform *Agrobacterium* EHA105 competent cells with this plasmid: Thaw a tube of 50 μL *Agrobacterium* EHA105 competent cells on ice. Then add about 0.5 μg plasmid (**step 3**) and mix them by gently pipetting up and down. Incubate the mixture on ice for 5 min, in liquid nitrogen for 5 min and then in a 37 °C water bath for 5 min. Subsequently, add 1 mL of LB liquid medium and incubate the tubes in a shaker at 28 °C for 2–4 h. Collect the bacteria by centrifuging at 894 × *g* for 3 min, remove most of the supernatant, leave only about 50 μL to resuspend the bacteria and plate them on LB agar plates containing 50 μg/mL kanamycin (for pYLCRISPR/Cas9P35S-B) and 30 μg/mL rifampicin (for EHA105). Incubate at 28 °C for 2–3 days.
5. Confirm the correct single colonies of *Agrobacterium* transformants by agarose gel analysis.

6. Select a single colony and inoculate it in a 10 mL LB liquid medium containing 50 μg/mL kanamycin and 30 μg/mL rifampicin. Shake the tube in a shaker at 28 °C and 200 rpm for 16 h, then preserve the bacteria culture in 50% glycerol at −80 °C.

### 3.2 *Agrobacterium*-Mediated Soybean Transformation and Regeneration of Transgenic Soybean Plants

This *Agrobacterium*-mediated soybean transformation and tissue culture procedures are modified from previously published protocols [27–29] (Fig. 2).

#### 3.2.1 Seed Sterilization by Chlorine Gas

1. Place two layers of mature soybean seeds in 15-mm × 90-mm plastic petri plates (*see* **Note 8**), and then place the plates and a 250 mL beaker containing 100 mL NaOCl in a desiccator in a fume hood.
2. Add 4 mL 12 N HCl to the beaker along the side. Cover the desiccator with a lid immediately.
3. After being sterilized for 16 h, release the chlorine gas, move seeds into a super-clean workbench to blow residual chlorine gas away.
4. Seal the petri plates and store the seeds for use (*see* **Note 9**).

#### 3.2.2 *Agrobacterium* Infection Solution Preparation

1. One day before explant infection, add 400 μL of the thawed bacteria culture (*see* Subheading 3.1.4, **step 6**) to 200 mL LB liquid medium containing 50 mg/L kanamycin and 30 mg/L rifampicin, incubate at 28 °C and 200 rpm until a certain density ($OD_{600}$ = 0.8–1.2).
2. Transfer the culture to several 50 mL centrifuge tubes, collect the bacteria by centrifuging at 894 × *g* for 10 min, resuspend the bacteria in RSM, and then adjust the bacterial density to an $OD_{600}$ of ~0.6.

#### 3.2.3 Explant Preparation

1. Incubate the sterilized seeds (16 seeds/plate) with the hilum facing the GM at 25 °C for 16–24 h under dark until their complete imbibition occurs.
2. Longitudinally dissect them into two equal halves along the hilum with a scalpel.
3. Remove the seed coat and the major part of the hypocotyl leaving only 3–5 mm close to the cotyledonary node.
4. Remove any buds attached to the cotyledonary node, and create 2–3 wounds perpendicular to the hypocotyl at the differentiation site of cotyledonary node (*see* **Note 10**). Half-seeds with a cotyledon and 3–5 mm hypocotyl segment are obtained as explants.

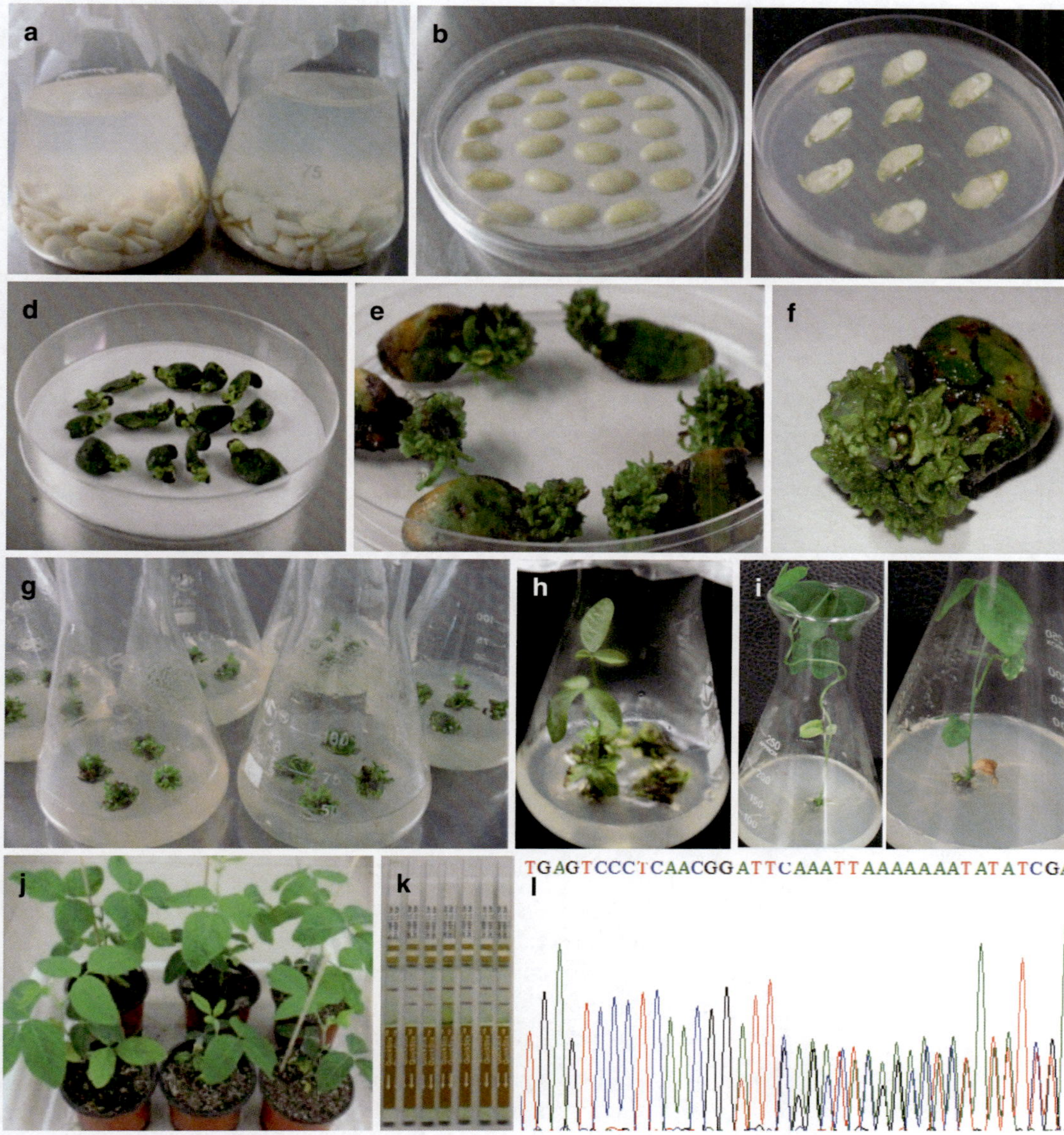


**Fig. 2** The processes of *Agrobacterium*-mediated soybean transformation and regeneration of soybean plantlets from cotyledonary node explants containing Cas9/sgRNA transgenes. (**a**) Infection of the explants with an *Agrobacterium* resuspension. (**b**) Cocultivation. (**c**–**f**) Induction and selection of adventitious shoots. (**g**, **h**) Shoot elongation. (**i**) Rooting. (**j**) Putatively transformed plantlets. (**k**) Strips detection for Bar protein in transformed plantlets. (**l**) Sequencing for detecting the target sites

*3.2.4 Infection and Cocultivation*

1. Infect the explants with *Agrobacterium* resuspension (*see* Subheading 3.2.2, **step 2**) by shaking in a shaker at 28 °C and 200 rpm for 30 min.

2. Pour out the resuspension and then remove the excess liquid from the explant surface by placing the explants on an autoclaved filter paper.
3. Place the explants into the CCM with the flat side facing the filter paper overlaying the CCM.
4. Incubate the explants at 25 °C for 3–5 days under dark.

*3.2.5 Induction and Selection of Adventitious Shoots*

1. After cocultivation, wash the explants with WM and lay them on an autoclaved filter paper to drain the excess liquid (*see* **Note 11**).
2. Cut off the hypocotyls grown during the stage of cocultivation with a scalpel, leaving only 3–5 mm hypocotyls near the cotyledonary node.
3. Obliquely insert the pruned explants into SIM in such a way that the hypocotyl and cotyledonary node are embedded in the SIM and the flat side is toward upside.
4. Incubate the explants under the conditions of 25 °C and 18-h light/6-h dark photoperiod for 2 weeks.
5. After 2 weeks, subculture the explants 1–2 times (at 2-week interval). At each time of subculturing, cut the very long shoots out of preformed shoots, abandon the dead explants and transfer the well-developed explants to fresh SIM and incubate them at the same conditions mentioned above.

*3.2.6 Shoot Elongation and Rooting*

1. Excise the cotyledons from the explants and then insert the remaining shoot clusters into the SEM for 2 weeks.
2. After 2 weeks, subculture the shoot clusters 3–4 times (at 2-week interval). At each time of subculturing, remove the outer browning tissues from the shoots and expose the inner green tissues to selective agents and nutrients. During this process, some of shoots survive and elongate.
3. When the shoots reach about 5 cm, cut them off from the base of shoot clusters and incubate them in RM without further selection. The culture conditions of shoot elongation and rooting are the same as that used for shoot induction.

*3.2.7 Plantlet Transplanting*

1. When the roots of the plantlets are sufficiently developed, wash off the medium from the roots, transplant them into pots (volume ratio of soil–vermiculite is 1:2) in the greenhouse, and cover them with a humid-keeping film (*see* **Note 12**).
2. After 3 days, uncover the film. The culture temperature of the greenhouse is 25 °C, and the photoperiod is 16-h light/8-h dark (*see* **Note 13**).

*3.2.8 Detection of Mutants*

When the plantlets vigorously and healthily grow, mutants can be detected by identifying their resistance to glufosinate, followed by determining whether the target genes are edited or not.

1. Detect the positive transformation events by using the Quick-Stix Strips according to the manufacturer's instructions. If two bands appear in the strip, it indicates the plant is positive.
2. Extract the genomic DNAs of the positive transformed plants as the templates for amplifying the target genes with the specific primer pairs located at 200–300 bp downstream and upstream of the target sites, respectively.
3. The PCR products can be subjected to a Sanger sequencing, and sequence analysis can be performed using the web tool CRISPR-GE (http://skl.scau.edu.cn/) [25].

## 4 Notes

1. These vectors [23, 24] were kindly provided by Professor Yaoguang Liu (South China Agricultural University). They can also be obtained from Addgene (http://www.addgene.org).
2. The default version of the target (reference) genome of soybean loaded by the web tool CRISPR-GE (http://skl.scau.edu.cn/) [25] is Glyma 1 version. Thus, the names of the target genes downloaded from Phytozome (Glyma 2 version gene model names) need to be converted into the Glyma 1 names of the correspondent genes.
3. It is recommended that two or more targets should be designed for a gene so that to have high chance to receive edited mutants. Especially, when the two target sites are very close to each other, large fragments may be deleted due to simultaneous targeting. In addition, it is desirable to have restriction enzyme (RE) sites in the target sequences, which may facilitate the identification of mutants using a PCR/RE assay.
4. In these vectors (pYLsgRNA-AtU3d/LacZ, pYLsgRNA-AtU3b, pYLsgRNA-AtU6-1, pYLsgRNA-AtU6-29), expression of sgRNA is driven by the *Arabidopsis thaliana* U3 and U6 promotes. The AtU3 and AtU6 promoters have a defined A and G as the first nucleotide of their transcript, respectively. Therefore, if the 5′ of the sgRNA 20-bp target sequences begin with A (AtU3 promoter used) and G (AtU6 promoter used), the overhangs of target adaptors can occupy this A and G, respectively; otherwise, the extra A and G must be added in the overhangs of target adaptors. In Table 1, the nucleotides in upper cases represent the target sequences, and the nucleotides

in lower cases represent overhangs that are used for matching with the nonpalindromic sticky ends of *Bsa*I-digested pYLsgRNA-AtU3/U6 intermediate vectors.

5. If a T4 DNA ligase buffer does not contain ATP, then ATP needs to be added separately to the final concentration of 0.5–1 mM. Make sure that the *Bsa*I-HF is not inactivated. Avoid overdigestion/ligation due to excess use of *Bsa*I/ligase, or longer incubation time.
6. It is best to use a DNA polymerase like KOD-Plus (TOYOBO) that has a higher fidelity. The extension speed of KOD-Plus is 1 min/kb, and the PCR product fragment sizes of LacZ-AtU3d-sgRNA, AtU3b-sgRNA, AtU6-1-sgRNA, and AtU6-29-sgRNA are 486 bp, 507 bp, 487 bp, and 502 bp, respectively [24]. Set a reasonable extension time according to the extension speed of the DNA polymerase.
7. The fragment size of pYLCRISPR/Cas9P35S-B is about 15 kb and the *Bsa*I-digested fragment sizes of LacZ-AtU3d-sgRNA, AtU3b-sgRNA, AtU6-1-sgRNA, and AtU6-29-sgRNA are 456 bp, 477 bp, 457 bp, and 472 bp, respectively [24]. Make sure that the molar ratio of vector to each inserted sgRNA expression cassette is 1:(4–6).
8. Healthy and vigorous seeds are the key to success of transformation experiments.
9. As a strong acid and chlorine gas are involved, proper personal protective tools should be used. Extended sterilization time (more than 24 h) may lead to lower tissue regeneration rate.
10. To avoid affecting differentiation, the wound cut at the effective differentiation site of cotyledonary node should not be too deep.
11. To avoid overgrowth of *Agrobacterium*, the step of rinsing the cotyledonary node explants 3–5 times with WM is recommended before inserting them into SIM.
12. A humid-keeping film is helpful for keeping moisture and protecting plantlets from environmental impact. However, the plantlets should not be covered too tightly to keep them aerated.
13. A long day-length is recommended to prolong vegetative growth period of the plantlets prior to flowering.

## Acknowledgments

The authors are thankful to Professor Yaoguang Liu from South China Agricultural University for providing the vectors pYLsgRNA-AtU3d/LacZ, pYLsgRNA-AtU3b, pYLsgRNA-AtU6-1, pYLsgRNA-AtU6-29, and pYLCRISPR/Cas9P35S-B.

## References


1. Hartman GL, West ED, Herman TK (2011) Crops that feed the world 2. Soybean-worldwide production, use, and constraints caused by pathogens and pests. Food Security 3(1):5–17
2. Homrich MS, Wiebke-Strohm B, Weber RLM, Bodanese-Zanettini MH (2012) Soybean genetic transformation: a valuable tool for the functional study of genes and the production of agronomically improved plants. Genet Mol Biol 35(4):998–1010
3. Li S, Cong Y, Liu Y, Wang T, Shuai Q, Chen N, Gai J, Li Y (2017) Optimization of *Agrobacterium*-mediated transformation in soybean. Front Plant Sci 8:246
4. Gregory P (2012) Gene overexpression: uses, mechanisms, and interpretation. Genetics 190 (3):841–854
5. Krysan PJ, Young JC, Sussman MR (1999) T-DNA as an insertional mutagen in *Arabidopsis*. Plant Cell 11(12):2283–2290
6. Kusaba M (2004) RNA interference in crop plants. Curr Opin Biotechnol 15(2):139–143
7. Ma X, Zhu Q, Chen Y, Liu Y-G (2016) CRISPR/Cas9 platforms for genome editing in plants: developments and applications. Mol Plant 9:961–974
8. Abdelrahman M, Al-Sadi AM, Pour-Aboughadareh A, Burritt DJ, Tran LSP (2018) Genome editing using CRISPR/Cas9–targeted mutagenesis: an opportunity for yield improvements of crop plants grown under environmental stresses. Plant physiology. Biochemistry 131:31–36
9. Bao A, Burritt DJ, Chen H, Zhou X, Cao D, Tran LSP (2019) The CRISPR/Cas9 system and its applications in crop genome editing. Crit Rev Biotechnol 39(3):321–336
10. Gao X, Chen J, Dai X, Zhang D, Zhao Y (2016) An effective strategy for reliably isolating heritable and Cas9-free *Arabidopsis* mutants generated by CRISPR/Cas9-mediated genome editing. Plant Physiol 171 (3):1794
11. Woo JW, Kim J, SIl K, Corvalán C, Cho SW, Kim H, Kim SG, Kim ST, Choe S, Kim JS (2015) DNA-free genome editing in plants with preassembled CRISPR-Cas9 ribonucleoproteins. Nat Biotechnol 33(11):1162–1164
12. Liang Z, Chen K, Zhang Y, Liu J, Yin K, Qiu JL, Gao C (2018) Genome editing of bread wheat using biolistic delivery of CRISPR/Cas9 *in vitro* transcripts or ribonucleoproteins. Nat Protoc 13(3):413–430
13. Malnoy M, Viola R, Jung MH, Koo OJ, Kim S, Kim JS, Velasco R, Kanchiswamy CN (2016) DNA-free genetically edited grapevine and apple protoplast using CRISPR/Cas9 ribonucleoproteins. Front Plant Sci 7(e188):1904
14. Cai Y, Chen L, Liu X, Sun S, Wu C, Jiang B, Han T, Hou W (2015) CRISPR/Cas9-mediated genome editing in soybean hairy roots. PLoS One 10(8):e0136064
15. Jacobs TB, Lafayette PR, Schmitz RJ, Parrott WA (2015) Targeted genome modifications in soybean with CRISPR/Cas9. BMC Biotechnol 15:16
16. Michno JM, Wang X, Liu J, Curtin SJ, Kono TJ, Stupar RM (2015) CRISPR/Cas mutagenesis of soybean and Medicago truncatula using a new web-tool and a modified Cas9 enzyme. GM Crops Food 6(4):243–252
17. Sun X, Hu Z, Chen R, Jiang Q, Song G, Zhang H, Xi Y (2015) Targeted mutagenesis in soybean using the CRISPR-Cas9 system. Sci Rep 5:10342
18. Li Z, Liu Z, Xing A, Moon BP, Koellhoffer JP, Huang L, Timothy Ward R, Clifton E, Carl Falco S, Mark Cigan A (2015) Cas9-guide RNA directed genome editing in soybean. Plant Physiol 169(2):960–970
19. Du H, Zeng X, Zhao M, Cui X, Wang Q, Yang H, Cheng H, Yu D (2016) Efficient targeted mutagenesis in soybean by TALENs and CRISPR/Cas9. J Biotechnol 217:90–97
20. Kanazashi Y, Hirose A, Takahashi I, Mikami M, Endo M, Hirose S, Toki S, Kaga A, Naito K, Ishimoto M (2018) Simultaneous site-directed mutagenesis of duplicated loci in soybean using a single guide RNA. Plant Cell Rep 37 (3):553–563
21. Cai Y, Chen L, Liu X, Guo C, Sun S, Wu C, Jiang B, Han T, Hou W (2018) CRISPR/Cas9-mediated targeted mutagenesis of *GmFT2a* delays flowering time in soya bean. Plant Biotechnol J 16(1):176–185
22. Cai Y, Chen L, Sun S, Wu C, Yao W, Jiang B, Han T, Hou W (2018) CRISPR/Cas9-mediated deletion of large genomic fragments in soybean. Int J Mol Sci 19:3835. https://doi.org/10.3390/ijms19123835
23. Ma X, Liu Y-G (2016) CRISPR/Cas9-based multiplex genome editing in monocot and dicot plants. Curr Protoc Mol Biol 115:31.6.1–31.6.21
24. Ma X, Zhang Q, Zhu Q, Wei L, Yan C, Rong Q, Wang B, Yang Z, Li H, Lin Y (2015) A robust CRISPR/Cas9 system for convenient, high-efficiency multiplex genome

editing in monocot and dicot plants. Mol Plant 8(8):1274–1284

25. Xie X, Ma X, Zhu Q, Zeng D, Li G, Liu Y-G (2017) CRISPR-GE: a convenient software toolkit for CRISPR-based genome editing. Mol Plant 10(9):1246–1249
26. Carola E, Romy K, Sylvestre M (2008) A one pot, one step, precision cloning method with high throughput capability. PLoS One 3(11): e3647
27. Chen L, Cai Y, Liu X, Yao W, Guo C, Sun S, Wu C, Jiang B, Han T, Hou W (2018) Improvement of soybean *Agrobacterium*-mediated transformation efficiency by adding glutamine and asparagine into the culture media. Int J Mol Sci 19(10):E3039. https://doi.org/10.3390/ijms19103039
28. Paz MM, Martinez JC, Kalvig AB, Fonger TM, Wang K (2006) Improved cotyledonary node method using an alternative explant derived from mature seed for efficient *Agrobacterium*-mediated soybean transformation. Plant Cell Rep 25(3):248–248
29. Paz MM, Shou H, Guo Z, Zhang Z, Banerjee A, Wang K (2004) Assessment of conditions affecting Agrobacteriummediated soybean transformation using the cotyledonary node explant. Euphytica 136:167–179

# Chapter 20

# Genome-Wide Mining of Disease Resistance Gene Analogs Using Conserved Domains

**Soodeh Tirnaz, Yueqi Zhang, and Jacqueline Batley**

## Abstract

The production of legume crop species is severely affected by disease, imposing a significant yield loss annually worldwide. Plant resistance gene analogs (RGAs) play specific roles in plant resistance responses, and their identification and subsequent application in breeding programs help to reduce this yield loss. RGAs contain conserved domains and motifs, which can be used for their identification and classification. Nucleotide-binding site-leucine-rich repeat (NLR), receptor like kinase (RLK), and receptor like protein (RLP) genes are the main types of RGAs. Computational identification and characterization of RGAs has been performed successfully among different plant species. Here, we explain the computational workflow for genome-wide RGA identification in legumes.

**Key words** Nucleotide-binding site-leucine-rich repeat (NLR), Receptor like kinase (RLK), Receptor like protein (RLP), Disease resistance, Domain structure

## 1 Introduction

Plant resistance gene analogs (RGAs), playing important roles in plant disease resistance, consist of genes with conserved domains and motifs [1]. There are three main classes of RGAs: NLRs, RLKs, and membrane-associated RLPs [1, 2]. Most resistance proteins identified to date belong to the NLR family, typically containing an N-terminal, an NBS-ARC domain APAF-1 R protein CED-4 domain (ARC), and a C-terminal LRR domain [1, 3–7]. This family can be further divided into two subtypes: TNL and nTNL, depending on the presence of a Toll/Interleukin-1 Receptor (TIR) domain at the N-terminus. Most nTNL, with a coiled-coil (CC) domain at the N-terminus, are termed CNL, and a small proportion with a Resistance to Powdery Mildew 8 (RPW8) domain at the N-terminus are called RNL. Another large resistance protein family is pattern recognition receptors (PRRs) including RLPs and RLKs [1, 8–10]. The two types of receptors have extra-cellular lysin motif (LysM) and LRR domains at the N-terminus for

Mukesh Jain and Rohini Garg (eds.), *Legume Genomics: Methods and Protocols*, Methods in Molecular Biology, vol. 2107, https://doi.org/10.1007/978-1-0716-0235-5_20, 

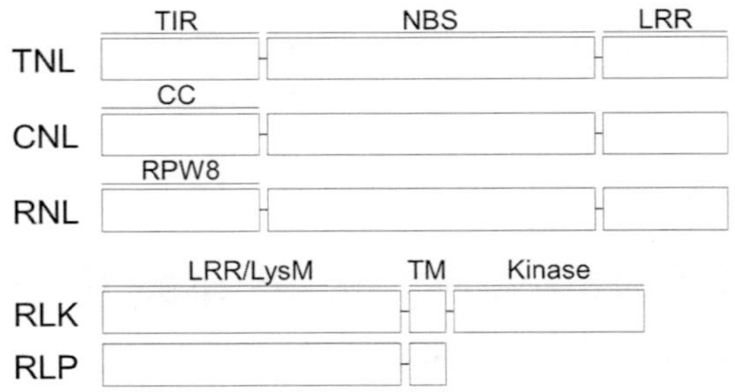


**Fig. 1** Schematic representation of domain and motif structures of four main classes of RGAs. *TIR* Toll/interleukin-1 receptor, *NBS* nucleotide-binding site, *CC* coiled-coil, *RPW8* Powdery Mildew Resistance Protein, *LysM* lysin motif, *TM* transmembrane, *LRR* leucine-rich repeats

the perception of the microbial patterns, and a transmembrane domain (TM), but only RLKs contain a cytoplasmic kinase domain [1]. PRRs are also important for regulating plant development [11]. RGAs are identified and classified by typical domains and motifs including CC and TIR, NB-ARC, LRR, TM, STTK, and LysM which can be predicted by a range of bioinformatics tools [4, 12, 13] (Fig. 1).

Legumes are threatened by a range of bacterial, viral, fungal, and oomycete pathogens and the plants employ mainly NLRs for resistance [14]. Some major RGAs conferring resistance to a range of pest and pathogens have been mapped and cloned in important legume species, as summarized in Table 1. The advances in genome sequencing and gene annotation help to further decipher plant–pathogen interactions and identification of resistance components in legumes [15]. Mostly TNL, CNL, RNL, atypical NLRs, and some RLKs have been identified in the genomes of pigeon pea, chickpea, soybean, bird's foot trefoil, barrel clover, and common bean using different methods (Table 2). Studies of the evolutionary relationships, structural and functional divergence of certain groups of RGAs, such as CNL in soybean and common bean, and TNL in soybean, have potential implications in crop improvements [14, 16, 17].

## 2 Materials

### *2.1 Genome Resources*

Reference genomes for a range of important legume species including cultivated and wild soybean, common bean, peanut, pigeon pea, chickpea, broad bean, mung bean, lentil, black-eyed pea, adzuki bean, narrow-leafed lupin, barrel clover, red clove, subterranean clover, and bird's foot trefoil are available. Of note, the availability and the quality of assembly and annotation of the sequenced genomes highly affect the prediction of resistance genes (*see* **Notes 1** and **2**).

**Table 1**
**Major RGAs conferring resistance to a range of pathogens that are mapped or cloned in important legumes**

| Legume species | Biotic stresses | RGAs mapped and cloned | References |
|---|---|---|---|
| *Lens culinaris* | Ascochyta blight (*Ascochyta lentis*) | *ral2* and *AbR 1* (cloned) | [34, 35] |
| *Cicer arietinum* | Fusarium wilt (*Fusarium oxysporum f. sp. ciceri*) | | [36] |
| *Phaseolus vulgaris* | Common bean rust (*Uromyces appendiculatus*) | *Ur-1* to *Ur-14* | [37] |
| *P. vulgaris* | Angular leaf spot (*Pseudocercospora griseola*) | *Phg* | [38] |
| *P. vulgaris* | Angular leaf spot (*P. griseola*) | *Phg-2* (cloned) | [39] |
| *Vicia faba* | Ascochyta blight (*A. fabae*) | *Af1-6* (cloned) | [40] |
| *Pisum sativum* | Fusarium wilt (*F. oxysporum f. sp. pisi*) | *Fw* | [41–43] |
| *P. sativum* | Powdery mildew (*Erysiphe polygoni*) | *Er* | [41, 44] |
| *P. sativum* | Pea common mosaic virus | *Mo* | [41] |
| *P. sativum* | Ascochyta blight (*Mycosphaerella pinodes*) | *Rmp1* to *Rmp4* | [45] |
| *Vigna unguiculata* | Fusarium wilt (*F. oxysporum f. sp. tracheifilum*) | *Fot3-1* | [46] |
| *V. unguiculata* | Cowpea rust (*U. vignae*) | *Ruv2*, *Ruv3* | [47] |
| *Glycine max* | Soybean aphid (*Aphis glycines* Matsumura) | *Rag1* to *Rag6* | [48–54] |
| *G. max* | Soybean mosaic virus | | [55, 56] |
| *G. max* | Phytophthora root rot (*Phytophthora sojae*) | *Rps1* | [57] |
| *G. max* | Soybean rust (*Phakopsora pachyrhizi*) | *Rpp1-5* | [58] |
| *G. max* | Phytophthora blight (*P. sojae*) | *Rps1-k* (cloned) | [59, 60] |
| *G. max* | Bacterial blight (*Pseudomonas syringae*) | *Rpg1-b* and *Rpg1-r* (cloned) | [61] |

## 3 Methods

RGA identification includes two main steps (Fig. 2). The first step is the detection of RGA-related domains (i.e., NB-ARC, LRR, STTK, LysM, CC, and TM), mainly based on the sequence similarity searches against different databases. The second step is classification of candidate genes based on their domain compositions. These two steps are explained in more detail, as follows:

**Table 2**
**Reported numbers of identified resistance genes analogs in legumes**

| Species | Tools | TNL | CNL | RNL | Atypical NLR | Total NLR | Reference | RLK | RLP | Reference |
|---|---|---|---|---|---|---|---|---|---|---|
| *Medicago truncatula* | HMMER/MEME | 86 | | | 247 | 333 | [20] | | | |
| *M. truncatula* | BLAST/Pfam | | | | | 764 | [62] | | | |
| *P. vulgaris* | HMMER/Pfam/COILS | 18 | 33 | | 272 | | [63] | | | |
| *G. max* | Pfam | 116 | 20 | | 183 | 319 | [64] | 467 | | [65] |
| *G. max* | HMMER/ | 117 | | | 26 | 143 | [14] | | | |
| *P. vulgaris* | InterProScan/ Geneious/MEME | 77 | | | 20 | 97 | | | | |
| *Lotus japonicus* | BLAST/InterPro | | | | | 229 | [66] | | | |
| *L. japonicus* | HMMER/BLAST/Pfam/COILS | 8 | 9 | | 66 | 158 | [67] | | | |
| *Cajanus cajan* | BLAST/HMMER/Pfam/ | 112 | 167 | 10 | | 289 | [68] | | | |
| *G. max* | SMART/ | 178 | 269 | 18 | | 465 | | | | |
| *M. truncatula* | MEME/COILS | 275 | 279 | 17 | | 571 | | | | |
| *P. vulgaris* | | 103 | 224 | 10 | | 337 | | | | |
| *C. arietinum* | BLAST | | | | | 187 | [69] | | | |
| *C. cajan* | BLAST/HMMER/Pfam/ | 47 | 37 | | 731 | 815 | [70] | | | |
| *C. arietinum* | SMART/ | 6 | 19 | | 202 | 227 | | | | |
| *G. max* | MEME/COILS | 67 | 68 | | 609 | 744 | | | | |
| *G. soja* | | 49 | 47 | | 856 | 952 | | | | |
| *L. japonicus* | | 16 | 11 | | 243 | 270 | | | | |
| *M. truncatula* | | 44 | 49 | | 677 | 770 | | | | |
| *P. vulgaris* | | 1 | 31 | | 327 | 359 | | | | |
| *P. vulgaris* | HMMER/Pfam/SMART | 98 | | | 803 | 901 | [31] | | | |
| *G. max* | | 254 | | | 1506 | 1760 | | | | |
| *M. truncatula* | | 361 | | | 2031 | 2392 | | | | |

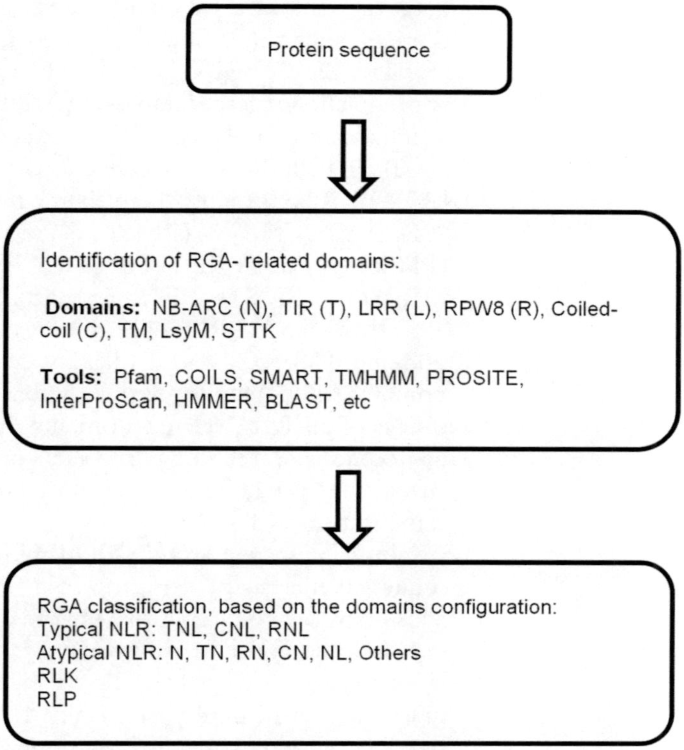


**Fig. 2** The workflow of resistance genes analogs identification in the legumes

### 3.1 Identification of RGA-Related Domains

The first step of genome-wide RGA identification is detection of RGA-related domains. The most common way to do this is searching for sequence similar to the sequence query in one or more databases. Basic Local Alignment Search Tool (BLAST) [18] and Hidden Markov Model (HMMER) [19] are the most common tools for similarity searches, and they have been commonly used for RGA identification in legumes (Table 2).

Similarity searches by HMMER and BLAST tools are performed by using the consensus sequence of specific domains, or genes, as a query and searching for similarity across the whole genome. For example, in many legume studies, listed in Table 2, the consensus sequence of NB-ARC domain (PF00931) is used as a query; however, Ameline-Torregrosa et al. (2008) used the consensus sequence of a complete TNL and CNL genes as a query for their identification in *Medicago truncatula* [20].

For similarity searches, the whole genome can also be used as a query to search for sequence similar to the sequence of RGA-related domains across different databases. These databases mainly use typical BLAST or HMM algorithms for searching similarity and domain detection. There are many domain annotation databases that are frequently used for detection of RGA-related domains including, but not limited to, Pfam [21], SMART [22], COILS [23], PRINTS [24], PROSITE [25], and ProDom [26]. Each of these databases has different content which makes them specific for classification of specific domains.

Pfam includes a wide collection of protein domains and families [27]. In RGA mining, Pfam is mainly used for detection of TIR (PF01582), TIR_2 (PF13676), RPW8 (PF05659), NB-ARC (PF00931), and LRR (PF00560, PF07723, PF07725, PF12799, PF13306, PF13516, PF13855, and PF14580) domains. However, TMHMM and PROSITE are available for detection of TM domains [28, 29] and COILS is available for detection of CC domains [23]. As there are no databases that can cover the identification of all RGA-related domains, it is necessary that the whole proteome is searched against each of these databases individually and at the end all data needs to be pooled for further analysis and classification.

InterProscan, a protein domain identifier, facilitates the protein sequence analysis by using up to 14 databases in one application. This helps to avoid repetitive information collected from individual searches and consequently making it less time consuming and less labor-intensive to analyze the data [30]. Moreover, InterProScan gives the user flexibility to use all 14 databases at once, or only one or some of them, and this allows users to optimize their searches by selecting the databases that specialize for the domain of interest [30]. Among all 14 database, Pfam, SMART and Gene3 show the best accuracy and performance for detection of RGA-related domains [12]. InterProscan has been frequently used in identification of RGAs in different plant species including legumes (Table 2). Of note, all the aforementioned tools are freely available and can be operated on command-line or on the user-friendly web interface, which facilitate their application.

Recently there have been many studies that also look at the integrated domains (ID) of RGAs (i.e., noncanonical domains), and it has been suggested that they are involved with the recognition of the pathogen. Many NLR-ID have been reported in legumes [31]. These domains can be detected using the Pfam database [31].

### 3.2 Classification of RGA

The second step of the RGA identification workflow, after detection of RGA-related domains, is analysis of their composition and final classification of genes. The classification is mostly done manually using custom scripts and is based on the domain configuration. The classification of NLR genes depends on the presence of NB-ARC domains, genes without NBS domains will be excluded from further classification. The genes carrying NBS domains will be classified based on the presence of the TIR, CC or RPW8 domains at the N-terminal and LRR domains at the C-terminal into TNL, CNL,

RNL, CN, TN, RN, N, NL, and Others categories. The Others category includes genes that do not fit into any of classes, such as genes with more than one domain in their N-terminus, for example genes containing both TIR and CC domains, and genes with disordered domains, such as genes with a TIR domain at the C-terminus (NLT).

The classification of RLK and RLP genes depends on the presence of a TM domain, genes without a TM domain will be excluded from further classification. Both RLKs and RLPs have LRR or LysM at their N-terminus. However the RLKs differ from the RLPs in having a STTK domain at their C-terminus. RLKs and RLPs can be further classified according to their extracellular N-terminal domain. Although more than twenty subclasses of RLK have been reported in *Arabidopsis thaliana*, LRR-RLK/RLP and LysM-RLK/RLP are the most common subclasses [32].

Genome-wide RGA identification among legume species, have been performed by application of different tools (Table 2) with separate domain detection and classification steps. These all together complicate downstream comparative analysis of RGAs (*see* **Notes 3** and **4**). Recently, Lie et al. (2016) introduced the RGAugury pipeline, which combined both RGA domain detection and classification steps into one application. They examined the pipeline across 50 sequenced plants genomes where the results were comparable with the case studies of the candidate genomes. RGAugury suits large-scale genome-wide RGA identification for comparative analysis within and between plants family with many members. RGAugury is capable of identification of RLK, RLP, TM-CC, and the 2 main classes of typical NLRs (i.e., TNL, CNL) and their atypical forms (i.e., TN, N, CN, TX (X represents unknown domain), Others). However, RNLs cannot be identified using RGAugury, and their classification needs to be performed manually using custom scripts. RGAugury is available in both command-line and web interface versions [12].

## 4 Notes

There are a number of factors and challenges that limit RGA identification, including but not limited to the following:

1. Genome-wide RGA prediction relies on the availability of the sequenced genomes. More than fifteen legume species have been sequenced, but still there are many species that have not been sequenced, and this can limit the comparative and evolutionary studies of RGA in legumes. However, with the decreasing trend of whole genome sequencing cost it is expected the genome sequence of many additional species will be available soon.

2. Apart from genome availability, the quality of assembly and annotation of already sequenced genomes can significantly affect the efficiency and accuracy of RGA identification. Bayer et al. (2018) show the prediction of RGAs is strongly affected by the repeat masking, in species from the Brassicaceae family. They reported that using different approaches of repeat masking causes significant differences of the RGA content between different genome assemblies of the same cultivars of *Brassica napus* and *B. rapa*. They also show many of the repeat sequences, such as transposons, that are stored in public databases and used for repeat masking contain some of the RGA-related domains, mainly TIR, CC, and NB-ARC. This means some of the true RGAs might unintentionally be categorized as repeats when using strong masking approaches, or some of the repeats might wrongly categorized as RGAs, when using less strong masking approaches [33]. Therefore, it is crucial to reevaluate repeat masking approaches in RGA studies.
3. Comparing the reported numbers of RGAs for the same legume species in different studies shows large differences (Table 2). The observed differences can be as a result of using different approaches and parameters for RGA identification. These differences make it almost impossible to compare their results and is one of the main barriers of mapping and cloning novel RGA.
4. As has been shown in Table 1, not all classes of RGAs have been identified in all studies and in particular there are not many studies in legume species that identify RLKs and RLPs. It is partially because researchers will study specific class(es) of RGAs based on their research interest and expertise. But it also can be as a result of the efficiency of available tools. Before the introduction of the RGAugury pipeline [12] there was not such a tool to identify all classes of RGA all together. However, with the rapid advances in optimization/introduction of current/novel bioinformatics tools it is expected the new tools for genome wide RGA mining will be available soon to facilitate their identification even more. All these advances will increase the number of RGA studies and will give the researchers the opportunity to look for new and different aspects of RGAs and hopefully all together ease their application in breeding programs toward less yield loss.

## References


1. Sekhwal MK, Li P, Lam I, Wang X, Cloutier S, You FM (2015) Disease resistance gene analogs (RGAs) in plants. Int J Mol Sci 16 (8):19248–19290
2. Zipfel C (2014) Plant pattern-recognition receptors. Trends Immunol 35(7):345–351
3. McHale L, Tan X, Koehl P, Michelmore RW (2006) Plant NBS-LRR proteins: adaptable guards. Genome Biol 7(4):212
4. Meyers BC, Dickerman AW, Michelmore RW, Sivaramakrishnan S, Sobral BW, Young ND (1999) Plant disease resistance genes encode members of an ancient and diverse protein family within the nucleotide-binding superfamily. Plant J 20(3):317–332
5. Baumgarten A, Cannon S, Spangler R, May G (2003) Genome-level evolution of resistance genes in *Arabidopsis thaliana* (author abstract). Genetics 165(1):309
6. Jeffery LD, Jonathan DGJ (2001) Plant pathogens and integrated defence responses to infection. Nature 411(6839):826. https://doi.org/10.1038/35081161
7. Cannon SB, Zhu H, Baumgarten AM, Spangler R, May G, Cook DR, Young ND (2002) Diversity, distribution, and ancient taxonomic relationships within the TIR and non-TIR NBS-LRR resistance gene subfamilies. J Mol Evol 54(4):548–562. https://doi.org/10.1007/s0023901-0057-2
8. Walker JC (1994) Structure and function of the receptor-like protein kinases of higher plants. Plant Mol Biol 26(5):1599–1609
9. Kruijt M, De MJD K, De PJGM W (2005) Receptor-like proteins involved in plant disease resistance, vol 6. Blackwell Science Ltd, Oxford, UK. https://doi.org/10.1111/j.1364-3703.2004.00264.x
10. Shiu S-H, Bleecker AB (2003) Expansion of the receptor-like kinase/Pelle gene family and receptor-like proteins in Arabidopsis. Plant Physiol 132(2):530–543
11. Morris ER, Walker JC (2003) Receptor-like protein kinases: the keys to response. Curr Opin Plant Biol 6(4):339–342
12. Li P, Quan X, Jia G, Xiao J, Cloutier S, You FM (2016) RGAugury: a pipeline for genome-wide prediction of resistance gene analogs (RGAs) in plants. BMC Genomics 17(1):852
13. Wan H, Yuan W, Ye Q, Wang R, Ruan M, Li Z, Zhou G, Yao Z, Zhao J, Liu S (2012) Analysis of TIR-and non-TIR-NBS-LRR disease resistance gene analogous in pepper: characterization, genetic variation, functional divergence and expression patterns. BMC Genomics 13 (1):502
14. Neupane S, Ma Q, Mathew FM, Varenhorst AJ, Andersen EJ, Nepal MP (2018) Evolutionary divergence of TNL disease-resistant proteins in soybean (*Glycine max*) and common bean (*Phaseolus vulgaris*). Biochem Genet 56 (4):397–422
15. Rubiales D, Fondevilla S, Chen W, Gentzbittel L, Higgins TJV, Castillejo MA, Singh KB, Rispail N (2015) Achievements and challenges in legume breeding for pest and disease resistance. Crit Rev Plant Sci 34 (1–3):195–236
16. Nepal MP, Benson BV (2015) CNL disease resistance genes in soybean and their evolutionary divergence. Evol Bioinforma 11:49–63. https://doi.org/10.4137/ebo.S21782
17. Benson BV (2014) Disease resistance genes and their evolutionary history in six plant species. Dissertation, South Dakota State University
18. Altschul SF, Madden TL, Schäffer AA, Zhang J, Zhang Z, Miller W, Lipman DJ (1997) Gapped BLAST and PSI-BLAST: a new generation of protein database search programs. Nucleic Acids Res 25(17):3389–3402
19. Finn RD, Clements J, Eddy SR (2011) HMMER web server: interactive sequence similarity searching. Nucleic Acids Res 39 (suppl_2):W29–W37
20. Ameline-Torregrosa C, Wang B, O'Bleness M (2008) Identification and characterization of nucleotide-binding site-leucine-rich repeat genes in the model plant *Medicago truncatula*. Plant Physiol 146:5–21. https://doi.org/10.1104/pp.107.104588
21. Bateman A, Birney E, Cerruti L, Durbin R, Etwiller L, Eddy SR, Griffiths-Jones S, Howe KL, Marshall M, Sonnhammer EL (2002) The Pfam protein families database. Nucleic Acids Res 30(1):276–280
22. Schultz J, Copley RR, Doerks T, Ponting CP, Bork P (2000) SMART: a web-based tool for the study of genetically mobile domains. Nucleic Acids Res 28(1):231–234
23. Lupas A, Van Dyke M, Stock J (1991) Predicting coiled coils from protein sequences. Science 252:1162–1164
24. Attwood TK, Croning MD, Flower DR, Lewis A, Mabey J, Scordis P, Selley J, Wright W (2000) PRINTS-S: the database formerly known as PRINTS. Nucleic Acids Res 28 (1):225–227
25. Hofmann K, Bucher P, Falquet L, Bairoch A (1999) The PROSITE database, its status in 1999. Nucleic Acids Res 27(1):215–219

26. Corpet F, Servant F, Gouzy J, Kahn D (2000) ProDom and ProDom-CG: tools for protein domain analysis and whole genome comparisons. Nucleic Acids Res 28(1):267–269
27. Bateman A, Coin L, Durbin R, Finn RD, Hollich V, Griffiths-Jones S, Khanna A, Marshall M, Moxon S, Sonnhammer EL (2004) The Pfam protein families database. Nucleic Acids Res 32(suppl_1):D138–D141
28. Apweiler R, Attwood TK, Bairoch A, Bateman A, Birney E, Biswas M, Bucher P, Cerutti L, Corpet F, Croning MD (2001) The InterPro database, an integrated documentation resource for protein families, domains and functional sites. Nucleic Acids Res 29(1):37–40
29. Krogh A, Larsson B, Von Heijne G, Sonnhammer EL (2001) Predicting transmembrane protein topology with a hidden Markov model: application to complete genomes. J Mol Biol 305(3):567–580
30. Quevillon E, Silventoinen V, Pillai S, Harte N, Mulder N, Apweiler R, Lopez R (2005) InterProScan: protein domains identifier. Nucleic Acids Res 33(suppl_2):W116–W120
31. Sarris PF, Cevik V, Dagdas G, Jones JD, Krasileva KV (2016) Comparative analysis of plant immune receptor architectures uncovers host proteins likely targeted by pathogens. BMC Biol 14(1):8
32. Shiu S-H, Bleecker AB (2001) Receptor-like kinases from Arabidopsis form a monophyletic gene family related to animal receptor kinases. Proc Natl Acad Sci 98(19):10763–10768. https://doi.org/10.1073/pnas.181141598
33. Bayer PE, Edwards D, Batley J (2018) Bias in resistance gene prediction due to repeat masking. Nat Plants 4(10):762
34. Chowdhury MA, Andrahennadi CP, Slinkard AE, Vandenberg A (2001) RAPD and SCAR markers for resistance to acochyta blight in lentil. Euphytica 118(3):331–337
35. Ford R, Pang ECK, Taylor PWJ (1999) Genetics of resistance to ascochyta blight (Ascochyta lentis) of lentil and the identification of closely linked RAPD markers. Theor Appl Genet 98 (1):93–98
36. Sharma KD, Muehlbauer FJ (2007) Fusarium wilt of chickpea: physiological specialization, genetics of resistance and resistance gene tagging. Euphytica 157(1-2):1–14
37. Souza TLPO, Faleiro FG, Dessaune SN, TJd P-J, Moreira MA, EGd B (2013) Breeding for common bean (Phaseolus vulgaris L.) rust resistance in Brazil. Trop. Plant Pathol 38 (5):361–374
38. Sartorato A, Nietsche S, Barros EG, Moreira MA (2000) RAPD and SCAR markers linked to resistance gene to angular leaf spot in common beans. Fitopatol Bras 25(4):637–642
39. Namayanja A, Buruchara R, Mahuku G, Rubaihayo P, Kimani P, Mayanja S, Eyedu H (2006) Inheritance of resistance to angular leaf spot in common bean and validation of the utility of resistance linked markers for marker assisted selection out side the mapping population. Euphytica 151(3):361–369
40. Rashid KY, Bernier CC, Conner RL (1991) Genetics of resistance in faba bean inbred lines to five isolates of Ascochyta fabae. Can J Plant Pathol 13(3):218–225
41. Dirlewanger E, Isaac PG, Ranade S, Belajouza M, Cousin R, De Vienne D (1994) Restriction fragment length polymorphism analysis of loci associated with disease resistance genes and developmental traits in Pisum sativum L. Theor Appl Genet 88(1):17–27
42. Coyne CJ, Inglis DA, Whitehead SJ, McClendon MT, Muehlbauer FJ (2000) Chromosomal location of Fwf, the Fusarium wilt race 5 resistance gene in Pisum sativum. Pisum Genetics 32:20–22
43. Mc Phee KE, Inglis DA, Gundersen B, Coyne CJ (2012) Mapping QTL for Fusarium wilt race 2 partial resistance in pea (Pisum sativum). Plant Breed 131(2):300–306
44. Nisar M, Ghafoor A (2011) Linkage of a RAPD marker with powdery mildew resistance er-1 gene in Pisum sativum L. Russ J Genet 47 (3):300
45. Clulow SA, Matthews P, Lewis BG (1991) Genetical analysis of resistance to Mycosphaerella pinodes in pea seedlings. Euphytica 58 (2):183–189
46. Pottorff M, Wanamaker S, Ma YQ, Ehlers JD, Roberts PA, Close TJ (2012) Genetic and physical mapping of candidate genes for resistance to Fusarium oxysporum f. sp. tracheiphilum race 3 in cowpea [Vigna unguiculata (L.) Walp]. PLoS One 7(7):e41600
47. Wu X, Wang B, Wu X, Lu Z, Li G, Xu P (2018) SNP marker-based genetic mapping of rust resistance gene in the vegetable cowpea landrace ZN016. Legum Res 41(2):222–225
48. Bales C, Zhang G, Liu M, Mensah C, Gu C, Song Q, Hyten D, Cregan P, Wang D (2013) Mapping soybean aphid resistance genes in PI 567598B. Theor Appl Genet 126 (8):2081–2091
49. Hill CB, Chirumamilla A, Hartman GL (2012) Resistance and virulence in the soybean-Aphis glycines interaction. Euphytica 186 (3):635–646
50. Jun T-H, Mian MAR, Michel AP (2012) Genetic mapping revealed two loci for soybean

aphid resistance in PI 567301B. Theor Appl Genet 124(1):13–22

51. Kim K-S, Bellendir S, Hudson KA, Hill CB, Hartman GL, Hyten DL, Hudson ME, Diers BW (2010) Fine mapping the soybean aphid resistance gene Rag1 in soybean. Theor Appl Genet 120(5):1063–1071
52. Kim K-S, Hill CB, Hartman GL, Hyten DL, Hudson ME, Diers BW (2010) Fine mapping of the soybean aphid-resistance gene Rag2 in soybean PI 200538. Theor Appl Genet 121 (3):599–610
53. Zhang G, Gu C, Wang D (2009) Molecular mapping of soybean aphid resistance genes in PI 567541B. Theor Appl Genet 118 (3):473–482
54. Zhang G, Gu C, Wang D (2010) A novel locus for soybean aphid resistance. Theor Appl Genet 120(6):1183–1191
55. Wu M, Wu W-P, Liu C-C, Liu Y-N, Wu X-Y, Ma F-F, Zhu A-Q, Yang J-Y, Wang B, Chen J-Q (2018) A bean common mosaic virus (BCMV)-resistance gene is fine-mapped to the same region as Rsv1-h in the soybean cultivar Suweon 97. Theor Appl Genet 131 (9):1851–1860
56. Yang QH, Gai JY (2011) Identification, inheritance and gene mapping of resistance to a virulent soybean mosaic virus strain SC15 in soybean. Plant Breed 130(2):128–132
57. Zhang J, Xia C, Wang X, Duan C, Sun S, Wu X, Zhu Z (2013) Genetic characterization and fine mapping of the novel Phytophthora resistance gene in a Chinese soybean cultivar. Theor Appl Genet 126(6):1555–1561
58. Goellner K, Loehrer M, Langenbach C, Conrath UWE, Koch E, Schaffrath U (2010) Phakopsora pachyrhizi, the causal agent of Asian soybean rust. Mol Plant Pathol 11(2):169–177
59. Gao H, Bhattacharyya MK (2008) The soybean-Phytophthora resistance locus Rps1-k encompasses coiled coil-nucleotide binding-leucine rich repeat-like genes and repetitive sequences. BMC Plant Biol 8(1):29
60. Gao H, Narayanan NN, Ellison L, Bhattacharyya MK (2005) Two classes of highly similar coiled coil-nucleotide binding-leucine rich repeat genes isolated from the Rps1-k locus encode Phytophthora resistance in soybean. Mol Plant-Microbe Interact 18 (10):1035–1045
61. Chen NWG, Sévignac M, Thareau V, Magdelenat G, David P, Ashfield T, Innes RW, Geffroy V (2010) Specific resistances against pseudomonas syringae effectors AvrB and AvrRpm1 have evolved differently in common bean (Phaseolus vulgaris), soybean (Glycine max), and Arabidopsis thaliana. New Phytol 187(4):941–956
62. Young ND, Debellé F, Oldroyd GE, Geurts R, Cannon SB, Udvardi MK, Benedito VA, Mayer KF, Gouzy J, Schoof H (2011) The Medicago genome provides insight into the evolution of rhizobial symbioses. Nature 480(7378):520
63. Wu J, Zhu J, Wang L, Wang S (2017) Genome-wide association study identifies NBS-LRR-encoding genes related with anthracnose and common bacterial blight in the common bean. Front Plant Sci 8:1398
64. Kang YJ, Kim KH, Shim S, Yoon MY, Sun S, Kim MY, Van K, Lee S-H (2012) Genome-wide mapping of NBS-LRR genes and their association with disease resistance in soybean. BMC Plant Biol 12(1):139
65. Zhou F, Guo Y, Qiu L-J (2016) Genome-wide identification and evolutionary analysis of leucine-rich repeat receptor-like protein kinase genes in soybean. BMC Plant Biol 16(1):58
66. Sato S, Nakamura Y, Kaneko T, Asamizu E, Kato T, Nakao M, Sasamoto S, Watanabe A, Ono A, Kawashima K, Fujishiro T, Katoh M, Kohara M, Kishida Y, Minami C, Nakayama S, Nakazaki N, Shimizu Y, Shinpo S, Takahashi C, Wada T, Yamada M, Ohmido N, Hayashi M, Fukui K, Baba T, Nakamichi T, Mori H, Tabata S (2008) Genome structure of the legume, Lotus japonicus. DNA Res 15(4):227–239. https://doi.org/10.1093/dnares/dsn008
67. Li X, Cheng Y, Ma W, Zhao Y, Jiang H, Zhang M (2010) Identification and characterization of NBS-encoding disease resistance genes in Lotus japonicus. Plant Syst Evol 289 (1-2):101–110
68. Shao Z-Q, Zhang Y-M, Hang Y-Y, Xue J-Y, Zhou G-C, Wu P, Wu X-Y, Wu X-Z, Wang Q, Wang B (2014) Long-term evolution of nucleotide-binding site-leucine-rich repeat genes: understanding gained from and beyond the legume family. Plant Physiol 166 (1):217–234
69. Varshney RK, Song C, Saxena RK, Azam S, Yu S, Sharpe AG, Cannon S, Baek J, Rosen BD, Tar'an B (2013) Draft genome sequence of chickpea (Cicer arietinum) provides a resource for trait improvement. Nat Biotechnol 31(3):240
70. Zheng F, Wu H, Zhang R, Li S, He W, Wong F-L, Li G, Zhao S, Lam H-M (2016) Molecular phylogeny and dynamic evolution of disease resistance genes in the legume family. BMC Genomics 17(1):1

# Chapter 21

# Determination of Bioactive Compounds of Fenugreek (*Trigonella foenum-graecum*) Seeds Using LC-MS Techniques

**Pratibha Singh, Vikas Bajpai, Vikaskumar Gond, Ashok Kumar, Narender Tadigoppula, and Brijesh Kumar**

## Abstract

*Trigonella foenum-graecum* (fenugreek), a legume has been used as spice throughout the world to enhance the sensory quality of foods, is considered to have a high nutraceutical value. The antidiabetic actions of *Trigonella* seeds have been speculated due to the presence of steroid saponins, alkaloids, and fiber content. Five bioactive compounds, namely, trigonelline, isoorientin, orientin, vitexin, and isovitexin, were detected and quantified by HPLC system. Further rapid detection of bioactive compounds was achieved using high-performance liquid chromatography hybrid electrospray quadrupole time-of-flight mass spectrometric (HPLC-ESI-QTOF-MS/MS) method. Twenty-five compounds were tentatively identified based on their retention time and measured accurate mass. The fragmentation patterns of known compounds were utilized to elucidate their structures. Six of these identified compounds, that is, 4-hydroxyisoleucine, trigonelline, isoorientin, isovitexin, pinitol, and sarsasapogenin, were simultaneously quantified by UHPLC-ESI-MS/MS method under the multiple reactions monitoring (MRM) mode using ultrahigh-performance liquid chromatography hybrid electrospray triple quadrupole linear ion trap mass spectrometer. The analytical method was validated and successfully applied for simultaneous determination of compounds in Fenugreek seeds, which indicated the suitability of the validated method for its quality control.

**Key words** HPLC/ESI-QTOF-MS/MS, UPLC/ESI-MS/MS Multiple reaction monitoring, *Trigonella foenum-graecum*

## 1 Introduction

Fenugreek (*Trigonella foenum-graecum*) commonly known as Methi is an annual herb belonging to family Fabaceae [1]. It is cultivated worldwide as a semiarid crop. Seeds of *T. foenum-graecum* are very common ingredient in dishes from the Indian sub-continent (Fig. 1) [2]. It is an herb similar to clover that is native to the Mediterranean region, southern Europe, and western Asia [3, 4]. The seeds are used in cooking to make medicine or to

Mukesh Jain and Rohini Garg (eds.), *Legume Genomics: Methods and Protocols*, Methods in Molecular Biology, vol. 2107, https://doi.org/10.1007/978-1-0716-0235-5_21, © Springer Science+Business Media, LLC, part of Springer Nature 2020

**Fig. 1** *Trigonella foenum-graecum* seed with leaf

supersede the taste of other medicines [5]. It has many potential medicinal applications in the health industry as well as food [6]. *T. foenum-graecum* is known to have antioxidant potency [7], hypocholesterolemic [8], digestive stimulant action and hepatoprotective effect [9]. It is an antibacterial [10], anticancer [11], antidiabetic [12] and antianorexia agent [13]. *T. foenum-graecum* seeds are taken by mouth for digestive problems such as loss of appetite, upset stomach, constipation and inflammation of the stomach [14]. These seeds are also used for painful menstruation [15], menopause [16], polycystic ovary syndrome [17], arthritis [18], poor thyroid function [19] and obesity [20]. It is also used for conditions that affect heart health such as atherosclerosis [21] and for high blood levels of certain fats including cholesterol and triglycerides [22]. *T. foenum-graecum* is used for kidney ailments [23], a vitamin deficiency disease called beriberi, mouth ulcers, boils, bronchitis, infection of the tissues beneath the surface of the skin (cellulitis), tuberculosis, chronic cough, chapped lips, baldness, Parkinson's disease and exercise performance [24]. Different classes of phytochemicals are identified from *T. foenum-graecum*. Major bioactive classes of compounds reported are sapogenins, flavonoids, alkaloids and steroids. Volatile compounds mainly anethole and stallone also possess bioactivity are identified and reported from *T. foenum-graecum* [25–27]. It is popularly used due to its pungent aromatic compounds in the seed that gives flavor, color and aroma to foods [28]. Due to presence of steroids, flavonoids and alkaloids it is used as hormonal and therapeutic drugs [29]. *T. foenum-graecum* has received much attention from pharmaceutical industries due to its extensive use in herbal dietary supplements. The quality control of its extracts in terms of bioactive constituents is essential to clinical efficacy and safety assurance. Therefore, it is important to simultaneously determine the bioactive constituents for quality control.

In recent years, several research groups have reported analytical methods using various chromatographic conditions and spectrophotometric technologies to develop quick and accurate analytical approaches for the identification, structural characterization and determination of chemical constituents of *T. foenum-graecum* seeds [30, 31].

Until now only a few constituents, namely, camboginol, garcinol, xanthochymol and isoxanthochymol, have been quantitatively determined by LC-MS/MS methods in *T. foenum-graecum* [32]. The present study deals with the application of a validated HPLC-ESI-MS/MS and UPLC-ESI-MS/MS method for the rapid detection and quantitation of multiclass bioactive constituents and a simple HPLC method for the identification and quantitation of four biomarkers in the seed extract of *T. foenum-graecum*.

## 2 Materials

### 2.1 Chemical and Solvents

1. Acetonitrile, methanol (LC-MS grade) and formic acid (analytical grade, from Fluka, Sigma-Aldrich) were used. Ultra-pure water was obtained from a Direct-Q 8 UV water purification system (EMD Millipore Corporation, Billerica, MA, USA) (*see* **Note 1**).
2. The plant material (seed) of *T. foenum-graecum* was purchased from local market of Lucknow, India.
3. The analytical standards isoorientin, isovitexin and sarsasapogenin (Sigma-Aldrich, purity ≥97%) were used. Trigonelline, pinitol and 4-hydroxyisoleucine (purity ≥99%) were isolated from *T. foenum-graecum* seed extract.
4. 0.22 μm syringe filter (Millex-GV, PVDF, Merck Millipore, Darmstadt, Germany) were used for sample filtration and ultrasonicator (Bandelin SONOREX, Berlin, Germany) for sonication.
5. Water (A) and acetonitrile (B) were used as the mobile phase in HPLC analysis while 0.1% formic acid with water (A) and methanol (B) in HPLC-QTOF-MS/MS and UPLC-$QqQ_{LIT}$-MS/MS analysis.

### 2.2 HPLC Instrumentation

1. The HPLC analysis was carried out using a Waters HPLC semi preparative system equipped with binary pump (515), auto sampler (717 plus) and a photo diode-array detector (2998).
2. The separation and detection of the compounds was carried out on a Lichrocart RP18 column (250 mm × 4.6 mm, 5.0 μm).

***2.3 HPLC-QTOF-MS/MS Instrumentation***

1. Agilent 1200HPLC system interfaced with Agilent 6520 hybrid quadrupole time of flight mass spectrometer (Agilent Technologies, Santa Clara, CA, USA) was used. It was equipped with a quaternary pump (G1311A), an aonline vacuum degasser (G1322A), an autosampler (G1329A) and a diode-array detector (G1315D).
2. Agilent Poroshell 120EC-C8 column (100 mm × 4.6 mm, 2.7 μm; was utilized for separation of compounds.

***2.4 UPLC-QqQ$_{LIT}$-MS/MS Instrumentation***

1. The UPLC-QqQ$_{LIT}$-MS/MS analysis was performed on a Waters Acquity UPLC™ system (Waters, Milford, MA, USA) connected to a hybrid linear ion trap triple-quadrupole mass spectrometer (API 4000 QTRAP™ MS/MS system from AB Sciex, Concord, ON, Canada) via an electrospray (Turbo V) ion source operated in positive and negative ionization mode (±), using polarity switching.
2. The LC-MS/MS system control, data acquisition and processing was done by Analyst software version 1.6.2 (AB Sciex).
3. The UPLC separation of selected analytes was achieved using an Acquity UPLC BEH $C_{18}$ column (50 mm × 2.1 mm id, 1.7 μm).

## 3 Methods

***3.1 Extraction and Sample Preparation***

1. Seed of *T. foenum-graecum* are washed thoroughly using tap water, air-dried at room temperature and ground.
2. The crushed seeds (100 g) are extracted with 400 mL of ethanol in an ultrasonic water bath (53 kHz) for 30 min and kept at room temperature (*see* **Note 2**).
3. After 24 h, the extract is filtered through Whatman filter paper and residues were further extracted thrice with fresh solvent following the same procedure. The combined filtrates are concentrated using rotary evaporator (Buchi Rotavapor-R2, Flawil, Switzerland) under reduced pressure of 50 kPa at 45 °C temperature.
4. Dried residue obtained after evaporation is weighed (1 mg) accurately, dissolved in methanol to prepare 1000 ppm solution, and vortexed (*see* **Note 3**).
5. The solution is filtered through 0.22 μm syringe filter. The prepared working solutions are vortexed for 30 s and injected into the LC-MS/MS system for analyses (*see* **Note 3**).

***3.2 Preparation of Standard Solution***

1. A mixed standard stock solution (1 mg/mL) of six reference compounds is prepared in methanol.

2. The working standard solutions are then prepared by appropriate dilution of the mixed standard solution with methanol to a series of concentration ranges from 0.1 to 1000 ng/mL.
3. The standard stock and working solutions are stored at −20 °C until use and vortexed for 30 s prior to injection (*see* **Note 4**).

### 3.3 HPLC Conditions

1. HPLC analysis via Lichrocart RP18 column is accomplished with gradient elution of 0.1% formic acid in water (A) and acetonitrile (B) as mobile phase at a flow rate of 1.0 mL/min with a sample injection volume of 10 μL.
2. The HPLC gradient elution is programmed as follows: 0–2 min, 5–15% B; 2–7 min, 15–15% B; 7–19 min, 15–25% B; 19–22 min, 25–75% B; 22–25 min, 75–50% B; 25–30 min, 50–5% B.
3. The diode array detector is set to monitor at 280 nm and the online UV spectra were recorded in the scanning range of 200–400 nm.
4. The accurate data of the molecular ions are processed using the Empower-2 Pro software.

### 3.4 Qualitative and Quantitative Analysis by HPLC System

1. HPLC method is among the most commonly used methods to verify quality of herbal drugs. It is cheaper than sophisticated LC-MS/MS systems, easy to handle and operate by trained technicians. Autosampler has made it cost-effective also.
2. HPLC also needs minimum maintenance and management at the same time it could be used for both qualitative and quantitative applications.
3. Recently, a simple HPLC-UV method for the standardization of new herbal supplement Kudzu (Japan) has been reported [33].
4. These methods can also be used to compare the fingerprint profiles of closely related species or the same species from different locations [34].
5. A suitable sensitive, selective and simple HPLC method has been developed for identification and quantification of bioactive molecules of *T. foenum-graecum*. Chromatographic conditions such as mobile phase composition, flow rate, injection volume and gradient program have been optimized for the separation of bioactive compounds of *T. foenum-graecum* seeds extract.
6. A mobile phase composed of 0.1% (v/v) formic acid aqueous solution (A) and acetonitrile (B) at 1.0 mL/min flow rate has been found optimum for detection of the desired peaks in seed extracts of *T. foenum-graecum* within 30 min.

**Table 1**
**Compounds identified in the seed extract of *Trigonella foenum-graecum* by HPLC technique**

| S. No. | Retention time (min) | Sample I[st] repeat area (%) | II[nd] repeat area (%) | III[rd] repeat area (%) | Average area (%) | Analytes |
|---|---|---|---|---|---|---|
| 1 | 2.40 | 22.08 | 22.40 | 22.47 | 22.32 | Trigonelline |
| 2 | 19.32 | 0.80 | 0.79 | 0.80 | 0.796 | Isoorientin |
| 3 | 20.25 | 2.16 | 2.21 | 2.15 | 2.17 | Orientin |
| 4 | 22.59 | 7.84 | 7.87 | 7.79 | 7.83 | Vitexin |
| 5 | 22.90 | 2.67 | 2.84 | 2.80 | 2.77 | Isovitexin |

7. Metabolic profiling of seed extract of *T. foenum-graecum* resulted in the identification and areawise quantification of five compounds including alkaloids and flavonoids (Table 1).
8. The experiment was done in three replicates and compounds were unambiguously identified by comparing their retention times with authentic standards.
9. The established HPLC analytical method was applied to determine five bioactive compounds, namely, trigonelline, isoorientin, orientin, vitexin and isovitexin in the ethanolic seed extract of *T. foenum-graecum* is summarized in Table 1.
10. The isovitexin and isoorientin were detected the most abundant while trigonelline was the second most abundant analyte in the analyzed sample (Table 1).
11. The chromatograms of standard compounds and seed extract of *T. foenum-graecum* are presented in (Fig. 2).

### 3.5 HPLC-QTOF-MS/MS Conditions

1. The mobile phase, consisted of 0.1% formic acid in water (A) and methanol (C) is delivered at a flow rate of 0.6 mL/min under a gradient program via Agilent Poroshell column operated at 25 °C.
2. The gradient system is 0.1–5 min, 10–20% C; 9–13 min, 30–35% C; 15–18 min, 35–36% C; 20–22 min, 10–10% C. The sample injection volume is 2 μL.
3. The diode array detector is set to monitor at 254 and 280 nm and the online UV spectra are recorded in the scanning range of 190–400 nm.
4. Mass spectrometer is operated in positive electrospray ionization mode and mass spectra are recorded by scanning the mass range of $m/z$ 50–1500 in MS mode.

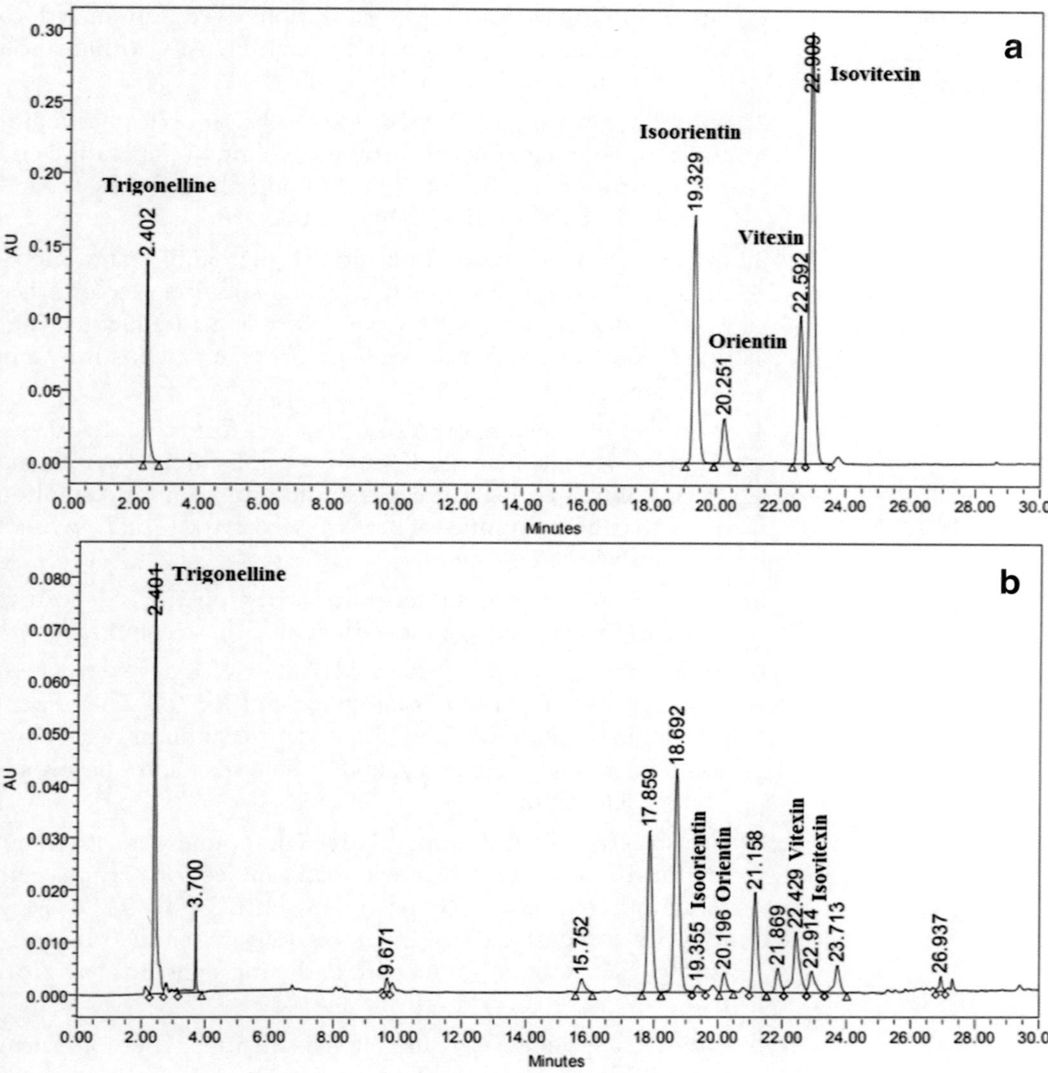


**Fig. 2** HPLC chromatograms of standards (**a**) and seed extract of *Trigonella foenum-graecum* (**b**)

5. Nitrogen is used as a drying, nebulizing and collision gas. Drying gas flow rate is 12 L/min.
6. The heated capillary temperature is set to 350 °C and nebulizer pressure at 45 psi.
7. The source parameters capillary voltage (VCap), fragmentor, skimmer and octapole voltages are set to 3500 V, 175 V, 65 V and 750 V respectively.
8. The accurate mass data of the molecular ions are processed through the Mass Hunter version B 04.00 (Agilent Technology Santa Clara, CA, USA).

### 3.6 Qualitative Analysis by HPLC-QTOF-MS/MS

1. The HPLC-QTOF-MS/MS conditions were optimized to obtain maximum resolution and signal intensity within minimum run time (*see* **Note 5**).
2. Several chromatographic conditions such as mobile phase composition, flow rate, injection volume, column temperature and gradient programs were studied and optimized for the separation of selected bioactive compounds.
3. Different mobile phase compositions (methanol–water, methanol–0.1% (v/v) formic acid aqueous solution, acetonitrile–water and acetonitrile–0.1% (v/v) formic acid aqueous solution) were tested in the gradient program at 0.4 mL/min flow rate.
4. A mobile phase composed of 0.1%(v/v) formic acid aqueous solution (A) and methanol (C) at 0.4 mL/min flow rate and 25 °C column temperature was found optimum for resolution of the maximum number of peaks in seed extracts of *T. foenum*-graecum within 22 min.
5. The MS analysis was successfully carried out in ESI positive mode which showed good sensitivity for these compounds.
6. Amino acids, alkaloids, flavonoids and glycosides were previously reported from *T. foenum*-graecum [35, 36]. The present metabolic profiling resulted in the identification of twenty-five compounds including alkaloids, steroids and flavonoids (Table 2) (*see* **Note 6**).
7. The structural identification of each compound was carried out on the basis of their accurate mass, molecular formula and retention time and MS spectra by HPLC-ESI-QTOF-MS/MS. Six compounds were unambiguously identified by comparing their retention time with authentic standards (*see* **Note 6**).
8. Other compounds were tentatively identified by comparing their MS spectra with available literature.
9. All the identified compounds, their retention time, molecular formulas, *m/z* calculated and observed, error ($\Delta_{ppm}$) and comparative profile of *T. foenum-graecum* are summarized in Table 2.
10. The base peak chromatograms (BPCs) of seed extract of *T. foenum-graecum* in positive mode are presented in (Fig. 3). Structures of all identified compounds are shown in Fig. 4.

### 3.7 UPLC-QqQ$_{LIT}$-MS/MS Conditions

1. UPLC analysis was accomplished with gradient elution of 0.1% formic acid in water (A) and acetonitrile (B) as mobile phase at a flow rate of 0.3 mL/min with a sample injection volume of 3 μL. The column temperature was maintained at 25 °C (*see* **Note 7**).

**Table 2**
**Compounds identified in the seed extract of *Trigonella foenum-graecum* in positive ionization mode by HPLC-QTOF-MS**

| Peak No. | $t_R$ (min) | $[M + H]^+$ (cal) | $[M + H]^+$ (obs) | Error ($\Delta_{ppm}$) | Molecular formula | Identification |
|---|---|---|---|---|---|---|
| 1 | 2.40 | 195.0863 | 195.0862 | 0.47 | $C_7H_{14}O_6$ | Pinitol |
| 2 | 2.50 | 138.0550 | 138.0550 | −0.47 | $C_7H_7NO_2$ | Trigonelline |
| 3 | 2.52 | 104.1070 | 104.1075 | 4.80 | $C_5H_{14}NO$ | Choline |
| 4 | 2.60 | 175.1190 | 175.1188 | 0.82 | $C_6H_{14}N_4O_2$ | Arginine |
| 5 | 2.72 | 118.0863 | 118.0867 | 3.39 | $C_5H_{11}NO_2$ | Betaine |
| 6 | 2.80 | 148.0968 | 148.0966 | 1.68 | $C_6H_{13}NO_3$ | 4-hydroxyisoleucine |
| 7 | 2.81 | 130.0863 | 130.0867 | 3.07 | $C_6H_{11}NO_2$ | Pipecolic acid |
| 8 | 2.82 | 268.1040 | 268.1042 | 0.75 | $C_{10}H_{13}N_5O_4$ | Adenosine |
| 9 | 3.00 | 180.0866 | 180.0867 | −0.19 | $C_6H_{13}NO_5$ | Tricine |
| 10 | 3.80 | 166.0863 | 166.0860 | −1.81 | $C_9H_{11}NO_2$ | Phenylalanine |
| 11 | 5.40 | 257.2475 | 257.2474 | 0.29 | $C_{16}H_{32}O_2$ | Palmitic acid |
| 12 | 6.30 | 205.0972 | 205.0970 | 0.57 | $C_{11}H_{12}N_2O_2$ | L-Tryptophan |
| 13 | 6.66 | 190.0498 | 190.0499 | 0.53 | $C_{10}H_7NO_3$ | Kynurenic acid |
| 14 | 7.42 | 295.1288 | 295.1284 | −1.36 | $C_{14}H_{18}N_2O_5$ | γ Glutamylphenyl-alanine |
| 15 | 8.30 | 595.1657 | 595.1667 | −1.66 | $C_{27}H_{30}O_{15}$ | Kaempferol-3-robinobioside |
| 16 | 8.50 | 595.1657 | 595.1665 | −1.28 | $C_{27}H_{30}O_{15}$ | Vitexin-7-*O*-glucoside |
| 17 | 9.30 | 449.1078 | 449.1083 | −1.04 | $C_{21}H_{20}O_{11}$ | Isoorientin |
| 18 | 9.70 | 449.1078 | 449.1079 | −0.07 | $C_{21}H_{20}O_{11}$ | Quercetin-3-*O*-L-rhamnoside |
| 19 | 9.90 | 449.1078 | 449.1078 | −0.12 | $C_{21}H_{20}O_{11}$ | Orientin |
| 20 | 10.50 | 433.1129 | 433.1127 | 1.29 | $C_{21}H_{20}O_{10}$ | Isovitexin |
| 21 | 10.70 | 433.1129 | 433.1129 | −0.16 | $C_{21}H_{20}O_{10}$ | Vitexin |
| 22 | 10.80 | 229.0859 | 229.0862 | 1.31 | $C_{14}H_{12}O_3$ | Resveratrol |
| 23 | 10.90 | 589.2280 | 579.2283 | 1.30 | $C_{30}H_{36}O_{12}$ | Vitexin-2′-*p*-coumarate |
| 24 | 15.00 | 417.3363 | 417.3366 | −1.52 | $C_{27}H_{44}O_3$ | Sarsasapogenin |
| 25 | 15.50 | 431.3156 | 431.3156 | −0.14 | $C_{27}H_{42}O_4$ | Lilagenin |

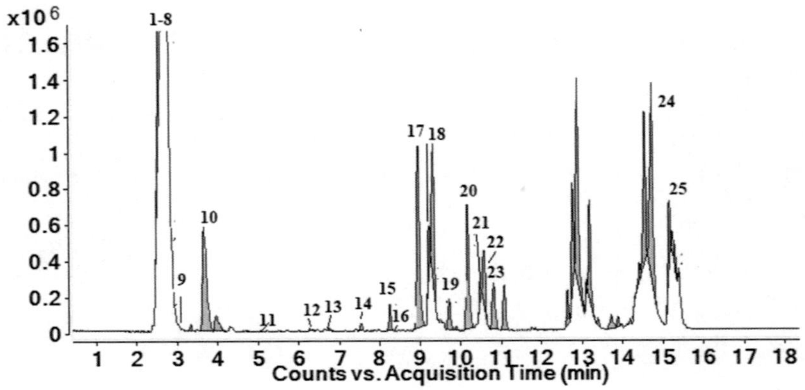


**Fig. 3** Base peak chromatogram (BPC) of extract of *Trigonella foenum-graecum*

2. The UPLC gradient elution was programmed as follows: 0–0.5 min, 5–5% B; 0.5–1.5 min, 5–5% B; 1.5–1.8 min, 5–10% B; 1.8–2.5 min, 10–12% B; 2.5–3 min, 12–15% B; 3–3.3 min, 15–18% B; 3.3–3.8 min, 18–45% B; 3.8–4.2 min, 45–95% B; 4.2–4.8 min, 95–95% B; 4.8–5 min, 95–5% B; 5–6 min, 5–5% B (*see* **Note 7**).
3. Quantitative analysis was performed using multiple reaction monitoring (MRM) acquisition mode at the unit resolution (*see* **Note 7**).
4. The optimized ion source parameters for positive mode were as follows: ion spray voltage, 5500 V; curtain (CUR) gas, 20 psi; nebulizer gas (GS1) and heater gas (GS2), 50 psi; ion source temperature, 450 °C; collision activated dissociation (CAD) gas, medium and the interface heater was on (*see* **Note 7**).
5. All source parameters for negative mode were same as for positive mode except ion spray voltage which was set at −4200 V (*see* **Note 7**).
6. The optimized compound dependent MRM parameters [declustering potential (DP), entrance potential (EP), collision energy (CE) and cell exit potential (CXP)] for precursor-to-product ion transition of each analyte are presented in (Table 3). The method validation parameters of each analyte are presented in (Table 4) (*see* **Note 7**).

### *3.8 Quantitative Analysis by UPLC-QqQ$_{LIT}$-MS/MS*

1. To the best of our knowledge there is no reported method for simultaneous quantification of flavonoids and isoflavonoids in *T. foenum-graecum*. Hence, we selected six bioactive compounds namely, 4-hydroxyisoleucine, trigonelline, isoorientin, isovitexin, pinitol and sarsasapogenin for this study (*see* **Note 7**).

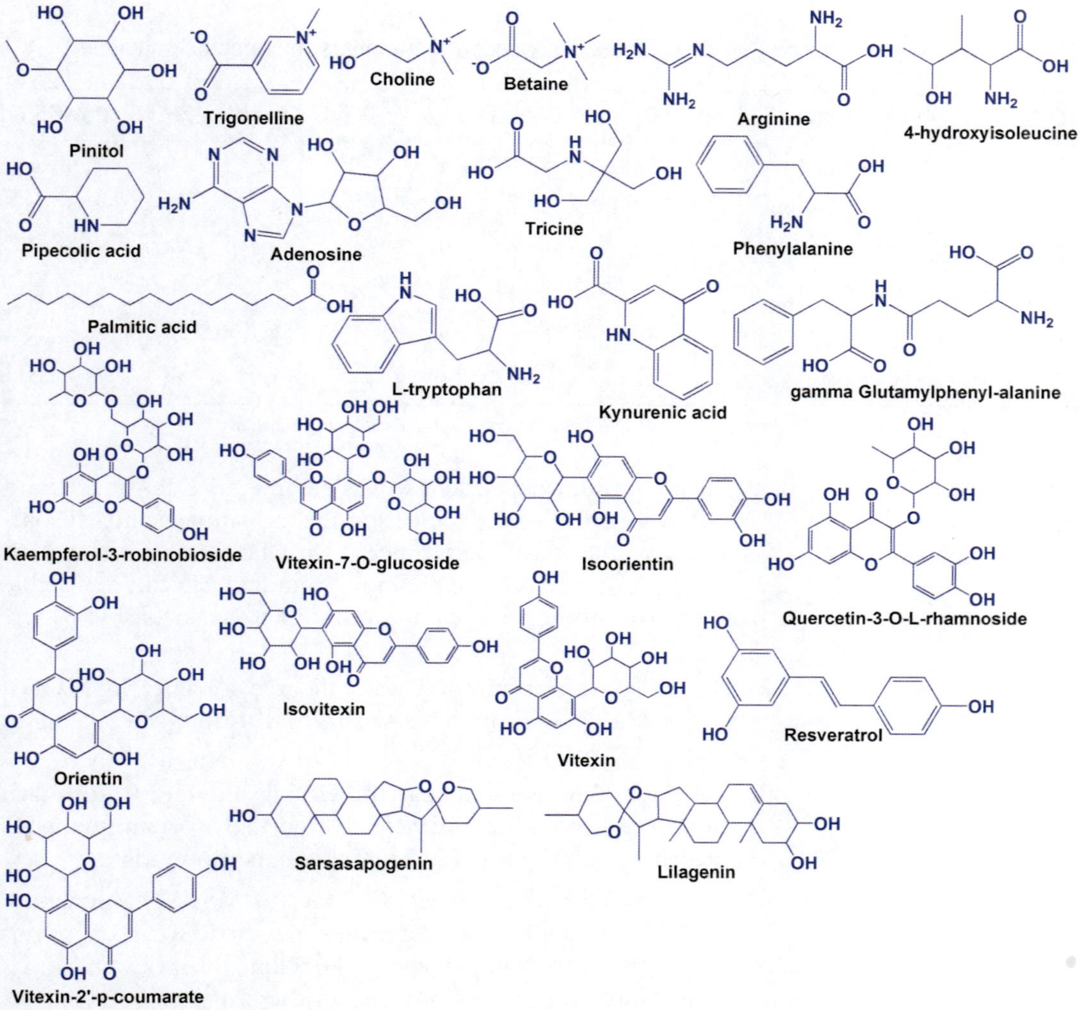


**Fig. 4** Structures of major compounds identified from *Trigonella foenum-graecum* by HPLC-QTOF-MS/MS analysis

2. All compounds were rapidly eluted with retention times at 0.53, 1.04, 3.65, 3.86, 4.32, and 5.32 min for 4-hydroxyisoleucine, trigonelline, isoorientin, isovitexin, pinitol and sarsasapogenin respectively (Fig. 5) (*see* **Note 8**). Extracted ion chromatogram (XIC) is shown in (Fig. 6).
3. This method could be used for quantitation of these bioactive compounds in the seed and also to study variations in its plant parts and products (*see* **Note 9**).

**Table 3**
**Multiple reaction monitoring (MRM), compound-dependent parameters for reference analytes**

| S. No. | $t_R$ (min) | Analytes | $Q_1$ (Da) | $Q_3$ (Da) | DP (V) | EP (V) | CE (eV) | CXP (V) | Polarity |
|---|---|---|---|---|---|---|---|---|---|
| 1 | 0.53 | 4-hydroxyisoleucine | 148.2 | 74.0 | 35 | 7.5 | 19 | 9.5 | Positive |
| 2 | 1.04 | Trigonelline | 137.9 | 92.0 | 70 | 7.2 | 31.5 | 13 | Positive |
| 3 | 3.65 | Isoorientin | 447.2 | 327.0 | −95 | −8 | −31 | −15 | Negative |
| 4 | 3.86 | Isovitexin | 431.2 | 311.1 | −91 | −6 | −26 | −7 | Negative |
| 5 | 4.32 | Pinitol | 195.2 | 163.0 | 34 | 9 | 13 | 7 | Positive |
| 6 | 5.32 | Sarsasapogenin | 417.6 | 273.4 | 24 | 7 | 27 | 13 | Positive |

*DP* declustering potential, *EP* entrance potential, *CE* collision energy, *CXP* cell exit potential

### 3.9 Method Validation and Application

1. The compound-dependent MRM parameters DP, EP, CE, and CXP were optimized by injecting the individual standard solution into the mass spectrometer with flow injection analysis (FIA) in order to achieve the most abundant, specific and stable MRM transition for each compound shown in (Table 3) (*see* **Note 9**).
2. The quantification method was validated for linearity, LODs, LOQs and precision. The satisfactory results are listed in (Table 4) (*see* **Note 9**).
3. The data demonstrated that the established UPLC approach was precise, accurate and sensitive enough for simultaneous quantitative determination of all the six compounds.
4. The established UPLC-ESI-$QqQ_{LIT}$-MS/MS analytical method was applied to determine six bioactive compounds namely, 4-hydroxyisoleucine, trigonelline, isoorientin, isovitexin, pinitol and sarsasapogenin in the ethanolic seed extract of *T. foenum-graecum* is summarized in (Table 5).
5. Compounds trigonelline (91.2 mg/g) and 4-hydroxyisoleucine (90.1 mg/g) were detected the most abundant constituents in the analyzed sample.

## 4 Notes

1. Mass spectrometry instruments were used due to sensitivity, selectivity, *and* accuracy. LC-MS grade solvents and ultrapure water were used to prevent the contamination.
2. Sonication and vortexing was done for the proper mixing of crude material and standards with solvent.
3. To prevent the degradation of the compounds, the standard stock and working solutions were stored at −20 °C.

**Table 4**
**Method validation parameters for six reference analytes**

| Parameters | Analytes | | | | | |
|---|---|---|---|---|---|---|
| | 4-Hydroxyisoleucine | Trigonelline | Isoorientin | Isovitexin | Pinitol | Sarsasapogenin |
| Regression equation | $1.19x + 1.97$ | $2.09x + 1.50$ | $9.93x + 5.75$ | $1.27x + 2.17$ | $9.23x + 1.22$ | $80.4x + 875$ |
| Correlation coefficient ($r^2$) | 0.9991 | 0.9997 | 0.9998 | 0.9994 | 0.9996 | 0.9992 |
| Linearity range (ng/mL) | 25–1000 | 25–1500 | 25–750 | 25–750 | 25–1200 | 22–1200 |
| LOD (ng/mL) | 1.9 | 4.60 | 5.69 | 1.93 | 24.96 | 0.30 |
| LOQ (ng/mL) | 6.04 | 13.9 | 17.27 | 5.85 | 75.65 | 0.91 |
| Precision RSD% (intra-day, $n = 6$) | 0.35 | 1.05 | 1.17 | 0.74 | 1.29 | 0.65 |
| Precision RSD% (inter-day, $n = 6$) | 1.02 | 1.21 | 1.48 | 1.21 | 1.04 | 1.11 |

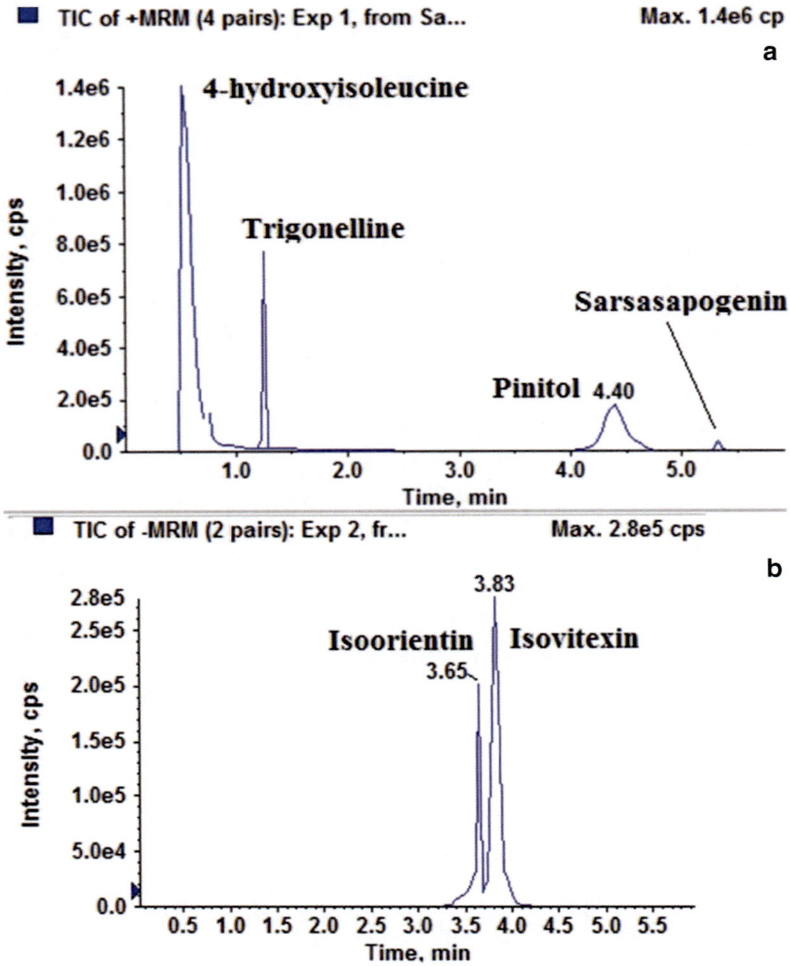


**Fig. 5** Total ion chromatograms (TICs) of compounds quantified from *T. foenum-graecum* (**a**): in positive ion mode and (**b**): in negative ion mode

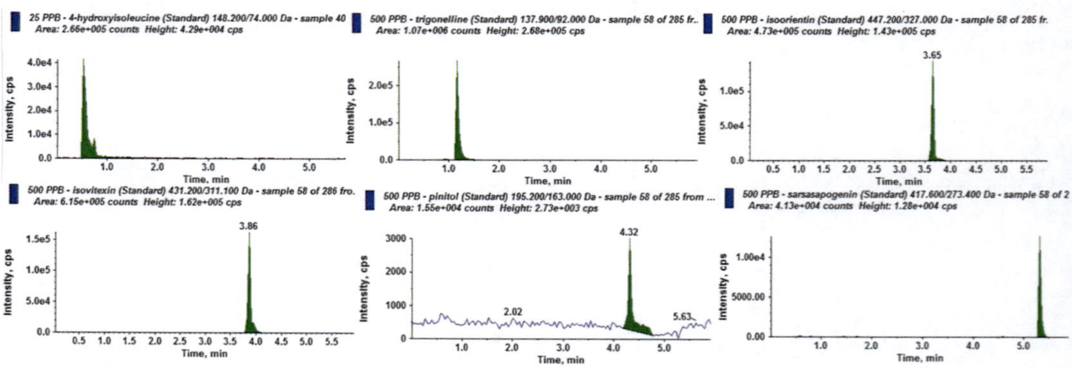


**Fig. 6** Extracted ion chromatograms (XICs) of compounds quantified from *Trigonella foenum-graecum*

**Table 5**
**Content (mg/g) of six bioactive compounds in the seed extract of *T. foenum-graecum***

| | Analytes (mg/g) | | | | | |
|---|---|---|---|---|---|---|
| **Sample** | **4-Hydroxyisoleucine** | **Trigonelline** | **Isoorientin** | **Isovitexin** | **Pinitol** | **Sarsasapogenin** |
| $T_{fg}$ | 90.10 | 91.20 | 1.12 | 3.18 | 45.80 | 1.35 |

*$T_{fg}$ Trigonella foenum-graecum*

4. The selectivity and specificity in phytochemical analysis have increased significantly through hyphenation of chromatographic separation and mass spectrometry detection of LC-MS.
5. This study elaborated a comprehensive qualitative and quantitative analysis of multiclass bioactive constituents including flavonoids, alkaloids, steroids, sapogenins, polyphenolics and amino acids in seed ethanolic extract of *T. foenum-graecum* using different analytical techniques like HPLC, HPLC-QTOF-MS/MS and UPLC-$QqQ_{LIT}$-MS/MS.
6. The UPLC-$QqQ_{LIT}$-MS/MS method was optimized and established for simultaneous determination of six compounds namely, 4-hydroxyisoleucine, trigonelline, isoorientin, isovitexin, pinitol and sarsasapogenin.
7. The UPLC system combined with mass spectrometry detection in MRM acquisition mode enabled significant reductions in separation time, solvent consumption and ensured excellent selectivity and sensitivity for quantitative analyses.
8. This method could be used for quality control of seed extracts and products according to the need.
9. The UPLC-$QqQ_{LIT}$-MS/MS quantification method was validated for linearity, LODs, LOQs and precision as per the International Conference on Harmonization Guidelines.

## Acknowledgments

The authors gratefully acknowledge the Sophisticated Analytical Instrument Facility, CSIR-CDRI, Lucknow, India, where all the mass spectrometric studies were carried out.

### References

1. Snehlata HS, Payal DR (2012) Fenugreek (*Trigonella foenum-graecum* L.) an overview. Int J Curr Pharm Rev Res 2(4):169–187
2. Hojjat SS, Hojjat H (2015) Effect of nano silver on seed germination and seedling growth in fenugreek seed. Int J Food Eng 1 (2):106–110

3. Acharya SN, Thomas JE, Basu SK (2008) Fenugreek, an alternative crop for semiarid regions of North America. Crop Sci 48 (3):841–853
4. Petropoulos GA (2003) Fenugreek: the genus Trigonella. CRC Press, Boca Raton, Florida
5. Krishnaswamy K (2008) Traditional Indian spices and their health significance. Asia Pac J Clin Nutr 17:265–268
6. Lobo V, Patil A, Phatak A, Chandra N (2010) Free radicals, antioxidants and functional foods: impact on human health Pharmacognosy reviews. Pharmacogn Rev 4(8):118
7. Reddy RR, Srinivasan K (2011) Hepatoprotective and antioxidant effect of fenugreek (*Trigonella foenum-graecum*) seeds in mice under lithogenic condition. J Food Biochem 35 (6):1619–1626
8. Ribes G, Sauvaire Y, Costa CD, Baccou JC, Loubatieres-Mariani MM (1986) Antidiabetic effects of subfractions from fenugreek seeds in diabetic dogs. Proc Soc Exp Biol Med 182 (2):159–166
9. Wani SA, Kumar P (2018) Fenugreek: a review on its nutraceutical properties and utilization in various food products. J Saudi Soc Agric Sci 17 (2):97–106
10. Dash BK, Sultana S, Sultana N (2011) Antibacterial activities of methanol and acetone extracts of fenugreek (*Trigonella foenum-gracum*) and coriander (*Coriandrum sativum*). Life Sci Med Res 27:1–8
11. Shabbeer S, Sobolewski M, Anchoori RK, Kachhap S, Hidalgo M, Jimeno A, Davidson NE, Carducci M, Khan SR (2009) Fenugreek: a naturally occurring edible spice as an anticancer agent. Cancer Biol Ther 8(3):272–278
12. Yoshinari O, Sato H, Igarashi K (2009) Antidiabetic effects of pumpkin and its components, trigonelline and nicotinic acid, on Goto-Kakizaki rats. Biosci Biotechnol Biochem 73(5):1033–1041
13. Khorshidian N, YousefiAsli M, Arab M, AdeliMirzaie A, Mortazavian AM (2016) Fenugreek: potential applications as a functional food and nutraceutical. Nutr Food Sci Res 3(1):5–16
14. Kunnumakkara AB, Koca C, Dey S, Gehlot P, Yodkeeree S, Danda D, Sung B, Aggarwal BB (2009) Traditional uses of spices: an overview. In: Molecular targets and therapeutic uses of spices: modern uses for ancient medicine, vol 2. World Scientific Publishing, Singapore, pp 1–24
15. Younesy S, Amiraliakbari S, Esmaeili S, Alavimajd H, Nouraei S (2014) Effects of fenugreek seed on the severity and systemic symptoms of dysmenorrhea. J. Reprod. Infertil. 15(1):41
16. Hakimi S, Mohammad Alizadeh S, Delazar A, Abbasalizadeh F, Bamdad Mogaddam R, Siiahi MR, Mostafa Garabagi P (2006) Probable effects of fenugreek seed on hot flash in menopausal women. J. Med. Plants 3(19):9–14
17. Bashtian MH, Emami SA, Mousavifar N, Esmaily HA, Mahmoudi M, Poor AHM (2013) Evaluation of fenugreek (*Trigonella foenum-graceum* L.), effects seeds extract on insulin resistance in women with polycystic ovarian syndrome. Iran J Pharm Res 12(2):475
18. Suresh P, Kavitha CN, Babu SM, Reddy VP, Latha AK (2012) Effect of ethanol extract of *Trigonella foenum graecum* (fenugreek) seeds on Freund's adjuvant-induced arthritis in albino rats. Inflammation 35(4):1314–1321
19. Arrak JK, Al-Zyadi AJ (2015) The therapeutic role of alcoholic extract of fenugreek seeds on hypothyroidism state induced by thiourea and some blood parameters in adult male rabbits. Iraqi J Vet Med 39(1):1–7
20. Avalos-Soriano A, De la Cruz-Cordero R, Rosado JL, Garcia-Gasca T (2016) 4-Hydroxyisoleucine from fenugreek (Trigonella foenum-graecum): effects on insulin resistance associated with obesity. Molecules 21 (11):1596
21. Roberts KT (2011) The potential of fenugreek (*Trigonella foenum-graecum*) as a functional food and nutraceutical and its effects on glycemia and lipidemia. J Med Food 14 (12):1485–1489
22. Sharma RD, Sarkar A, Hazra DK, Misra B, Singh JB, Maheshwari BB, Sharma SK (1996) Hypolipidaemic effect of fenugreek seeds: a chronic study in non-insulin dependent diabetic patients. Phytother Res 10(4):332–334
23. Olaiya CO, Soetan KO (2014) A review of the health benefits of fenugreek (*Trigonella foenum-graecum* L.): nutritional, biochemical and pharmaceutical perspectives. Am J Soc Issues Humanties 4(2):3–12
24. Charles DJ (2012) Fenugreek. In: Antioxidant properties of spices, herbs and other sources. Springer, New York, NY, pp 295–303
25. Bahmani M, Shirzad H, Mirhosseini M, Mesripour A, Rafieian-Kopaei M (2016) A review on ethnobotanical and therapeutic uses of fenugreek (*Trigonella foenum-graceum* L). J Evidence-Based Complementary Altern Med 21(1):53–62
26. Belguith-Hadriche O, Bouaziz M, Jamoussi K, El Feki A, Sayadi S, Makni-Ayedi F (2010) Lipid-lowering and antioxidant effects of an ethyl acetate extract of fenugreek seeds in

high-cholesterol-fed rats. J Agric Food Chem 58(4):2116–2122

27. Nagulapalli Venkata KC, Swaroop A, Bagchi D, Bishayee A (2017) A small plant with big benefits: fenugreek (Trigonella foenum-graecum Linn.) for disease prevention and health promotion. Mol Nutr Food Res 61(6):1600950
28. Mishra P (2009) Isolation, spectroscopic characterization and molecular modeling studies of mixture of *Curcuma longa*, ginger and seeds of fenugreek. Int. J. PharmTech Res. 1(1):79–95
29. Yadav UC, Baquer NZ (2014) Pharmacological effects of *Trigonella foenum-graecum* L. in health and disease. Pharm Biol 52(2):243–254
30. Taylor WG, Zaman MS, Mir Z, Mir PS, Acharya SN, Mears GJ, Elder JL (1997) Analysis of steroidal sapogenins from amber fenugreek (*Trigonella foenum-graecum*) by capillary gas chromatography and combined gas chromatography/mass spectrometry. J Agric Food Chem 45(3):753–759
31. Chopra S, Ahmad FJ, Khar RK, Motwani SK, Mahdi S, Iqbal Z, Talegaonkar S (2006) Validated high-performance thin-layer chromatography method for determination of trigonelline in herbal extract and pharmaceutical dosage form. Anal Chim Acta 577(1):46–51
32. Faeste CK, Christians U, Egaas E, Jonscher KR (2010) Characterization of potential allergens in fenugreek (*Trigonella foenum-graecum*) using patient sera and MS-based proteomic analysis. J Proteome 73(7):1321–1333
33. Bebrevska L, Bravo L, Vandervoort J, Pieters L, Vlietinck A, Apers S (2007) Development and validation of an HPLC method for quality control of *Pueraria lobata* flower. Planta Med 73:1606–1613
34. Tistaert C, Dejaegher B, Vander Heyden Y (2011) Chromatographic separation techniques and data handling methods for herbal fingerprints: a review. Anal Chim Acta 690 (2):148–161
35. Vígh S, Cziaky Z, Sinka LT, Pribac C, Moş L, Turcuş V, Remenyik J, Mathe E (2017) Analysis of phytoconstituent profile of fenugreek-- Trigonella Foenuem-graecum L.-seed extracts. Stud Univ Babes-Bolyai, Chem 62:145–166
36. Benayad Z, Gómez-Cordovés C, Es-Safi N (2014) Characterization of flavonoid glycosides from fenugreek (Trigonella foenum-graecum) crude seeds by HPLC–DAD–ESI/MS analysis. Int J Mol Sci 15 (11):20668–20685

# Chapter 22

# Quantitative Phosphoproteomic Analysis of Legume Using $TiO_2$-Based Enrichment Coupled with Isobaric Labeling

**Pragya Barua, Nilesh Vikram Lande, Sunil Kumar, Subhra Chakraborty, and Niranjan Chakraborty**

## Abstract

Phosphorylation of proteins is the most dynamic protein modification, and its analysis aids in determining the functional and regulatory principles of important cellular pathways. The legumes constitute the third largest family of higher plants, Fabaceae, comprising about 20,000 species and are second to cereals in agricultural importance on the basis of global production. Therefore, an understanding of the developmental and adaptive processes of legumes demands identification of their regulatory components. The most crucial signature of the legume family is the symbiotic nitrogen fixation, which makes this fascinating and interesting to investigate phosphorylation events. The research on protein phosphorylation in legumes has been focused primarily on two model species, Medicago tr*u*ncatula and *Lotus japonicus*. The development of reciprocal research in other species, particularly the crops, is lagging behind which has limited its beneficial uses in agricultural productivity. In this chapter, we outline the titanium dioxide-based enrichment of phosphopeptides for nuclear proteome analysis of a grain legume, chickpea.

**Key words** False discovery rate, Legumes, Nuclear fraction, Phosphopeptide enrichment, Phosphoproteome, Protein phosphorylation, Stress response, Titanium dioxide

## 1 Introduction

Protein phosphorylation is recognized as a keystone in accelerating our understanding of stress responses and cell defense [1]. A functional protein molecule is not essentially the translated product of the RNA transcribed from the coding gene. Addition of series of covalent modifications in amino acid side chains, during or following translation, can significantly alter the nature of protein [2]. Phosphorylation is one such post-translational modification (PTM) and used as a chemical switch to regulate important processes such as ubiquitin-mediated proteolysis, mRNA splicing and regulation of signal transduction, among others [3]. Histone phosphorylation-mediated modification of chromatin structure and alterations in protein–protein and DNA–protein interactions

Mukesh Jain and Rohini Garg (eds.), *Legume Genomics: Methods and Protocols*, Methods in Molecular Biology, vol. 2107, https://doi.org/10.1007/978-1-0716-0235-5_22, 

are the classical examples, highlighting the importance of phosphorylation in cellular context [4]. Phosphorylation is an esterification reaction in which chemical addition of a phosphoryl group ($PO_3^-$) takes place with the hydroxyl (–OH) group of a serine, threonine or tyrosine side chain of a polypeptide [5]. Addition or deletion of phosphate groups alters surface charge distribution leading to change in structural conformation, enzyme activity, and protein–protein interaction. Therefore, identification of these modifications are imperative in understanding the multiple functional forms of a protein and their role in systemic pathways.

Global phosphoproteomic analysis in legumes, particularly in crop species, has been strikingly limited with even lesser studies on stress response. Interestingly, the change in phosphorylation status does not always involve protein level alterations. Under unfavorable conditions, the degree of phosphorylation might control protein folding, enzyme activity as well as protein–protein interaction. Also, some residues might be more preferentially regulated than the others. The frequency of phosphorylation of serine residues are highest followed by threonine and the least for tyrosine residues [6]. However, the identification of altered phosphorylation status at tyrosine residue holds greater significance, owing to its limited occurrence and thus specificity in regulatory context [7]. Till date, several phosphoproteomic studies have been carried out in different legumes at various stages using different tissue types [8–21]. These studies primarily included immobilized *m*etal *a*ffinity *c*hromatography (IMAC) enrichment and *m*etal *o*xide *a*ffinity *c*hromatography (MOAC), as the key steps for enrichment of phosphorylated peptides. In contrast to IMAC, where immobilized positively charged metals are used to trap the negatively charged phosphate groups, in MOAC the matrix itself is composed of metal oxides or hydroxides, eliminating the need of resin [22]. A number of metal oxides can be used for enrichment, such as zirconium dioxide ($ZrO_2$), gallium oxide ($Ga_2O_3$), ferric oxide ($Fe_3O_4$), niobium oxide ($Nb_2O_3$), stannic oxide ($SnO_2$), hafnium dioxide ($HfO_2$), and tantalum oxide ($Ta_2O_5$). Notably, titanium dioxide ($TiO_2$) enrichment has been the current method of choice owing to its better binding specificity.

The large-scale phosphoproteomic analysis requires a relatively higher amount of proteins to be able to enrich and detect the phosphorylation events because only one-third of the total cellular proteome becomes phosphorylated at any instant. Moreover, the intensity varies, making detection of such peptides to be extremely difficult without performing enrichment. Recent advances in the sensitivity and online fractionation techniques coupled with mass spectrometry have helped in the detection of even limited number of phosphopeptides. However, the choice of instrument and even the peptide fragmentation method all contribute toward the outcome. Electron-based fragmentation methods are often employed for mapping phosphorylation sites, as they generally produce full spectrum of sequence-informative ions without neutral loss of modifying functional groups [23–28]. Electron transfer dissociation (ETD) is generally chosen over collision-activated dissociation (CAD) for phosphopeptide detection as the loss of phosphate groups is prevalent in the later [29]. In this chapter, we elaborate the steps involved in $TiO_2$-based phosphopeptide enrichment in nuclear fraction. Use of highest quality reagents must be ensured as the quality of the reagents used can significantly affect experimental outcome. Also, the physical properties of the $TiO_2$ particles have to be considered. Different manufacturers [30] and/or different crystallinity [31] of these particles result in different degrees of specificity. Such differences could lead to differing results being obtained in different laboratories, and thus to diverse interpretations. It is advantageous to use mesoporous particles instead of smooth ones since they have a higher surface-to-volume ratio and thus greater binding capacity [32]. Once the correct type of particles is chosen, it is important to apply the optimal ratio for the $TiO_2$ particles and the peptide amount [33].

## 2 Materials

### 2.1 Equipment

1. Mortar and pestle.
2. SpeedVac concentrator.
3. Hitachi (ELITE LaChrom) high-performance liquid chromatography (HPLC).
4. C18 StageTips.
5. Vortex mixer.
6. Probe sonicator.
7. Sonication bath.
8. Microcentrifuge.
9. Bioruptor.
10. Mass spectrometer.

### 2.2 Buffers and Solutions

It is important to obtain the highest purity of all chemicals listed below.

#### 2.2.1 Nuclei Isolation

1. Polyvinylpolypyrrolidone (PVPP).
2. Homogenization buffer (HB): 10 mM Trizma (Tris base), 80 mM KCl, 10 mM EDTA, 1 mM spermidine, 1 mM spermine, 0.5 M sucrose, pH 9.5. After adjusting pH, add 0.15% β-mercaptoethanol and 0.5% Triton X-100. Add protease inhibitor cocktail (Sigma-Aldrich) and phosphatase inhibitors (PhosSTOP, Sigma-Aldrich).
3. Nuclei Wash buffer (NWB): HB without Triton X-100.

*2.2.2 Protein Extraction, Quantitation, Digestion, and Labeling of Peptides*

1. 1% sodium dodecyl sulfate (SDS).
2. Bicinchoninic acid (BCA) kit (Thermo Pierce, Waltham, USA).
3. 5 mM dithiothreitol (DTT).
4. 20 mM iodoacetamide (IAA).
5. Acetone.
6. Triethylammonium bicarbonate (TEABC).
7. Trypsin Gold (Worthington Biochemical Corp.).
8. Tandem Mass Tag (TMT) 6-plex kit (Thermo Fisher Scientific).
9. Acetonitrile anhydrous.
10. 5% Hydroxylamine.
11. Solvent B (10 mM TEABC in 90% acetonitrile, pH 8.5).

*2.2.3 Phosphopeptide Enrichment and LC-MS/MS Analysis*

1. $TiO_2$ beads (GL Sciences product number 5010-21315).
2. Loading buffer (80% ACN, 6% TFA).
3. TFE digestion buffer: 10% 2,2,2-trifluroethanol in 100 mM ammonium bicarbonate.
4. Phosphoenrichment buffer: 300 mM KCl, 5 mM $KH_2PO_4$, 50% acetonitrile (ACN), 6% trifluoroacetic acid.
5. Wash buffer: 60% ACN, 1% TFA.
6. Transfer buffer: 80% ACN, 0.5% acetic acid.
7. 2,5-Dihydroxybenzoic acid (DHB; Fluka)
8. Phthalic acid (Fluka).
9. Acetonitrile, HPLC grade (Bie & Berntsen A/S).
10. Formic acid (Mass spectrometry grade).
11. Trifluoroacetic acid (TFA), protein sequencer grade (Applied Biosystems).
12. Ammonia solution, 25% (Merck).
13. Orthophosphoric acid (Bie & Berntsen A/S).
14. Mobile phase A: 0.1% formic acid.
15. Milli-Q water (Millipore; UHQ water).
16. Solvent C: 80% acetonitrile in 0.1% formic acid.

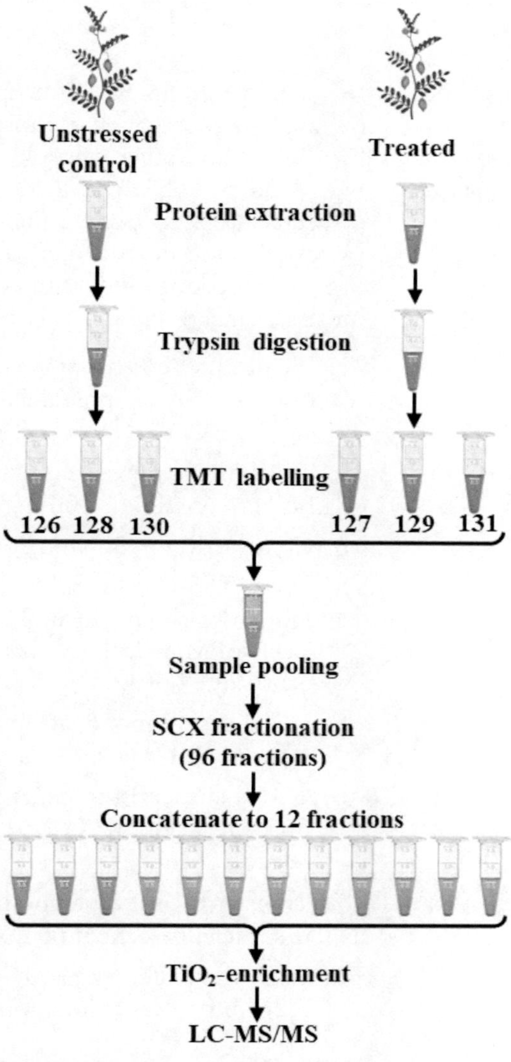


**Fig. 1** Diagrammatic representation of experimental workflow for carrying out quantitative phosphoproteome analysis. Extract organelle of interest from control and treated samples. Perform protein extraction and trypsin digestion. Label the digested peptides. Pool and fractionate the labeled peptides. Subject each fraction to $TiO_2$-based phosphopeptide enrichment followed by LC-MS/MS analysis

## 3 Methods

***3.1 Isolation of Nucleus from Unstressed Control and Treated Chickpea Seedlings***

In general, enrichment of phosphopeptide is critical, particularly for the analysis of organellar phosphoproteome. A generalized workflow of quantitative phosphoproteome analysis is presented in Fig. 1. Notably, the phenotypic plasticity of plants, in response to subjected stress, is a consequence of altered proteome status. Here we briefly outline the method for isolation of nuclei [34], the organelle hosting and regulating gene expression. All the steps must be carried out at 4 °C in chilled containers.

1. Weigh and crush approximately 40 g of tissue using liquid nitrogen in a prechilled mortar and pestle, with 0.3% (w/w) PVPP.
2. Transfer the powdered tissue immediately into an ice-cold 1 L beaker containing 400 mL of ice-cold HB (*see* **Note 1**).
3. Gently stir for 30 min for complete lysing of the organellar membranes.
4. Filter the slurry through four layers of cheesecloth and two layers of Miracloth (Merck, Germany) into an ice-cold 500 mL centrifuge bottle.
5. Pellet the homogenate by centrifugation using a fixed-angle rotor (1200 × *g*) at 4 °C for 20 min.
6. Gently suspend the pellet in 30 mL ice-cold NWB and centrifuge (57 × *g*) at 4 °C for 2 min to remove intact cells and tissue residues.
7. Transfer the supernatant into a fresh centrifuge tube, and pellet the nuclei by centrifugation (1200 × *g*) at 4 °C for 15 min.
8. Wash the pellet two additional times by resuspension in NWB followed by centrifugation (1200 × *g*) at 4 °C for 15 min.

***3.2 Extraction and Solubilization of Proteins from Unstressed Control and Treated Samples***

1. Separately suspend the nuclei enriched pellets from both unstressed control and treated samples in 1% SDS and incubate at 25 °C for 10 min.
2. Sonicate each pellet (20% amplitude, pulse of 10 s followed by 5 s pause) on ice for 20 min using a probe sonicator to extract proteins.
3. Incubate the samples at 95 °C for 10 min and centrifuge at 10,000 × *g* for 10 min.
4. Transfer the supernatants in a fresh tube and carefully determine protein concentration using BCA assay (*see* **Note 2**).
5. Subject 3 mg protein, each from unstressed control and treated samples, to reduction using 5 mM DTT for 20 min.

6. Following reduction, alkylate the samples using 20 mM IAA for 15 min in dark.
7. Precipitate the proteins overnight using five volumes of acetone at −20 °C.
8. Retrieve the protein pellets through centrifugation at 10,000 × *g* for 10 min and dissolve in 50 mM TEABC (pH 8.4) (*see* **Note 3**).
9. Digest the proteins by trypsin at 37 °C (1:10 trypsin–protein), overnight.
10. Check the pre- and post-digest (10 μg) by one-dimensional gel electrophoresis (1-DE) to confirm digestion status (*see* **Note 4**).

### 3.3 Labeling and Fractionation of the Digested Peptides

Spit the tryptic peptides of each time point into three parts, to serve as technical replicates for labeling using TMT 6-plex kit. Carry out the labeling as follows:

1. Label the unstressed control replicates with channels 126, 128 and 130; and treated replicates with 127, 129, and 131, respectively, according to manufacturer's instructions.
2. Allow the vials containing the labels 126, 127, 128, 129, 130, and 131 to reach room temperature (*see* **Note 5**).
3. Also, allow the digested peptides to come to room temperature (*see* **Note 6**).
4. Add anhydrous acetonitrile or ethanol (41 μL to 0.8 mg vials; 256 μL to 5 mg vials) to each vial. Allow the reagent to dissolve with occasional vortexing for 5 min. Centrifuge the tubes briefly to spin down the solution.
5. Carefully add 41 μL of TMT label reagents to each 25–100 μg sample. Alternatively, transfer the digested peptides to the TMT reagent vial.
6. Incubate the reaction at room temperature for 1 h.
7. Terminate the reactions were quenched by adding 8 μL of 5% hydroxylamine (provided in TMT kit) and incubated at room temperature for 15 min.
8. Pool the peptides of each of the time points (unstressed control and treated) and dry using a SpeedVac concentrator.
9. Reconstitute the pooled peptides in 1 mL of 10 mM TEABC (pH 8.4) and fractionate using basic reverse phase liquid chromatography.
10. Inject the peptides using a manual injector onto an SCX column connected to HPLC system.
11. Separate peptides using a gradient of 0–100% solvent B in 120 min into 96 fractions.

12. Collect the fractions, pool, and concatenate into 12 fractions.
13. Dry the peptides and desalt using C18 StageTips and subject each fraction to phosphopeptide enrichment.

### 3.4 Phosphopeptide Enrichment by $TiO_2$

For proteomic analysis, keep one-tenth volume of each fraction and evaporate to dryness in microcentrifuge tubes. Process the remaining samples for $TiO_2$-based phosphopeptide enrichment as described earlier [35].

#### 3.4.1 Preparing $TiO_2$ Beads

1. Weigh $TiO_2$ beads at a ratio of 10:1 to protein.
2. Suspend the beads in Loading buffer at a concentration of 100 μL Loading buffer per sample.
3. Incubate the $TiO_2$ bead suspension Bioruptor at 4 °C for 1 min to disaggregate beads.

#### 3.4.2 $TiO_2$-Based Enrichment of Phosphopeptides

1. Reconstitute the dried peptides of each fraction in 500 μL of TFE digestion buffer (*see* **Note 7**).
2. Add the components of phosphoenrichment buffer to a final volume of 1.6 mL in the following order:
   (a) 150 μL 3.2 M KCl.
   (b) 55 μL 150 mM $KH_2PO_4$.
   (c) 800 μL ACN.
   (d) 95 μL TFA.
3. Carefully mix the peptide solution at room temperature.
4. Add $TiO_2$ beads at a ratio of 10:1 (beads–peptides) and incubate at 40 °C for 5 min with mild shaking.
5. Centrifuge the beads for 1 min at 3500 × *g* at room temperature to pellet beads and discard the supernatant by aspiration (non phosphopeptides).
6. Remove the nonspecifically bound peptides by pelleting beads 4 additional times with 1 mL wash buffer, incubating each wash for 30 s.
7. After the final wash, 100 μL transfer buffer and incubate at RT for 30 s in ThermoMixer at 800 rpm to resuspend beads.
8. Elute the enriched phosphopeptides thrice using 40 μL of 2% ammonia solution into microfuge tubes containing 10 μL of 20% TFA on ice (*see* **Note 8**).
9. Dry the eluted peptides and suspend in 30 μL of 0.1% TFA.
10. Elute phosphopeptides twice with 100 μL of 5% $NH_4OH$ and combine eluents.
11. Subject the eluate to a second round of $TiO_2$ enrichment using the same protocol and fresh $TiO_2$ beads (**steps 1–10**).
12. Desalt using C18 StageTips.

13. Acidify the eluate with 30 μL of formic acid (pH should be <4) (*see* **Note 9**).
14. Vacuum dry to near completion and store at −80 °C.
15. Resuspend the phosphopeptides in 25 μL mobile phase A prior to LC-MS/MS analysis.

### 3.5 Mass Spectrometry and Data Acquisition

Mass spectrometric analysis can be carried out as follows:

1. Reconstitute the desalted peptides in 0.1% formic acid and loaded onto trap column nanoViper 2 cm (3 μm C18 Aq; Thermo Fisher Scientific).
2. Separate peptides using EASY-Spray C18 analytical column (15 cm, 75 μm PepMap C18, 2 μm C18 Aq; Thermo Fisher Scientific) set at 35 °C. The following solvent gradients are widely used for peptide separation: linear gradient of 5–35% solvent C over 90 min with a total run time of 120 min and flow rate set to 300 nL/min.
3. Data can be acquired in positive scan mode using a data-dependent acquisition method wherein MS1 survey scans are carried out in 400–1600 $m/z$ range (120,000 mass resolution at 200 $m/z$).
4. The most intense precursor ions are selected with maximum cycle time of 3 s (high-energy collisional dissociation fragmentation, collision energy 33%, mass resolution 30,000) for peptide fragmentation.
5. The charge states for peptides can be set to 2–6, and dynamic exclusion to 30 s along with an exclusion width of ±20 ppm.
6. Internal calibration must be carried out using lock mass option.

### 3.6 Processing of Raw Data Obtained by Mass Spectrometry Analysis

1. Mass spectrometry data can be processed to generate peak list files using dedicated software such as Proteome Discoverer software (Thermo Fisher Scientific).
2. The data is searched against appropriate database (e.g., NCBI *Cicer arietinum* proteome database, containing 33,000 protein sequences).
3. Peptide identifications can be carried out using search algorithms such as SEQUEST and Mascot. Usage of multiple search algorithms is preferred for more confident identifications.
4. The search parameters include trypsin (or the enzyme used) as the proteolytic enzyme with maximum of two allowed missed cleavages.
5. The dynamic modifications are set as oxidation of methionine and acetylation of protein at the N-terminus, whereas

carbamidomethylation of cysteine and TMT 6-plex modification at peptide N-terminus, and lysine are set as static modifications.

6. Phosphorylation of serine, threonine, and tyrosine are set as additional dynamic modifications to analyze phosphoproteome.
7. Precursor and fragment mass tolerance are set to 10 ppm and 0.05 Da, respectively.
8. The false discovery rate is calculated by enabling the peptide sequence analysis using a decoy database and a cutoff of 1% peptide spectral match.
9. For protein identification, 1% peptide level false discovery rate (FDR) is used.
10. The reporter ions quantifier node of Proteome Discoverer can be used for relative quantitation, and PhosphoRS3.1 node can be used for calculating the probability of phosphorylation at each residue.
11. The phoshopeptides with >75% localization probability are considered for analysis.
12. Proteins and phoshopeptides are considered as differentially regulated after assigning suitable cutoff value.

## 4 Notes

1. The inclusion of protease and phosphatase inhibitors in homogenization buffers is necessary to prevent sample degradation and dephosphorylation. Kinase activity also has to be blocked to prevent nonbiological events of protein phosphorylation and exploration of artificial phosphorylation pattern.
2. All measurements must be done accurately in order to avoid any false estimation. Equal amount of protein from unstressed control and treated sample must be taken.
3. It is normal for proteins not to completely solubilize in the TEABC, samples can appear translucent, and become more transparent 1–2 h after commencing trypsin digestion. Prior to initiating digestion, samples must appear as a smooth, homogenous suspension, with no fragments visible, and should not settle within ~5 min.
4. It is important to confirm that the digestion was successful before proceeding further into the experiment. Trypsin activity is pH dependent and lack of enzyme activity might lead to partial or no digestion.

5. The TMT reagents are moisture-sensitive. Vials must be equilibrated to room temperature before opening to avoid moisture condensation onto the product.
6. For optimal results, use 25–100 μg of peptide for each labeling reaction.
7. When purifying phosphopeptides using the $TiO_2$ method, it is very important to mix the sample thoroughly with the DHB before loading the sample onto the $TiO_2$ microcolumn.
8. When eluting the phosphopeptides from the $TiO_2$ microcolumn, it is extremely important to use sufficient elution buffer, since some phosphopeptides, especially multiphosphorylated peptides bind strongly to the $TiO_2$.
9. Peptides are better dissolved at near neutral pH.

## Acknowledgments

This work was supported by grants from the Department of Science and Technology (DST)-SERB (EMR/2015/001870), India. The authors thank Department of Biotechnology (DBT) and Council of Scientific and Industrial Research (CSIR), India for providing research fellowship to PB, NVL, and SK, respectively.

### References

1. Silva-Sanchez C, Li H, Chen S (2015) Recent advances and challenges in plant phosphoproteomics. Proteomics 15:1127–1141
2. Mann M, Ong SE, Grønborg M, Steen H, Jensen ON, Pandey A (2002) Analysis of protein phosphorylation using mass spectrometry: deciphering the phosphoproteome. Trends Biotechnol 20:261–268
3. Bah A, Vernon RM, Siddiqui Z, Krzeminski M, Muhandiram R, Zhao C et al (2015) Folding of an intrinsically disordered protein by phosphorylation as a regulatory switch. Nature 519:106
4. Houben A, Demidov D, Caperta AD, Karimi R, Agueci F, Vlasenko L (2007) Phosphorylation of histone H3 in plants-a dynamic affair. Biochim Biophys Acta 1769:308–315
5. Li XS, Yuan BF, Feng YQ (2016) Recent advances in phosphopeptide enrichment: strategies and techniques. Trends Anal Chem 78:70–83
6. Reiland S, Messerli G, Baerenfaller K, Gerrits B, Endler A, Grossmann J et al (2009) Large-scale Arabidopsis phosphoproteome profiling reveals novel chloroplast kinase substrates and phosphorylation networks. Plant Physiol 150:889–903
7. Delom F, Chevet E (2006) Phosphoprotein analysis: from proteins to proteomes. Proteome Sci 4:15
8. Meyer LJ, Gao J, Xu D, Thelen JJ (2012) Phosphoproteomic analysis of seed maturation in Arabidopsis, rapeseed, and soybean. Plant Physiol 159:517–528
9. Gupta R, Min CW, Meng Q, Agrawal GK, Rakwal R, Kim ST (2018) Comparative phosphoproteome analysis upon ethylene and abscisic acid treatment in Glycine max leaves. Plant Physiol Biochem 130:173–180
10. Nanjo Y, Skultety L, Ashraf Y, Komatsu S (2010) Comparative proteomic analysis of early-stage soybean seedlings responses to flooding by using gel and gel-free techniques. J Proteome Res 9:3989–4002
11. Nguyen THN, Brechenmacher L, Aldrich J, Clauss T, Gritsenko M, Hixson K et al (2012) Quantitative phosphoproteomic analysis of soybean root hairs inoculated with *Bradyrhizobium japonicum*. Mol Cell Proteomics 11:1140–1155
12. Yin X, Komatsu S (2015) Quantitative proteomics of nuclear phosphoproteins in the root tip of soybean during the initial stages of flooding stress. J Proteome 119:183–195

13. Yin X, Sakata K, Komatsu S (2014) Phosphoproteomics reveals the effect of ethylene in soybean root under flooding stress. J Proteome Res 13:5618–5634
14. Subba P, Barua P, Kumar R, Datta A, Soni KK, Chakraborty S, Chakraborty N (2013) Phosphoproteomic dynamics of chickpea (Cicer arietinum L.) reveals shared and distinct components of dehydration response. J Proteome Res 12:5025–5047
15. Kumar R, Kumar A, Subba P, Gayali S, Barua P, Chakraborty S, Chakraborty N (2014) Nuclear phosphoproteome of developing chickpea seedlings (Cicer arietinum L.) and protein-kinase interaction network. J Proteome 105:58–73
16. Barua P, Lande NV, Subba P, Gayen D, Pinto S, Keshava Prasad TS, Chakraborty S, Chakraborty N (2019) Dehydration-responsive nuclear proteome landscape of chickpea (Cicer arietinum L.) reveals phosphorylation-mediated regulation of stress response. Plant Cell Environ 42:230–244
17. Grimsrud PA, den Os D, Wenger CD, Swaney DL, Schwartz D, Sussman MR et al (2010) Large-scale phosphoprotein analysis in Medicago truncatula roots provides insight into in vivo kinase activity in legumes. Plant Physiol 152:19–28
18. Trapphoff T, Beutner C, Niehaus K, Colditz F (2009) Induction of distinct defense-associated protein patterns in *Aphanomyces euteiches* (oomycota)–elicited and–inoculated Medicago truncatula cell-suspension cultures: a proteome and phosphoproteome approach. Mol Plant-Microbe Interact 22:421–436
19. Yang ZB, Eticha D, Führs H, Heintz D, Ayoub D, Van Dorsselaer A et al (2013) Proteomic and phosphoproteomic analysis of polyethylene glycol-induced osmotic stress in root tips of common bean (Phaseolus vulgaris L.). J Exp Bot 64:5569–5586
20. Ino Y, Ishikawa A, Nomura A, Kajiwara H, Harada K, Hirano H (2014) Phosphoproteome analysis of *Lotus japonicus* seeds. Proteomics 14:116–120
21. Sun H, Xia B, Wang X, Gao F, Zhou Y (2017) Quantitative phosphoproteomic analysis provides insight into the response to short-term drought stress in Ammopiptanthus mongolicus roots. Int J Mol Sci 18:2158
22. Leitner A (2010) Phosphopeptide enrichment using metal oxide affinity chromatography. Trends Anal Chem 29:177–185
23. Meng F, Forbes AJ, Miller LM, Kelleher NL (2005) Detection and localization of protein modifications by high resolution tandem mass spectrometry. Mass Spectrom Rev 24:126–134
24. Chi A, Huttenhower C, Geer LY, Coon JJ, Syka JE, Bai DL et al (2007) Analysis of phosphorylation sites on proteins from Saccharomyces cerevisiae by electron transfer dissociation (ETD) mass spectrometry. Proc Natl Acad Sci U S A 104:2193–2198
25. Khidekel N, Ficarro SB, Clark PM, Bryan MC, Swaney DL, Rexach JE et al (2007) Probing the dynamics of O-GlcNAc glycosylation in the brain using quantitative proteomics. Nat Chem Biol 3:339
26. Molina H, Horn DM, Tang N, Mathivanan S, Pandey A (2007) Global proteomic profiling of phosphopeptides using electron transfer dissociation tandem mass spectrometry. Proc Natl Acad Sci U S A 104:2199–2204
27. Wiesner J, Premsler T, Sickmann A (2008) Application of electron transfer dissociation (ETD) for the analysis of posttranslational modifications. Proteomics 8:4466–4483
28. Chalkley RJ, Thalhammer A, Schoepfer R, Burlingame AL (2009) Identification of protein O-GlcNAcylation sites using electron transfer dissociation mass spectrometry on native peptides. Proc Natl Acad Sci U S A 106:8894–8899
29. Mikesh LM, Ueberheide B, Chi A, Coon JJ, Syka JE, Shabanowitz HDF (2006) The utility of ETD mass spectrometry in proteomic analysis. Biochim Biophys Acta 1764:1811–1822
30. Cantin GT, Shock TR, Park SK, Madhani HD, Yates JR (2007) Optimizing $TiO_2$-based phosphopeptide enrichment for automated multidimensional liquid chromatography coupled to tandem mass spectrometry. Anal Chem 79:4666–4673
31. Thingholm TE, Jørgensen TJ, Jensen ON, Larsen MR (2006) Highly selective enrichment of phosphorylated peptides using titanium dioxide. Nat Protoc 1:1929
32. Larsen MR, Thingholm TE, Jensen ON, Roepstorff P, Jørgensen TJ (2005) Highly selective enrichment of phosphorylated peptides from peptide mixtures using titanium dioxide microcolumns. Mol Cell Proteomics 4:873–886
33. Li QR, Ning ZB, Tang JS, Nie S, Zeng R (2009) Effect of peptide-to-$TiO_2$ beads ratio on phosphopeptide enrichment selectivity. J Proteome Res 8:5375–5381
34. Choudhary MK, Basu D, Datta A, Chakraborty N, Chakraborty S (2009) Dehydration-responsive nuclear proteome of rice (Oryza sativa L.) illustrates protein network, novel regulators of cellular adaptation, and evolutionary perspective. Mol Cell Proteomics 8:1579–1598
35. Humphrey SJ, Azimifar SB, Mann M (2015) High-throughput phosphoproteomics reveals in vivo insulin signaling dynamics. Nat Biotechnol 33:990

# INDEX


## A

Abiotic stresses ........ 20, 21, 24–28, 99–120, 254, 269, 277, 279, 317
*Agrobacterium*-mediated transformation ........ 349–351
Annotations ........ 8, 20–22, 28, 35, 36, 43–45, 54, 82, 84–91, 93, 110, 112–114, 117, 141, 154–156, 197, 208, 237, 253–259, 366, 369, 372
Antibiotic screening ........ 277, 279
*Arachis hypogaea* miniature inverted-repeat transposable element 1 (*AhMITE1*) ........ 237–242, 244
*A. rhizogenes* ........ 336–338, 340–342, 350
Assembly ........ 2, 20, 35–45, 49–94, 101, 138, 154, 178, 255, 272, 344, 366

## B

Bcftools ........ 149, 150, 155, 179, 226
Biochemical ........ 50, 56, 264, 277, 279, 398
Bioinformatics ........ 37–38, 49, 60, 62, 79, 83–91, 93, 100, 101, 141, 149, 176–180, 203, 204, 206, 221–227, 262, 263, 366, 372
Bisulphite sequencing ........ 127–143

## C

Cellular processes ........ 262, 277
Chickpea (*Cicer arietinum*) ........ 1, 10–12, 15, 19–29, 99, 100, 110, 129, 142, 147, 159, 160, 191, 192, 254, 269–275, 278, 318, 335–346, 366, 400
Chloroplasts ........ 6, 49–54, 56–58, 63, 65, 72, 73, 75, 79, 80, 83, 86–88, 90, 91, 135, 139, 142, 280, 294
Composite plants ........ 336–339, 342–343
Co-transformation ........ 336, 337
Cotyledonary nodes ........ 328, 358–360, 362
CRISPR/Cas9 ........ 336, 349–362

## D

ddRAD-seq, *see* Double-digest restriction site-associated DNA sequencing (ddRAD-seq)
De novo ........ 3, 4, 6, 7, 9, 12, 13, 15, 36, 38, 41–42, 44, 49, 56, 60, 66, 68, 72, 73, 75, 77, 78, 86, 114–115, 177, 178, 182, 237, 255
DGE, *see* Differential gene expression (DGE)
Differential gene expression (DGE) ........ 27, 29, 101, 103, 115–116, 120
Differentially methylated regions (DMRs) ........ 28, 128–132, 136–141
Disease resistance ........ 244, 365–366
DNA ........ 6, 21, 35, 50, 104, 127, 160, 189, 199, 233, 261, 278, 306, 317, 340, 349, 395
Double-digest restriction site-associated DNA sequencing (ddRAD-seq) ........ 159–183, 200

## E

Enzyme Kinetics ........ 280, 283, 291
Error rates ........ 77, 128, 135, 139, 142

## F

False discovery rate (FDR) ........ 120, 129, 130, 151, 404
FDR, *see* False discovery rate (FDR)

## G

Gene editing ........ 349–362
Genetic diversities ........ 3, 4, 9, 29, 160, 178, 203, 243, 317
Genomes ........ 1–13, 19, 35, 49, 100, 127, 153, 159, 189, 199, 233, 261, 306, 335, 350, 366
Genome-wide association study (GWAS) ........ 3, 9, 25, 147–157, 190, 194, 197
Genomic selection (GS) ........ 189–197
Genotyping-by-sequencing (GBS) ........ 5, 6, 13, 25, 160, 170, 200, 202, 207
G-quadruplex ........ 261–267
GUS analysis ........ 320, 321, 325, 328

## H

Hairy root transformation ........ 335–346
HPLC/ESI-QTOF-MS/MS ........ 379, 380, 383


Mukesh Jain and Rohini Garg (eds.), *Legume Genomics: Methods and Protocols*, Methods in Molecular Biology, vol. 2107, https://doi.org/10.1007/978-1-0716-0235-5, © Springer Science+Business Media, LLC, part of Springer Nature 2020

**I**

In vitro transformation ........ 318, 337

**L**

Legumes ........ 1–13, 19, 21, 22, 29, 35–45, 99–120, 127–143, 159, 160, 199–228, 254, 256, 261–267, 269–275, 277, 279, 314, 317, 318, 335, 336, 349, 366–372, 395–405
Linkage maps ........ 8, 11, 12, 20, 21, 24, 25, 29, 160, 244
Long non-coding RNAs (lncRNAs) ........ 27, 29, 253–259

**M**

Machine learning ........ 253–259
Mapping ........ 4, 11, 12, 21, 22, 24, 25, 29, 35–45, 62, 78, 83, 85, 94, 112, 113, 115, 128–131, 135–139, 142, 143, 155, 160, 190, 223–226, 237, 243, 244, 269, 372, 397
Methylated cytosine ........ 138
Mitochondria ........ 54, 55, 59, 79, 91, 280, 305, 306, 308–313
Molecular markers ........ 20–25, 29, 159, 244
Morphological ........ 3, 24, 273, 296, 326
Mung bean ........ 306, 308–310, 312, 313, 366

**N**

Next-generation sequencing (NGS) ........ 4, 21, 27, 37, 49, 50, 52, 56, 60, 78, 79, 87, 100, 101, 108, 111, 112, 119, 128, 129, 132, 134, 199, 203, 221, 237, 253
Non-coding RNA ........ 27, 253–259
Nuclear fraction ........ 397
Nucleotide-binding site-leucine-rich repeat (NLR) ........ 365, 368, 370

**P**

Pangenome ........ 4, 35–45
Percoll gradient (PG) ........ 58, 309, 310
PG, *see* Percoll gradient (PG)
Phenotyping ........ 150, 270–272, 274, 294
Phosphopeptide enrichment ........ 397–399, 402–403
Phosphoproteome ........ 399, 400, 404
Physiological ........ 101, 261, 262, 277, 279
Pigeonpea regeneration ........ 318, 323
Plant breeding ........ 189
Plant Long Non-Coding RNA Prediction by Random forests (PLncPRO) ........ 27
Plumular meristem transformation ........ 317–332
Prediction accuracy ........ 189, 190, 192, 194–197
Probe design ........ 202, 203, 206–209
Probe hybridization ........ 205, 206
Protein phosphorylation ........ 395, 404
Pseudomolecule ........ 11, 150

**Q**

Quantitative RT-PCR (qRT-PCR) ........ 100, 104, 106–109, 118, 119
Quantitative trait loci (QTLs) ........ 3, 25, 28, 29

**R**

Receptor like kinases (RLK) ........ 118, 368, 371
Receptor like proteins (RLP) ........ 368, 371
Resequencing ........ 2–5, 9, 11, 13, 15, 25, 40, 160, 201, 335
Restriction enzyme (RE) ........ 161, 171, 182, 200, 202, 234, 239, 327, 361
RLK, *see* Receptor like kinases (RLK)
rMVP ........ 148–153
RNA-sequencing (RNA-seq) ........ 8, 26, 27, 69, 72, 78, 101, 104, 108, 110–115, 117, 254, 255
Root nodules symbiosis (RNS) ........ 336
Root traits ........ 269–275

**S**

Semihydroponic system ........ 270
Sequencing ........ 1–13, 20, 21, 26, 27, 35–37, 39, 40, 43, 44, 49–94, 108, 110–113, 115, 117, 119, 127–143, 155, 159–183, 199–228, 233, 235, 237, 239–241, 264, 356, 357, 359, 361, 366
Shoot grafting ........ 318
Single nucleotide polymorphism (SNP) ........ 5, 6, 8, 11, 22, 25, 38, 78, 100, 148, 149, 151, 153, 155–157, 160, 176–179, 182, 183, 192, 193, 197, 199, 200, 202–204, 225, 244
Soybean ........ 1–3, 8, 13, 38, 39, 44, 99, 100, 147–157, 159, 160, 203, 235, 254, 255, 278, 349–362, 366, 367
Stress response ........ 27, 100, 118, 119, 253, 280, 396
Stress tolerance ........ 101, 118, 119, 147, 253, 277, 279
Suppression PCR ........ 235, 239, 240

**T**

Target enrichment sequencing (TES) ........ 199–228
Tissue culture ........ 318–320, 323–328, 337, 358
Titanium dioxide ($TiO_2$) ........ 396
Transcriptome ........ 20, 21, 24, 26–29, 100, 101, 104, 108–117, 119, 202, 234, 253, 255, 256

Transformation frequency .......................... 318, 320, 328
Transposable elements (TEs) marker
system .......................................................... 233–245
Transposon display (TD) ...................................... 235, 237
*Trigonella foenum-graecum* .......................................... 377, 378, 390

## U

UPLC/ESI-MS/MS multiple reaction
monitoring ........................................................ 379

## V

Visualization ............................................... 21, 28, 90–91, 117, 130, 141, 142, 263

## W

Web resources ..................................................... 21, 28, 29
Whole-genome bisulphite sequencing
(WGBS) ........................... 128, 129, 132, 133, 141
Whole-genome sequencing
(WGS) ............................................ 2, 6, 9–11, 13, 85–87, 202, 207, 371

FSC
www.fsc.org
MIX
Papier | Fördert gute Waldnutzung
FSC® C083411

Kolibri 360 GmbH
Ferdinand-Jühlke-Straße 7
99095 Erfurt, Deutschland
produktsicherheit@kolibri360.de